General Genetics

A SERIES OF BOOKS IN BIOLOGY

Editors: Douglas M. Whitaker, Ralph Emerson, Donald Kennedy, George W. Beadle (1946–1961)

ADRIAN M. SRB
CORNELL UNIVERSITY

RAY D. OWEN
CALIFORNIA INSTITUTE OF TECHNOLOGY

ROBERT S. EDGAR
CALIFORNIA INSTITUTE OF TECHNOLOGY

General Genetics

SECOND EDITION

W. H. FREEMAN AND COMPANY *San Francisco & London*

Preface to the Second Edition

In the preface to the first edition of this book, we evaluated the science of genetics at that time as being in a state of exceptionally vigorous growth. If the evaluation was faulty, it erred on the side of underestimate. In the years since 1952, remarkable advances have been made, particularly in the understanding of the nature and activities of genetic material in physical and chemical terms. Over the same period, other areas of genetics, most of them already soundly based some decades ago, have not stood still. And the interactions of genetics with other natural sciences have continued to increase, so that a more profound understanding of genetics has become concomitant with greater insight into all the life sciences.

Given the many advances in genetics during the past dozen years, and its ever broader implications, the present book inevitably contains a great deal of new material. Anyone who compares this edition with the earlier will note significant changes. Among those that are immediately striking is the increased emphasis on studies with bacteria, bacteriophage, and certain of the fungi. Organisms that once were exotics in reference to genetic study are now common research objects whose special experimental attributes have permitted many of the major advances of recent years. The genetics of man, always interesting but once not highly rewarding as an area of investigation, can now be used to provide the student with information and ideas that go beyond the realm of human interest and into a realm of breadth and real substance. In the first edition, it was said that "certain characteristics of the nucleic acids, *as yet not understood,* confer upon genes and viruses their essential faculty for specific duplication." Now these characteristics are relatively well understood, and can be defined in the highly specific language of the chemist. Moreover, the messages carried by genes are beginning to be decoded by chemical techniques, and the processes by which these messages are translated in the organism are understood in principle, and indeed in some detail. Those forms of inheritance that are called *extranuclear,* although still mysterious, have now been found so widely that they demand increased attention.

With the inevitable and totally desirable addition of important new material, we have had to face a substantial pruning of some of the old, if the book were to be kept to usable size for teaching purposes. The omission of sound material that appeared in the first edition is perhaps some cause for regret, since we do not judge the omitted material by any means valueless. But to provide students the basis for a fair and working knowledge of genetics as a whole, we have had to decide what must be included, and what cannot be, in a book of limited size. These judgments in fact have dictated the retention of a very substantial portion of what is sometimes called classical genetics. Our decision to introduce students to genetics by way of the contributions of Mendel was not merely a matter of graceful tribute on the occasion of the centennial of his work. It is a recognition that his pioneering contributions framed the terms of reference of typical genetic analysis and thinking, even as they are carried out today.

We have condensed and reorganized the structure of the book. Many of the chapters are admittedly too long and cover too much ground to serve as the accompaniments of single class lectures. In different courses selections from the material offered in the text will doubtless be made in different ways. But our aim has been to present genetics in coherent blocks of material whose comprehension will enable students to grasp what is now a vast and rather sprawling although interrelated set of subject matters.

One of the satisfying attributes of genetics, and also one of the evidences of its maturation as a science, is that in many of its facets the principles are so well understood that a student does not need to be belabored with great detail or multiple examples. The order of presentation, like the selection of material, is again a matter of our judgment, with which, of course, there can be legitimate difference. We have not attempted consciously to compromise between the extremes of divergent points of view. We have simply written the book in a form that we as authors could accept as a reasonably orderly and representative way in which to present genetics while keeping its constituent areas in context.

Among the consequences that the rapid proliferation and major advances in genetics have had for this book is that the authors of the first edition invited Dr. Robert S. Edgar to join them in preparation of the second. His participation provides increased professional scope and proficiency, particularly in certain of the newer areas of genetics where progress is being made with great speed and where current activity is great. Again, the authors have undertaken the work on a roughly equivalent basis, and the order of their names on the book is not a correlate of effort or responsibility.

We have retained those particular features of the first edition that appear to have been especially useful to students and teachers. These include the annotated references, which represent various degrees of sophistication and have been expanded and brought up to date. We continue to believe in the importance of problem-solving as a way to facilitate the student's mastery of genetics, and have included the better problems from the first edition as well as adding new ones. The figures and tables are meant to be an integral part of our presentation of genetics; they should be considered in context and not as isolated embellishments.

We wish to acknowledge gratefully the help of a large number of persons. Not all of them can be mentioned individually here, but perhaps first among them are our wives, Jo, June, and Lois, who have helped directly, and have exercised patience beyond the dictates of marital duty. The entire text has been read and reread by J. F. Crow, whose comments and criticisms have been perceptive and helpful. Our professional associates have aided us cheerfully, by reading and commenting on parts of the manuscript, by providing special materials, or by giving us the advantage of their special knowledge and skills in a number of other ways. These include Colin Driscoll, S. Emerson, N. H. Horowitz, Guido Pincheira, B. G. Sanders, H. T. Stinson, Jr., C. H. Uhl, B. Wallace, and S. Zahler.

We are pleased also to recognize here the generosity of publishers and of individuals who have permitted us the use of many of the illustrative materials and data on which the book depends. Specific acknowledgments are made in connection with these materials as they appear throughout the book. Furthermore, the debts of gratitude expressed in the preface to the first edition (printed also in this second edition) still persist. Although much of the present book is new, or has been rewritten or reorganized from the first, a good deal is not. And in any case the second edition is an evolution of the first. Therefore, those who contributed so helpfully in 1952 should not be forgotten.

June 1965   ADRIAN M. SRB
                                              RAY D. OWEN
                                              ROBERT S. EDGAR

Preface to the First Edition

The science of genetics is relatively young and in a state of exceptionally vigorous growth. It has branched into many fields, deriving sustenance and stimulus from its close interrelationships with cytology, evolution, biochemistry, physiology, morphogenesis, and practical agriculture; its development has depended in significant ways on the application of appropriate mathematical techniques; and, at many levels, genetics has important implications for the welfare of the human individual. These interrelationships are not sidelights but are part of genetics itself. For this reason, genetics offers great opportunities for integration of the student's knowledge of various fundamental aspects of biology. At the same time it offers great challenges to teachers of general genetics and to the writers of textbooks.

We have felt the urge to meet this challenge in our own way. In presenting the subject, we believe it most important that the general biological implications of genetics should never be lost from view. To accomplish this end, we have attempted from the outset to present genetics in terms of the effects of hereditary units, in dynamic interplay with environment, on the development and function of organisms. We have also thought it important to draw examples from a wide variety of living forms, ranging from microorganisms through diverse plants and animals to man. This has meant the abandonment of anything like a systematic historical approach to the subject. On the other hand, we have the strong conviction that genetics, as one of the great experimental sciences of our era, should be viewed as a series of problems and challenges, some of which have been successfully met and been supplanted by others. This means that genetic ideas may be developed along a logical sequence, from the simple to the complex. In following such a sequence, the book should be compatible with the overall organization of most introductory courses in genetics as they are presently being taught.

Different courses in introductory genetics provide more or less time for presentation of the subject; and instructors may wish to emphasize or neglect certain areas of the field as appropriate to the needs of their students. We have attempted to plan a certain amount of flexibility for this text in anticipation of varying situations. The first sixteen chapters provide elementary coverage of the so-called "classical" areas of genetics. These areas are included in almost every general course in the subject. Chapters 17, 18, and 19 are concerned with some of the details of genic effects that have emerged from recent investigations. Where there is a limitation on time, instructors in certain introductory courses, for example those serving as a prerequisite to work in applied genetics in agricultural schools, may choose to omit portions of these three chapters from formal assignment. On the other hand, students majoring in a basic biological science will find these chapters fundamental. Chapters 20 and 21 present population genetics in fair detail. Here again there is opportunity for selection of material in the light of particular course needs. The final group of three chapters emphasizes application of genetic principles. Chapters 22 and 23 should be of special interest and use to students in colleges of agriculture. But this material may be conveniently omitted, without loss in continuity, if it is inappropriate to a particular course. Chapter 24 summarizes aspects of genetics that relate to the welfare of the human individual. Examples of inheritance in man are found throughout the book.

The illustrations are meant to be an integral part of our presentation of genetics. They should be considered in context. The references at the end of each chapter are no doubt too extensive to be read in entirety by the ordinary undergraduate student. Admittedly, some of this reference material may be overly difficult for a beginner in the field. But the references are meant to serve a variety of needs, including those of students with special interests and capabilities and those of relatively advanced students who are reviewing general genetics. The annotations of references should help the individual student and teacher to select reading suitable to his purposes and interest.

Since genetics is essentially a "problem-solving" kind of science, questions and problems are a major accessory to a genetics text. We have been particularly concerned to confront the student with actual experimental situations for his interpretation. The problems following the different chapters are to some degree graded in difficulty; those toward the end of each sequence are designed to challenge the serious student of genetics.

The order of author names on this book in no way signifies a major and a minor effort or responsibility. We have undertaken the work on as equivalent a basis as possible and must share equally the responsibility for its merits and deficiencies. But for whatever merits there are, we wish to acknowledge gratefully the contributions of many other persons. First of all, our wives, Jo and June, have contributed substantial aid, as well as an uncommon measure of patience and understanding. G. W. Beadle, Ralph Emerson, and D. M. Whitaker, Editors of the W. H. Freeman and Company Biology Series, have provided critical comment on the manuscript. Additional critical reading furnished by the publishers was done by G. H. Mickey. Among our professional associates, many have read various portions of the text and passed on to us their helpful comments. These include D. W. Bishop, I. Blumen, R. W. Bratton, R. L. Cushing, Zlata Dayton, W. T. Federer, E. Hadorn, C. R. Henderson, N. H. Horowitz, E. B. Lewis, H. M. Munger, Margery Shaw, H. H. Smith, Curt Stern, A. H. Sturtevant, A. Tyler, C. H. Uhl, and R. P. Wagner. Claude Hinton, Dan Lindsley, Earl Patterson, Carlos Schlottfeldt, and Val Woodward have helped in reading proof and in other ways. Mrs. Alfred B. Clark kindly typed portions of the manuscript.

We also owe numerous debts of gratitude for special materials that have gone into making up the book. The W. B. Saunders Company and A. H. Sturtevant and G. W. Beadle generously loaned us use of the Drosophila eye color plate. Materials for the corn color plate were made available to us and grown especially for our purposes by E. G. Anderson. Color transparencies for the fox color plate were prepared for our use by R. M. Shackelford, of the University of Wisconsin. We are also indebted to A. B. Chapman for the materials illustrating the Wisconsin Swine Selection Cooperative Program, of which we made extensive use in Chapter 23. R. A. Brink provided the photograph of semisterile corn. And H. M. Munger has permitted us to utilize illustrations from an unpublished outline to his course in plant breeding at Cornell. A condensation of an extensive review of the action of colchicine on dividing cells was made available to us by O. J. Eigsti. We are particularly indebted to Dr. Marta S. Walters and Dr. Spencer W. Brown for the photomicrographs reproduced in Figure 7-6. These are selected from a series of twenty large prints that may be obtained from Dr. Walters, 3844 Lincoln Road, Santa Barbara, California. The series shows a mitotic sequence in great clarity and detail. We are indebted to Professor R. A. Fisher, Cambridge, and to Messrs. Oliver and Boyd Ltd., Edinburgh, for permission to reprint Table 3-5 from their book, *Statistical Methods for Research Workers.*

The illustrations of Evan Gillespie speak eloquently for themselves, but we wish to pay personal tribute to our illustrator's insight and imaginative collaboration.

March 1952   ADRIAN M. SRB
                                                         RAY D. OWEN

Contents

CHAPTER 1

Mendelism

Through simple observation, we have all become aware that living things have individuality. One human being is easily distinguished from another in a variety of ways, unless our acquaintance includes certain kinds of twins of like sex. Similarly, dogs can be told apart, even if they are of the same breed. An expert can distinguish among certain colonies of a simple organism like bakers' yeast. At the same time we all know by experience that the variation among living things is not haphazard. Although dogs differ more or less markedly from one another, they do not differ in ways that permit them to be confused with people, or with some other kind of animal, or with plants. Another way of saying this, again based on what every one knows, is that when dogs have offspring these are always dogs, never cats or carnations. Furthermore, most of us are aware, for example, that the babies of two white-skinned people will be white- and not black-skinned. And if we have observed just a bit more closely we know that some individuals seem to represent recombinations of characteristics of their parents. We have heard many times statements of the kind, "Johnny has his father's nose but his mother's eyes." Observations of this kind, even though unsystematic and sometimes wrongly interpreted, lead one who has thought about it to the correct general conclusion: biological individuality has patterns, limitations, and characteristics that depend at least in part on the parents of an individual.

The science primarily concerned with precise understanding of biological properties that are transmitted from parent to offspring is called *genetics*. Through experimentation and study geneticists have established certain rules that characterize the transmission of biological heredity. They have also concerned themselves with the mechanisms of hereditary transmission. An exciting and meaningful aspect of genetics is analysis of the physical-chemical properties of the hereditary determinants that bridge the gap between one generation of individuals and the next. Geneticists are also concerned with finding out how hereditary "information," transmitted from one generation to the next, is translated into the various characteristics that define the similarities and differences among living things. Early in the development of the science of genetics, investigators realized that genetic systems need to be understood in relation to the environments in which organisms exist; therefore, the study of biological heredity includes a consideration of biological environments. Changes in genetic systems, at given points in time and through the course of time, are likewise part of genetics. And, of course, genetics, like any science, has seen application of its knowledge to practical problems of concern to man. These and other aspects of biological heredity will be considered in the following pages, which are written in an attempt to introduce you to an important and fascinating science. In a general way, the importance of this science of genetics is that it leads to a profound understanding of the world of living things. Much of its fascination is that it is a particularly dynamic science, interacting with all the other biological sciences and with the physical sciences. The result of this interaction has been a constant infusion of new techniques and concepts, that lead continuingly to additional understanding and new challenges.

One could begin the study of genetics by consideration of any of a variety of the aspects of biological inheritance. We shall begin by considering the first clearly understood aspect of genetics—the transmission of biological properties. We begin in this way not only to preserve some sense of the historical development of genetics but also because the methods and reasoning used in the study of hereditary transmission are those that give genetics the particular flavor that distinguishes it from the other sciences with which it interacts and overlaps.

The hereditary transmission of biological properties is obviously an aspect of the reproduction of organisms. A brief introductory statement about reproduction will enable us at once to consider hereditary transmission in relatively specific ways.

Sexual Reproduction. Throughout our introductory discussion of genetics we shall be primarily concerned with sexual reproduction. This is the type of reproduction in which the individual begins life as a fertilized egg, or *zygote*. More specifically, sexual reproduction describes a life cycle that involves an alternation of gametic union and meiosis. New individuals can originate, of course, by means of other kinds of reproduction. Strawberries, for example, like many other higher plants, may multiply by sending out "runners," which can take root and function as independent plants. Among other plants, new individuals can be developed by means of cuttings, or by grafting, or by budding. Many unicellular plants and animals may reproduce simply by dividing into "daughter cells." Molds and other organisms may produce asexual spores—

specialized pieces of the parent organism capable of germinating to form new individuals. In honey bees the males usually develop from unfertilized eggs. Similar *parthenogenetic* development of the egg is not infrequent in other animals and plants. These and additional examples that might be cited illustrate *asexual* or *vegetative* reproduction. In none of them does the individual originate as a zygote, formed by the fusion of egg and sperm. Parthenogenesis is not strictly asexual, inasmuch as one of the elements of sexual reproduction, the egg, is involved. Neither is it sexual reproduction in a complete sense, because the egg is not fertilized.

Our initial emphasis is placed on sexual reproduction for several good reasons. One is the fact that our own species, and the higher animals in which we are generally most interested, almost invariably reproduces by means of fertilized eggs. Sexual processes are also most important in the propagation of higher plants, and occur in many lower plants and animals. Among viruses, bacteria, and some fungi, genetic processes exist that are like sexual reproduction in that they bring together hereditary elements from different cell lines but that do not fit in other ways the definition of sexual reproduction given above. At the moment we will not discuss further these so-called *parasexual* systems. But an understanding of sexual reproduction in relation to heredity will give the basis for later consideration of parasexuality.

Another reason for emphasizing sexual reproduction is that it provides better opportunities for the analysis of inheritance than does vegetative propagation. When individuals multiply asexually, their offspring are almost literally "chips off the old block." An investigator cannot combine or extract inherited differences in asexually reproduced generations, and the lack of inherent variation in such a line of descent leaves few "handles" for the experimenter to grasp. In sexually reproducing material, on the other hand, differing individuals can be mated, or *crossed*, and the consequences of such combinations can be observed through successive generations. Thus sexual reproduction offers techniques useful for the investigation of the processes of inheritance. Knowledge accumulated through such studies in turn helps in understanding the sexual reproduction that made the investigations possible. We will get some idea, as we progress, why sexual reproduction is so important and so widespread in plants and animals; and through familiarity with the transmission of inherited potentialities in sexual reproduction, we will find it easy to understand also the bases of asexual multiplication.

Sexual reproduction is so familiar a method of organic multiplication that few of us have had occasion to consider how remarkable it is. The sperm, the egg, and the product of their fusion, the zygote, are notable for their small size and for the precision of their organization and activity. It has been estimated that the sperm that fertilized all the eggs from which the present human population of the world developed could be packed into a container smaller than an eraser on a pencil. The eggs, since they store materials for early development, are somewhat larger; nevertheless they are also small by everyday standards. A gallon jug would readily hold all the eggs from which humans living in the world today originated. All of the biologically inherited qualities of human beings—the similarities as well as the differences that distinguish one human from another and from all other living things—have their basis in a minute mass of sperm and less than a gallon of eggs! You will learn later how the chemical nature of hereditary determinants permits coding such vast amounts of information into the abbreviated messages carried from one generation of living things to the next.

The minute and delicately organized material that forms the physical basis for the transmission of inherited qualities has been called *germ plasm*. It occupies only a fraction of the volume of the cells containing it. We shall be concerned with the architecture and the activity of this germinal material.

Mendel, and the Beginning of the Science of Genetics. Few sciences have so clear-cut a beginning as genetics. Common observational information of the kind to which we referred in the introductory paragraph of this chapter goes back far into man's history. And at various times in the past men made studies or carried out experiments suggested by such observations. But experimental genetic science with real meaning for the present began in the middle of the nineteenth century with the work of Gregor Johann Mendel, an Austrian monk.

Mendel's definitive experiments, carried out in a monastery garden, were done with peas, a sexually reproducing plant. The flower of a pea plant is so constructed that pollen from other flowers is generally excluded. As a result, peas are naturally self-fertilizing. That is, the gametes that unite to form a new individual under natural conditions are produced by a single parent plant. On the other hand, an investigator can open the bud of a pea flower, emasculate it before male gametes are produced, and fertilize the flower with pollen obtained from some other plant. Thus, if precautions are taken against other fertilizations by stray foreign pollen, a controlled cross, or cross-fertilization, is possible. Mendel's experiments were controlled crossing experiments.

Before considering certain of these experiments specifically, we will say a few things about them in general terms, to give a context for appreciating Mendel's work as well as learning from it. Others before Mendel had made controlled crosses, or matings, within various species of organisms. Why did Mendel's crosses, rather than those of some earlier worker, provide the basis for the science of genetics? Above all, perhaps, Mendel had a brilliant analytical mind that enabled him to interpret his results in ways that defined the principles of heredity. But Mendel was good at experimentation as well as interpretation. He knew how to conduct an investigation in such a way as to maximize the chances that meaningful results would be obtained. He knew how to simplify, and how to reduce relatively meaningless complexities. As parents for his crosses, he chose individuals that differed by sharply contrasting alternative characteristics.

Moreover, these parents represented true-breeding lines of plants. That is, the parental types, if allowed to self-fertilize normally, had the simplest predictable hereditary pattern, giving rise only to individuals like themselves. You will see later that such true-breeding lines are a consequence of continued self-fertilization.

Among other elements in Mendel's success are the simple, logical sequence of crosses that he made and the careful numerical counts of his progenies that he kept with reference to the readily definable characteristics on whose inheritance he focused attention. In retrospect, certain present-day biologists have argued that Mendel also had a bit of luck riding with him, especially in the choice of materials whose high degree of suitability was not entirely foreseeable in advance of the experiments. Whatever the role of luck in the foundation of our present knowledge of genetics, Mendel's sound and painstaking work is highly exemplary. His innovations of thought and experiment, summarized in a paper published in 1866, rank among the admirable accomplishments of the rational human mind.

Some Mendelian Studies. Let us now consider a few representative experiments carried out by Mendel, and their results and the interpretation of the results. One of the pairs of alternative characters studied by Mendel was round versus wrinkled seeds. When Mendel crossed plants from a round-seeded line with plants from a wrinkled-seeded line, all of the first filial generation, abbreviated as F_1, had round seeds. The characteristic of only one of the parental types was represented in these *hybrids*. (Hybrids are the progeny of a cross between inherently unlike strains of individuals.) In the next generation, the F_2, achieved by self-pollination of F_1 individuals, each of the alternatives that differentiated the original parental types reappeared. Mendel's count of the two types in the F_2 was 5,474 round and 1,850 wrinkled, a ratio of 2.96 to 1.

Mendel found it notable that the same general result occurred when he made crosses between plants from lines differing by other alternative characteristics. Another example, expressed in genetic shorthand, is that crossing yellow cotyledons with green gave all yellows in F_1, and a ratio of about 3 yellows to

Table 1-1. RESULTS OF SOME OF MENDEL'S CROSSES OF GARDEN PEAS.

Phenotypes of Parents	F_1 Phenotype	Numbers of F_2 Progeny, by Phenotype		F_2 Ratio, Dominant to Recessive
round × wrinkled (seeds)	round	round: 5,474	wrinkled: 1,850	2.96 to 1
yellow × green (cotyledons)	yellow	yellow: 6,022	green: 2,001	3.01 to 1
long × short (stems)	long	long: 787	short: 277	2.84 to 1
axial × terminal (inflorescence)	axial	axial: 651	terminal: 207	3.14 to 1

1 green in F_2. Analogous findings were obtained when long-stemmed plants were crossed with short-stemmed plants, and when plants with axial inflorescence were crossed with plants having terminal inflorescence. In all, Mendel studied seven pairs of alternative characters and found a remarkable uniformity of type of result when crosses were made. Some of his pertinent results are summarized in Table 1-1. (Before reading on, list the generalizations that can be drawn from the information given in the table.)

What generalizations can be drawn from the experimental findings?
1. In F_1 only one of a pair of alternative characters that typify the different parental lines appears.
2. In F_2 both of the alternatives that define the parental lines appear.
3. In F_2 the character that appeared in F_1, to the exclusion of its alternative, is found about three times as frequently as its alternative.

What interpretations can be made? What did Mendel make of his generalized experimental findings? One of the keys to the situation is recognition that in F_1 the hereditary basis for the character that fails to appear is not lost. The character appears again in F_2. One can infer in the crossing sequence the persistence of hereditary elements derived from each of the original parents. Recognizing this, Mendel was able to visualize the simple mechanism accounting for the numerical ratio of 3 to 1 in F_2.

Assume that for a given pair of alternative characters F_1 individuals have received an appropriate genetic element from each parent. Let X be the element representing the character that is expressed in F_1 and x be the element that is the basis for the character that is not expressed in F_1. The F_1 individuals may then be designated Xx. Suppose that the pollen and egg cells produced by these individuals contain only one of the elements, X or x, and that these two kinds of elements are represented equally among gametes. If the genetically different kinds of pollen fertilize the genetically different kinds of egg cells with equal likelihood, the resultant combinations may be predicted as shown below.

| | POLLEN | |
	X	x
X	XX	Xx
x	Xx	xx

EGG CELLS

Summarizing, we expect to find F_2 combinations of genetic elements in the proportion $1XX:2Xx:1xx$. If in F_2 the relationship of element X to x is as it was in F_1, Xx individuals can be expected to express the characteristic expressed in F_1 individuals. This means that in F_2 one expects XX and Xx individuals to show the same characteristic and *xx* individuals to show the alternative. With reference to alternative characteristics, then, F_2 individuals are expected to occur in a proportion of three ($1XX + 2Xx$) to one ($1xx$).

Now let us apply such a system to one of Mendel's specific experiments. Figure 1-1 specifies the generalized Mendelian interpretation in the case of a cross between long- and short-stemmed peas. Note that the assumption that hereditary elements exist in pairs applies to the parents as well as to F_1 and F_2 individuals.

If you have Table 1-1 in mind, you are aware, of course, that Mendel's actual counts do not fit precisely the 3:1 ratio designated for F_2 in Figure 1-1. You will learn in the next chapter that some deviation from the predicted ratio is to be expected simply as a result of chance. If, for the time being, you accept this as true, the Mendelian interpretation obviously fits the results given so far. How can we substantiate the correctness of this interpretation? One approach is to use the interpretation as the basis for making predictions of results to be obtained from further crosses of particular kinds. Mendel made further crosses.

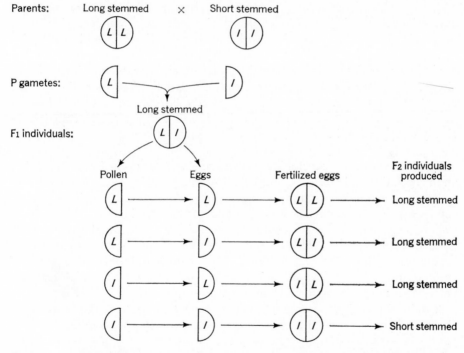

Figure 1-1. *Mendelian ratios in F_2, as observed following crosses between long- and short-stemmed peas, have a simple basis in the particulate determinants of heredity. These determinants exist in pairs in an individual, but are separated singly into different gametes.*

The results fulfill the only tenable predictions that can be made on the basis of his interpretations of hereditary transmission.

Consider, for example, the F_2 individuals deriving from crosses between long- and short-stemmed plants. According to the Mendelian interpretation this F_2 population, which includes about 3 long-stemmed for every short-stemmed individual, should be described as including three genetic constitutions in the frequency, $1LL:2Ll:1ll$. It follows that self-fertilization of short-stemmed plants (*ll*) should give rise in the following generation only to short-stemmed plants. Experimental results confirm this expectation. Self-fertilization of long-stemmed plants should produce an F_3 generation that includes both long- and short-stemmed plants. More precisely, one-third of the F_2 long-stemmed plants, being *LL*, should give rise only to long-stemmed plants. Two-thirds, being *Ll*, should behave like the F_1 hybrids, and give long- and short-stemmed plants in a proportion of $3:1$. Experiments confirm this prediction.

Other kinds of crossing combinations can be made to test Mendel's interpretation of his results. He carried out many appropriate hybridizations. But rather than describe these here we will introduce you to similar work done with organisms other than garden peas, to emphasize that the Mendelian principles have wide application. Investigations carried out after Mendel's time have shown clearly that his pioneering work can be generalized. This kind of verification of his findings, and subsequent development of knowledge of genetics, occurred only after a curious gap in time. Mendel's work was ignored and virtually forgotten for thirty-four years after its publication. Its rediscovery in 1900, by three different biologists independently, burst upon a biological world that had, in the interval since the publication of the paper, progressed sufficiently to appreciate it.

Terminology. Before developing further the elementary principles of genetics, an interpolation is desirable. Genetics is a problem-solving kind of science, and we hope to minimize its memorizing aspects. But it would be inefficient to continue using sentences or long phrases for which single-word terms are available and commonly encountered. Accordingly, you should pause at this point to equip yourself with the rudiments of a genetic vocabulary. The following are not intended as definitions to be memorized, but as descriptions of useful terms. With *germ plasm, cross, hybrid, zygote,* F_1, and F_2 you have already begun to accumulate genetic terminology.

A unit of heredity, for example, the element that controls the stem-length character in peas, is called a *gene*. You will have many opportunities to refine your understanding of the nature of genes.

The members of a pair of such units, like *L* and *l* controlling the long-short stemmed alternative, are called *alleles*.

An individual's genetic constitution is called its *genotype*. Thus F_1 hybrids after crosses between long- and short-stemmed plants have the genotype *Ll*.

When, as in such hybrids, the members of a pair of alleles are unlike, the individuals are described as *heterozygous*. When members of a given allelic pair are alike (for example, *LL* or *ll*), the individual is *homozygous* for these alleles. The corresponding nouns are often used. The F_1 hybrids can be called *heterozygotes*. Parents from the true-breeding lines of garden peas can be called *homozygotes*.

When Mendel crossed peas characterized by different alternative characters, he found that the F_1 individuals resembled only one of the parental types. The character that appeared in F_1 he called (in translation) *dominant*, its alternative *recessive*. This terminology is retained in current genetic usage. We can also say that the *L* gene (for long stems) is dominant to its alternative *l* (for short stems). The *l* allele is recessive.

Because of dominance, *LL* and *Ll* plants look alike, even though they differ in genotype. We say that they have the same *phenotype*, a word that refers to the appearance of an individual. (You have learned that long-stemmed pea plants may have the same phenotype but differing genotypes. Could the same statement be made correctly about short-stemmed plants?)

In the examples given so far, dominant genes are represented by capital letters of the alphabet, their recessive alleles by corresponding lower-case letters. Somewhat more precise and meaningful conventions used by geneticists in assigning gene symbols will be introduced in Chapter 2.

Mendelian Inheritance in Rabbits. In wild rabbits the fat beneath the skin is white. Certain domestic breeds have yellow fat. When rabbits from true-breeding strains of the two types are crossed, F_1 individuals all have white fat. If F_1 males are mated with F_1 females to produce an F_2 generation, white- and yellow-fat individuals are found in a proportion of 3 white to 1 yellow. The result is like that obtained by Mendel in garden peas, and a similar interpretation can be made. In a way, of course, the similarity of result gives us increased confidence in the Mendelian interpretations. At least we see that Mendel's results with peas are not unique. But instead of being satisfied with the mere similarity of result, let us use the rabbits in another way, to test the principles emerging from Mendel's work.

Suppose that the white-fat F_1 hybrids are *backcrossed* to yellow-fat parents, a backcross being a mating of a hybrid with one of its parents or with an individual of parental type. The actual result is that white- and yellow-fat progeny are obtained in equal frequency (not in a 3:1 ratio!). Figure 1-2 shows that the system of interpretation used for Mendel's results with F_2 populations covers this situation nicely. The F_1 white-fat rabbits, when bred, are found to carry a recessive allele (*y*).

What happens when the F_1 hybrids are backcrossed to white-fat parents? Because the parents displaying the dominant character are from true-breeding lines, their genotypes are *YY*. Eggs or sperm from these parents can carry only

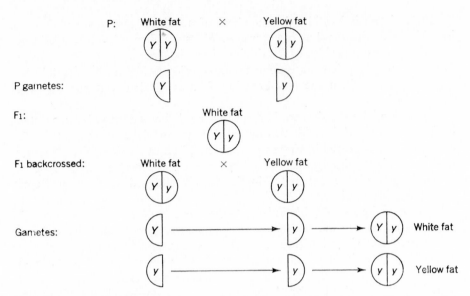

Figure 1-2. *When rabbits heterozygous for the fat-color alternative are backcrossed to the recessive parent, the particulate basis of this alternative is revealed.*

gene *Y*. Since *Y* is dominant, as shown by earlier crosses, one expects to find no yellow-fat progeny when the backcross to the white-fat parents is made. This is confirmed by experimental findings. The genetic situation is summarized in Figure 1-3.

Plant and animal breeders, as well as genetic investigators, often wish to know whether an individual carrying a dominant character is homozygous or heterozygous. An appropriate backcross will usually give the answer. (Which kind of backcross is the appropriate one, that shown in Figure 1-2 or that shown in Figure 1-3? Can you make a general statement that covers the situation? Mendel used backcrosses in his experimental series. On the basis of what you

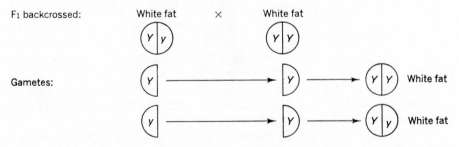

Figure 1-3. *Progeny of the backcross to the dominant parent do not segregate phenotypically.*

know of Mendel's other experiments, diagram a backcross he might have made; predict and account for appropriate results.)

Physiological and Environmental Considerations. Something has now been said about transmission of genes and about characteristics determined by genes. The characteristics discussed have been directly observable. The genes that account for these characteristics have been inferred from what happens when individuals having these characteristics produce offspring. So far you have been given no clues that would enable you to answer questions like the following. How do genes act to give rise to the characters they determine? In heterozygotes, how can one allele act in such fashion as to obscure the expression of its alternative? Or, in other words, how can one visualize a basis for dominance and recessiveness? The answers to questions of this sort are complicated enough that thorough treatment of them should be deferred until a later chapter. In fact, totally unequivocal answers cannot be given, even at the end of the book. Geneticists still have important problems to work on. Meantime some additional knowledge about yellow- and white-fat rabbits will suggest plausible partial answers to the questions posed above, answers to be supported later when gene action is discussed in detail.

When a wild rabbit eats green plants, one of the digestive processes that goes on is the breakdown of certain yellow components (xanthophylls) of this food. An enzyme found in the liver controls this process. When a rabbit lacks the enzyme, the breakdown of xanthophylls to colorless derivatives does not occur. Instead, the xanthophylls are stored intact in the animal's fat, giving it a yellow color. This difference, presence or absence of the enzyme, is the immediate, physiological cause of the difference between white-fat and yellow-fat rabbits. The presence or absence of the enzymatic function is determined by the allelic alternatives Y and y.

Consideration of the Y and y alleles in relation to enzyme activity suggests a reasonable interpretation of their dominance relationship. An enzyme is an organic catalyst; a relatively small amount of the appropriate enzyme probably is all that is required to break down the xanthophyll in a rabbit's normal diet. When we observe that the two genotypes YY and Yy give identical phenotype, white fat in each instance, we are observing that a single "dose" of the Y allele is sufficient to produce all the enzyme needed by a rabbit to break down all the xanthophyll he is likely to eat. This reasonable interpretation implies that rabbits of genotype Yy may indeed produce less enzyme than rabbits of genotype YY, perhaps only half as much; rabbits of both genotypes have white fat because even the lesser enzyme activity in the heterozygotes does not limit the breakdown of xanthophyll. However, in such a case, if we were to consider enzyme production as a phenotypic attribute, we would conclude that gene Y is not dominant over its allele y, because genotypes YY and Yy are distinguishable as producing different amounts of enzyme. This apparent complication illustrates a generalization: when several aspects of a genetically controlled attribute can be evaluated, our judgments as to dominance may vary, depending on the aspect of phenotype being observed.

Earlier we pointed out in a very general way that genetic systems interact with environment. Yellow-fat rabbits illustrate one aspect of genetic-environmental interaction. Suppose that no xanthophylls were included in the diet of a rabbit of genotype yy, ordinarily yellow-fat. (The situation can be arranged rather easily by an investigator.) On a xanthophyll-free diet even animals lacking the enzyme for xanthophyll destruction have white fat. This points to an important principle: Genes determine potentialities; the realization of these potentialities depends on the environment in which the genes perform their functions.

At least one of the allelic pairs studied by Mendel appears to exemplify the kind of physiology of gene action described for genes Y and y in the rabbit. The round-seeded pea stores the bulk of its reserve carbohydrate as starch. The alternative genetic character, wrinkled seed, stores far less starch but accumulates considerable stachyose, a simpler carbohydrate. The simplest inference is that in homozygous wrinkled an enzyme concerned in starch synthesis is deficient. Whether or not this is correct, gene R (for round) gives biochemical competence for starch synthesis that gene r (for wrinkled) is unable to provide.

Incomplete Dominance. The genetic situations we have encountered so far involve pairs of characters that exist as more or less unequivocal alternatives. Life is not always so simple, as one sees when he looks into the genetics of the frizzle fowl. These chickens present a curious phenotype. They look as one might imagine the proverbial hen to have looked when she got off the hot griddle. Their feathers are brittle and curly or even corkscrew in shape; they wear off so easily that the birds may be nearly naked for intervals in their lives. Microscopic examination of frizzle feathers reveals extreme abnormalities. Some of these characteristics of frizzle fowls are shown in Figure 1-4. Frizzle fowls have a variety of other physical and of physiological abnormalities, secondary consequences of the fact that these birds have difficulties in maintaining their body temperature. Their hearts are enlarged, as are the spleen, crop, gizzard, pancreas, adrenals, and kidney. All in all they are rather sad birds, although at high temperature, and under experimental conditions in which rapid heat loss is necessary, frizzles actually get along better than normal chickens. It is worth remembering that alternative genetic types may be favored or discouraged by a given environment.

Frizzle fowls mated to other frizzle fowls have only frizzle progeny. Therefore we can presume that the frizzle character indeed has a genetic basis. When frizzles are mated to normal birds, neither parental type is represented in F_1. Instead one finds a third kind of bird, one that we shall call "mild frizzle," a sort of intermediate between the parental types. The feathers of the F_1 birds

Figure 1-4. *Frizzle fowls have abnormal feathers.* A: *Feathers from normal birds under low magnification show a closely interwebbed structure.* B: *Frizzle feathers are weak and stringy.* C: *Frizzle fowls at the end of the breeding season are almost naked.* D: *Even when fully feathered, a frizzle has poor insulation against heat loss.* (A *and* B *after Landauer, p. 131, "Temperature and Evolution," volume VI of* Biological Symposia, *The Ronald Press Company, 1942.* C *and* D *courtesy of F. B. Hutt,* J. Genetics, **22:**126, 1930.)

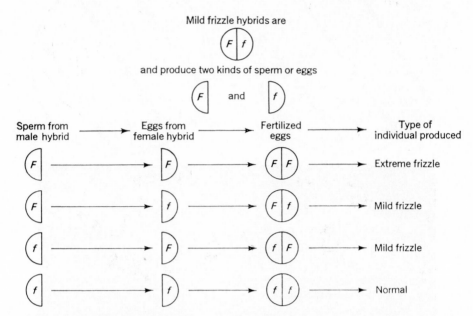

Figure 1-5. *An understanding of the particulate basis of heredity enables us to predict the results of crossing mild frizzles together.*

curl to some degree, but are less brittle and less modified than those of their frizzle parents, whom by contrast we can call "extreme frizzles." The other abnormalities of extreme frizzles are correspondingly milder in the hybrids.

Let us predict what the F_2 would be like if the genotypic (not phenotypic) situation were the same for the extreme frizzle-normal alternative as for white-yellow fat in rabbits. Figure 1-5 makes this prediction, assigning the genotype FF to extreme frizzles and ff to normal fowl. The prediction says that the F_1 mild frizzles have a genotype Ff, and that in F_2 the genotypes are anticipated

Table 1-2. RESULTS OF INTERCROSSING MILD FRIZZLE, FIRST-GENERATION HYBRIDS.

	Observed Numbers	*Predicted* (1:2:1)
extreme frizzle	23	23
mild frizzle	50	46
normal	20	23
total	93	92†

† Avoiding the absurdity of predicting fractions of chickens.
Source: Landauer and Dunn, *J. Heredity*, **21**:300, 1930.

in a proportion 1 *FF*:2 *Ff*:1 *ff*. If the genotype *Ff* gives mild frizzle phenotype as defined by F₁ birds, the phenotypic ratios for F₂ should be 1 extreme frizzle:2 mild frizzle:1 normal. Table 1-2 gives the results of actual crosses between mild frizzles of an F₁ generation. These results confirm the prediction made on the basis of a simple allelic difference between normals and extreme frizzles. We need to make only one adjustment to what we have learned before about Mendelian genetics. We find that the rule of dominance does not always hold. This finding, viewed in terms of our discussion of the dominance of white over yellow fat in rabbits, should not be unexpected to you.

We can solidify our confidence in the foregoing interpretation of Mendelian heredity with incomplete dominance by turning again to the backcross for further information. Figure 1-6 shows what happens when F₁ mild frizzles are backcrossed to normal and to extreme frizzle. The results confirm that mild frizzles are heterozygous for an allelic pair. These hybrids produce in equal

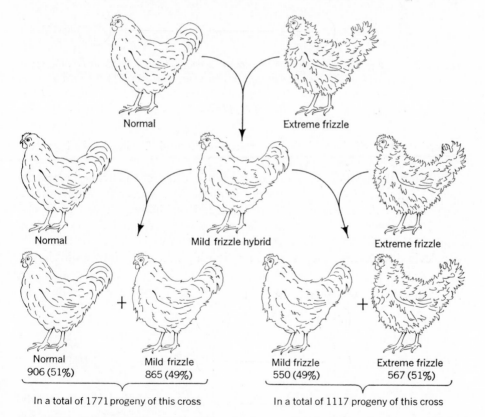

Normal

Extreme frizzle

Normal

Mild frizzle hybrid

Extreme frizzle

Normal
906 (51%)

+

Mild frizzle
865 (49%)

Mild frizzle
550 (49%)

+

Extreme frizzle
567 (51%)

In a total of 1771 progeny of this cross   In a total of 1117 progeny of this cross

Figure 1-6. *When mild frizzle hybrids are backcrossed to the parental types, the kinds of progeny and their frequencies confirm that mild frizzles are heterozygous for an allelic pair.*

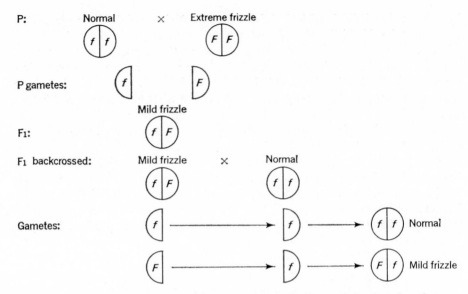

Figure 1-7. *The conclusion drawn from the results seen in Figure 1-6 is based on the same kind of analytical reasoning that permits interpretation of other instances of Mendelian heredity. Results of the backcross of mild frizzle to normal are readily understood in terms of simple rules of particulate heredity.*

proportions two distinct kinds of germ cells, *F* and *f*. The detailed genetic interpretation is shown in Figures 1-7 and 1-8.

Although the phenomenon of dominance is often encountered in genetic studies, instances of incomplete or partial dominance are far from infrequent. Depending on the characteristics involved, and on the ways they can be described or measured, dominance relations may appear to be more or less complex. You will find in the literature of genetics a variety of terms that have been devised to describe these various relationships. At this point we will simply

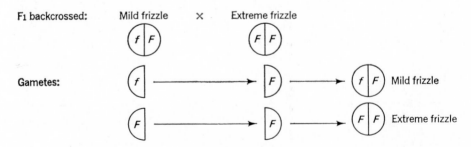

Figure 1-8. *An interpretation of the backcross of mild frizzle hybrids to extreme frizzles completes the explanation of the results seen in Figure 1-6.*

give a few illustrations of the phenomenon. Blue Andalusian chickens, for example, are heterozygous for a pair of alleles that in homozygous condition produce white-splashed and black Andalusians, respectively. Roan shorthorn cattle are heterozygous for an allelic pair in which the corresponding homozygotes are, respectively, red or white. Red-flowered snapdragons crossed with white-flowered yield F_1 hybrids with pink flowers. A rare allele in man, producing a severe anemia (thalassemia major) in the homozygous condition, causes only a mild anemia (thalassemia minor) when heterozygous.

Independent Assortment of Allelic Pairs. Mendel's contribution to genetics was not confined to elucidation of the principles that govern the inheritance of simple alternatives. He also dealt with the attributes of situations in which two pairs of alternatives are involved at the same time in a particular mating sequence. For example, when one parent had round seeds and yellow cotyledons and the other wrinkled seeds and green cotyledons, the F_1 was uniformly round, yellow. The F_2 population was

> 315 round, yellow
> 101 wrinkled, yellow
> 108 round, green
> 32 wrinkled, green

If you consider the different characteristics singly, you will see that the summarized proportions of round to wrinkled and of yellow to green are each approximately 3:1, a genetic situation now familiar. Mendel saw this, and saw also that his F_2 population as given above fell into a ratio of 9:3:3:1, the double dominants being the most frequent class and the double recessives the least frequent. Most important, he perceived that this more complex *dihybrid* ratio is precisely what is expected if members of the two pairs of genetic alternatives are assorted independently of one another. We shall now examine such a situation in detail, but using tomatoes rather than garden peas to provide data for analysis.

A Dihybrid Cross in Tomatoes. First, we shall consider two pairs of unit differences in tomatoes (Fig. 1-9).

Tall Stature Versus Dwarf. Dwarf tomatoes differ from the familiar tall type in having short, thick stems and deeper green, puckered, and somewhat curved leaves.

Cut Versus Potato Leaves. The leaves of most tomatoes typically have cut margins, but some varieties are called potato-leaf types because their leaves are broad and entire, like the leaves of potato plants. J. W. MacArthur, a Canadian geneticist, published extensive studies of these and many other characteristics of tomatoes. We shall utilize his data and conclusions to provide an excellent example of Mendelian genetics.

Figure 1-9. A: *Standard (tall) tomato plant.* B: *Dwarf plant, the same age as* A *and grown under similar circumstances, but differing from it in a unit of inheritance.* C: *Cut leaf.* D: *Potato leaf in the tomato differing from* C *in a unit of inheritance. (Photographs by R. H. Burnett.)*

Considered separately, the two pairs of differences are inherited in a pattern with which you are now well acquainted:

P	tall × dwarf	cut × potato
F₁	tall	cut
F₂	3 tall:1 dwarf	3 cut:1 potato

When the tall F₁ is backcrossed to dwarf, one observes a ratio of 1 tall:1 dwarf in the progeny. Similarly, F₁ cut backcrossed to potato gives 1 cut:1 potato. Evidently, the difference between tall and dwarf behaves as a unit difference in inheritance, tall being dominant; and the difference between cut and potato leaves also is a unit difference, cut being dominant. (You can fill in genotypes, germ cells, and genotypic ratios for review. In doing so, let D = tall, d = dwarf; and C = cut, c = potato.)

Our problem is to consider how these two different pairs of alleles behave with respect to each other in inheritance. A way of studying this is to cross individuals that differ in both characteristics:

tall, cut × dwarf, potato

When this cross is made, the F₁ shows both dominant characteristics, tall and cut. It is heterozygous for two pairs of alleles. Such an individual is called a *dihybrid.*

A good method of testing the behavior of these two pairs of alleles with respect to each other is to backcross the F₁ to the dwarf, potato-leaf parent plant type.

Actual data obtained from this backcross are as follows:

77 tall, cut
62 tall, potato
72 dwarf, cut
73 dwarf, potato

Four types occur among the backcross progeny—the original parental types and two new combinations, tall, potato and dwarf, cut. These four types appear in very nearly equal numbers.

Figure 1-10. *When a dihybrid forms germ cells, there are two equally likely ways in which the germinal elements may line up. This results in the formation of four kinds of germ cells, in equal numbers.*

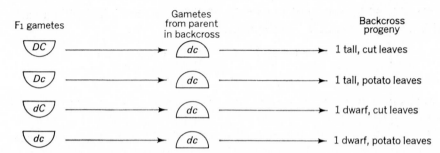

Figure 1-11. *The four kinds of germ cells produced by the dihybrid result in four distinct and equally frequent phenotypes among the progeny of a backcross to the double recessive parent type.*

We can conclude that when the F_1 forms germ cells, the two different lineups shown in Figure 1-10 occur with equal frequency. Subsequent fertilizations give the observed ratio in the backcross (Fig. 1-11).

Segregation and Independent Assortment. Two important implications emerge from the dihybrid example just discussed. These implications were seen clearly by Mendel, and his formulations of them into principle are part of the firm basis he gave to the science of genetics.

First, the members of a given pair of alleles separate from each other when an individual forms germ cells. We observed this in connection with single allelic pairs, and our present consideration of the dihybrid offers no occasion for modifying the rule. The backcross in the example above shows that about half of the progeny received the *D* allele from the dihybrid parent, and the other half the *d* allele. (The actual figures are $77 + 62 = 139D, 72 + 73 = 145d$.) Similarly, about half received the *C* allele, and half the *c*. (The figures are $77 + 72 = 149C$, and $62 + 73 = 135c$.) In working with and understanding genetics, it is important to observe this *principle of segregation* first of all: *The members of a pair of alleles separate cleanly from each other when an individual forms germ cells.*

The second rule is that *the members of different pairs of alleles assort independently of each other when germ cells are formed.*

In the dihybrid *DdCc*, *D* is just as likely to be included in a germ cell with *c* as it is to be with *C*, and *c* is equally likely to be with *D* or *d*. Another way of looking at the behavior of the *D-d* and the *C-c* allelic pairs as independent phenomena is suggested in Figure 1-12.

Later we will encounter some regular modifications of the principle of independent assortment, and we will see that studies of such modifications have provided much information about the structure of germ plasm. But for the present we will be concerned with the great variety of instances in which this principle holds.

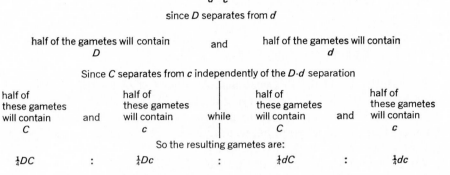

In an individual of genotype

$$\frac{D}{d}\ \frac{C}{c}$$

since D separates from d

half of the gametes will contain
D

and

half of the gametes will contain
d

Since C separates from c independently of the D-d separation

| half of these gametes will contain C | and | half of these gametes will contain c | while | half of these gametes will contain C | and | half of these gametes will contain c |

So the resulting gametes are:

$\frac{1}{4}DC$: $\frac{1}{4}Dc$: $\frac{1}{4}dC$: $\frac{1}{4}dc$

Figure 1-12. *The separation of the C-c allelic pair, and of the D-d allelic pair, respectively, can be treated as independent events.*

The Testcross. We cannot regard the principle of independent assortment to be established by a single test of the principle; we need to test it further. Of several possible tests, the first might involve this reasonable extension: The combinations in which two pairs of alleles are introduced in a cross should not influence the kinds or proportions of germ cells formed by the dihybrid. In other words, if we set up the cross:

tall, potato × dwarf, cut,

the F_1 should produce the same four kinds of germ cells in equal numbers as did the dihybrid from the cross:

tall, cut × dwarf, potato.

(List parental genotypes and germ cells; F_1 phenotype, genotype, and germ cells from both crosses for practice and to convince yourself that this is indeed expected on the basis of independent assortment.)

Many tests of this sort have given results consistent with the predictions.

Similar situations already described (p. 10) should make it obvious to you that in all such tests the F_1 should be crossed with individuals homozygous for both recessive characteristics. If, in the cross:

tall, potato × dwarf, cut

the F_1 is simply backcrossed to the tall parent, all the offspring will be phenotypically tall, because the tall parent contributes the dominant D allele to each backcross offspring. The segregation and assortment of the tall-dwarf alternative is therefore obscure. Similarly, a simple backcross to the cut parent will give only the cut phenotype in the progeny. Crossing to the double-recessive type (dwarf, potato) has the virtue of *making the segregation and assortment of both allelic pairs immediately apparent in the phenotypes of the progeny of the cross.*

The cross with a double-recessive type is called a *testcross* to distinguish it from a *backcross*, which is a cross of the F_1 with either of the parent types. Some backcrosses are testcrosses, and some testcrosses are backcrosses, but the two terms are not always equivalent.

Dihybrid Ratios in the F_2 Generation. Another test of the principle of independent assortment involves carrying a dihybrid cross to the F_2 generation. The F_1 individuals are assumed to produce four kinds of germ cells in equal numbers. Under this assumption, what ratios do we predict among the progeny when two such individuals are crossed?

An easy way to illustrate the possible fertilizations is through use of a "checkerboard," in which the four kinds of germ cells from one parent are listed in a row along the top, and the four kinds from the other parent are put in a column down the left side. The 16 combinations are then obtained by "filling in" the checkerboard:

POLLEN FROM DIHYBRID MALE

	$D\,C$	$D\,c$	$d\,C$	$d\,c$
$D\,C$	$\dfrac{D\ C}{D\ C}$	$\dfrac{D\ C}{D\ c}$	$\dfrac{D\ C}{d\ C}$	$\dfrac{D\ C}{d\ c}$
$D\,c$	$\dfrac{D\ c}{D\ C}$	$\dfrac{D\ c}{D\ c}$	$\dfrac{D\ c}{d\ C}$	$\dfrac{D\ c}{d\ c}$
$d\,C$	$\dfrac{d\ C}{D\ C}$	$\dfrac{d\ C}{D\ c}$	$\dfrac{d\ C}{d\ C}$	$\dfrac{d\ C}{d\ c}$
$d\,c$	$\dfrac{d\ c}{D\ C}$	$\dfrac{d\ c}{D\ c}$	$\dfrac{d\ c}{d\ C}$	$\dfrac{d\ c}{d\ c}$

(EGGS FROM DIHYBRID FEMALE)

Collecting like genotypes, we predict the following genotypes and phenotypes:

$$1/16\,\frac{D\ C}{D\ C} + 2/16\,\frac{D\ C}{d\ C} + 2/16\,\frac{D\ C}{D\ c} + 4/16\,\frac{D\ C}{d\ c} = 9/16 \text{ tall, cut}$$

$$1/16\,\frac{D\ c}{D\ c} + 2/16\,\frac{D\ c}{d\ c} = 3/16 \text{ tall, potato}$$

$$1/16\,\frac{d\ C}{d\ C} + 2/16\,\frac{d\ C}{d\ c} = 3/16 \text{ dwarf, cut}$$

$$1/16\,\frac{d\ c}{d\ c} = 1/16 \text{ dwarf, potato.}$$

Table 1-3 shows an actual result of this cross as compared with the prediction.

The observation is close to the prediction and confirms the hypothesis that these two pairs of alleles are independently transmitted. If one wanted to check

Table 1-3. RESULTS IN F_2 OF A CROSS BETWEEN TALL, POTATO TOMATOES AND DWARF, CUT TOMATOES.

	Tall, Cut	Tall, Potato	Dwarf, Cut	Dwarf, Potato
observed	926	288	293	104
expected (9:3:3:1)	906	302	302	101

Source: MacArthur, *Trans. Royal Can. Inst.*, **18**:8, 1931.

the matter even further, the genotypic ratio in the F_2 could be tested by further matings. This has been done in many cases, and the results confirm again the hypothesis. (What further matings can you suggest that would be efficient in checking the F_2 genotypic ratio? Go back to p. 17, which shows an F_2 population after a dihybrid cross by Mendel. If a testcross were made of the F_1, what phenotypes would be expected in the testcross progeny, and in what proportions?)

Interactions in Phenotypic Expression. The preceding paragraphs bring out an important principle: different pairs of alleles may be independent of each other in their segregation and recombination patterns—in their patterns of genetic transmission from one generation to the next. Independence of gene transmission, however, does not necessarily imply independence of gene action. In fact, in terms of its final expression in the phenotype of the individual, *no gene acts by itself.* A difference between two individuals may be traced to a particulate difference in their germ plasm. But the individuals themselves, complex as they are in their development and functions, are the results of innumerable integrated reactions. Much of the rest of this book will support the point that has just been made. We can find evidence for it immediately by considering modifications of the dihybrid phenotypic ratio of 9:3:3:1.

Cyanide in White Clover. Some strains of white clover are high in cyanide content, while others are low. Rather surprisingly if one thinks of the usual toxicity of cyanide, its presence in clover eaten by cattle apparently does not hurt the animals. On the contrary, high-cyanide lines of white clover, because of their generally richer vegetative growth, may be agriculturally desirable.

Usually, when a high- and a low-cyanide strain are crossed, the F_1 is high, and the F_2 gives a ratio of 3 high:1 low, indicating that a single pair of alleles controls this difference. This suggests that low-cyanide strains, being homozygous recessives, should always give low-cyanide progeny when intercrossed. However, in rather rare instances the following result has been noted:

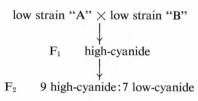

low strain "A" × low strain "B"

F_1 high-cyanide

F_2 9 high-cyanide : 7 low-cyanide

This result is at first unexpected, but with a little thought you will be able to explain it.

The F_2 ratio in this instance is a dihybrid ratio, modified so that the phenotype having both dominants, with a frequency of $\%_{16}$, is high; the remaining $\%_{16}$, having either or both recessives homozygous, are low in cyanide content. To explain this, assume that two different allelic pairs each affect cyanide in such a way that:

$L\text{-} = \text{high} \qquad H\text{-} = \text{high}$
$ll = \text{low} \qquad hh = \text{low}$

P $LLhh \times llHH$
                    low   low

F_1 $LlHh$
                     high

F_2 $L\text{-}H\text{-} = \text{high (9)}$
          $llH\text{-} = \text{low (3)}$
          $L\text{-}hh = \text{low (3)} \left.\begin{array}{c}\\\\\\\end{array}\right\} = 7 \text{ low}$
          $llhh = \text{low (1)}$

The observed 9:7 ratio in the F_2 and all other aspects of this cross are consistent with this explanation.

Now, something is known about the chemistry of cyanide production in white clover. Cyanide (or, more properly, hydrocyanic acid, HCN) is liberated under the action of a specific enzyme from a specific source or substrate known as a *cyanogenic glucoside*. The reaction can be represented as follows:

$$\text{substrate} \xrightarrow{\text{enzyme}} \text{cyanide}$$

One of the low-cyanide lines (*LLhh*) in the cross can easily be shown to have very little, if any, of the enzyme. Therefore, gene *H* determines presence of the enzyme, and we can diagram:

gene *H*
↓
$$\text{substrate} \xrightarrow{\text{enzyme}} \text{cyanide}$$

But, in the absence of *H* (that is, when its allele, *h*, is homozygous), there is little or no enzyme, and little or no cyanide is formed.

It can be shown that the other low-cyanide parental line (*llHH*) has plenty of the enzyme, but lacks the substrate. Therefore, gene L determines presence of the substrate, and we can diagram:

$$\text{gene } H \atop \downarrow$$

$$\underset{\xrightarrow{\phantom{\text{gene } L}}}{\text{gene } L} \text{ substrate} \xrightarrow{\text{enzyme}} \text{cyanide}$$

But, in the absence of L (that is, when its allele, l, is homozygous), there is little or no substrate, and little or no cyanide is formed.

It is likely that gene L also governs the specificity of an enzyme concerned with the conversion of some precursor to the substrate upon which the H enzyme works. For the sake of unity, we will diagram this partly speculative relationship:

$$\text{gene } L \qquad\qquad \text{gene } H$$
$$\downarrow \qquad\qquad\quad \downarrow$$
$$\text{enzyme} \qquad\quad \text{enzyme}$$
$$\xrightarrow{} \text{precursor} \xrightarrow{} \text{substrate} \xrightarrow{} \text{cyanide}$$

Plant leaf extracts can be tested for cyanide, and by adding substrate and enzyme separately the four types of F_2 can be demonstrated (Table 1-4).

Table 1-4. TESTS OF LEAF EXTRACTS FOR CYANIDE CONTENT.

	Phenotype	Leaf Extract Alone	Leaf Extract and Substrate	Leaf Extract and Enzyme H
(9)	L-H-	+	+	+
(3)	L-hh	0	0	+
(3)	llH-	0	+	0
(1)	llhh	0	0	0

"+" indicates a positive test for cyanide.
"0" indicates little or no cyanide.

One observes phenotypic ratios of 9:7, or 12:4 (two different ways), or 9:3:3:1 (taking all tests into account) in this dihybrid F_2 depending on how the plants are tested and classified.

This consideration of cyanide in white clover suggests a basis that is quite common for gene interaction. One gene may depend for its expression on the product of the action of another gene. There are, of course, other bases for other types of interaction. You may enjoy speculating about the possible material bases, in terms of gene action and interaction, of some of the other modified dihybrid ratios that emerge in the questions at the end of this chapter.

General Aspects of Independence and Interaction. In some ways the interaction between two different genes, in which one gene affects the expression of another, will remind you of the phenomenon of dominance. The two phenomena are essentially different, however. *Dominance* always refers to the modification of the expression of one member of a pair of alleles by the other, never to an interaction between different genes; *epistasis* is the term generally used to describe effects of nonallelic genes on each other's expression. The questions at the end of this chapter will confront you with a variety of epistatic effects revealed by modifications of the Mendelian dihybrid ratio.

Keys to the Significance of This Chapter

The differences between simple alternative genetic characters depend on particulate elements, called genes, in the germ plasm. These are paired in an individual, one pair member having come from the paternal, the other from the maternal parent of the individual. When an individual forms germ cells, the members of an allelic pair, of paternal and maternal origin, respectively, separate cleanly without having influenced each other, and go into different germ cells. Consideration of the transmission of two or more pairs of alleles at a time leads to the conclusions that members of different pairs can assort into the germ cells independently of each other, and that they recombine at random at fertilization.

In hybrids one member of an allelic pair may be dominant over the other. Careful discrimination between genotype and phenotype is, therefore, necessary. Environment and genotype interact to determine the phenotype of an individual.

REFERENCES

Auerbach, C., *The Science of Genetics*. New York: Harper & Bros., 1961. (A simple, lucid treatment of the elements of genetics.)

Emerson, R. A., "A Fifth Pair of Factors, *Aa*, for Aleurone Color in Maize, and its Relation to the *Cc* and *Rr* Pairs." *Memoir 16, Cornell Univ. Agr. Exp. Sta.,* 1918. (An early classic on phenotypic interactions.)

Iltis, H., *Life of Mendel*, trans. by E. and C. Paul. New York: W. W. Norton & Co., 1932. (Interestingly written, thorough biography.)

Landauer, W., "Form and Function in the Frizzle Fowl: The Interaction of Hereditary Potentialities and Environmental Temperature." *Biol. Symposia*, **6**:127–166, 1942. (Broad treatment of experimental studies with the frizzle fowl.)

Mendel, G., "Versuche über Pflanzen Hybriden," 1866. (This classic of genetics is available in the original German as Vol. 42, No. 1, of the *Journal of Heredity* and in English translation in Sinnott, E. W., Dunn, L. C., and Dobzhansky, Th., *Principles of Genetics*. New York: McGraw-Hill, 1958.)

Roberts, H. F., *Plant Hybridization before Mendel*. Princeton: Princeton University Press, 1929. (This historical account will reinforce your appreciation of Mendel.)

Stern, C., editor, "The Birth of Genetics," Supplement to *Genetics*, **35**:No. 5, Part 2, 1950. (Includes Mendel's letters to Carl Nageli, 1866–1873, and the papers by DeVries, Correns, and von Tschermak reporting the rediscovery of Mendelism in 1900.)

Tschermak-Seysenegg, E. von, "The Rediscovery of Mendel's Work." *J. Heredity*, **42**:163–171, 1951. (A fascinating firsthand report by one of the rediscoverers of Mendel's work.)

QUESTIONS AND PROBLEMS

Thalassemia is a type of human anemia rather common in Mediterranean populations, but relatively rare in other peoples. The disease occurs in two forms, called minor and major; the latter is much more severe. Severely affected individuals are homozygous for an aberrant gene; mildly affected persons are heterozygous. Persons free of this disease are homozygous for the normal allele. The following four questions concern thalassemia and heredity.

1-1. A man with thalassemia minor marries a normal woman. With respect to thalassemia, what types of children, and in what proportions, may they expect? Diagram the germ-cell unions producing children in this marriage, letting T = the allele for thalassemia, t = its normal alternative. *1:1*

1-2. Both father and mother in a particular family have thalassemia minor. What is the chance that their baby will be severely affected? Mildly affected? Normal? Diagram the possible germ-cell unions in this family. *1:2:1*

1-3. An infant has thalassemia major. From the information given so far, what possibilities might you expect to find if you checked the infant's parents for anemia? *7*

1-4. Thalassemia major is almost always fatal in childhood. How does this fact modify your answer to Question 1-3?

1-5. Purebred Holstein-Friesian cattle are black and white. A recessive allele that, when homozygous, results in red and white is present but rare in this breed. Red-and-white calves are barred from registration, and therefore it is economically important to avoid using for breeding purposes black-and-white individuals that carry the undesirable recessive allele hidden in the heterozygous condition. How might you detect such heterozygosity in a bull to be used extensively in artificial insemination? *mate w/ red cow*

Hans Nachtsheim, of the Free University, Berlin-Dahlem, over a period of several years investigated an inherited anomaly of the white blood cells of rabbits. This *Pelger anomaly*, in its usual condition, involves an arrest of the typical segmentation of the nuclei of certain of the white cells. The rabbits do not appear to be seriously inconvenienced by this anomaly.

1-6. When rabbits showing the typical Pelger anomaly were mated with rabbits from a true-breeding normal stock, Nachtsheim counted 217 offspring showing the Pelger anomaly to 237 normal progeny. What appears to be the genetic basis of the Pelger anomaly? *dominant allele*

1-7. When rabbits with the Pelger anomaly were mated to each other, Nachtsheim found 223 normal progeny, 439 showing the Pelger anomaly, and 39 *extremely abnormal progeny*. Besides having defective white blood cells, these very abnormal progeny showed severe deformities of the skeletal system, and almost all of them died soon after birth. What (in genetic terms) do you suppose these extremely defective rabbits represented? Why do you suppose that there were only 39 of them? *1:2:1*

1-8. What additional experimental evidence might you collect to support or disprove your answers to Question 1-7?

1-9. About one human being in a thousand (in Berlin) shows a Pelger anomaly of white blood cells very similar to that described in rabbits. It is also inherited, in man, as a simple dominant, but in man the homozygous type has not been observed. Can you suggest why, if you are permitted an analogy with the condition in rabbits?

1-10. Again by analogy with rabbits, what genetic situations might be expected among the children of a man and wife each showing the Pelger anomaly?

1-11. From the text, you are already familiar with the alternative characteristics cut leaf and potato leaf found in tomatoes. Another set of alternatives is purple stem versus green stem. A dominant gene A determines purple stem; plants of genotype aa have green stems. When MacArthur crossed a purple-stemmed, potato-leafed true-breeding strain with a true-breeding green-stemmed, cut-leafed strain, he observed a 9:3:3:1 ratio in F_2. Diagram a cross of the F_1's to the purple, potato parent; to the green, cut parent. Diagram a testcross of these same F_1's. What will be the phenotypic ratio in each cross?

1-12. In the table below, results are given for five separate matings of tomato-plant phenotypes. What are the most probable genotypes for the parents in each instance? (These are not experimental data, but are postulated for illustrative purposes.)

	NUMBER OF PROGENY			
PHENOTYPES OF PARENTS	*purple, cut*	*purple, potato*	*green, cut*	*green, potato*
a) purple, cut × green, cut	321	101	310	107
b) purple, cut × purple, potato	219	207	64	71
c) purple, cut × green, cut	722	231	0	0
d) purple, cut × green, potato	404	0	387	0
e) purple, potato × green, cut	70	91	86	77

Example: In cross (a) we know that the purple, cut parent had the dominant allele of each pair, but we do not know from its phenotype whether it was homozygous or heterozygous. It may therefore be written: *A ?C ?*. Similarly, the other parent is *aaC ?*, and our problem is to remove the question marks.

The simplest way to do this is to note that the green, potato progeny (*aacc*) must have received the recessive allele of *each* pair from *each* parent. The parental genotypes therefore must be *AaCc* and *aaCc*. You could reach the same conclusion in other ways. As a check, note that the expected ratio, 3:1:3:1, in this cross agrees well with the observed result.

1-13. A true-breeding tall, purple-stemmed, cut-leafed strain is crossed with a dwarf, green-stemmed, potato-leafed strain. What types of germ cells, and in what proportions, will be produced by the F_1 plants, assuming independent assortment of all three allelic pairs?

1-14. What proportion of the F_2 progeny will be genotypically like the dwarf, green, potato parent strain? Phenotypically? Genotypically like the tall, purple, cut strain? Phenotypically?

1-15. What does epistasis mean? Dominance? How do the two differ?

Many kinds of wild mammals have a peculiar distribution of pigment in their hair. The hair is mostly black or dark brown, but each hair has, just below the tip, a yellow band. This color pattern, called the *agouti* pattern after a wild animal displaying it, gives wild mice, rats, and rabbits, for example, their peculiar "mousy" color, almost indescribable in ordinary color terms.

1-16. In black mice and other black animals the subapical yellow band is not present; the hair is all black. This absence of the *wild* agouti pattern is called *nonagouti*. When mice of a true-breeding agouti strain are crossed with nonagoutis, the F_1's are all agouti, and in F_2 three agoutis appear to one nonagouti. Diagram this cross, letting A = agouti and a = nonagouti, and giving: parental phenotypes, genotypes, and germ cells; F_1 phenotype, genotype, and germ cells; and F_2 genotypes and phenotypes.

1-17. Another inherited color deviation in mice substitutes brown for the black color in the wild-type hair. Brown-agouti mice are called "cinnamons," a good descriptive term for their color. When wild-type mice are crossed with cinnamons, the F_1's are all wild type, and the F_2 consists of three wild type to one cinnamon. Diagram this cross as in Question 1-16, letting B = the black of the wild type, b = the brown of the cinnamon.

1-18. When mice of a true-breeding cinnamon strain are crossed with mice of a true-breeding nonagouti strain (black), the F_1 are all wild type. Explain this "reversion" to wild type by means of a genetic diagram. $aB \times Ab$

1-19. In the F_2 of the cross in Question 1-18, besides the parental types (cinnamon and nonagouti-black), and the wild type of the F_1, a fourth color, called chocolate, shows up. Chocolates are a solid, rich-brown color. What do the chocolates represent genetically? $aa\ bb$

1-20. Assuming that the A-a and the B-b allelic pairs assort independently of each other, what would you expect to be the relative frequencies of the four color types in the F_2? Diagram the crosses of Questions 1-18 and 1-19, showing phenotypes, genotypes, and germ cells. $9:3:3:1$

1-21. What phenotypes, and in what proportions, would be observed in the progeny of a backcross of the F_1's in Question 1-18 to the cinnamon parent stock? To the nonagouti-black parent stock? Diagram these backcrosses.

1-22. Diagram a testcross for the F_1's of Question 1-18. What colors would result and in what proportions?

1-23. Albino (pink-eyed white) mice are homozygous for the recessive member of an allelic pair, C-c, independent of the A-a and B-b pairs in its genetic transmission. Suppose that you had four different highly inbred (and therefore presumably homozygous) albino lines, and that you crossed each of them with a true-breeding wild-type strain and raised a large F_2 progeny in each case. What genotypes for the albino lines would you deduce from the following F_2 ratios:

	NUMBERS OF F_2 PROGENY				
F_2 OF LINE	*wild type*	*nonagouti-black*	*cinnamon*	*chocolate*	*albino*
1	87	0	32	0	39
2	62	0	0	0	18
3	96	30	0	0	41
4	287	86	92	29	164

Handwritten annotations in left margin:
$AA\ bb\ cc$
$AABB\ cc$
$aa\ BB\ cc$
$Aa\ Bb\ cc$

Example: F_1 mice in all crosses are known to have received *ABC* from their wild-type parent and *c* from their albino parent. Crosses to produce the F_2 can therefore be written *A ?B ?Cc* × *A ?B ?Cc*, in which the males and females in any given F_1 have the same genotype.

In the F_2 of line 1, the cinnamon mice must have received the *b* allele from each parent. All of the colored mice, however, have the agouti color pattern. The F_1's of line 1 must therefore have been *AABbCc*. Since these received *ABC* from their wild-type parent, they must have received *Abc* from their albino parent, and since the albino of line 1 is presumed to be homozygous, its genotype must be *AAbbcc*.

As a check, note that the particular 9:3:4 ratio expected in such an F_2 agrees well with the observed result. You could get the answer more quickly by noting that this is a modified dihybrid ratio (9:3:4) involving heterozygosity for albinism and for the *B-b* pair of alleles.

The following three questions refer to general genetic situations, in which the particular phenotypes are not specified.

1-24. A particular cross gives in F_2 a modified dihybrid ratio of 9:7. What phenotypic ratio would you expect in a testcross of the F_1's?

1-25. What phenotypic ratio would you expect from the testcross of an F_1 giving a 13:3 ratio in F_2? A 9:3:4 ratio? A 12:3:1 ratio? A 15:1 ratio? A 9:6:1 ratio? A 1:2:2:4:1:2:1:2:1 ratio? A 3:6:3:1:2:1 ratio? A 9:3:3:1 ratio?

1-26. Indicate the basis of each of the nine dihybrid ratios listed in Questions 1-24 and 1-25.

Example: The 13:3 ratio results when a dominant allele of one gene produces a phenotypic effect indistinguishable from that produced by the homozygous recessive allele of an independent gene. Thus, the dihybrid phenotypic ratio of 9*A-B-*:3*A-bb*: 3*aaB-*:1*aabb* becomes 13:3, since only the third phenotype is different; the first, second, and fourth are alike.

In Duroc-Jersey pigs the coat color is usually red. However, two different true-breeding sandy-colored lines (A and B) are known. If sandy pigs from either of these lines are mated with the normal reds, the F_1 is red and F_2's come out in a ratio of 3 red to 1 sandy.

1-27. If a sandy from line A is mated with a sandy from line B, the F_1 is all red. Make a hypothesis to cover this situation, using appropriate gene symbols and assigning plausible genotypes to the parents and to the F_1 animals.

1-28. If the F_1's in question 1-27 are carried into an F_2 generation one obtains a proportion of color types that is 9 red:6 sandy:1 white. Make a plausible interpretation of this situation. See whether this interpretation is consistent with your hypothesis for Question 1-27. If not, give a new, more appropriate, answer to 1-27.

In the fowl, genotype *rrpp* gives single comb, *R-P-* walnut comb, *rrP-* pea comb, and *R-pp* rose comb.

1-29. What comb types will appear and in what proportions, in F_1 and in F_2, if single-combed birds are crossed with birds of a true-breeding walnut-combed strain?

1-30. What are the genotypes of the parents in a walnut × rose mating from which the progeny are 3 rose:3 walnut:1 pea:1 single?

1-31. What are the genotypes of the parents in a walnut × rose mating from which all the progeny are walnut?

1-32. How many genotypes will produce a walnut phenotype? Write them out.

1-33. F_2 phenotypic ratios of 27:37 are known. What is their basis? Hint: You may find it useful to think in terms of trihybrids.

1-34. Does the genetic explanation of a 27:37 ratio have more in common with that of a 9:7 ratio or of a 15:1 ratio? Explain.

In cattle, *RR* individuals have a red coat, *R'R'* a white coat, and the heterozygotes are roan. The polled characteristic is determined by a dominant gene *P*, the horned alternative by genotype *pp*.

1-35. What will be the phenotype of F_1 individuals after the mating of a white, horned animal with a red that is homozygous for polled? What phenotypes will be found in F_2 and in what proportions?

1-36. If the F_1 individuals in 1-35 are backcrossed to white, horned animals, what phenotypes will be produced and in what proportions?

CHAPTER 2

Sex Chromosomes and Sex Linkage; Probability

The principles of Mendelism were ignored for many years after the publication of Mendel's findings. Not until 1900 was Mendelism rediscovered; it was then immediately appreciated. Why should the world have become so ready to accept it in the interval?

Two main components of progress in biology were probably significant. First, cytologists, looking at cells through microscopes, had accumulated observations sufficient to explain the transmission and behavior of the abstract Mendelian "factors." Second, biologists had begun to appreciate mathematics as an important tool in biology. Mendel's own mathematical inclinations, which had led him to deal easily with the analysis of data recorded as numbers and with chance combinations, had been relatively rare for the biologist of his day. Biology broadened along these lines toward the end of the nineteenth century.

Both of these categories of advance, the observation of the vehicles of inheritance and the mathematical treatment of probabilities, will be introduced in this chapter. They are as important today for understanding heredity as they were at the time of the rediscovery of Mendelism.

Chromosomes. During the latter part of the nineteenth century, and particularly during the remarkable decade of 1880–90, the chromosomes of cell nuclei, now familiar even to one who has studied only the most elementary biology, were identified and examined in detail. The rediscovery of Mendelism coincides with the publication of the second edition of E. B. Wilson's *The Cell*, a book that for many years was the standard work in cytology. If you look back at this edition, you will see that at the turn of the century the behavior of chromosomes in cell division and the essential processes of fertilization of the egg were fairly well understood. The behavior of the chromosomes in the formation of germ cells had been described, but there were many disputed interpretations and some of the views we now regard as established were at that time discredited. It was not accepted, for example, that chromosomes exist in pairs, that they pair and then separate in the formation of germ cells, or even that there are differences among the chromosomes of different pairs. It was not until 1903 that W. S. Sutton gave the first modern interpretation of the relationship between genes and chromosomes, with a cytological explanation of segregation and independent assortment. Details of chromosome behavior will be discussed in the following chapter; for the present we will consider only a particular kind of chromosomes, the *sex chromosomes*, and their behavior with respect to genes displaying a particular pattern of inheritance, *sex linkage*.

Sex Chromosomes. In 1901 and 1902, C. E. McClung, a cytologist working with grasshoppers and other Orthoptera, suggested that a particular chromosome, which he called the "accessory chromosome," was concerned with the determination of sex at fertilization. Although the principle was right, McClung at first had the details wrong because of a peculiarity of his material. They were set straight in 1905 by Miss Nettie Stevens and by Wilson. Looking backward the story seems as simple as could be. In some of the Orthoptera, the female has an even number of chromosomes (*N* pairs = 2*N* chromosomes). The male has an odd number; one of the chromosomes paired in the female has only a single representative in the male. This was McClung's "accessory chromosome"; Miss Stevens dubbed it the X-chromosome. When the female forms eggs, the members of each pair of chromosomes separate from each other; one member of each pair goes into the egg, so that all of the eggs have an X-chromosome. As Miss Stevens said of the normal spermatids, however, "one-half of them must contain the element X, the other half not." When an egg is fertilized by an X-bearing sperm, the zygote is female (XX); when it is fertilized by a sperm lacking the sex chromosome the zygote is male, having only one X, derived from the egg. The sex of the individual that develops from the fertilized egg is therefore determined at fertilization by a chance event, that is, by which type of sperm has functioned.

In other insects, and especially clearly in the common mealworm *Tenebrio molitor*, Miss Stevens observed that the male did not lack a mate for the X-chromosome, but that the partner was a very small chromosome quite different from the X (Fig. 2-1). It was natural to call this unlike partner of the X the Y-chromosome. When the male formed sperm it divided into two classes,

Figure 2-1. *A pair of dissimilar chromosomes, called the X-Y pair, in the mealworm provided a key to sex determination. Note the unequal pair. (From Stevens in* Studies in Spermatogenesis, Carnegie Institution of Washington, *1905, p. 31.)*

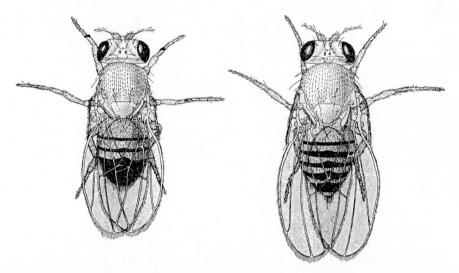

Figure 2-2. Drosophila melanogaster *has contributed greatly to our knowledge of heredity.* Left, *male.* Right, *female.* (*After Sturtevant and Beadle,* An Introduction to Genetics, *W. B. Saunders Co., 1940, frontispiece.*)

one carrying the X-chromosome, the other the Y. During the next two or three years, Miss Stevens, Wilson, and others noted similar situations in other species. One that was soon to prove most important in genetic research was a small fly, *Drosophila melanogaster* (Fig. 2-2). Here the Y-chromosome is not smaller than the X but is different in shape (Fig. 2-3). The pattern of transmission of the X- and Y-chromosomes from one generation to another will be obvious to you (Fig. 2-4).

Sex Linkage. In 1910, Thomas Hunt Morgan published a short paper in *Science* that began: "In a pedigree culture of Drosophila which had been running

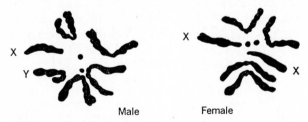

Male   Female

Figure 2-3. *A female* D. melanogaster *has four pairs of chromosomes in her body cells, including a pair of X-chromosomes. A male has an X-Y pair instead of X-X, but is otherwise like the female in his chromosomal makeup.* (*After Dobzhansky, in Morgan,* The Scientific Basis of Evolution, *W. W. Norton, 1932, p. 80.*)

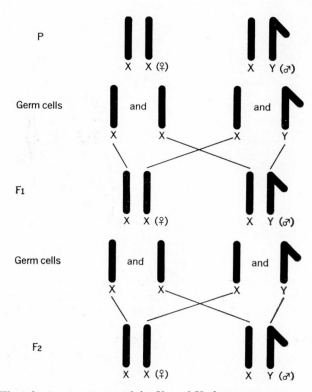

P

X X (♀)   X Y (♂)

Germ cells   and   and

X   X   X   Y

F₁

X X (♀)   X Y (♂)

Germ cells   and   and

X   X   X   Y

F₂

X X (♀)   X Y (♂)

Figure 2-4. *The inheritance pattern of the X- and Y-chromosomes.*

for nearly a year through a considerable number of generations, a male appeared with white eyes. The normal flies have brilliant red eyes." (See Color Plate I.) Morgan mated the white-eyed male to his red-eyed sisters and found that the offspring had red eyes. White eye was therefore evidently recessive. In the F_2 generation there were 3,470 red-eyed flies and 782 white-eyed ones. The data fit rather badly the ratio we might expect, but for the moment we might accept the numbers as representing a Mendelian 3:1 ratio with some shortage in the recessive class (see Prob. 2-33). Morgan noted a surprising feature of this generation: all of the white-eyed flies were males. He then crossed a white-eyed male with some of his red-eyed daughters, and produced:

> 129 red-eyed females
> 132 red-eyed males
> 88 white-eyed females
> 86 white-eyed males

Again, there is some deficiency in the recessive class, but the ratio approximates 1:1 for red eye versus white eye in both sexes, and, as Morgan wrote, "the results show that the new character, white eyes, can be carried over to the females

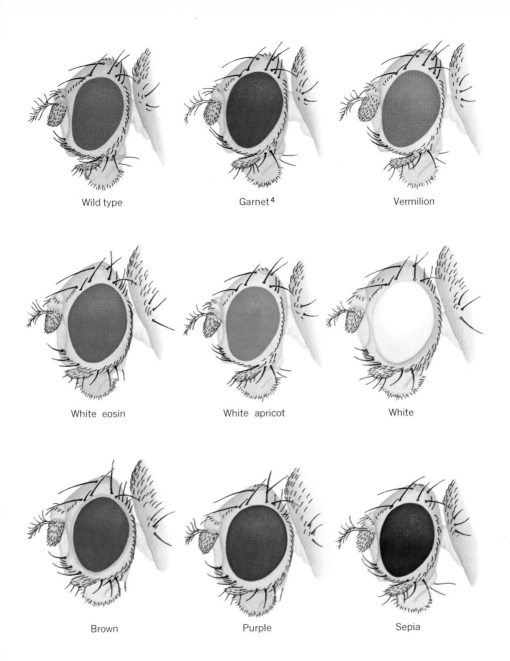

Plate I. *Some eye colors in* Drosophila melanogaster. *(After E. M. Wallace, in* An Introduction to Genetics *by Sturtevant and Beadle, Saunders, 1938.)*

Plate II. *Three-dimensional model of the myoglobin molecule. The molecule consists of a polypeptide chain containing 150 amino acids. Attached to it is a heme group with an iron atom that functions to bind oxygen. (From Kendrew in* Scientific American, *December 1961, pp. 98, 99.)*

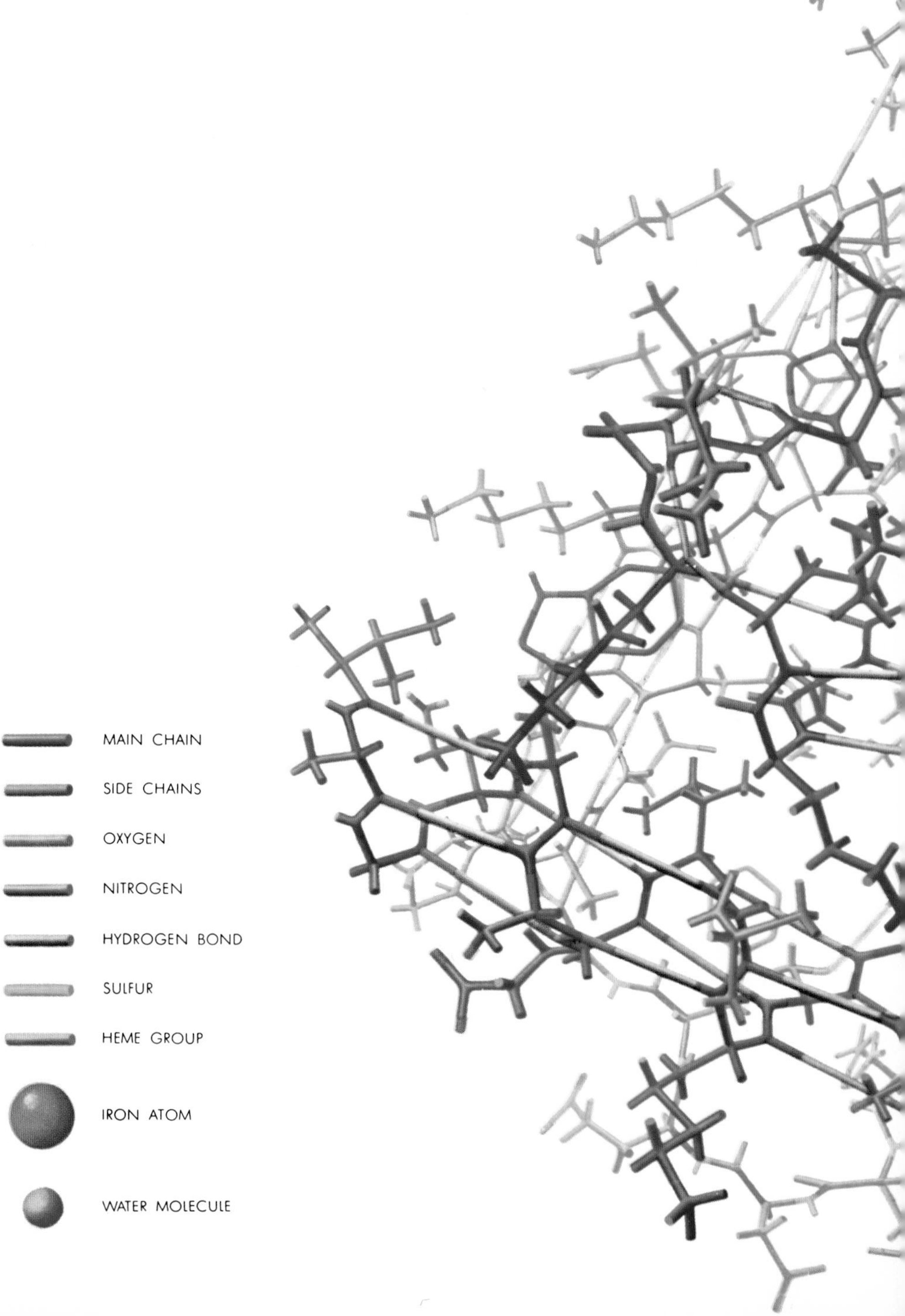

MAIN CHAIN

SIDE CHAINS

OXYGEN

NITROGEN

HYDROGEN BOND

SULFUR

HEME GROUP

IRON ATOM

WATER MOLECULE

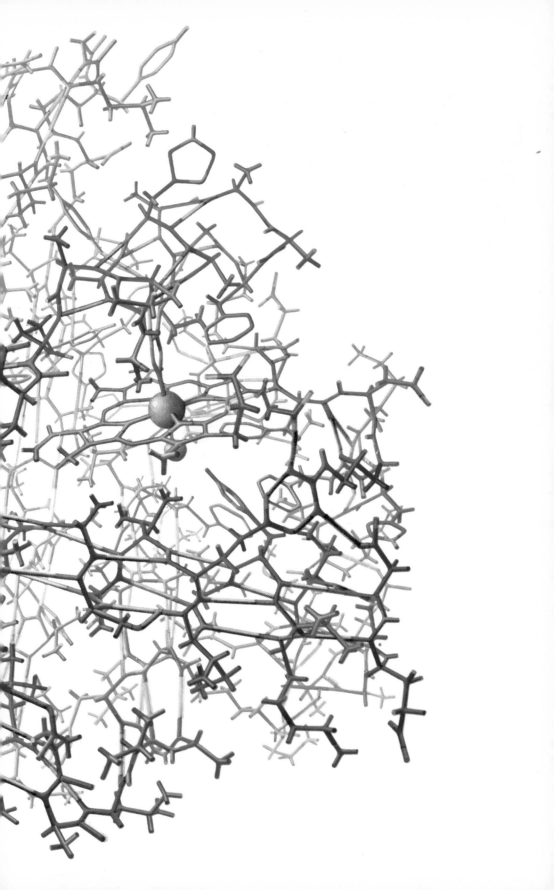

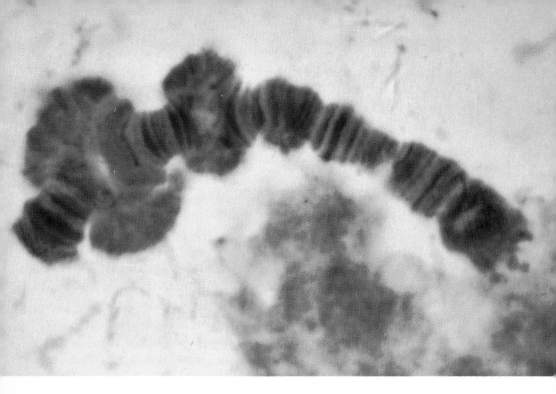

Plate III. *Puffs on chromosome IV in the salivary gland of a midge,* Chironomus tentans. *In addition to the normal salivary chromosome banding, the figures show puffs as protuberances that appear at particular times and places on the chromosome during development, then disappear. In the upper figure, DNA has been stained brown and protein, green. In the lower figure, RNA is stained reddish-violet and DNA, blue. (From Beermann and Clever in* Scientific American, *April 1964, p. 51.)*

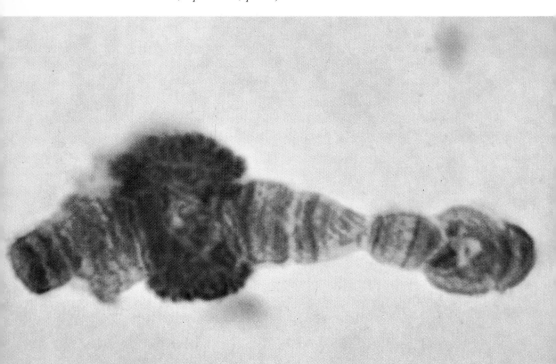

by a suitable cross, and is in consequence in this sense not limited to one sex." Why, then, had all the white-eyed flies in the F₂ been males?

Assume that the gene for this eye-color alternative is located in the X-chromosome, and is not represented at all in the Y. The white allele is recessive. Now, the original white-eyed male must have had the recessive allele for white (*w*) in his single X-chromosome (Fig. 2-5). This male was mated to females having the dominant allele for red eye (*w*⁺) in both X chromosomes. The F₁ males received their mother's X through the egg, and were therefore red-eyed; the F₁ females inherited the dominant *w*⁺ from their mothers, but carried the recessive *w* in their other X, from their fathers. In F₂, only the males who received the X bearing *w* from their mothers were white-eyed. (Work through a diagram similar to Fig. 2-5 for Morgan's backcross, the data for which are given in the preceding paragraph. What would you expect if white-eyed females were crossed with red-eyed males?)

Characteristics dependent on genes that follow the pattern of the X-chromosome in inheritance are known as *sex-linked* characteristics, and the controlling genes are sex-linked genes.

Attached-X. Soon after the publication of Morgan's paper, numerous other sex-linked genes were detected in Drosophila, and in other species. One of the most useful of these to the geneticist involved a marked reduction in the number

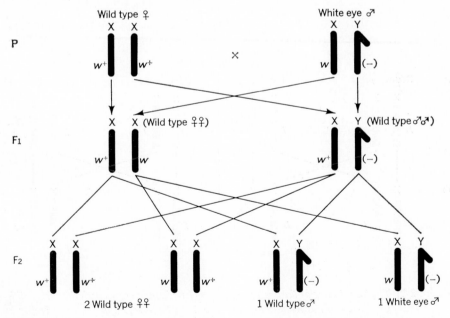

Figure 2-5. *White eye color in Drosophila is a sex-linked recessive characteristic.*

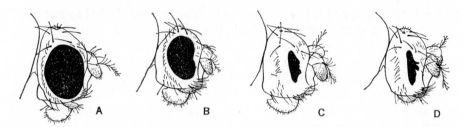

Figure 2-6. *Wild-type Drosophila* (A) *have round eyes, but the eyes of bar-eye females* (C) *and males* (D) *are small and narrow because of a greatly reduced number of facets. Only females may show the intermediate "wide-bar" phenotype* (B). (*After Morgan*, The Theory of the Gene, *Yale University Press, 1926, p. 87.*)

of facets in the fly's normally large, round compound eyes. This difference from *wild type* is called *bar eye*, because when it is obtained in true-breeding stocks, the eyes of both sexes are reduced to narrow red bars (Fig. 2-6C and D). When flies of the bar-eye stock are crossed with wild type, the following results are obtained in F_1 (σ = male, φ = female. $\sigma\sigma$ and $\varphi\varphi$ are conventional plural symbols, like "pp" for pages.):

round eye φ × bar eye σ (wild type)		bar eye φ × round eye σ (wild type)	
F_1:	$\sigma\sigma$ round eye $\varphi\varphi$ wide bar eye	F_1:	$\sigma\sigma$ bar eye $\varphi\varphi$ wide bar eye

You will note that this pattern of inheritance is consistent with an assumption that eye shape is controlled by a pair of sex-linked alleles without dominance, so that heterozygous females have wide bar eyes (Fig. 2-7).

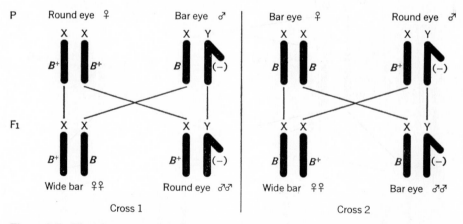

Figure 2-7. *The inheritance of the bar-eye characteristic behaves as though the responsible gene were in the X-chromosome. The Y-chromosome appears to be empty in this regard.*

How can this be checked? One way is to predict on this basis what kind of an F_2 would be produced in each cross, and then compare what is actually obtained with this prediction. A checkerboard for cross 1 of Figure 2-7 is shown in Figure 2-8. We would predict a ratio of approximately 1 round ♀ :1 wide bar ♀ :1 round ♂ :1 bar ♂ in the F_2 of this cross. This is the result actually observed. (Show that the observed ratio of 1 bar ♀ :1 wide bar ♀ :1 round ♂ :1 bar ♂ in the F_2 of cross 2 is also in agreement with expectation.)

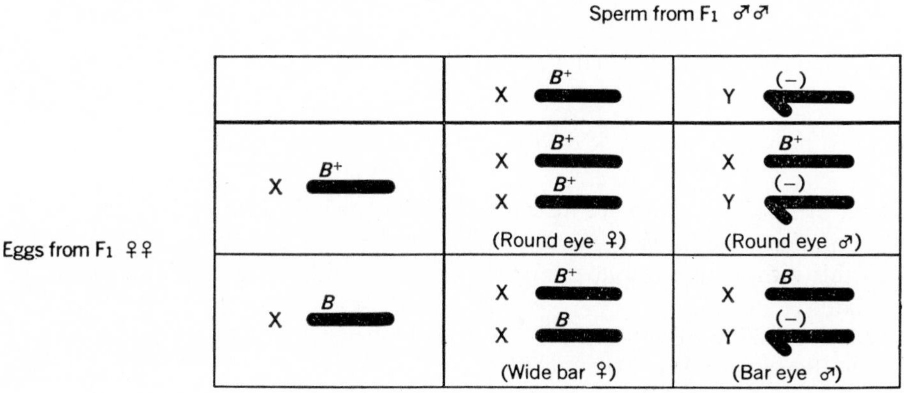

Sperm from F₁ ♂ ♂

	X B^+	Y (−)
X B^+	X B^+ X B^+ (Round eye ♀)	X B^+ Y (−) (Round eye ♂)
X B	X B^+ X B (Wide bar ♀)	X B Y (−) (Bar eye ♂)

Eggs from F₁ ♀ ♀

Figure 2-8. *The F_2 prediction from the hypothesis in Figure 2-7 is subject to experimental test.*

Other predictions based on the same hypothesis (backcross results, for example) have also been tested repeatedly. The uniform agreement between observed and expected results provides convincing evidence that the bar-round alternative is controlled by a pair of alleles in the X-chromosome.

One day several decades ago, an unusual fly turned up in the Drosophila cultures of L. V. Morgan (Mrs. T. H. Morgan). While the fly was being examined by some of the workers in the laboratory, the ether anesthesia under which it was held wore off, and with little warning it left the stage of the microscope and flew away. There were other flies about, and there seemed little chance of recapturing this particular one. But it was a queer-looking female specimen, and considerable effort was made to find her. She was finally caught on a window pane.

This story may seem trivial. But in the development of knowledge of heredity an individual fly has often been more important than a cow. And this particular fly was an unusually important one, because she was the beginning of an exceptional stock. Incidentally, the fact that she looked queer had nothing to do with her eventual contribution to genetic research.

A typical experimental result with females of this stock is this: When bar-eye males are crossed with females of the exceptional stock, the offspring are just

Figure 2-9. *A female Drosophila homozygous for* B⁺, *whose X-chromosomes were attached to each other, would form two kinds of eggs.*

the opposite of the normal and usual result of a cross between round-eye females and bar-eye males:

exceptional ♀♀ × bar eye ♂♂ → bar eye ♂♂, round eye ♀♀
(round eye)

Here, the *males* seem to get their allele of the bar pair *from their fathers*, while the *females are like their mothers*—the reverse of ordinary sex linkage. How can this remarkable result, uniformly obtained with the exceptional stock, be explained?

After thinking about her experimental observations, Mrs. Morgan arrived at the hypothesis that the X-chromosomes of the exceptional females might be attached to each other, so that they failed to separate when germ cells were produced. Such females would then form two kinds of eggs, as shown in Figure 2-9. Their predicted behavior in a cross with a bar-eye male is shown by the checkerboard in Figure 2-10.

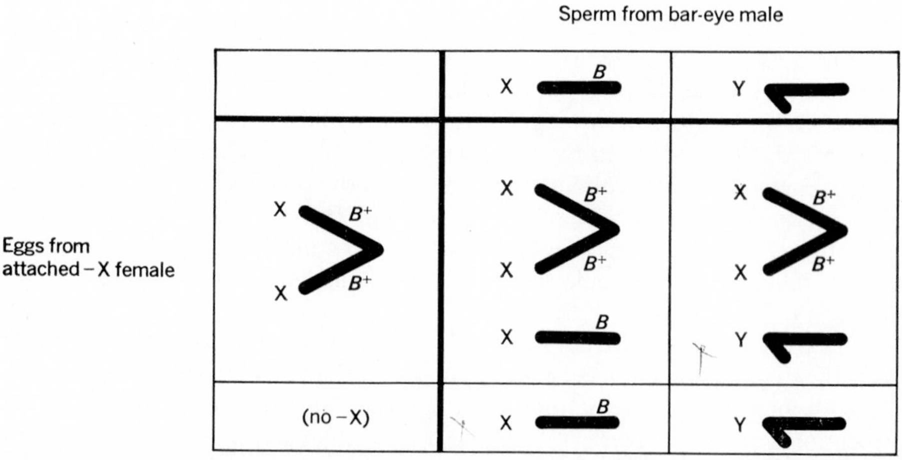

Sperm from bar-eye male

	X ▬▬ B	Y ◀
Eggs from attached–X female X ▷ B⁺ / X B⁺	X ▷ B⁺ / X B⁺ X ▬▬ B	X ▷ B⁺ / X B⁺ Y ◀
(no –X)	X ▬▬ B	Y ◀

Figure 2-10. *An attached-X female homozygous for* B⁺, *when crossed with a bar-eye male, would produce round-eye female and bar-eye male progeny, and two additional exceptional classes as well.*

The checkerboard shows some important points in common with the actual breeding results. It provides for XX progeny lacking the bar allele; these might well be the round-eye females in the cross. And it provides for one-X progeny with the bar allele from their fathers; these may be the bar-eye male progeny. It also raises some obvious questions. If the hypothesis is correct, what happens to the expected "three-X" and "no-X" progeny? And how does the Y-chromosome fit into the picture?

At any rate, the hypothesis showed sufficient promise to be worth checking. The confirmation was straightforward: microscopic observation of sex chromosome behavior in germ-cell formation by the exceptional females showed without question that the two X-chromosomes were attached to each other. Besides the attached-X's, the exceptional females had a Y-chromosome (like those predicted in the upper-right block of our checkerboard). We need to revise our checkerboard slightly in line with this observation (Fig. 2-11).

Knowing that the checkerboard is based on fact, it is now possible to dispose of the questions it raises.

1. Individuals lacking an X-chromosome (lower-right corner of checkerboard) die before hatching from the egg.

2. Individuals with three X's (upper left) usually die. The few that survive are so different from other flies that they are easily identified. Such individuals

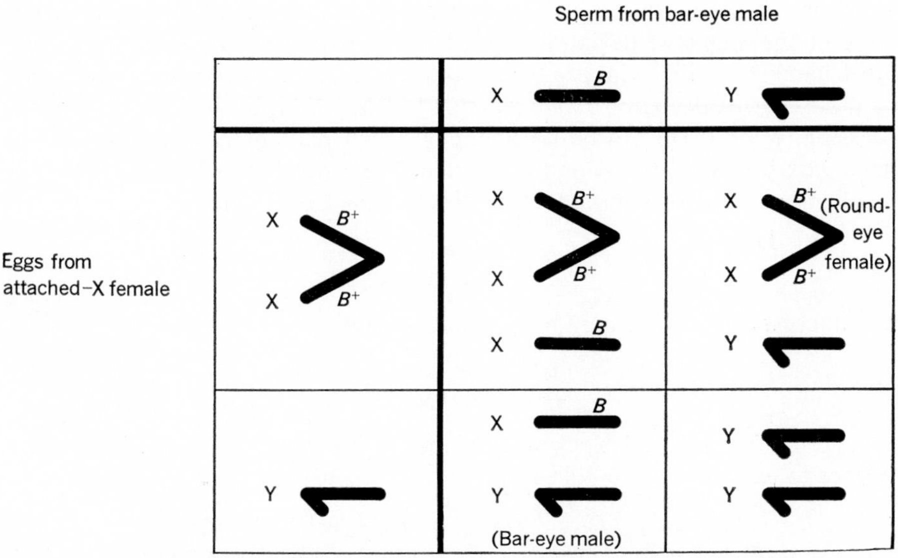

Figure 2-11. *The attached-X condition in the exceptional stock, postulated from genetic evidence, is confirmed by cytological observation. Attached-X females also have a Y-chromosome, which modifies the hypothetical diagram in Figure 2-10 slightly.*

have contributed to genetic knowledge. We shall refer to them again in Chapter 12.

3. The presence of a Y-chromosome in an attached-X (symbol: $\widehat{XX}$) female does not affect her sexual characteristics at all. It has also been shown, in other cases, that one-X flies lacking a Y are normal males in appearance, although they are sterile. It is evident that it is the "one-X or two-X" alternative that is essential in the normal determination of sex in Drosophila; the presence or absence of the Y is largely irrelevant.

Observations like those on $\widehat{XX}$ provide the most convincing kind of confirmation that the allele for bar eye and its normal alternative are actually located in the X-chromosomes. We have already found that this pair of alleles coincides with the X-chromosomes in its normal pattern of inheritance. When a visible and regular abnormality in the distribution of the X-chromosomes can be shown always to be associated with an identical abnormality in the distribution of the bar gene, the relationship becomes an established fact.

Genetic Symbols. In Drosophila genetics, the use of a "+" to distinguish the wild-type allele of any gene, as in the white- and bar-eye examples, automatically identifies a given allele as that found in the standard type. *By using a capital letter to designate a dominant deviation from type, and a small letter to designate a recessive one*, the dominance relationships of allelic forms of a gene are clear from the symbolism. For example B^+ is the symbol for the wild-type allele of the *dominant* deviation from the standard type, bar eye (B). On the other hand, b^+ is the symbol for an entirely different gene, the wild-type allele of the *recessive* deviation from the standard type, black body color (b). On occasion, when there can be no doubt of which particular wild-type allele is under consideration, the symbol is shortened to a simple "+." Thus, in crosses involving only the bar-normal alternative, we can conveniently use:

$$B = \text{bar eye}$$
$$+ = \text{wild type,}$$

and in crosses involving only body color, we could use

$$b = \text{black body color}$$
$$+ = \text{wild type.}$$

When both bar eye and black body color are involved in a cross, however, it is safer to use:

B = bar eye	b = black body color
B^+ = wild-type eye shape	b^+ = wild-type body color.

In fact, whenever two or more different *genes* are concerned in a cross, it is generally safer to specify which plus-allele you are talking about by using the appropriate symbol, rather than using a simple "+."

The letter used for the symbol of the gene conventionally comes from the *name of the characteristic, different from the standard, controlled by a deviant form of the gene*. Thus, a gene controlling eye color, giving a white-eye/red-eye alternative, has *w* as its basic symbol, since *white* is the deviant type; the symbol *r* would be wrong here, because *red* is the standard characteristic. Consider the information, then, provided by the symbol *w*+: It represents the wild-type allele of a recessive deviation from type, and we can easily remember what this deviation is, since *w* stands for white.

Many geneticists have adopted the Drosophila conventions in assigning symbols to genes in other animals and in plants. However, many other geneticists retain older and less informative conventions. Recall the cut-leaf/potato-leaf alternative for tomatoes, in Chapter 1. Here, the standard symbolism does not include the "+" convention at all; the symbol *C* comes from the *standard type* (cut leaf), and *c* stands for a recessive deviation from type called potato leaf.

You may at first find the convention in which a capital letter represents a dominant, and the corresponding small letter its recessive allele, and in which the particular letter chosen is arbitrary, the least confusing method of assigning symbols to genes. It undertakes to give less information, and it therefore appears less complex. If you continue to study genetics, however, you will increasingly appreciate the Drosophila conventions, and in reading other genetic literature you will find it important to be able to follow both types of representation with ease. For these reasons, we will use both methods in this book.

Multiple Alleles. Implied in our introductory sentence to the section on attached-X, "numerous other sex-linked genes were detected in Drosophila," was the fact that the X-chromosome bears many distinct genes. Early examples included different genes affecting eye color in Drosophila; for example, when flies with the recessive *vermilion* eye color were crossed with those with the recessive *white* eyes, the female hybrids had wild-type *red* eyes (see Color Plate I). The preceding section on genetic symbols will help you to visualize the situation: the vermilion stock has X-chromosomes (w^+v), where w^+ is the wild-type allele of the gene for white eyes, and *v* is the recessive allele of another gene, for vermilion; and the white stock has X-chromosomes (wv^+). The female hybrids are heterozygous for both genes, and show the dominant effect of the wild-type allele for each allelic pair.

A different situation was observed by Morgan in 1912, and interpreted by Sturtevant in 1913, for the eye colors white and *eosin* (see Color Plate I). Here eosin, like white, was recessive to the wild-type red. But when eosin and white stocks were crossed, the female progeny did not "revert" to wild-type red, but were eosin in eye color. As Sturtevant noted, "Red eye is a dominant to eosin and to white, and eosin is also a dominant to white." The *eosin* and *white* alleles could be considered to be variants of a single wild-type gene; again the genetic symbolism will help to make the situation clear. Here, we let *w* represent the gene and distinguish three alleles: w^+ = wild-type red, w^e = eosin, and w = white. Assuming dominance in the order given, and remembering that females have only two X chromosomes and can therefore have only two representatives of the gene, one on each chromosome, while males can have only one, you can easily work through problems involving three or more alleles of this sex-linked gene. For example, eosin females heterozygous for white are w^ew; when mated to wild-type (w^+) males they are expected to produce equal numbers of eosin (w^e) and white (w) sons, but all their daughters are red-eyed, half of them heterozygous for eosin (w^+w^e) and the other half heterozygous for white (w^+w). You will find that the common genetic conventions of using a basic symbol for a gene, distinguishing allelic forms of the gene by superscripts, and using different basic symbols for different genes, provide clear and useful tools for thinking about genetic problems.

The number of possible allelic forms of a gene is not, of course, limited to three; very long multiple-allelic series representing hundreds of different forms are known for some genes. And multiple-allelic series are not limited to sex-linked genes; other genes also exist, in populations, in many allelic forms. Some of the problems at the end of this chapter will give you an opportunity to practice with multiple-allelic situations. When you come to them, remember that one chromosome can normally carry only one representative of a given gene; even though many alleles of a gene may exist in a population, in an individual the number is limited by the number of chromosomes of the appropriate pair—typically, *two* for most chromosomes and *one* for situations like the single X of male Drosophila.

Multiple alleles have provided material for fine genetic dissections of the genetic material, and sometimes apparent multiple alleles turn out to be more complex than they seemed to be at first. These subjects will be discussed later in this book, particularly in Chapter 9.

Sex Linkage in Other Organisms. Sex determination by sex chromosomes, and the location of genes in these chromosomes, are similar in a variety of species to the situation described for Drosophila. In man, for instance, many of the known, simply inherited characteristics are controlled by sex-linked genes. Examples are the sex-linked recessive gene for ordinary red-green color blindness, and the "bleeder's disease," *hemophilia*, in which the blood fails to clot normally—also a sex-linked recessive.

Knowledge of the genetic pattern for such a defect often helps in making definite predictions about the children of concerned individuals. It is clear, for example, that a man whose blood clots normally cannot transmit to his children the allele for hemophilia, no matter how high the incidence of this disease may be in his family. A man has only one X-chromosome; if the allele for hemophilia is present in that chromosome, the man is a "bleeder," and if he is not a "bleeder," then his X-chromosome does not carry the defective allele. A normal woman whose father was hemophilic, on the other hand, can expect that about half of her sons will be hemophilic. (Chance, of course, enters into the experience of any particular family in this respect.)

It is of interest to note that a few dogs show a condition similar to human hemophilia, and that here too the allele responsible is a sex-linked recessive. Sex linkage has been infrequently observed in other mammals. In spite of long and extensive studies of inheritance in mice, for example, it was only recently that the first X-borne gene was identified. Curiously, a number of others were then soon recognized, and by now more than 20 sex-linked mutations have been noted in the mouse.

One interesting example in cats has been known for many years. The black and yellow pigments in the coats of cats seem to be controlled by a sex-linked pair of alleles in such a way that the heterozygote is the familiar "tortoiseshell," having areas of black and areas of yellow in its coat. We can list (note that Y stands for a gene in the X-chromosome here, not for the Y-chromosome):

FEMALES	MALES
yy = yellow	$y(-)$ = yellow
Yy = tortoiseshell	$Y(-)$ = black
YY = black	

You are therefore justified in giving good odds in a wager that the calico cat in the next block is a female. Winning your bet would not be quite certain, however, since tortoise tomcats do very rarely appear. The fact that they are almost always sterile shows that there is something abnormal about their sexual development.

Plants like corn and some other common garden vegetables have both sexes represented in the same individual. There is no sex difference between individuals; no sex chromosomes exist, and no sex linkage is possible. But some plant species are bisexual (*dioecious*, "two houses"). Hemp, date palms, and willows are examples. In some dioecious plants, sex chromosomes have been identified, and sex linkage is known.

Birds differ from the Drosophila type of sex determination in one important respect: The *male* has two like sex chromosomes, and it is the female that has an unpaired chromosome or an unlike-membered pair. The effect of this is to reverse the pattern of sex-linked inheritance now familiar to us in Drosophila. To see the difference illustrated, observe the inheritance of the sex-linked dominant for barred feathers in poultry—a color pattern characteristic of the Barred Plymouth Rock breed (Fig. 2-12).

Besides birds, the moths and butterflies, all reptiles that have been carefully examined, and some kinds of fish and amphibia, have the XX = male, XY or XO = female type of sex determination. Other kinds of animals, and many dioecious plants for which a sex-chromosome mechanism of sex determination holds, have the Drosophila type.

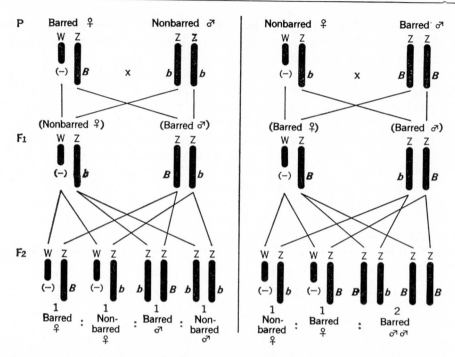

Figure 2-12. *Barred feathers, a sex-linked dominant color pattern in the fowl, shows the pattern of inheritance characteristic of forms in which it is the female that has a single X-chromosome.*

Sex Determination in Mice and Men. The Y-chromosome in Drosophila seems unimportant in sex determination; XO flies (lacking a Y-chromosome) are normal males in appearance, although they are sterile. It is the 1-X or 2-X alternative that is essential in the normal determination of sex in Drosophila; the presence or absence of the Y-chromosome is largely irrelevant. For many years, it was assumed that the same situation held for mammals, even though it was known that in other forms (for example, in the plant Melandrium) the male-specific Y-chromosome played an essential part in the determination of the male sex. In 1959, geneticists discovered that the Y-chromosome plays a definite role in sex determination of mice and men. In mice, certain females showed a recessive phenotype that should have been obscured by a dominant allele if they had inherited an X-chromosome from their mothers. These females were deduced to be XO in chromosomal constitution, their single X-chromosome being of paternal origin and their femaleness being due to lack of a Y-chromosome. Cytological studies proved this hypothesis to be true. Later, XXY mice were shown to be male in phenotype; again, the Y-chromosome played a determining role in sex. Coincidentally, certain human intersexes that are predominantly female in phenotype (Turner's syndrome) were shown to be

XO in chromosomal constitution, while aberrant males (Klinefelter's syndrome) were shown to have two or more X-chromosomes but also to have a Y. We will leave to a later chapter a consideration of the mechanisms by which such aberrant types arise.

The recognition of the active role of the Y-chromosome in mammals led to the solution of a long-standing puzzle. We noted earlier that tortoiseshell tomcats do very rarely appear and suggested that since they are usually sterile, there is something abnormal about their sexual development. It has recently been found that a tortoiseshell tomcat has the sex chromosome constitution XXY.

Probability

An appreciation of the laws of chance is obviously basic to understanding the transmission of hereditary qualities. Sex ratios in families provide a convenient introduction to this subject. To illustrate, we can ask: "Among families with two children, what proportion will have two boys? A boy and a girl? Two girls?"

There will be four kinds of families of two children, and if we can assume for the present that equal numbers of girls and boys are born (a sex ratio of 1:1), these four types of families will be equal in frequency. Considering the children in order of birth, and letting B stand for boy and G for girl, the families will be:

$$1BB:1BG:1GB:1GG$$

If only sex ratios are considered, disregarding birth order, the families will be distributed:

$$1BB:2BG:1GG, \text{ or } \tfrac{1}{4}BB:\tfrac{1}{2}BG:\tfrac{1}{4}GG$$

Another way of arriving at this is more direct. If the probability of a boy on any trial is $\frac{1}{2}$, then the probability of boys on each of two independent trials is $\frac{1}{2} \times \frac{1}{2} = \frac{1}{4}$. Similarly, the probability of two girls is $\frac{1}{2} \times \frac{1}{2} = \frac{1}{4}$. But there are two ways one could get one boy and one girl—namely, *boy then girl* and *girl then boy*. So this probability is $2 \times \frac{1}{2} \times \frac{1}{2} = \frac{1}{2}$. The probability distribution is therefore:

$$(\tfrac{1}{2})^2 \text{ BB}:2(\tfrac{1}{2})(\tfrac{1}{2})\text{BG}:(\tfrac{1}{2})^2 \text{ GG}$$
$$\tfrac{1}{4} \qquad \tfrac{1}{2} \qquad \tfrac{1}{4}$$

This checks with our previous result.

How about families of three children? On the same assumption, the possible combinations are:

	GBB	BGG	
BBB	BGB	GBG	GGG
	BBG	GGB	

1	:	3	:	3	:	1

(3 boys) (2 boys, 1 girl) (1 boy, 2 girls) (3 girls)

$\frac{1}{8}$ $\frac{3}{8}$ $\frac{3}{8}$ $\frac{1}{8}$

or, more directly:

$$(\tfrac{1}{2})^3 \; + \; 3(\tfrac{1}{2})^2(\tfrac{1}{2}) \; + \; 3(\tfrac{1}{2})(\tfrac{1}{2})^2 \; + \; (\tfrac{1}{2})^3$$

(3 boys)   (2 boys, 1 girl)   (1 boy, 2 girls)   (3 girls)

$\frac{1}{8}$ $\frac{3}{8}$ $\frac{3}{8}$ $\frac{1}{8}$

You can work out the distribution in families of four children, and confirm that it would be:

$$(\tfrac{1}{2})^4 \; + \; 4(\tfrac{1}{2})^3(\tfrac{1}{2}) \; + \; 6(\tfrac{1}{2})^2(\tfrac{1}{2})^2 \; + \; 4(\tfrac{1}{2})(\tfrac{1}{2})^3 \; + \; (\tfrac{1}{2})^4$$

(4 boys)   (3 boys, 1 girl)   (2 boys, 2 girls)   (1 boy, 3 girls)   (4 girls)

$\frac{1}{16}$ $\frac{4}{16}$ $\frac{6}{16}$ $\frac{4}{16}$ $\frac{1}{16}$

Such computations become tedious. Fortunately, there is a convenient shortcut. The frequencies of the various combinations correspond to the coefficients of the binomial expansion:

$$(a + b)^2 = a^2 + 2ab + b^2,$$
$$(a + b)^3 = a^3 + 3a^2b + 3ab^2 + b^3,$$
$$(a + b)^4 = a^4 + 4a^3b + 6a^2b^2 + 4ab^3 + b^4,$$

and so on.

The *exponents* of a and b in these expansions correspond to the *number of children of each type* in the family to which the coefficient applies. We can make this clearer by illustrating its use.

The general formula is

$$(a + b)^n,$$

where a is the probability of a boy ($\tfrac{1}{2}$), b is the probability of a girl ($\tfrac{1}{2}$), and n is the number of children per family in the family size under consideration.

Now we can ask: What is the probability that in a family of five children, there will be two boys and three girls? In this example:

$$(a + b)^5 = a^5 + 5a^4b + 10a^3b^2 + 10a^2b^3 + 5ab^4 + b^5$$

We want the value of $10a^2b^3$ (that is, 2 boys, 3 girls). This is:

$$10(\tfrac{1}{2})^2(\tfrac{1}{2})^3 = \tfrac{10}{32} = \tfrac{5}{16}.$$

In other words, among sixteen families of five children each, about five such families on the average should have two boys and three girls.

Regularities of the Binomial Expansion. Two symmetries in the binomial expansion make its use easy.

1. Coefficients. The coefficients of successive powers of the binomial can be arranged in a regular triangle (*Pascal's triangle*).

POWERS			COEFFICIENTS						
(*n*)									
1	1	1							
2	1	2	1						
3	1	3	3	1					
4	1	4	6	4	1				
5	1	5	10	10	5	1			
6	1	6	15	20	15	6	1		
7	1	7	21	35	35	21	7	1	
8	1	8	28	56	70	56	28	8	1

You will note that the horizontal rows are symmetrical, and that in the triangle each number is the sum of the numbers immediately above and one place to the left above.

2. Exponents. The exponents in the binomial expansion also follow regular series. For example, in

$$(a + b)^6 = a^6 + 6a^5b + 15a^4b^2 + 20a^3b^3 + 15a^2b^4 + 6ab^5 + b^6,$$

you will note that the exponents of *a* begin at the power to which the binomial is raised and decrease regularly to 0, while the reverse is true of the exponents of *b*.

Other Similar Uses of the Binomial. The utility of the binomial is not limited to calculations involving 1:1 ratios. The probabilities can be any fractions whose sum equals unity. For example, among marriages between heterozygotes the probability of a child showing the dominant characteristic is $\frac{3}{4}$, and the probability of a child showing the recessive characteristic is $\frac{1}{4}$.

Suppose we ask: Among families of eight children whose parents are both heterozygous, what proportion will have exactly six children showing the dominant and two the recessive characteristic? To answer this kind of question, remember the following rule, which is based on the expression for the binomial coefficients.

If the probability of occurrence of an event *A* is *p*, and the probability of occurrence of the alternative event *B* (or the nonoccurrence of *A*) is *q*, then the probability that in *n* trials event *A* will occur *s* times and event *B* will occur *t* times is given by

$$\frac{n!}{s!t!} p^s q^t,$$

where the symbol *n*! is read "*n* factorial," and represents the product of all the integers from 1 to n ($8! = 1 \times 2 \times 3 \times \cdots \times 8$). *Note:* You may sometimes need to know that $0! = 1$. Note also that $s + t = n$, and $p + q = 1$.

Rather than write factorial expressions in full, it is handier to write them in such a way that cancellations can be used to minimize arithmetic:

$$\frac{n!}{s!t!} = \frac{8!}{6!2!} = \frac{6! \times 7 \times 8}{6! \times 2!} = 28.$$

The term needed is therefore

$$28p^6q^2 = 28(\tfrac{3}{4})^6(\tfrac{1}{4})^2 = 0.31.$$

This is a rather academic example, but it illustrates one interesting point: Among families of eight, only about one-third will show the ideal ratio of 6:2, and the other families will deviate from this expected value in one direction or the other.

We could also ask questions along this line: How often, in families of eight, would we expect the deviation from the expected 6:2 ratio to be as much as or more than two? In other words, how often would there be eight children showing the dominant, or four or less showing the dominant? We could get the answer by summing the appropriate terms of the binomial.

Probabilities and Independent Events. It is sometimes confusing to note that two apparently similar cases have different probabilities if more information is available on one than on the other. For example, as a prediction about an anticipated two-child family, we could say that there is one chance in four that both children will be boys. After one son is born, however, there is one chance in two that the birth of a second child will produce a two-boy family. This may seem inconsistent; in the first situation the chance for two boys is only $\frac{1}{4}$, while in the second it is $\frac{1}{2}$.

The difference, of course, is that the probabilities are based on different possibilities in the two situations. In the first, having two boys is one of four equally possible eventualities:

$$\text{BB} \quad \text{BG} \quad \text{GB} \quad \text{GG}$$

But in the second, the last two of these eventualities have been eliminated. Now there are only two possible eventualities;

$$\text{BB and BG}$$

and the probability of BB has increased to $\frac{1}{2}$.

The problems we have been considering all depend on the *independence* of the successive events whose probabilities are computed. Even if a family has nine boys in a row, we assume that the chance a tenth baby will be a girl is $\frac{1}{2}$.

Simple probability considerations can provide shortcuts for the calculation of dihybrid F_2 ratios, and similar problems. The shortcuts are based on one of the laws of chance:

If two events are independent, the chance that they will occur together is the product of their separate probabilities.

It will be simpler to illustrate an application of this law than to engage in further definition. With regard to the tall-dwarf alternative in the tomato examples of Chapter 1, the expected $3:1$ phenotypic ratio in the F_2 generation tells us that:

The chance (probability) that an individual plant will be tall $= \frac{3}{4}$; the chance that an individual plant will be dwarf $= \frac{1}{4}$.

Similarly, with regard to the cut-potato leaf alternative:

The chance that an individual plant will be cut $= \frac{3}{4}$; the chance that an individual plant will be potato $= \frac{1}{4}$.

Assuming independence between these two pairs of alternatives, we can compute the chance that an individual plant will be:

$$\text{tall, cut} = \frac{3}{4} \times \frac{3}{4} = \frac{9}{16}$$
$$\text{tall, potato} = \frac{3}{4} \times \frac{1}{4} = \frac{3}{16}$$
$$\text{dwarf, cut} = \frac{1}{4} \times \frac{3}{4} = \frac{3}{16}$$
$$\text{dwarf, potato} = \frac{1}{4} \times \frac{1}{4} = \frac{1}{16}$$

These results are the same as those from the checkerboard.

Table 2-1. DISTRIBUTION OF GENOTYPES IN F_2.

Tall-Dwarf Alternative		Cut-Potato Alternative	
Genotype	*Probability*	*Genotype*	*Probability*
DD	$\frac{1}{4}$	*CC*	$\frac{1}{4}$
Dd	$\frac{1}{2}$	*Cc*	$\frac{1}{2}$
dd	$\frac{1}{4}$	*cc*	$\frac{1}{4}$

Table 2-1 shows that essentially the same method can be used to compute genotypic ratios. You can compute the expected frequency for any genotype separately, without filling in and collecting genotypes for the whole checkerboard. For example, the genotype $DDcc$ occurs with the frequency $\frac{1}{4} \times \frac{1}{4} = \frac{1}{16}$, the genotype $DdCc$ with the frequency $\frac{1}{2} \times \frac{1}{2} = \frac{1}{4} (= \frac{4}{16})$, and so on. These computations agree with the laborious collections of the same genotypes from the checkerboard.

Chance Deviations from Expected Ratios. Another aspect of the operation of chance in genetic systems must be considered. We have had repeated occasion to compare predicted ratios with those observed experimentally. Those chosen for illustrative purposes have seemed close enough to expectation, when the expectation was properly computed. But two questions have perhaps bothered you: Why does the observed not turn out to be *identical* with the expected if the hypothesis on which the expected distribution is based is correct? Just how large a difference can we permit between the observed and the expected before we suspect that the hypothesis being tested is false or inadequate?

An answer to the first question will have occurred to you if you have flipped pennies. The working hypothesis in this game is that there is equal likelihood a penny will fall heads or tails. Or, to put the matter more formally, the probability that a penny will fall heads is equal to $\frac{1}{2}$, and the chance that it will fall tails is also $\frac{1}{2}$. Then, if you toss a penny 100 times, you expect that it will fall heads 50 times ($\frac{1}{2} \times 100$) and tails 50 times. But if you undertake this experiment in probability and toss a penny 100 times, you will find that the ideal expectation of 50 heads:50 tails is not often realized. It is much more likely that you will observe a distribution somewhere between 45 heads:55 tails on the one hand, and 55 heads:45 tails on the other. The ideal expectation, based on probability considerations, is an average expectation in an infinite number of trials, but in any finite number of trials it is unlikely to be realized exactly.

Now, we observe that these same laws of probability operate when a mixture of pollen types fertilizes the female parts of flowers, or when sperm fertilize eggs. For example, when pollen from a plant of genotype *Dd* is placed on flowers of a homozygous recessive plant *dd*, two types of pollen grains, each present in large numbers, are available for the fertilization of a finite number of female germ cells. We can predict, if no consistent bias of any kind favors one type of pollen over the other, that about half the embryos will come from eggs fertilized by *D* pollen, and half from eggs fertilized by *d* pollen. But we cannot expect that this prediction will be exactly realized any more than in the similar case of the tossed penny. Nor will any repetition of the experiment involving the same number of trials be likely to give numbers in each class identical with those observed in the first experiment. Each "sample" of the ideal, infinite population will give somewhat different values. Each is subject to *sampling errors*—to chance deviations from the ideal expected values.

This brings us to our second question regarding chance deviations from expected ratios: how large a difference can we permit between the observed and the expected before we suspect that something other than chance alone is involved in the deviations observed? This is a more difficult question, because any answer must be arbitrary. If, in a game of penny tossing, your friend tossed a penny 100 times and threw a head every time, you would probably ask to see the penny, suspecting that it must be so abnormal as to have heads on both sides. If he threw 90 heads and 10 tails you would probably consider this ratio so far off as to suggest that the penny might not be properly balanced, or you would watch the tosses carefully to be sure that the penny turned over in the air. Would you accept the ratio of 80:20 as a chance deviation? 70:30? 60:40? Just where would your skepticism stop?

The question can be rephrased for a concrete example. Take the ratio between tall and dwarf tomatoes observed in the backcross discussed in the previous chapter:

	TALL	DWARF	TOTAL
observed	139	145	284
expected (1:1)	142	142	284
deviation	3	3	

We are really wondering *how often*, in backcrosses involving 284 progeny, we would expect chance alone to produce a deviation as large as three plants in each class. Would it happen in nine such trials out of ten? Then certainly we could accept this particular trial as only a chance deviation. Would it happen only once in ten trials? Then we might be rather doubtful, but, being conservative, we would probably accept this as a chance deviation. If it would happen only once in twenty trials, we might well be skeptical, and suspect that something other than chance alone might be involved in the deviation observed in this particular trial. In other words, few of us would risk a positive conclusion on a "twenty-to-one shot." And if the odds were a hundred or more to one we would regard it as a poor gamble indeed.

Statisticians, in general, accept these reasonable standards of evaluation of chances: If a deviation can be shown to occur more often than once in twenty trials, on the basis of chance alone, the observations are conventionally accepted as a satisfactory fit to the expected. If the chance is less than twenty to one ($\frac{1}{20} = 0.05$), the deviation is regarded as "significant;" that is, something other than chance is suspected to be operating. If the probability is less than a hundred to one (0.01), the deviation is "highly significant," and it is considered very unlikely that the difference between observed and expected is due to chance alone.

The Chi-Square Test. It should be clear that the significance of a given deviation is related to the size of the sample. If we expect a 1:1 ratio in a test involving six individuals, an observed ratio of 4:2 is not at all bad. But if the test involves six hundred individuals, an observed ratio of 400:200 seems rather far off. Similarly, if we test 40 individuals and find a deviation of 10 in each class, this deviation seems serious:

observed	30	10
expected	20	20
deviation	10	10

But if we test 200 individuals, the same numerical deviation seems reasonably enough explained as a purely chance effect:

observed	90	110
expected	100	100
deviation	10	10

The statistical test most commonly used on such problems is simple in design and application. Each deviation is *squared*, and each squared deviation is then divided by the expected number in its class. The resulting quotients are then all added together to give a single value, called the *chi-square* (χ^2), for the distribution. To substitute symbols for words, let d represent the respective deviations, e the corresponding expected values, and the Greek letter Σ "the sum of." Then

$$\chi^2 = \Sigma \left(\frac{d^2}{e} \right).$$

We can calculate χ^2 for the two arbitrary examples above to show how this value relates the magnitude of the deviation to the size of the sample (see Table 2-2).

Table 2-2. CALCULATION OF CHI-SQUARE VALUES.

	Sample of 40 Individuals		Sample of 200 Individuals	
observed	30	10	90	110
expected (e)	20	20	100	100
deviation (d)	10	10	10	10
d^2	100	100	100	100
$\dfrac{d^2}{e}$	5	5	1	1
$\chi^2 = \Sigma \left(\dfrac{d^2}{e} \right)$	$\chi^2 = 10$		$\chi^2 = 2$	

You will note that the value of χ^2 is much larger for the smaller population, even though deviations in the two populations are numerically the same. In view of our earlier common-sense comparison of the two, this is a practical demonstration that the calculated value of χ^2 is related to the significance of a deviation. It has the virtue of reducing many different samples, of different sizes and with different numerical deviations, to a common scale for comparison.

The chi-square test can also be applied to samples including more than two classes. For example, Table 2-3 shows the value of chi-square for the testcross ratio introduced in Chapter 1.

The number of classes upon which the χ^2 value is based must be considered in evaluating its significance. A value for a two-class distribution includes only two squared deviations. But a value for a four-class distribution is based on four

Table 2-3. CALCULATION OF CHI-SQUARE FOR A DIHYBRID
TESTCROSS RATIO.

	Tall, Cut	*Tall, Potato*	*Dwarf, Cut*	*Dwarf, Potato*
observed	77	62	72	73
expected (e)	71	71	71	71
deviation (d)	6	9	1	2
d^2	36	81	1	4
$\dfrac{d^2}{e}$	0.51	1.14	0.01	0.06

$$\chi^2 = \Sigma \left(\frac{d^2}{e} \right) = 1.72$$

Note: Cases in which the expected numbers are all the same are more quickly
calculated by adding the squared deviations and making a single division by
the expected number. We have retained the longer method of computation in
order to avoid confusion in extending the calculation to other ratios, like the
9:3:3:1, in which the expected numbers differ in respective classes.

squared deviations, and it seems reasonable that here a larger value of χ^2 should
be permitted before we question the hypothesis that chance alone explains the
deviation. Conventionally, the effect of number of independent classes is
recognized in the phrase *degrees of freedom.*

The number of degrees of freedom in tests of genetic ratios is almost always
one less than the number of classes. Thus in tests of 1:1 or 3:1 ratios there is
one degree of freedom. A test of 1:2:1 ratio would have two degrees of freedom.
A test of a 1:2:2:4:1:2:1:2:1 would have eight degrees of freedom. The general
idea of degrees of freedom is like the situation encountered by a small boy when
he puts on his shoes. He has two shoes, but only one degree of freedom. Once
one shoe is filled by a foot, right or wrong, the other shoe is automatically
committed to being right or wrong too. Similarly, in a two-place table, one
value can be filled arbitrarily, but the other is then fixed by the fact that the
total must add up to the precise number of observations involved in the experiment, and the deviations in the two classes must compensate for each other.
When there are four classes, any three are usually free, but the fourth is fixed.
Thus, when there are four classes, there are usually three degrees of freedom.

We are now ready for the final step in the application of the χ^2 test. This is a
very simple step in practice, although the mathematical processes and assumptions upon which it is based are rather complex.

Once the value of χ^2 is computed, and the number of degrees of freedom
noted, the probability for a strictly chance deviation *as large as or larger than*
that obtained can be read directly from a chart of χ^2 (Fig. 2-13). It is convenient
to use a straightedge set perpendicular to the axis of the chart at the calculated

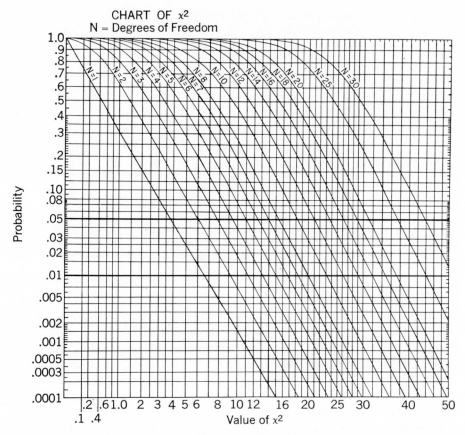

Figure 2-13. *Chart of chi-square.* (*After* Crow, Genetics Notes, *Burgess Publishing Co.,* *5th ed., 1963, p. 151.*)

value of χ^2, and to note where the straightedge intersects the curve for the proper number of degrees of freedom (N). The probability is read at the left, opposite the point of intersection you have noted.

In Table 2-2 we calculated $\chi^2 = 10$ for one degree of freedom for the sample of 40 individuals. You will note on the chart that the probability decreases as the value of χ^2 increases. In this case chance alone would be expected in considerably less than one in a hundred independent trials to produce as large a deviation as that obtained. We cannot reasonably accept chance alone as responsible for this particular deviation; it represents an event that would occur, on a chance basis, much less often than the one-time-in-twenty that we have agreed on as our point of rejection; this event would occur less often than even the one-time-in-a-hundred that we decided to regard as highly significant.

In the same table, we calculated $\chi^2 = 2$ for the sample of 200 individuals.

Here the probability, according to the χ^2 chart, is near 0.15. Independent repetitions of *this* experiment would produce chance deviations as large as those observed about three times in twenty trials. We can reasonably regard this deviation as simply a sampling, or chance, error. A result like this, of course, does not establish that the hypothesis being tested is *correct;* it only indicates that the data provide no statistically compelling argument *against* it.

In Table 2-3 we calculated $\chi^2 = 1.72$ for a dihybrid testcross. Because there are four classes, the chart is read from the intersection with the curve for three degrees of freedom; the probability is between 0.6 and 0.7. Clearly, the deviations can be accepted as resulting from chance. The dihybrid F_2 ratio of Table 1-3, page 23, can be used to illustrate the entire chi-square test (Table 2-4).

Table 2-4. CHI-SQUARE TEST OF DIHYBRID F_2 DATA.

	Tall, Cut	Tall, Potato	Dwarf, Cut	Dwarf, Potato
observed	926	288	293	104
expected 9:3:3:1 (e)	906.3	302.1	302.1	100.7
deviation (d)	19.7	14.1	9.1	3.3
d^2	388.09	198.81	82.81	10.89
$\dfrac{d^2}{e}$	0.43	0.66	0.27	0.11

$$\chi^2 = \Sigma \left(\frac{d^2}{e} \right) = 1.47; \ p \text{ is between 0.6 and 0.7}$$

Between six and seven times in ten, chance alone would produce a deviation as great as or greater than that observed by MacArthur in this F_2. The data fit the expectations well.

Improper Use of Chi-Square. The two most important reservations regarding ordinary use of the χ^2 method in genetics are:

1. Chi-square can usually be applied only to numerical frequencies themselves, not to percentages or ratios derived from the frequencies. For example, if in an experiment one expects equal numbers in each of two classes, but observes 8 in one class and 12 in the other, he might express the observed numbers as 40 percent and 60 percent, and the expected as 50 percent in each class. A χ^2 value computed from these percentages could not be used directly for the determination of p. When the classes are large, a χ^2 value computed from percentages can be used, if it is first multiplied by $n/100$, where n is the total number of individuals observed.

2. Chi-square cannot properly be applied to distributions in which the frequency of any class is less than 5. In fact, some statisticians suggest that a particular correction be applied if the frequency of any class is less than 50.

However, the approximations involved in χ^2 are close enough for most practical purposes when there are more than 5 expectations in each class.

Keys to the Significance of This Chapter

The subject matter of this chapter falls into two related parts, each of which can be summarized briefly.

Sex is generally controlled by a pair of *sex chromosomes. Sex-linked genes* exhibit distinctive patterns of inheritance, correlated with the transmission of the sex chromosomes. They suggest the fundamental fact that *genes are located in chromosomes.* More than two alternative forms of a gene (*multiple alleles*) often exist in a population.

An appreciation of the mathematics of probability is essential to understanding genetics, and probability computations can serve as significant time-savers in calculating the frequencies of various phenotypes and genotypes. Binomial distributions describe the characteristics of families of given sizes, segregating in clear-cut ratios for specific attributes such as sex. It is important to have at hand arbitrary techniques for evaluating the statistical significance of any deviations observed in an experiment testing a Mendelian hypothesis. The chi-square test has been discussed and applied for this purpose.

REFERENCES

Bamber, R. C., "Genetics of Domestic Cats." *Bibliographia Genetica*, **3**:1–86, 1927. (A comprehensive review, including the tortoiseshell cat example discussed in this chapter.)

Bridges, C. B., and Brehme, K. S., "The Mutants of *Drosophila melanogaster*." *Carnegie Inst. Wash. Pub.* 552, 1944. (An exhaustive summary of reported hereditary variants in this species.)

Hollander, W. F., "Auto-Sexing in the Domestic Pigeon." *J. Heredity*, **33**:135–140, 1942. (An interesting account of a method of breeding pigeons whose sex will be evident from their color.)

Hutt, F. B., Rickard, C. G., and Field, R. A., "Sex-Linked Hemophilia in Dogs." *J. Heredity*, **39**:2–9, 1948.

Morgan, L. V., "Non-Criss-Cross Inheritance in *Drosophila melanogaster*." *Biol. Bull.*, **42**:267–274, 1922. (Reporting the discovery and genetic behavior of attached-X.)

Peters, James A., editor, *Classic Papers in Genetics*. Englewood Cliffs: Prentice-Hall, 1959. (Included, and particularly relevant here, are papers by Mendel, 1865, Sutton, 1903, and Morgan, 1910—the sex-linkage paper.)

Stern, C., *Principles of Human Genetics*. San Francisco: W. H. Freeman, 2nd ed., 1960. (Especially, at this point, Chap. 5, "Probability," Chap. 8, "Genetic Counseling," and Chap. 9, "Genetic Ratios," for extensions of probability considerations to problems of human heredity; and Chap. 13, "Sex Linkage.")

Sturtevant, A. H., *Genetics and Evolution*. San Francisco: W. H. Freeman, 1961. (A reprinting of Sturtevant's selected papers; you will be interested in the 1913 paper on "The Himalayan Rabbit Case, with Some Considerations on Multiple Allelomorphs.")

Wilson, E. B., *The Cell in Development and Inheritance*. London: Macmillan, 2nd ed., 1900. (Until 1928, when the third edition was published, this edition was the standard work in this field.)

QUESTIONS AND PROBLEMS

2-1. What are sex chromosomes? What does the symbol $\widehat{XX}$ stand for?

2-2. In terms of standard Drosophila gene symbolism, what can you deduce from the symbol y^+ if you know that an inherited deviation from type in this species is yellow body color?

2-3. The standard type of tomato plant has cut leaves, is tall (not dwarf), has an anthocyanin pigment that makes the stems purple, and the ripe fruit has red flesh. Unit deviations from type in these characteristics include, respectively, potato leaves, dwarfism, green stems (lacking the anthocyanin), and yellow fruit-flesh color. Standard symbols for the allelic pairs are: C-c for cut vs potato leaves; D-d for tall vs dwarf; A-a for the presence vs the absence of the anthocyanin; and R-r for red vs yellow fruit-flesh color. The deviations from type are recessive to the type allele in each case.

    a. How would this symbolism be modified if the Drosophila system for designating alleles prevailed in tomato genetics?

    b. What are advantages and disadvantages of each convention of nomenclature?

2-4. How would your answer to Question 2-3a be changed if the conventional standard type of tomato had green, rather than purple, stems?

In man, *hemophilia*, the "bleeder's disease" in which the time required for blood to clot is greatly prolonged, depends on the recessive allele of a sex-linked gene. There are at present about 40,000 cases of the disease in the United States. In the following questions, let $h^+ =$ the allele for normal clotting time; $h =$ the allele for hemophilia.

Note: In all problems involving sex linkage, list the phenotypes of sons and daughters separately.

2-5. A man whose father was hemophilic, but whose own blood-clotting time is normal, marries a normal woman with no record of hemophilia in her ancestry. What is the chance of hemophilia in their children?

2-6. A woman whose father was hemophilic, but who is not herself a "bleeder," marries a normal man. What is the chance of hemophilia in their children?

?-7. What is the chance of hemophilia among the sons of a daughter of the marriage in Question 2-6 if she marries a normal man?

2-8. It has been reported that women heterozygous for hemophilia may be distinguishable from homozygous normal women; the clotting time of the heterozygotes appears to be longer, but not sufficiently long to constitute a hazard to survival. How might you apply this information to Problem 2-7?

The probable sex linkage of alleles for yellow and black pigments in cats was described on page 45.

2-9. A calico cat has a litter of eight kittens: one yellow male, two black males, two yellow females, and three calico females. Assuming there is a single father for the litter, what is his probable color?

2-10. A black cat has a litter of seven kittens: three black males, one black female, and three calico females. Comment on the probable paternity of this litter.

2-11. A yellow cat has a litter of four kittens: one yellow and three calico. Assuming there is a single father for the litter, what is the probable sex of the yellow kitten?

2-12. Suppose you set out to ascertain whether, in cats as in mice, XO individuals are phenotypically female. What female-kitten colors, with respect to yellow, tortoiseshell, and black, would you look for, in what parental color crosses?

2-13. Very rarely, a tortoiseshell tomcat has been reported to be fertile. What genetic data would you require to suggest that such tomcats are $\widehat{XX}$?

Vermilion eye color in Drosophila (Plate I) is inherited as a sex-linked recessive.

2-14. What phenotypes would be found in the progeny of a cross between a vermilion female and a wild-type male?

2-15. Suppose that the vermilion female in Question 2-14 were $\widehat{XX}$. What phenotypes would the surviving progeny of this cross show? Diagram the cross, showing parental genotypes, phenotypes, and gametes; F_1 genotypes and phenotypes.

2-16. Like vermilion eye color, white eye color (w) in Drosophila depends on a sex-linked recessive. Suppose that a particular white-eyed female, when crossed with a wild-type (w^+) male, gave white-eyed daughters and red-eyed sons. How might this result be explained? How would you check your explanation?

Recall the discussion of the inheritance of comb shape in poultry in the Questions and Problems section of Chapter 1, and Figure 2-12 showing the barred feather pattern.

2-17. A homozygous barred walnut-comb cock is mated with a nonbarred single-comb hen, and 320 eggs from this cross hatch. How many chicks would you expect in each phenotypic class?

2-18. Progeny of the cross in Problem 2-17 are mated to produce an F_2 generation. If 320 F_2's are produced, how many would you expect in each phenotypic class?

Rarely, a hen's ovary may lose its function (presumably as a result of a local infection), and a testis develops instead. Such "sex-reversed" hens have even become fathers of chicks. No change in the chromosomal makeup of the hen occurs under these conditions.

2-19. What types of sperm, with regard to X-chromosome constitution, would a sex-reversed hen be expected to produce?

2-20. What sex ratio would you expect among the progeny of a sex-reversed hen and a normal hen?

2-21. Suppose that the barred walnut "cock" in Question 2-17 were really a sex-reversed hen. What answer would you give to Question 2-17?

2-22. The parents in Problem 2-6 have a normal (nonhemophilic) son. What is the probability that their next son will also be normal?

2-23. Suppose that a family like the one in Problem 2-6 has two sons. What is the probability that they will both be normal?

2-24. If a family like the one in Problem 2-6 has six sons, what is the probability that three will be normal and three hemophilic?

2-25. The family in Problem 2-6 is expecting a child. What is the probability that it will be normal? That it will be hemophilic?

2-26. The family in Problem 2-6 has six children. What is the probability that two are normal and four hemophilic?

Data in the following problems are from MacArthur's studies of tomatoes.

Let: A = purple stem R = red fruit flesh C = cut leaf
      a = green stem r = yellow fruit flesh c = potato leaf

2-27. True-breeding, purple-stem tomatoes crossed with green-stem plants gave all purple-stem F_1's. When these F_1's were backcrossed to green-stem plants, the progeny were: 482 purple-stem, 526 green-stem. Diagram the crosses, showing all pertinent genotypes, phenotypes, and germ cells.

2-28. Verify that, in the backcross ratio of Problem 2-27, $\chi^2 = 1.92$, and p is greater than 0.05. Can this deviation be accepted as a sampling error?

2-29. The F_2 of the cross in Question 2-27 consisted of 3084 purple-stem and 1093 green-stem plants. Does this fit the expected ratio satisfactorily? (Verify that p is greater than 0.05.)

2-30. Purple-stem, cut-leaf plants ($AACC$) crossed with green-stem, potato-leaf plants would give what genotype and phenotype in F_1? What would be the backcross ratio when these F_1's are crossed with the green-stem, potato-leaf parent strain if the A-a and C-c allelic pairs assort independently?

2-31. In an F_2 grown from the seeds of the F_1's in Question 2-30, MacArthur counted:

purple, cut	purple, potato	green, cut	green, potato
1790	620	623	222

Is this consistent with our postulate of independent assortment of these two allelic pairs? (Verify that p here is a little less than 0.5.)

2-32. When a true-breeding purple-stem, potato-leaf strain was crossed with a true-breeding green-stem, cut-leaf strain, MacArthur observed in F_2:

purple, cut	purple, potato	green, cut	green, potato
247	90	83	34

Calculate the expected number of purple, cut F_2's in this cross (454 total F_2 progeny), using (a) the checkerboard method and (b) the probability, or fraction, method described on page 51.

2-33. Genetic ratios are often complicated by differences in viability of particular genotypes; for example, white-eye Drosophila are less likely to survive to be counted than are wild-type flies. Perform χ^2 tests on both sets of Morgan's data on page 36, and comment on the statistical aspects of these experiments—which were the basis for the original recognition of sex linkage.

Of the Drosophila eye colors illustrated in Color Plate I, apricot (w^a) may be assumed, for our present purposes, to be an allele of white (w) and of eosin (w^e), while purple (pr), which is recessive to wild type (pr^+), is not sex-linked.

2-34. Give genotypes for:
   a. A female fly with wild-type phenotype heterozygous for white and purple.
   b. An apricot male homozygous for purple.

2-35. What progeny genotypes, and in what proportions, would you expect from crossing the male and female in Problem 2-34?

In mice, there is a set of multiple alleles of the gene for albinism. Four of these alleles, listed in order of decreasing amount of color in the hair of homozygotes, are

$$C = \text{full color (wild type)}$$
$$c^{ch} = \text{chinchilla}$$
$$c^d = \text{extreme dilution}$$
$$c = \text{albino.}$$

This gene is not sex-linked. Assume that each allele is dominant to those below it in the list.

2-36. Diagram a cross between a wild-type mouse heterozygous for extreme dilution and a chinchilla mouse heterozygous for albinism.

2-37. Another non-sex-linked gene affecting hair color in mice has alleles B for black hair (wild type) and b for brown (recessive). Expand your diagram in Problem 2-36 to include the information that both of the mice crossed are heterozygous for brown (Bb).

CHAPTER 3

The Vehicles of Inheritance

In the preceding chapter we encountered a visible, material basis for the behavior of the units of heredity that had, up to that point, a logical but only a
rather abstract reality. Yet we do not need to know about chromosomes to
feel sure of the existence of genes. Mendel established the particulate and
duplicate basis of inheritance without knowing anything about chromosomes—
in fact, several years before chromosomes had been named or described in any
detail. Even the fact that a single sperm fertilizes a single egg to give rise typically to a single individual had not been established beyond doubt at the time
when Mendel did his work, although the assumption that this is true seems
implicit in some of his reasoning. The Mendelian approach was, and to a large
extent still is, based on breeding experiments. The units of heredity share the
kind of reality of the atoms of chemistry or the particles of physics: they have
not been directly observed, but their structure and properties have been deduced
from indirect observation.

Similarly, it is not necessary to know about Mendelian units to be reasonably
sure that chromosomes are intimately concerned with inheritance. During the
thirty-four years that Mendel's paper lay unnoticed, chromosomes were described, and their behavior was observed with great care and enthusiasm. It
was appreciated that in their regular and precise duplication and distribution
at ordinary cell division, their neat parcelling-out to the germ cells, and their
behavior during fertilization and development, the chromosomes are uniquely
important to the cell. And chromosome theories of heredity had been suggested
while Mendelism was still virtually unknown.

Shortly after the rediscovery of Mendel's work, during the period when
Mendelism was being enthusiastically extended to many plants and animals,
it was realized that what was known about chromosomes fit remarkably well

with what was being learned about genes. Within a brief period of time, these two previously independent subjects of investigation became fused along their common boundary, and today we recognize as the science of *cytogenetics* a field that is an integral part both of the study of heredity and of the study of cells. This hybrid field has enriched, and has been enriched by, both of its parent disciplines. It is imperative that we, as students of heredity, become familiar with the nature and behavior of chromosomes, the vehicles of inheritance.

The Structure of Chromosomes. On the level of visual observation, a good deal is known about the structure of chromosomes. They change in form and character from stage to stage in the continuous processes of cell growth and multiplication. These changes are in general cyclic; as the cell prepares to divide, and proceeds through division, the chromosomes pass through a regular progression of structural alterations until, at the end of the division cycle, the characteristics of the original cell are again restored in two daughter cells.

The fundamental element of the chromosome as observed in the light microscope is a thread called the *chromonema* (plural: *chromonemata*). Studies with the electron microscope reveal that chromosomes are made up of very fine fibers, which are not visible, individually, in the light microscope. Thus a chromonema must consist of a bundle of such threads, or one such thread coiled and supercoiled. The chromosomes usually change in length through the division cycle, in part due to the coiling or uncoiling of the chromonemata. Several different types of coiling have been described, and cytologists speak of standard, relic, or relational coils and others as well. These terms are necessary to the cytologist, but need not be defined here.

Chemically, chromosomes are composed of a substance called *deoxyribonucleic acid* (DNA for short) and a special class of proteins called *histones* that form aggregates with DNA. The DNA-histone complexes are referred to as nucleo-proteins. Chromosomes also contain other types of proteins and RNA (*ribonucleic acid*). (These various chemical substances are discussed more fully in Chap. 5.) Since DNA and histones are major components of chromosomes and are always present in them they probably are the essential ingredients of chromosomes. The way in which the chemical components of the chromosomes are associated is not yet known. A chromosome may consist, for example, of a single long molecule of DNA complexed with histone or a series of DNA molecules "glued" end-to-end by histone.

As we attempt to describe, from what we can see with the aid of an electron microscope, how one chromosome is different from another, we disclose various facts about the structure and behavior of chromosomes.

The Centromere. Each chromosome has (in most cases) a single differentiated region somewhere along its length that seems to act as the point at which force is exerted in the separation of dividing chromosomes. The structure has been called by a number of names, among which the terms *centromere* and *kinetochore* are used most generally. It has also been called the *spindle-fiber attachment,* or the *attachment region.* The force that accounts for the movement apart of a dividing chromosome is associated with a visible cellular structure, which, because of its shape, is called the *spindle.* The spindle, when stained, appears to be made up of fibers; hence the term *spindle-fiber attachment* for the centromere, the point at which the spindle fiber appears to affect the movement of the chromosome.

The position of the centromere along the length of a chromosome contributes in a direct manner to the shape of the chromosome as the nucleus divides. If the centromere is near the middle, the chromosome becomes V-shaped as this central spot leads the way to the poles of the cell, the "arms" of the chromosome trailing behind. If the centromere is nearer one end, the chromosome becomes J-shaped, since the arms are of unequal length. If the centromere is very near one end, the chromosome moves as a straight rod; there is some doubt that a centromere is ever truly terminal under normal conditions.

The centromere is typically a permanent, well-defined organelle of the chromosome, and not a vague, transient, or generalized region. In most cells it is lost to view through changes in the staining qualities and general character of the nucleus during part of the cell's history. But there is every reason to believe that in such instances the "loss" is purely a visual illusion. For example, when a centromere is really lost, through the breaking off of a part of a chromosome, it cannot be regenerated by the other part. Moreover, in favorable material it can be demonstrated all through the cell's division cycle. Like other structures and regions of the chromosome, continuity of pattern from cell generation to cell generation is characteristic of the centromere. The chromosome duplicates all its parts with almost undeviating fidelity and only rarely ventures into novelty.

Chromomeres, Knobs, and Constrictions. In many species, a chromosome is seen to be not a smooth or regularly coiled thread, but to have bumps and constrictions along its length. When this is true, the specific pattern of such specialized regions is reasonably constant for any given chromosome from cell to cell. The smaller "beads on the string" are called *chromomeres;* the larger ones are called *knobs.* This structural differentiation of the chromosome along its length is probably caused in part by differential coiling of the chromonema in different regions.

Nucleolus-Forming Regions. Within a nucleus, there are often one or more spheres of darkly staining material called *nucleoli* (singular: *nucleolus*). The organization of a nucleolus seems usually to be the function of a specific point on a particular chromosome, and when a nucleolus is visible it can be seen to be attached to this *nucleolus-forming region* or *nucleolus organizer.* Chromosome 6 in corn, for instance, can usually be identified in certain stages of cellular division through its possession of a nucleolus organizer.

Satellites. In regions such as the nucleolus organizer region, a chromosome is often constricted to so fine a thread that it may be difficult to see. If another sec-

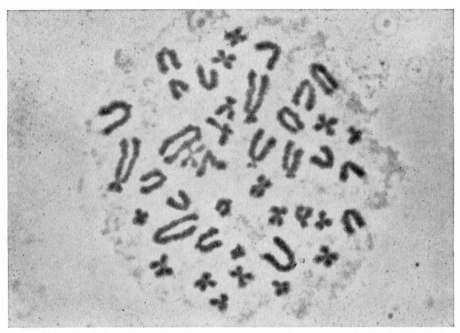

Figure 3-1. *The chromosome complement of the rat. Cell division has been arrested at metaphase by treatment with colchicine. All chromosomes have doubled but chromatids are held together at the centromere. Note that some chromosomes have centrally located centromeres (these bivalents have X configurations); others have nearly terminal centromeres (U configurations). Forty-two chromosomes are visible. (A micrograph by Klaus Bayreuther.)*

tion of the chromosome extends beyond this point, it appears as an "appendage" to the chromosome, and is called a *satellite.*

Heteropyknosis. Certain chromosomes or parts of chromosomes become darkly staining earlier in the division cycle and retain their dense appearance longer than do others. This dense-staining property is called *heteropyknosis* (literally: "difference in density"). Major parts of the sex chromosomes in many species are heteropyknotic, and this, together with the fact that the sex chromosomes often are the first to move apart on the spindle, has been useful in the tentative identification of the sex chromosomes in these species. In humans, one of the two X-chromosomes in the female is positively heteropyknotic (condensed) in resting nuclei and is visible as a darkly staining blob. This fact makes possible sexing the nuclei of human cells, as the single X-chromosome in the male is not heteropyknotic.

Some chromosomes or parts of chromosomes appear to be condensed at all or almost all stages of cell division. These regions are often referred to as *heterochromatic* (literally: "different colored") to contrast them with the *euchromatic* chromosomes or chromosome parts that go through the regular changes in morphology associated with cell division. In *Drosophila melanogaster* for example the Y-chromosome and the regions of the other chromosomes adjacent to the centromeres are heterochromatic. It has been suggested that heterochromatic regions are composed of *heterochromatin* differing both in structure and chemical composition from *euchromatin*.

These and other characteristics of chromosome structure make it possible to identify particular chromosomes in favorable material. In corn, for example, it has been possible to prepare charts of the relative lengths of the ten different chromosomes, the relative lengths of the arms (fixed by the positions of the centromeres), and the patterns of chromomeres, constrictions, knobs, nucleolus organizers, and satellites, that enable cytologists to identify not only particular chromosomes but even specific regions of chromosomes. The accomplishments of cytologists in this direction are admirable, since they are sometimes observing objects so small as to be near the resolving limits of their chief tool, the light microscope.

The Diptera (two-winged insects, including Drosophila) have a remarkable advantage for cytogenetic work in the giant chromosomes of their larval salivary glands. These will be described, in connection with their chief uses, in Chapter 7.

The Chromosome Complement. We have been discussing the characteristics of individual chromosomes; we can now turn to a consideration of the nature and behavior of the chromosomes as a group.

The number of chromosomes per nucleus is as a rule constant for all the individuals of a species, and varies from one species to another. Man has 46, the fox 34, the rabbit 44, the rat 42, the mouse 40, red clover 14, garden peas 14, corn 20, tomatoes 24, and so on. The number of chromosomes has no general significance relative to the evolutionary achievement of any species. Within a series of related plant species, however, regular multiples of the same basic chromosome number often appear to be related to evolutionary changes. This phenomenon will be discussed in Chapter 7.

The chromosomes are present in pairs, and, with the exception of certain sex chromosomes, the members of a pair of chromosomes cannot normally be distinguished under the microscope. It is often convenient to speak of the chromosome number of a species in terms of the number of *pairs* of chromosomes present; thus human beings have 23 pairs, foxes 17, red clover 7, corn 10, and so on. When we casually stated that corn has 20 chromosomes we actually meant that a corn plant has ten different *pairs* of chromosomes in each somatic cell; the members of each pair are alike, but the different pairs are distinguishable. The different chromosomes are symbolized by numbers, from chromosome 1 (the longest) to chromosome 10 (the shortest) (Fig. 3-2). A nucleus of a root cell, for example, will generally contain two number 6 chromosomes, two number 9 chromosomes, and so on.

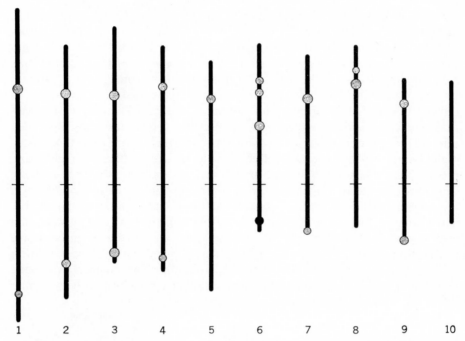

Figure 3-2. *The ten different chromosomes of corn are distinguishable in terms of their length (chromosome 1 is the longest, 10 the shortest); the relative lengths of their arms (the centromere, shown as a dash across the chromosome, divides the chromosome into two arms); the nucleolar-forming region of chromosome 6; and knobs. (After Longley,* Botan. Rev., 7:266, 1941.*)*

There are at least six kinds of exceptions to the statement that a constant number of chromosomes is present in pairs in the cells of any individual of a species. The first of these is very important.

1. The mature germ cells (*gametes*) of a sexually reproducing individual contain only half the usual number of chromosomes—one member of each pair. In this respect, the germ cells are different from the body cells, or *somatic cells*, of the individual. Conventionally, the gametes are described as *haploid* in chromosome number, and the somatic cells as *diploid*. The symbol n is used to signify the haploid chromosome number; you will note that n also describes

ORGANISM	CHROMOSOMES IN GAMETES ($= n$)	CHROMOSOMES IN SOMATIC CELLS ($= 2n$)	PAIRS OF CHROMOSOMES IN SOMATIC CELLS ($= n$)
corn	10	20	10
man	23	46	23

the number of *pairs* of chromosomes in diploid tissues. The table above will make the relationship clear.

We will be concerned, during much of the remainder of this chapter, with the mechanisms and the significance of the maintenance of these regularities.

2. The importance of the second exception to the statement that a constant number of pairs of chromosomes is present in the cells of any individual is difficult to evaluate in terms of our present knowledge. The fact is, however, that occasional deviations from the rule occur in somatic tissues; some body cells can sometimes be shown to have less and others more than the normal diploid number of chromosomes. Occasionally, these deviations are regular and predictable. In legumes like peas and clover, for example, the root nodules, which are the site of activity of the nitrogen-fixing bacteria so important in the maintenance of soil fertility, contain plant cells whose chromosome number is consistently doubled (Fig. 3-3). In other tumorlike growths, in cancers, and even in certain normal organs like animal liver, an increase of chromosome number in multiples of the basic number (called *polyploidy*) also appears with some degree of regularity. A diminution in chromosome number in somatic tissues may also occur. Rarely, this diminution can be shown to be a regular

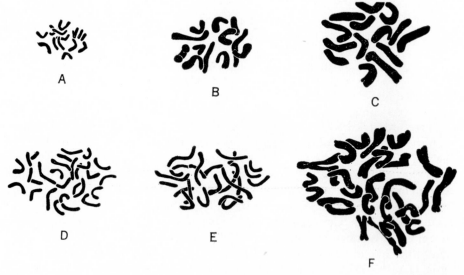

Figure 3-3. *The root nodules of legumes contain cells with twice the normal chromosome number.* A: *red clover, ordinary root tip cell; count* 2n = *14.* D: *nodule tissue of red clover; count 28 chromosomes.* B: *common vetch, ordinary root tip cell; count* 2n = *12.* E: *nodule of common vetch; count 24 chromosomes.* C: *garden pea, ordinary root tip cell; count* 2n = *14, two beginning to appear double.* F: *nodule of garden pea; count 28 chromosomes, many longitudinally double.* (*From Wipf and Cooper,* Proc. Nat. Acad. Sci . **24:**88. 1938.)

process; in certain organisms, for example, particular chromosomes are limited to the "germ line" and are uniformly lost in the formation of somatic tissue.

3. The third kind of exception to the statement that a constant number of pairs of chromosomes is present in the cells of any individual of a species is almost certainly of little general significance. For example, some types of corn (particularly Black Mexican Sweet Corn) have extra chromosomes in addition to the normal ten pairs. These extra chromosomes are called *B-type* chromosomes. They seem to have no essential genetic activity; they can be either entirely absent or present in rather large numbers in a plant without appreciably affecting the plant's characteristics. The B-type chromosomes have provided the basis for some very interesting and significant work in experimental cytology, but in terms of normal development they are unimportant.

4. An important part of the seed of higher plants regularly has three, rather than two, sets of chromosomes. This structure, the *endosperm*, plays a dynamic part in seed development, and it later contributes to the nourishment of the young seedling. Its origin will be described in Chapter 4.

5. Variations in number of chromosomes or sets of chromosomes among different individuals of the same species occur and are of considerable importance with regard to the mechanisms of evolution, the improvement of plants, and the understanding of gene action in development.

6. Finally, as we have previously noted, the sex difference within a species is often associated with a difference in chromosome number. One sex may be XX and the other XO; or one sex may be diploid, and the other haploid, resulting from the development of unfertilized eggs, as in the honey bee and other Hymenoptera.

Mitosis

We turn now to a consideration of the mechanisms through which the regularities in chromosome complement are maintained. The problem has two facets. First, how is the regular diploid complement of chromosomes kept constant through the successive nuclear divisions involved in the growth and development of a multicellular individual from a single cell, the fertilized egg? And, second, what special phenomena in germ-cell formation result in haploid gametes, each with one member of each pair of chromosomes, so that diploidy is restored at fertilization? The process responsible for the maintenance of the first of these two bases of regularity is ordinary somatic nuclear division, and is called *mitosis*. It will provide a frame of reference for the second process— the formation of germ-cell nuclei, or *meiosis*.

We can begin with the kind of nuclear division through which the fertilized egg of a human being, for example, gives rise to the 10^{14} or so cells of the adult human body. The process of mitosis is uniform in its essentials through all the somatic divisions that occur. And the essential characteristics of mitosis are very simple: *Each chromosome duplicates itself*, and the duplicates are separated from each other at cell division, one going into the nucleus of one daughter cell, and its "twin" into the other. The daughter cells are therefore identical with each other and with their parent cell in chromosome constitution. Figure 3-4 diagrams this essential characteristic of mitosis. The following more detailed account of the process should not mask or confuse its essential simplicity.

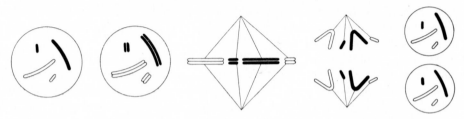

Figure 3-4. *In mitosis, each chromosome duplicates itself. The duplicates separate as the nucleus divides, so that the daughter nuclei are identical in chromosomal constitution. The process is shown in greater detail in Figures 3-5 and 3-6. (By permission after* Fundamentals of Cytology, *by Sharp, Copyright 1943, McGraw-Hill Book Co., p. 64.)*

It has already been emphasized that nuclear division is a continuous and a cyclic process. A cell proceeds smoothly through a series of changes that result in the formation of two daughter cells, and a very precisely synchronized part of this process is the division and distribution of the chromosomes of the nucleus to the daughter cells. The daughter cells may then follow the same cycle, giving rise in turn to two cells each, and so on. Cell multiplication is therefore a geometric progression: $1 \longrightarrow 2 \longrightarrow 4 \longrightarrow 8 \longrightarrow 16 \longrightarrow 32 \longrightarrow 64$, and so on. At this rate, it would take less than 50 consecutive divisions for a single egg to produce 10^{14} cells. Since the division rates vary in different tissues during development, this average figure has little actual significance except as an order of magnitude.

Because the process of nuclear division is a continuous series of changes, only by arbitrary distinctions can we break up the process into "stages" or "phases" for separate consideration. We can enter the division cycle at any point, and the cyclic nature of the process will bring us back to that point at the completion of our consideration of one nuclear division. It is convenient to enter between divisions, when the active cell is metabolizing but is not dividing. (See the diagram of Fig. 3-5 and the photomicrographs of dividing plant cells in Fig. 3-6.)

The "Resting" Nucleus. The nucleus of the metabolic stage is often called the "resting" nucleus—a misnomer in a general sense, because the nucleus may, in fact, be functionally very active in this stage. It is, however, "resting" from the standpoint of division; it may remain in this stage for long periods of time without evident structural change.

The resting nucleus (Figs. 3-5,A and 3-6,A) is a fine, relatively uniform network (*reticular* in structure), in which individual chromosomes are usually very difficult to distinguish. One or more nucleoli are often conspicuous; their number is not always constant because of the tendency of separate nucleoli to fuse into single, darkly staining spheres.

Prophase. As the nucleus prepares to divide, the reticular network resolves itself into visible chromonemata, which stand out as double threads. (Figs. 3-5,B–*D* and 3-6,B–D). They have already duplicated themselves; the remainder of the division is concerned with separating these sister threads. The threads shorten and the chromosomes become distinct. Each chromosome as a whole appears double. The nucleoli shrink and fade during this period, and by late prophase they have disappeared, as has the nuclear membrane.

Metaphase. The chromosomes come to the center of the cell (Figs. 3-5,E and 3-6,E). Their centromeres are distributed as if on a plate at the cell's equator (the "equatorial plate"), and the spindle can be seen when it is properly stained. Careful examination in very favorable material shows that at this stage each longitudinal half-chromosome (*chromatid*) contains chromonemata that are already double. Each chromonema, therefore, may have duplicated itself more than a whole mitotic cycle before the stage at which the sister threads so formed are to separate. It is by no means certain that the duplication always occurs at the same stage in all cells.

Anaphase. The divided chromosomes separate on the spindle, moving toward the poles of the cell (Figs. 3-5,F and 3-6,F–G).

Telophase. The netlike character of the metabolic nucleus begins to develop. Nucleoli again grow into prominence (Figs. 3-5,G–I and 3-6,H). Two cells, with identical metabolic nuclei, are now present where one existed before.

We have considered only the nuclear behavior in division. Late in anaphase or early in telophase the cells themselves divide. This process is typically somewhat different in plants and in animals. In plants, a plate develops between the daughter cells; in animals the cells "pinch apart." In some cells nuclear division may take place unaccompanied by cellular division.

General Aspects of Mitosis. The precision of mitosis makes it a remarkable phenomenon. A more direct mechanism than this one, to guarantee the equal distribution of essential materials to daughter cells, is difficult to conceive. If we can assume that the chromonemata are essentially the visible evidence of linear strings of genes (and we will be able to marshal convincing evidence for this assumption), then the point-for-point reduplication of these linear strings, and their longitudinal separation into the daughter nuclei, represents a beautiful solution to the problem of providing each daughter cell with an identical set of genes.

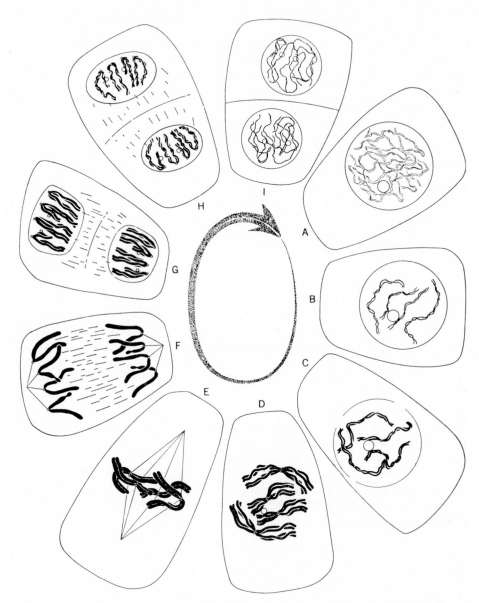

Figure 3-5. *In mitosis, a resting nucleus* (A) *undergoes a continuous sequence of changes called prophase* (B, C, D), *metaphase* (E), *anaphase* (F), *and telophase* (G, H, I). *Either or both of the two daughter nuclei so formed may then enter the division cycle again.* (By permission after Fundamentals of Cytology, *by Sharp, Copyright 1943, McGraw-Hill Book Co., p. 60.*)

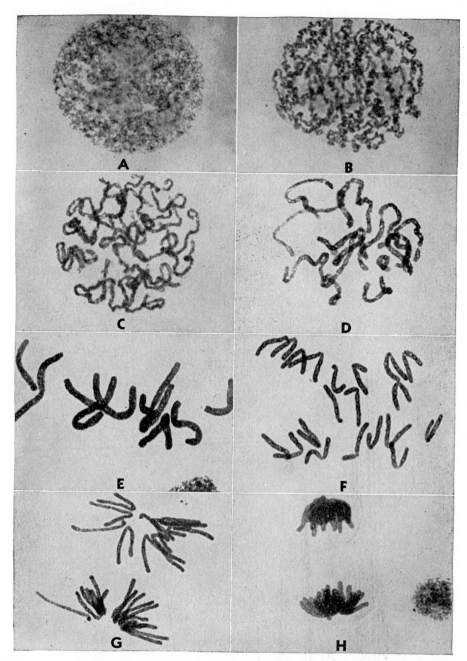

Figure 3-6. *Photomicrographs of dividing tapetal cells of young anthers in the Californian coastal peony show the chromosomes enlarged about 900 diameters. Compare with the diagram of Figure 3-5. A, resting nucleus. B, C, D, early, middle, and late prophase. E, metaphase. F, G, middle and late anaphase. H, telophase. (Courtesy of M. S. Walters and S. W. Brown.)*

But the very precision of this process raises some difficult questions. If, as a result of mitotic division, all or almost all of the cells of the body have the same set of genes, then what makes cells in different tissues different both in structure and in function? This is probably the most elusive riddle in biology today, and we cannot solve it in this book. We will, however, be able to bring some further information to bear upon it, and will return to its systematic consideration in Chapters 11 and 12.

Meiosis

The special kind of nuclear division that results in the formation of haploid gametes, each with one member of each pair of chromosomes, now demands our attention. Meiosis is not really a single division, but two successive ones, the first and the second meiotic divisions. The process of meiosis in its essentials is simple and straightforward. The members of each pair of chromosomes come to lie side by side in the nucleus. Each chromosome is double, so that four strands are associated in a *tetrad;* there are therefore n (the haploid number) tetrads. The tetrads are separated into *dyads* at the first meiotic division, and the dyads into single chromosomes by the second meiotic division. The resulting nuclei are the nuclei of the germ cells; each, you will observe, must have one member of each pair of chromosomes.

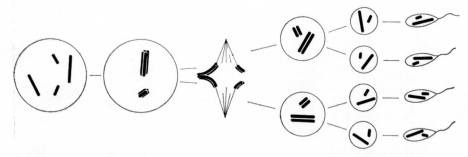

Figure 3-7. *In the formation of sperm, duplicated members of each pair of chromosomes come to lie side by side in four-strand configurations. Two successive nuclear divisions then result in the formation of four sperm, each with one member of each pair of chromosomes. (By permission after* Introduction to Cytology, *by Sharp, Copyright 1934, McGraw-Hill Book Co., p. 251.)*

Figure 3-7 diagrams the essentials of the process of meiosis. If you study the figure at this time, the more detailed consideration of meiosis below should not obscure its true simplicity.

The Germ Line. As an embryo grows and differentiates, certain cells are sooner or later set aside as potential gamete-forming tissues. The location, nature, and time of formation of these tissues vary greatly from one kind of plant or animal to another. In chickens and other birds, for example, there is good evidence that the primordial germ cells originate in an area outside the embryo itself, very early in development, and that they later migrate into the developing sex organs (gonads). In some of the lower animals, the cells of the germ line are clearly set apart from ordinary somatic cells at extremely early stages in development. In plants, similar extreme variations occur in the time at which the germ line becomes distinct from somatic tissues.

But whenever, however, and wherever the cells of the germ line originate, their behavior in the process of gamete formation is remarkably similar in all sexually reproducing forms. We can take a male mammal as a standard, and list the major deviations from this model after the model has been described.

Spermatogonia. Arranged around the basement membranes of the tubules of the testes are rather large cells called *spermatogonia*. They are the direct descendants (by mitosis) of the primordial germ cells, and are eventual sources of gametes. The spermatogonia multiply by mitosis, and in this process give rise to somewhat smaller, different cells called *primary spermatocytes* (Fig. 3-8,A–B).

Primary Spermatocytes: The First Meiotic Division

Prophase I. The reticulum of the nucleus of the primary spermatocyte resolves itself, in early prophase of the first meiotic division, into the diploid number of long, very slender threads. These threads sometimes appear to be already double or in the process of doubling. This stage is called *leptotene* (Fig. 3-8,C).

The slender threads come to lie side by side, in units corresponding to the number of pairs of chromosomes (Fig. 3-8,D). The process of side-by-side conjunction is called *synapsis;* the stage at which it occurs is called *zygotene*.

The paired threads shorten and thicken somewhat, and each thread is now clearly double, except that the centromere does not divide. Four strands are associated in a *tetrad* of four *chromatids* for each original pair of chromosomes. This stage is called *pachytene* (Fig. 3-8,E).

In each tetrad, two chromatids fall apart from the other two, so that each chromatid has a single pairing partner in each region. But the chromatids often seem to exchange pairing partners along their length, so that cross-shaped figures are formed (Fig. 3-8,F–G). These cross-shaped figures, or *chiasmata* (singular: *chiasma*), may constitute cytological evidence for genetic phenomena with which we shall be concerned in Chapters 6 and 7. The stage in which the chiasmata are first evident is called *diplotene*.

The chiasmata seem to progress toward the ends of the chromatids (that is, to *terminalize*) at least in many plants and animals, as the chromatids continue

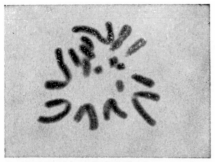

A. *Spermatogonial metaphase.*

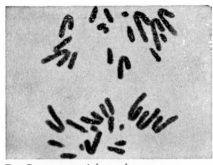

B. *Spermatogonial anaphase.*

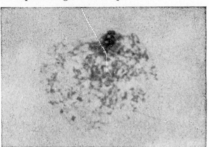

C. *Leptotene. Note positively heteropycnotic X-chromosome at 12 o'clock.*

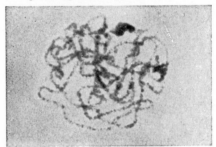

D. *Zygotene. Note chromomeric organization of paired homologues, and the X-chromosome at 1 o'clock.*

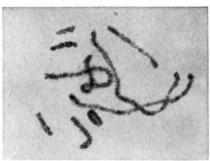

E. *Late pachytene. The X-chromosome is bent on itself and thus U shaped.*

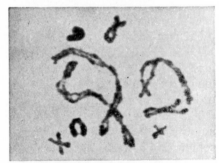

F. *Early diplotene. The homologues are beginning to separate.*

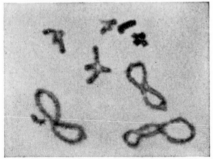

G. *Diplotene. Sister chromatids and chiasmata are now visible.*

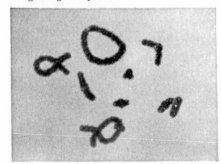

H. *Early diakinesis. Difference in condensation between the X (at 4 o'clock) and the autosomes is now less distinct.*

Figure 3-8. *Pre-meiotic mitosis and meiosis in the male of the common British grasshopper,* Chorthippus brunneus. *The complement consists of eight pairs of*

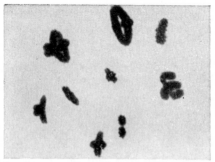

I. First metaphase. Equatorial plate in polar view.

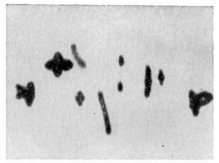

J. First metaphase. Equatorial plate in side view.

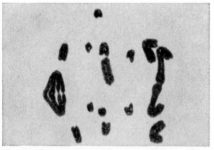

K. Early anaphase. The two chromatids of the X-chromosome (center) can now be resolved.

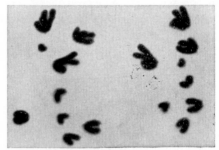

L. Late anaphase.

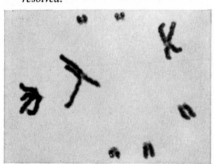

M. Second metaphase. 8 chromosomes (autosomes only).

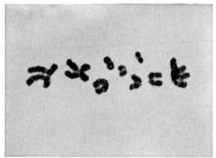

N. Second metaphase. 9 chromosomes (X + 8 autosomes).

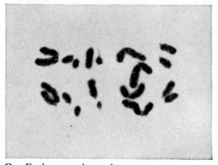

O. Early second anaphase.

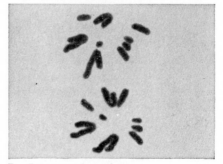

P. Late second anaphase.

autosomes and a single X-chromosome (an XO sex-determining mechanism). (*Courtesy of B. John and K. R. Lewis*).

to shorten and thicken (3-8,H). This stage, called *diakinesis*, is an important one, because the number of chromosomes is often easier to count at this stage than at any other, and our information about chromosome numbers in many plants and animals is based on diakinesis figures.

Metaphase I. The tetrads take their usual positions at the cell's equatorial plate (Fig. 3-8,I–J).

Anaphase I. The two chromatids of a tetrad that are still connected at their centromere separate from the other two of the tetrad, and these *dyads* go to opposite poles of the cell (Fig. 3-8,K–L). With regard to each original pair of chromosomes, the separation of centromeres has now been accomplished; the centromere of each chromosome has been separated from that of its pairing partner. We shall see later that, because of exchanges between pairing partners at intervals along their length, the separation of other regions of the original pair of chromosomes is not effected until the second meiotic division. The chiasmata often seem to offer some resistance to the separation of the dyads, but eventually they pull free, and the two chromatids of each dyad then spread away from each other. They usually remain connected at their centromeres.

Telophase I. There is considerable variation, from one species to another, in the nature and duration of the subsequent telophase. This period between the two meiotic divisions is designated by the special term *interkinesis*. It may be rather long, but typically it is very short, and the cells press immediately into the second meiotic division.

Secondary Spermatocytes: The Second Meiotic Division

Prophase II. In the usual prophase of the second meiotic division, the dyads of the preceding telophase are still associated at their centromeres, but they diverge widely elsewhere. The members of the dyads themselves can sometimes be observed to be already double or doubling, in evident preparation for the next cell division, which in animals comes after fertilization.

Metaphase II. The dyads line up at the equatorial plate again, with their centromeres at the equator (Fig. 3-8,M–N).

Anaphase II. With the division of the centromeres, and the separation of the chromatids on the spindle, each chromatid now becomes a complete and separate chromosome (Fig. 3-8,O–P).

Telophase II. The nuclei, four from each original primary spermatocyte, now each contain one member of each pair of chromosomes. These are the nuclei of the spermatids, which, without further division, reorganize to form the major part of the head of the sperm cell. This process of reorganization of spermatids into sperm cells includes not only nuclear changes but cytoplasmic modifications as well, leading to the formation of the sperm with its tip, head, midpiece, and tail. The general process of sperm-cell formation, including the meiotic divisions, is called *spermatogenesis*.

Germ-Cell Formation in Female Animals. As indicated earlier, the processes of meiosis are similar in all sexually reproducing plants and animals of either sex. The following tabulation of comparative terms emphasizes the similarities between male and female animals:

MALE	FEMALE
spermatogonium	oögonium
primary spermatocyte	primary oöcyte
secondary spermatocyte	secondary oöcyte
spermatid	oötid (or ovum)
sperm cell	egg cell

There is, however, one important difference between male and female gamete formation. In the male, four functional sperm cells are formed from each primary spermatocyte. The egg must often provide stored food material for the developing embryo, and the division of this stored material, accumulated by the egg during its development, among four functional female gametes would have real disadvantages. Accordingly, a regular modification of cell division has been fixed among females by evolution. Three of the four products of meiosis in females are small, abortive cells called *polar bodies*. These contain little cytoplasmic material. They simply "bud off" from the egg as meiosis proceeds (Fig. 3-9). Thus, in females one primary oöcyte gives rise to only a single egg cell.

The meiotic divisions in the eggs of some species occur *after* the sperm has entered the egg. In other species, sperm entrance occurs after the first, but before the second, meiotic division. In others, meiosis is completed before fertilization.

In many animal species, in addition to the food material stored as yolk in the egg cell, additional material is added outside the cell as it passes down the oviduct. Birds carry this trend to something of an extreme; albumin is secreted around the egg cell, and the whole is packaged within a membrane and a porous shell—a neat arrangement for an independently developing embryo.

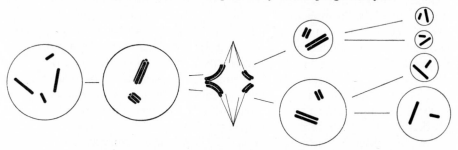

Figure 3-9. *Meiosis in a female animal gives rise to only one functional egg from each primary oöcyte. (By permission after* Introduction to Cytology, *by Sharp, Copyright 1934, McGraw-Hill Book Co., p. 251.)*

Gamete Formation in Plants. The germ-cell-forming tissues in plants behave in meiosis in much the same way as do those of the corresponding sexes in animals. In plants, however, the situation is somewhat complicated by the occurrence of two or more *mitotic* divisions of the haploid products of meiosis before fertilization occurs. The subject will be discussed in detail in the next chapter.

Genes and Chromosomes

Autosomal Inheritance. The implication has been clear throughout this chapter that all the chromosomes of an individual, not just his sex chromosomes, carry genes. But our demonstration, in the preceding chapter, that genes are in fact located in chromosomes was limited to sex-linked genes in the X-chromosomes.

The chromosomes that are not sex chromosomes are called *autosomes*. It should be obvious that the autosomes provide a reasonable basis for the inheritance of characteristics that are not sex-linked, just as the sex chromosomes can be shown to be involved in sex-linked inheritance. The autosomes are present in pairs, just as are the non-sex-linked Mendelian units. They maintain their individual identity, just as do genes. The members of a pair separate when germ cells are formed, as do the members of a pair of alleles. And the independent behavior of the members of different pairs of chromosomes provides a clear physical basis for the independent assortment of different pairs of alleles. Still more convincing evidence that the autosomes are the vehicles of non-sex-linked inheritance will be forthcoming when we begin to consider correlations between abnormalities of chromosomal number, behavior, and structure, and abnormalities of gene distribution (Chap. 7).

Genes, Meiosis, and Mitosis. If we accept the conclusion that the autosomes as well as the sex chromosomes carry genes, we can consider further the genetic significance of the two types of nuclear division we have been discussing. The cytologist is often limited by the fact that the members of a pair of autosomes are not distinguishable; with his methods they appear to be identical. But if one chromosome of a pair carries a given gene, and its homologue carries an allelic form of this gene, then the members of this pair of chromosomes are distinguishable to the cytogeneticist. Figure 3-10 illustrates mitosis and meiosis, labeling three pairs of chromosomes on this basis.

You will observe that mitosis (Fig. 3-10,I) produces daughter cells that are genetically identical with the parent cell and with each other. Meiosis, on the other hand, results in the formation of different types of haploid gametes; it involves the segregation and assortment of genes. Subsequently, fertilizations result in a variety of new combinations.

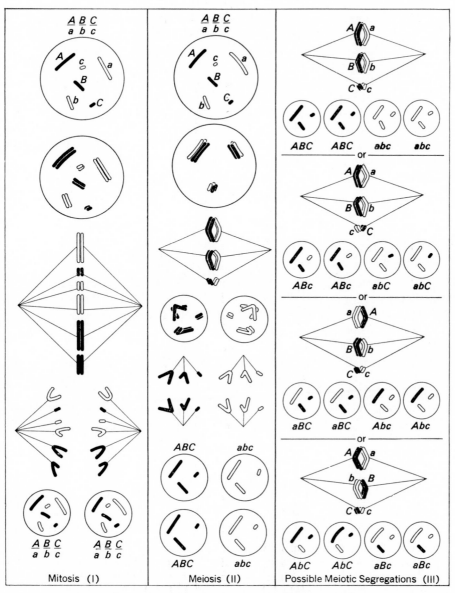

Figure 3-10. *Mitosis* (I) *in a trihybrid, when the three allelic pairs mark three different pairs of chromosomes, results in daughter cells identical with each other and with the parent cell. Meiosis* (II) *provides the basis for segregation and assortment of alleles. The different possible alignments* (III) *at the first meiotic metaphase result in the production of eight kinds of gametes, in equal proportions. (By permission, based in part on* Introduction to Cytology, *by Sharp, Copyright 1934, McGraw-Hill Book Co., p. 253.)*

In a male, a given meiotic division will produce, under the conditions of Figure 3-10,II, two different kinds of sperm. You can easily see, from the abbreviated diagram in Figure 3-10,III, how different alignments of the paired chromosomes at the first meiotic metaphase would result in all eight kinds of gametes actually produced by the trihybrid illustrated. In the female, since only one functional egg is formed in each meiotic series, it would take at least eight different primary oöcytes to produce the eight possible kinds of eggs, depending on the chance alignment of chromosome pairs when the first polar body is formed. But the qualitative consequences of meiosis are alike in the two sexes— eight kinds of germ cells are produced by the trihybrid, in approximately equal numbers.

Keys to the Significance of This Chapter

A constant number of pairs of chromosomes is in general present in each of the somatic cells of all individuals of a species. The members of different pairs of chromosomes differ in constant structural details, and often they can be visibly distinguished from each other. But the members of a pair of chromosomes, except for the sex chromosomes, are normally alike.

Ordinary somatic nuclear division (*mitosis*) results in the formation of two daughter cells, identical with each other and with the parent cell in chromosomal and genetic constitution.

The *meiotic* divisions, which result in the formation of haploid germ cells, provide a physical basis for the segregation and independent assortment of genes.

REFERENCES

Darlington, C. D., and Ammal, E. K. J., *Chromosome Atlas of Cultivated Plants.* London: G. Allen & Unwin, Ltd., 1945. (A thorough compilation of reported chromosome numbers, with useful introductions, discussions, and bibliography.)

Huskins, C. L., and Cheng, K. C., "Segregation and Reduction in Somatic Tissues." *J. Heredity*, **41**:13–18, 1950. (One of a series of related articles, with references to others.)

Longley, A. E., "Chromosome Morphology in Maize and Its Relatives." *Botan. Rev.*, **18**:399–413, 1952. (A thorough review of this subject, with many references.)

Makino, Sajiro, *An Atlas of the Chromosome Numbers in Animals.* Ames: Iowa State College Press, 1951. (A compilation for animals similar to Darlington's work on plant chromosome numbers, cited above.)

Morgan, T. H., Sturtevant, A. H., Muller, H. J., and Bridges, C. B., *The Mechanism of Mendelian Heredity.* New York: Henry Holt & Co., 1915. (A classic. A later edition of this book, published in 1923, when compared with this first edition shows the rapid accumulation of genetic information during the interval between editions.)

Schultz, J., and St. Lawrence, P., "A Cytological Basis for a Map of the Nucleolar Chromosome in Man." *J. Heredity,* **40**:30–38, 1949.

Sharp, L. W., *Fundamentals of Cytology.* New York: McGraw-Hill Book Co., 1943. (Concise and interestingly presented, the chapters on mitosis [Chap. 5], chromosomes [Chap. 7], and meiosis [Chap. 8] are particularly pertinent here.)

Swanson, C. P., *Cytology and Cytogenetics.* Englewood Cliffs: Prentice-Hall, 1957. (An excellent advanced book which treats the genetic aspects of cytology.)

Wipf, L., and Shackelford, R. M., "Chromosomes of a Fox Hybrid." *Proc. Nat. Acad. Sci.,* **35**:468–472, 1949. (Source of the material in Problems 3-13 and 3-14.)

QUESTIONS AND PROBLEMS

3-1. What do the following terms signify?

anaphase	heteropyknosis	satellite
autosomal	interkinesis	secondary spermatocyte
centromere	knob	somatic
chiasma	leptotene	spermatid
chromatid	meiosis	spermatogenesis
chromomere	metaphase	spermatogonia
chromonemata	mitosis	spindle
cytogenetics	nucleolus	synapsis
diakinesis	oötid	telophase
diplotene	polar body	tetrad
euchromatin	primary spermatocyte	zygotene
heterochromatin	prophase	

Characterize the four chromosomes carried by an X-bearing Drosophila sperm, in terms of the centromeres of these chromosomes, as X_m, 2_m, 3_m, 4_m (letting m stand for a centromere of *male* origin). Similarly, label the centromeres of a Drosophila egg as X_f, 2_f, 3_f, 4_f; and those of a Y-bearing sperm as Y_m, 2_m, 3_m, and 4_m.

3-2. In terms of the above symbolism, what will be the "chromosomal centromere formula" of a fertilized egg that will develop into a male? A female? Of male somatic cells generally, assuming no abnormality in somatic mitosis? Female somatic cells?

3-3. What is the centromere formula of a spermatogonium? A primary spermatocyte? (Remember that the centromeres do not usually divide until the second meiotic division.) Give a formula for any particular secondary spermatocyte. A spermatid. A sperm.

3-4. What is the centromere formula of an oögonium? A primary oöcyte? Any particular secondary oöcyte? An egg, after meiosis, but unfertilized? A polar body?

3-5. What is the chance that a particular sperm has all four centromeres from the father of the individual producing the sperm? From the mother?

3-6. What is the chance that a fertilized egg that will develop into an F_2 has all eight of its centromeres from the male in the original parental cross?

3-7. Mankind has 22 pairs of autosomes plus an XX pair in women and an XY pair in men. What proportion of a man's sperm will contain all of the centromeres he received from his father?

3-8. Assume an ancestral, sexually reproducing organism in which there was originally only one pair of chromosomes but that produced gametes without meiosis, so that the germ cells in each generation were diploid. What consequences would this have in terms of the chromosome number of successive generations? In terms of genetic variation?

3-9. In a normal male, how many functional sperm will be expected from 100 primary spermatocytes? 100 secondary spermatocytes? 100 spermatids?

3-10. How many functional eggs will be expected from 100 primary oöcytes? 100 secondary oöcytes? 100 oötids?

3-11. What advantage for the species has the process of polar body formation in females, compared with the possibility of forming four functional eggs from a primary oöcyte?

3-12. Assume a plant with six chromosomes—a pair of rods, a pair of V's, and a pair of dots. After three successive generations of self-fertilization, what proportion of the population will have all V-shaped chromosomes? What proportion will be like the original plant, with a pair of rods, a pair of V's, and a pair of dots?

Fox breeders have sometimes hoped to improve the size, productivity, and color variety of the red fox (*Vulpes vulpes*) by hybridizing it with the Arctic fox (*Alopex lagopus*). Numerous such hybrids have been produced, but they are always completely sterile.

3-13. The red fox has 17 pairs of relatively large, long chromosomes. The arctic fox has 26 pairs of shorter, smaller chromosomes. What do you expect to be the chromosome number in somatic tissues of the hybrid?

3-14. The first meiotic division in the hybrid shows a mixture of paired and single chromosomes. Why do you suppose this occurs? Can you suggest a possible relationship between this fact and the sterility of the hybrid?

3-15. Before the rediscovery of Mendelism, Galton formulated a "Law of Ancestral Inheritance" according to which a person shows about half of the hereditary characteristics of each parent, a quarter of those of each grandparent, and so on. The law was based on studies of such characteristics as height in man. Can you account for the "law," in terms of chromosomes and genes?

Assume a corn plant that is missing the knob at the end of one of its ninth chromosomes, and has a knob at the end of one of its tenth chromosomes (*cf.* Fig. 3-2). Letting K stand for *knob*, the plant's chromosomal formula may be designated $\left(\dfrac{9K}{9}, \dfrac{10K}{10}\right)$. In corn, as in animals, the "male" tissues form four functional meiotic products, while in the "female" tissues three degenerate and only one is functional.

3-16. Suppose that you examined cytologically the meiotic products in the male tissues of this exceptional plant. What proportion would you expect to find with both knobs ($9K$, $10K$)? What other types would you expect to find, and in what proportions?

3-17. Answer the questions in 3-16 with regard to the chromosomal constitution of the functional products of meiosis in the female tissues.

3-18. If this exceptional plant is self-fertilized, what proportion of the zygotes would you expect to be like the standard type of corn $\left(\dfrac{9K}{9K}, \dfrac{10}{10}\right)$?

3-19. Discuss the value of material like that in the preceding three questions with regard to the visible recognition of the independent assortment of chromosomes.

3-20. In Drosophila, an allelic pair concerned with body color

> y, a sex-linked recessive for yellow body color, and
> y^+, the wild-type allele

is known to be located very near the end of the X-chromosome (the opposite end from the centromere). If we had used this pair of alleles to "mark" the X-chromosomes of a female fly, rather than using the centromere, how would your answers to Question 3-4 have been different?

3-21. Diagram the process of egg formation in a female Drosophila, with X-chromosomes marked as follows:

(X$_m$ and X$_f$ refer to the centromeres.) Show tetrad formation, but do not at this time show chiasmata or exchanges between chromatids. Show how two different kinds of functional eggs may be formed under these circumstances.

3-22. Now, diagram the process as in Problem 3-21, but show an exchange between two of the four strands of the tetrad, between the position of the y gene and the centromere. What new types of eggs may result from this kind of exchange?

3-23. Rabbit eggs have been stimulated to develop without fertilization, and certain of the parthenogenetic rabbits so formed have been reported to be heterozygous for certain allelic pairs. How might this be explained?

CHAPTER 4

Life Cycles and Reproduction

The preceding chapters have included, either expressly or implicitly, certain principles that provide a foundation upon which much of your further understanding of genetics will rest. They may be summarized briefly as follows:

1. Life, as we know it, is derived only from previously existing life.
2. Each of the different kinds of living things gives rise to other living things of the same general kind.
3. Many of the similarities and differences of organisms have a biologically heritable basis residing in units of heredity called genes.
4. Genes are located in chromosomes in the nuclei of cells.
5. Two well defined kinds of nuclear division—mitosis and meiosis—are known to have special and far-reaching implications for the life histories of organisms.

When expanded and qualified, these genetic principles tell us a good deal about life as a general phenomenon. But to bring them into sharp focus, and to be able to apply them, you will need to consider them in relation to particular organisms. Because all these principles relate in some degree to reproduction, it is the *reproduction* of particular organisms that must be clearly understood. This chapter will give some idea of the diversity of reproductive processes in various forms of life.

The Summarization of Reproductive Processes in Life Cycles. The significant stages in the processes by which an organism gives rise to others of its kind make up the *life cycle* of that organism. Biologists often represent a particular life cycle by diagraming in a circle the salient events of a completed life history. Thus a simplified life cycle of man might be shown as in Figure 4-1. Schematic

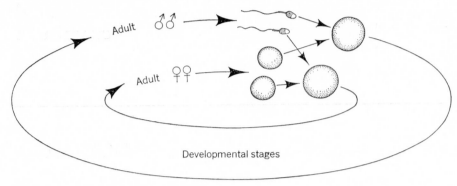

Figure 4-1. *A simplified life cycle of man.*

representations of this kind provide simple, flexible frameworks for summarizing significant knowledge that may be difficult to apply unless remembered in a meaningful, sequential arrangement. You will find, however, that strictly sequential events are not characteristic of the reproduction of every species of organism.

Chromosome Cycles and Life Cycles. Chromosomes play an unseen but highly important role in life cycles. In sexually reproducing organisms, life begins as a single cell formed from the union of two parental germ cells, or *gametes*. This fusion cell has the combined chromosome complements of the germ cells and, therefore, normally has twice the chromosome number of either gamete alone. The subsequent growth and development of a mature multi-cellular organism from a fertilized egg cell, or *zygote*, are based on cellular divisions in which the nuclei divide mitotically, so that somatic cells typically have the same chromosome complement as the fertilized egg. Reproduction by the new adult, however, is preceded by a reduction of chromosome number through meiosis of the nuclei in certain cells. This whole series of events repeated through succeeding reproductive cycles insures that, generation after generation, the adults of a species have the same number of chromosomes and produce gametes with half that number. With few known exceptions, chromosome cycles in higher animals follow this pattern. The life cycle of man, already presented briefly, can now be made more complete by adding the information about haploid and diploid stages in the cycle, and showing where reduction in chromosome number and resumption of diploidy occur. For details of spermatogenesis and oögenesis in higher animals, refer to Chapter 3.

In many groups of plants, the life cycle involves a regular alternation of a sexual with an asexual generation. This is true of angiosperms, gymnosperms, ferns, mosses, liverworts, and certain members of lower groups. In the typical situation, a zygote develops into a *sporophytic* plant. The *sporophyte*, through meiosis, gives rise to spores, each of which by itself may produce a *gametophytic* plant. *Gametophytes* are plants that form gametes. With the union of gametes, a zygote is produced, and from the zygote a new sporophyte comes into being. The alternation of a gamete-producing generation with a spore-producing generation in plants coincides with an alternation in chromosome number in these forms. For ease of reference, the typical relation of chromosome complement to prominent stages in the life cycles of higher animals and plants can be summarized in the following way:

	HAPLOID CHROMOSOME COMPLEMENT	DIPLOID CHROMOSOME COMPLEMENT
most animals	gametes	zygote, animal body
many plants	spores, gametophyte, gametes	zygote, sporophyte

The reproduction of many groups of microorganisms cannot be summarized in so neat a fashion. Indeed in some instances one is not sure that the hereditary material is organized into chromosomes.

The Life Cycles of Particular Organisms

In examining the genetically significant aspects of the reproduction of particular kinds of organisms, we must limit ourselves to a few examples. Our choice of examples is admittedly somewhat arbitrary but has a rationale of which you should be aware: (1) the organisms chosen have life histories that either are widely representative or represent some significant departure from generally familiar schemes of reproduction; (2) the organisms chosen are important objects of genetic research, either because they are widely studied, or because they offer exceptional opportunities for investigation, or both. The omission of animals other than Paramecium does not imply that man, Drosophila, mice, and foxes are uninteresting or unfruitful objects of genetic study. Quite the contrary. But the general scheme and some of the typical details of life cycles of animals should be familiar to you from the previous chapter and the first part of this one. You will see at once that we have selected microorganisms for most of the examples; the greater variability in schemes of reproduction for microorganisms, as compared with those for higher organisms, and the fact that microorganisms offer unusual advantages for many kinds of genetic investigations have determined this selection.

THE LIFE CYCLE OF CORN

Reproduction in corn (*Zea mays*) is typical enough of what happens in most seed plants to permit broad applications from the outlines of its life history to analogous events in different members of the group. A more particular reason for concentrating attention on corn is that of all plants its genetics has probably been investigated most thoroughly. Why have geneticists spent so much time on corn? Its agronomic importance, the fact that a single ear of corn represents a large progeny, the early discovery of interesting deviant characters, and the relative ease of controlled matings are all involved. In addition, corn chromosomes are of a size and morphology that make them well suited for cytogenetic

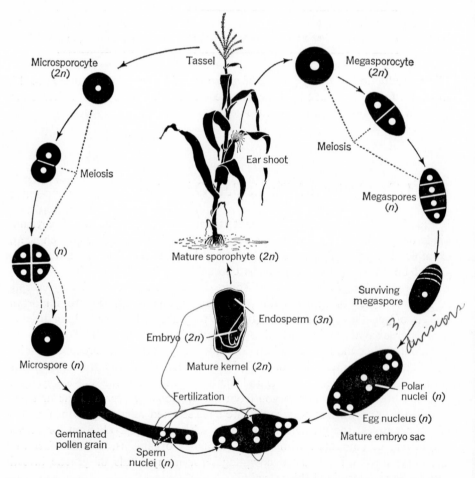

Figure 4-2. *A knowledge of the life cycle of corn* (Zea mays) *is basic to an understanding of the many elegant genetic studies that have been carried out with this plant.*

study. These advantages of corn as an object of genetic research have far outweighed such disadvantages as its comparatively long life cycle. Finally, it should be remarked that some exceptionally gifted investigators have been fascinated by the genetics of corn, and have devoted their time and thought to its study.

Sporophytic Generation and Spore Production. At least for preliminary purposes, we can identify the sporophyte of corn as the large green corn plant familiar to everyone. Corn is *monoecious*, which means that both *pistillate* ("female") and *staminate* ("male") flowers occur on the same plant. The pistillate flowers are borne in the *ears* of corn; the staminate flowers are in the *tassel* at the top of the stalk (see Fig. 4-2). In the flowers are formed spores that give rise to the gametophytic generation.

Megaspores are produced in the ovules of pistillate flowers. The cells that undergo meiosis to produce megaspores are called *megasporocytes*. Each megasporocyte gives rise to four megaspores, three of which degenerate. The remaining megaspore develops into the female gametophyte. Notice that the fate of the four products of meiosis of a megasporocyte nucleus corresponds to the fate of the meiotic products of a primary oöcyte in animals.

Microspores are produced from *microsporocytes* (pollen mother cells) within the anthers of the staminate inflorescence. Each microsporocyte gives rise by meiosis to four functional microspores.

The Gametophytic Generation and Gamete Production. As is characteristic of angiosperms, the gametophytic generation of corn is inconspicuous and much reduced. The female gametophyte is produced by three divisions of a megaspore. The eight nuclei thereby derived lie at first in a common cytoplasm, and are said to make up the *embryo sac*. This is the female gametophyte. The nuclei all are haploid. One is the egg nucleus. Two others, the polar nuclei, have a supporting role in the fertilization process.

The male gametophyte also is uncomplicated, consisting merely of the pollen grain (and later the pollen tube) containing three nuclei. Two of these are *sperm nuclei*, or gametes. The other is the *tube nucleus*.

Fertilization. As pollen matures, it comes free of the anthers that contain it. Under natural conditions, its fate thereafter is governed largely by gravity and by the whim of whatever air currents are flowing at the time. The pollen may either travel far or settle near home. Much pollen never functions in fertilization, either because it comes to rest in the wrong place or because it dries out before arriving at the right place. However, when a mature, physiologically reactive pollen grain comes in contact with a silk of an ear of corn, the pollen grain sends out a long tube that grows down through the silk and eventually pushes into the embryo sac. In the embryo sac, one sperm nucleus fuses with the egg, forming a zygote. Meantime, the other sperm nucleus and the two polar nuclei of the embryo sac unite in a triple fusion. This fusion nucleus has three sets of chromosomes, and hence is said to be *triploid*. Through subsequent mitotic divisions, it gives rise to the *endosperm*, a tissue that plays a dynamic role in the development of the seed and serves as a nutritional storehouse for the young sporophyte.

Sporophyte Again. The zygote, through cell division and differentiation, develops into an embryo. When a kernel of corn germinates, the embryo enlarges rapidly and eventually becomes a mature sporophyte with roots, stem, and leaves. The endosperm is consumed early in the history of the growing seedling plant.

Controlled Matings in Corn. Genetic analysis in corn or any other organism is facilitated if the investigator is not forced to deal with instances of unknown or uncontrolled parentage. To control matings in most animals is relatively simple, and may be accomplished merely by isolating individuals from members of the opposite sex and then pairing off males and females at an appropriate time. The problem of controlling matings in higher plants may be somewhat different, especially in those species where both male and female sex organs occur in single individuals.

In order to control matings in corn, two kinds of precautionary measures must be taken: (1) The pistillate inflorescence must be protected from pollen other than that the investigator wishes to apply. (2) The desired pollen must be collected and applied in such a way that no mixture with other pollen occurs. Separation of the staminate and pistillate inflorescences in different parts of the corn plant makes these objectives readily attainable. A paper bag tied over the ear shoot before the silks appear will prevent chance fertilization, and the pollen to be used in the cross can be collected by enclosing an appropriate tassel in a different paper bag. Since corn pollen is fairly short-lived, bagging a tassel the day before its pollen is to be used is adequate insurance against active contamination by foreign pollen that may have settled on the tassel. The controlled pollination process can be accomplished by removing the ear-shoot bag and pouring pollen from the tassel bag directly onto the silks. Then the ear-shoot covering must be replaced promptly. Such techniques permit corn, at the will of an investigator, either to be entirely self-pollinated or to be cross-pollinated from a given source. Under uncontrolled conditions (open-pollination), a few of the kernels on an ear of corn usually have resulted from self-pollination and the majority from cross-pollination.

Problems of controlling matings in seed plants may differ considerably from group to group, depending in a large part on the kind of inflorescence encountered. When both stamens and pistils are within the same flower (a *perfect flower*), self-fertilization may be the rule under natural conditions. Controlled cross-pollination must then be preceded by timely emasculation of the flower to be pollinated, in order to prevent self-pollination. Tomatoes, for example, can be emasculated simply by removing the male organs of the flower with forceps. In some other plants, this straightforward approach may not be practical. Techniques for genetic investigation must be adapted to the characteristics of the particular kind of plant to be studied. There are good reasons, then, for geneticists to know a great deal about the actual details of reproductive processes.

Genetic Applications. You understand, of course, that we cannot attempt to review the entire genetics, or even any considerable portion of it, for the organisms whose life cycles we are examining. But it should be profitable to consider a few of the genetic-research challenges and opportunities found in the particulars of each life cycle under consideration.

Genetics of the Male Gametophyte. You have learned that a meiotic division in a plant heterozygous for an allelic pair *Aa* gives rise to two nuclei of genotype *A* and two nuclei of genotype *a*. This is seldom observed directly. The rule is usually inferred from microscopic observations of chromosomal distribution at meiosis and from the fact that Mendelian ratios presuppose equal numbers of the different genotypes occurring in gametes produced by a heterozygote. There are two chief impediments to obtaining direct evidence for the basic 1:1 segregation at meiosis in organisms heterozygous for a single allelic pair: (1) Meiotic products such as pollen grains seldom have phenotypic characters that permit determination of their genotype. Usually, a meiotic product must participate in fertilization before its genetic constitution can be identified. (2) It is difficult in most organisms to obtain for analysis all the products of a single meiosis. This is certainly true of corn, where on the female side three of the four products of each meiosis degenerate, and where on the male side the four pollen grains arising from a single meiosis intermingle with other male gametophytes.

Nevertheless, we can in general be certain that there really is a 1:1 segregation of members of allelic pairs at meiosis. Phenotypic ratios obtained among the progeny of hybrids (*Aa*) prove that in large populations of pollen there must be roughly equal members of *A* and *a* pollen grains. Similarly, Mendelian ratios among segregating characters in sporophytes are consistent with the supposition that there are equal chances of either the *A* or the *a* allele being present in the single nucleus that survives after meiosis of a megasporocyte.

In a few special instances it is possible to detect segregation directly in pollen grains and in female gametophytes. A clear example is found in the segregation of alleles for the characters *waxy* and *nonwaxy* in corn. These alternative phenotypes depend on a difference in the chemical nature of starch reserves that can be distinguished by treating with an iodine solution. Pollen with the *wx* gene is red in iodine; pollen carrying the normal allele, *Wx*, turns blue. If pollen from a plant of the genetic constitution *Wxwx* is stained with iodine, about half is red and half blue. In an experiment of this kind by Demerec, 3,437 pollen grains were found to stain blue, and 3,482 were red-staining. The results provide direct evidence for the basic 1:1 segregation expected after meiosis in a heterozygote. Since the pollen collected in such an experiment is only a sample of an entire population, chance deviations from a 1:1 ratio are expected and do occur. Such chance deviations in gametic ratios account in part for deviations from the ideal Mendelian ratios predicted for the diploid generation following. Figure 4-3 is a photographic demonstration that corn heterozygous for the waxy gene gives pollen approximately half of which stains blue, indicating the allele for *nonwaxy*, and half of which stains red, indicating *waxy*. Female gametophytes from a *Wxwx* hybrid also may be subjected to the iodine treatment. If this test is applied to a good sample of gametophytes, approximately the same number of individuals stain blue as stain red. The result is a straightforward indication that alleles *Wx* and *wx* have equal chances of being in the surviving member of the quartet of megaspores formed by a meiosis.

Endosperm Genetics. All the nuclei in a male gametophyte of corn are genotypically alike, and the same is true for a female gametophyte. Therefore an

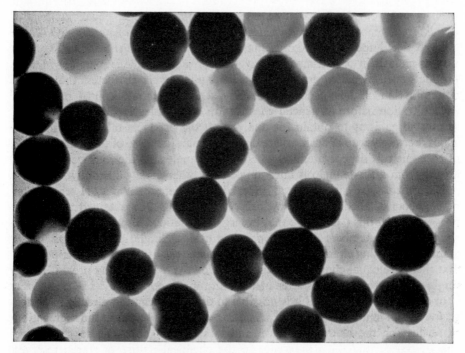

Figure 4-3. *The iodine test applied to pollen from a corn plant heterozygous for the* waxy *gene. A count of the darkly and of the lightly stained pollen grains provides a straightforward demonstration of the 1:1 segregation of members of an allelic pair into gametes. (Pollen supplied by O. E. Nelson, Jr.; preparation and photography by G. Pincheira.)*

Sugary ♀ ✕   Starchy ♂
     su su   *Su Su*
     (egg) *su*   ✕   (sperm) *Su* → (embryo) *Su su*
(polar nuclei) *su + su*   ✕   (sperm) *Su* → (endosperm) *Su su su*

Starchy ♀ ✕   Sugary ♂
     Su Su   *su su*
     (egg) *Su*   ✕   (sperm) *su* → (embryo) *Su su*
(polar nuclei) *Su + Su*   ✕   (sperm) *su* → (endosperm) *Su Su su*

Figure 4-4. *Reciprocal crosses in corn involving the alternative characters* starchy *and* sugary *illustrate that the endosperm derives two identical alleles from the maternal parent and a third allele from the paternal parent.*

embryo and its surrounding endosperm have the same alleles represented. But because the endosperm arises from a triple rather than a double fusion of gametic nuclei, the allelic balance in the endosperm differs from that in the embryo. Heterozygous endosperm always has two of one allele and one of the other. This permits a view of dominance relations different from that usually observed in the sporophyte. The difference between allelic balance in endosperm and that in embryo can be seen in Figure 4-4, which shows reciprocal crosses involving the characters *sugary* and *starchy* endosperm. In both the hybrids indicated in the figure, the endosperm turns out to be starchy, no matter which allele enters through the female line. One *Su* gene, then, is dominant over two *su* alleles.

Other alleles, however, may not show the same kind of dominance relations in the endosperm. This can be seen in certain genotypes that result from crosses involving *floury* (*f*) and *flinty* (*F*) endosperm. The genotype *Fff* is found to determine the floury character; *FFf* endosperm is flinty.

An advantage of endosperm genetics is that the investigator can keep one jump ahead in the game of genetic analysis as it is ordinarily played. In the usual investigation of inheritance of a sporophytic character, seed must be planted for the parental generation, and then again for the F_1, and a third time for the F_2. In the study of endosperm characters, however, kernels that form on the parent plants show the F_1 hybrid characteristics. One more planting, and the F_2 endosperm characteristics appear. Two plantings, then, give the same amount of information about endosperm characters that three plantings give about sporophytic characters. Geneticists, of course, have not concentrated entirely on the study of endosperm characters in corn, since the characters of the sporophyte are much more numerous and, on the whole, more important.

The value of knowing life histories becomes quite obvious when you consider the genetic intricacies of the apparently simple kernel of corn. Covering the kernel is a protective layer called the *pericarp*. This consists of maternal tissue, and its characteristics are determined by the genotype of the mother plant. Beneath it is the endosperm, including its outer layer, the *aleurone*, whose characteristics appear soon after pollination. Still deeper inside is the embryo, having a genotype one generation removed from that of the pericarp—a genotype whose phenotypic expression is largely deferred until germination and the emergence of the new sporophyte takes place. If the origin of structures in the kernel of corn were not known and properly fitted into the life cycle, genetic interpretation of the inheritance of kernel characteristics might go rather far astray.

THE LIFE CYCLE OF NEUROSPORA

Neurospora is a bread mold belonging to the group of fungi called Ascomycetes. It was once known as a pest in bakeries, but it also has some reputation as a necessary participant in the manufacture from peanut meal of a Javanese delicacy called *ontjom*. We are interested in Neurospora because it offers exceptional opportunities for genetic study. If we were to try to imagine an ideal organism for genetic research, this synthetic creature would need to have many of the qualities already found in a simple mold.

What are these qualities? Neurospora has a brief life cycle, and is well adapted for investigation on a mass-production scale. Since it may be readily propagated asexually, unlimited populations of a given genotype may be had for little trouble. Moreover, it is easily kept in pure culture on a chemically defined medium—a great advantage for physiological research. Finally, the nature of its life cycle permits unusually close genetic analysis.

Reproduction in *Neurospora crassa*. There are several different species of the genus Neurospora. We are to discuss *N. crassa*, the species on which the bulk of genetic investigation has been carried out. When simply the generic name Neurospora is employed in genetic writing, the species *crassa* is usually meant, as Drosophila, unless further qualified, refers to *D. melanogaster*.

Figure 4-5 summarizes the life cycle of Neurospora. You will note that reproduction of this organism may be sexual or asexual. Asexual reproduction occurs by means of spores called *conidia* or simply by propagation through unspecialized fragments of mycelium. The vegetative hyphae of Neurospora are segmented, with each segment normally containing many haploid nuclei. Asexual reproduction, then, is based on mitotic divisions of haploid nuclei.

Neurospora has two mating types, designated *a* and *A*, which are determined by members of an allelic pair. Sexual reproduction occurs only when cells of

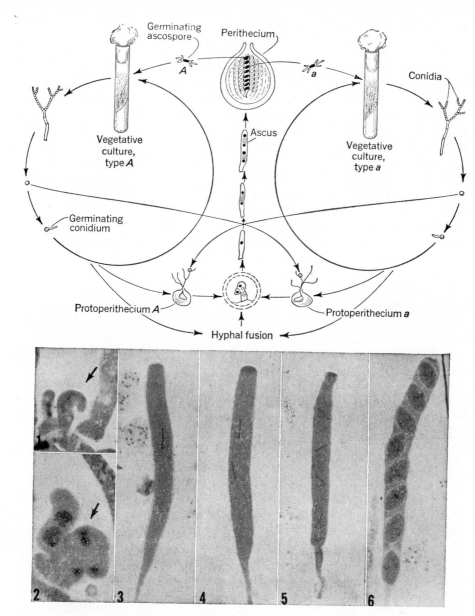

Figure 4-5. *Below a diagrammatic representation of the life cycle of* Neurospora crassa *is a photographic sequence of the major events in the development of the ascus. In the photographs, you see, from the left, the cell in which nuclei fuse to form the zygote, then stages of meiosis I, meiosis II, and of mitosis of each of the meiotic products, and finally eight spores in an ascus. (Diagram of the life cycle after Beadle, from Hardin,* Biology: Its Human Implications, *W. H. Freeman and Co., 1949, p. 377. Photographs courtesy of G. Pincheira.)*

opposite mating type unite. The fusion nucleus that forms is the sole diploid stage in this organism. The zygote quickly undergoes meiosis in a saclike structure called the *ascus*. As is represented in our diagram of the life cycle of Neurospora, the narrow contours of the ascus hold the division figures so that their long axes coincide with the long axis of the sac. The ascus is so narrow that divisions following the first division of meiosis occur in tandem, and the resultant nuclei do not slip past each other. The meiotic divisions are typical of their kind, and give rise to the usual four nuclei. Each product of meiosis, still in place, then divides once mitotically, giving rise to eight nuclei in linear order in the ascus. Heavy spore walls form around each of these nuclei, and eventually the eight nuclei are contained in eight *ascospores*. Starting at either end of an ascus one may count off four pairs of spores, each pair representing one of the products of meiosis. The members of a pair of ascospores are genetically alike.

Eventually, an ascus of Neurospora ruptures. Then the ripe ascospores that were within mingle with other ascospores from different sources. In order to identify individuals arising from crosses between strains of Neurospora, ascospores must be isolated and cultured separately. The germination of an ascospore begins a new generation of haploid mycelium of the genetic constitution determined by the shuffling of genotypes at meiosis.

Genetic Applications. Until an ascus breaks up, the eight linearly arranged ascospores provide an accurate diagrammatic record of what has happened at meiosis. This is because: (1) All of the products of a single meiosis are preserved within one structure and cannot be confused with products of other meiotic divisions; (2) the position of a particular ascospore within its ascus can be referred to the actual position of a nucleus in meiosis, as determined by the orientation of separating chromosomes on a division spindle.

The eight ascospores within an ascus, then, represent an *ordered tetrad* of duplicate meiotic products, and therefore provide exceptional information. Interpretation is possible when ascospores are removed singly from an ascus and kept in their original order. Dissection of ascospores in order is a task that requires pains and skill because a single spore is only about 0.028 mm long, but with the use of a microscope and a fine glass needle the job can be done. After isolations are made in order, the ascospores may be placed in separate culture tubes and germinated. Then the genetic attributes of cultures so derived may be referred to segregations of genes at meiosis. In summary, if eight such culture tubes are lined up in the same order as that of their corresponding ascospores in the ascus, one can observe directly which alleles went to which nuclei at meiotic segregation. This kind of situation is illustrated in Figure 4-6, which shows the segregation of a wild-type gene as against an allele that affects the morphology of the mold. The variant-type morphology is characterized by

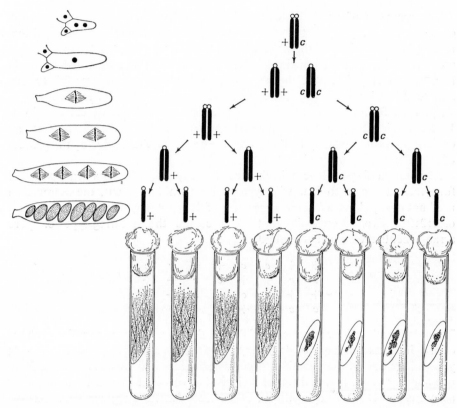

Figure 4-6. *The basis of first-division segregation within an ascus of Neurospora. The sequence at the left shows the general nuclear behavior synchronized with a detailed sequence for a particular chromosomal pair shown at the right. Gene c determines a colonial type growth, while its wild-type allele determines the usual spreading mycelial growth habit. The centromeres of the chromosomes are represented by clear circles. (After Beadle; Baitsell,* Science in Progress, *Fifth Series.* Yale University Press, 1947, p. 180.)

buttonlike, colonial growth on agar medium as contrasted with the weblike, spreading growth typical of the wild type under similar cultural conditions.

You will note that after a meiosis such as has been illustrated in Figure 4-6, the first four spores in the line would germinate to give wild-type mycelium and that the next four would produce colonial cultures. This spore order may be represented simply as $+ + + + c c c c$. An order $c c c c + + + +$ would be equally possible. This kind of difference in orientation depends on whether the chromosome carrying gene c lines up to the right or to the left of its homologue at the first division of meiosis.

The recovery of precisely four variant and four wild-type spores from a complete ascus is just what is expected when a single allelic pair is segregating. Remember that when all the products of a meiosis are recovered, genetic ratios are mechanically, not statistically, determined. Naturally, observed ratios in Neurospora vary from the theoretical ratios if certain ascospores fail to germinate. Also, if ascospores were isolated at random, instead of in order out of single asci, following the cross $+ \times c$, deviations from the expected ratio of 1 colonial:1 wild-type mycelium would be expected to occur. Some of these deviations would certainly be due to the operation of chance during the process of selecting ascospores to isolate and germinate.

Although a single allelic pair characteristically gives rise to a segregation of four variant spores to four wild-type spores within an ascus, the segregation need not always have a *linear arrangement* of 4:4, but may be some combination of 2:2:2:2. From a cross like that just discussed, the following types of spore arrangements are sometimes found:

$$
\begin{array}{cccccccc}
+ & + & c & c & + & + & c & c \\
c & c & + & + & c & c & + & + \\
+ & + & c & c & c & c & + & + \\
c & c & + & + & + & + & c & c \\
\end{array}
$$

To understand how arrangements of this kind occur, it is necessary first to have clearly in mind the basis of 4:4 segregations. If you look back to Figure 4-6, you will see that 4:4 segregations occur when a gene and its allele separate from one another at the first division of meiosis. Now look at Figure 4-7 to see what accounts for 2:2:2:2 segregations, called *second-division segregations*. Observe that second-division segregations are accounted for by actual physical exchanges of corresponding segments of chromatid strands of homologous chromosomes. This process of exchange of chromosomal material is called *crossing-over*. When crossing-over occurs between a gene and the centromere of its chromosome, segregation of the alleles does not occur at the first division of meiosis but is deferred to the second division. The centromeres of homologous chromosomes seem always to segregate from one another at the first division of meiosis. In Neurospora, segregation within an ascus showing a 2:2:2:2 spore arrangement indicates crossing-over between the segregating gene and the centromere of its chromosome.

Heterokaryosis in Neurospora. The cells of a Neurospora mycelium are typically multinucleate. If the mycelium has arisen from a single isolated ascospore, the nuclei are genotypically identical. When mycelia of different genetic origin are in proximity, however, vegetative fusions of the hyphae may take place, and nuclei may be interchanged. A mycelium containing nuclei of diverse genotypes within the common cytoplasm of a single cell is called a *heterokaryon*. Such a condition of heterokaryosis has obvious analogies with the heterozygous

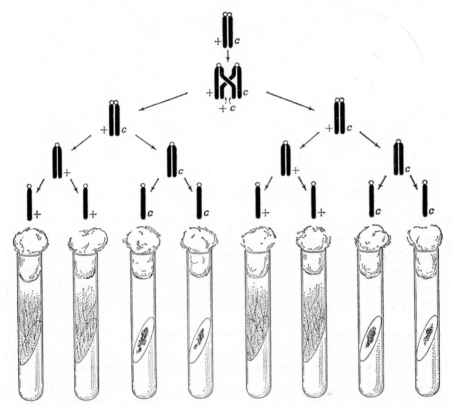

Figure 4-7. *Exchange between chromatids that occurs between gene locus and the centromere results in second-division segregation of members of an allelic pair. (After Beadle; Baitsell,* Science in Progress, Fifth Series, *Yale University Press, 1947, p. 182.)*

state. You will see at once the chief difference between the two (Fig. 4-8). In heterokaryosis the different allelic representatives of a particular locus are separated by nuclear membranes; in heterozygosity the different allelic representatives are contained within the same nucleus. With reference to gene dosage, heterokaryosis is more versatile, since the ratios of different kinds of nuclei within a heterokaryon are not mechanically fixed. Heterokaryosis is not a phenomenon confined to Neurospora but is found also in a good many other filamentous fungi, for example, Aspergillus and Penicillium.

Where heterokaryosis occurs, it can be used as a tool for genetic investigations. The most elegant of such studies, as exemplified in work carried out by G. W. Beadle and V. Coonradt, utilize genetic deviants that differ from the standard type in nutritional properties. These workers utilized deviant strains of Neurospora characterized by deficiencies in the capacity to synthesize certain vitamins

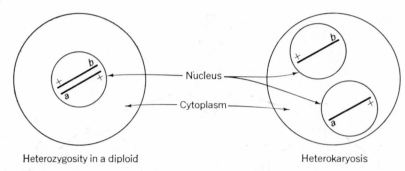

Figure 4-8. *In a heterozygote, the different alleles are in the same nuclei; in a hetero-karyotic cell, the different alleles are in separate nuclei.*

that the wild-type mold can make from simpler compounds. Two such strains are the *nicotinicless* and the *pantothenicless*. The former is unable to grow unless the vitamin nicotinic acid is included among nutrients available to it; the latter has a dietary requirement for pantothenic acid, another vitamin. In contrast, the *minimal medium* for Neurospora, the simplest chemically defined medium on which the wild type will grow, includes neither nicotinic nor pantothenic acid.

Attempts to culture the pantothenicless and nicotinicless strains individually on the minimal medium for Neurospora are unsuccessful. However, if the two strains are inoculated together into the minimal medium, a luxuriant mycelium arises. The explanation is simple and convincing. Inoculation of the two mutant strains side-by-side is followed by heterokaryon formation. The pantothenicless nuclei can direct the synthesis of nicotinic acid, because they have the appropriate wild-type allele. Similarly, the nicotinicless nuclei can direct the formation of pantothenic acid. The two nuclei *complement* each other's activities and taken together have as much genetic competence as the wild type. Working in concert in a kind of intracellular, internuclear symbiosis, the two different genotypes within the heterokaryon are able to do everything required for vigorous growth. The situation is summarized in Figure 4-9.

If you reorient your thinking slightly, you will see how heterokaryosis provides a basis for a powerful tool for genetic analysis, a tool called the *complementation* test. Consider again this experimental observation: a normal phenotype is produced when the nicotinicless and pantothenicless variants are brought together in a heterokaryon. The result can be explained only on the basis that the variants complement each other; put in other words, the variants are not deficient in the same genetic function. Therefore, one can conclude that the genes accounting for the pantothenicless and nicotinicless attributes are non-allelic. This conclusion would in any case have seemed probable for the example given, on the grounds that the two variant phenotypes suggest deficiencies in very different genetic functions.

Consider now two nicotinicless variants of different origin, nicotinicless-1 and nicotinicless-2. The phenotypes, in each instance a requirement for nicotinic acid in the growth medium, do not suggest deficiency in different functions. But when an experiment was done by Beadle and Coonradt, in which *nic-1* and *nic-2* were inoculated together into a minimal medium, a heterokaryon was formed that grew in the absence of nicotinic acid. The results justify the conclusion that each variant strain includes a normal genetic component absent in the other. Thus one might symbolize the genotypes of the two variant strains as *nic-1* + and + *nic-2*. If the experimental result had been absence of growth in minimal medium by the heterokaryon, one might suspect, although the case would not be conclusively proven, that *nic-1* and *nic-2* represent lesions in the same genetic function.

In general, then, the complementation test is a way for fractionating germ plasm into meaningful components. We will say more about complementation in Chapter 9, when we have a more sophisticated idea of the nature of the gene. Meantime, two or three broad statements may be helpful to your thinking. (1) The formation of an appropriate heterozygous diploid nucleus can be the basis for a complementation test in the same way as is heterokaryosis. You will see this readily if you re-examine Figure 4-8. (2) If the genetic variants under consideration are dominants, the distinction between complementarity and noncomplementarity is not meaningful. (3) A negative test, that is, absence of complementation, is not in itself proof of identical genetic function. For example, in Neurospora some strains of the mold refuse to accept nuclei from certain other strains; in such instances, inability to grow on the minimal medium would be irrelevant to considerations of the complementarity of a pair of nutritionally deficient variants.

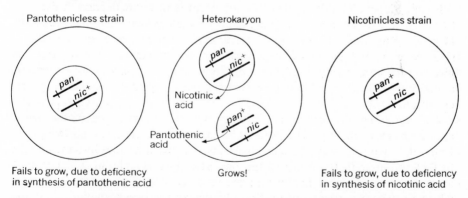

Figure 4-9. *Variant strains of Neurospora unable to synthesize particular vitamins cannot grow on media lacking these vitamins. However, strains deficient for different vitamins may form heterokaryons when inoculated together on these same media. In the heterokaryon the different nuclei complement each other's activities.*

THE LIFE CYCLE OF PARAMECIUM

Paramecium is a Protozoan belonging to the group called ciliates. We will be concerned particularly with the species *Paramecium aurelia*. Paramecia of this species are shaped somewhat like a cigar, with dimensions of about 150 by 50 microns. Those of you who have observed Paramecium under the microscope know this one-celled animal to have a complex structural organization. In function, also, Paramecium is complex, and its reproductive processes are not only complex but highly unusual, when one considers organisms in general. Indeed not all these processes are completely understood, and we will deal only with those in which a combination of genetic and cytological evidence has led to basic understanding.

Why focus attention on an organism that seems to represent a very special case? The answer is that the unusual reproductive properties of Paramecium provide unusual possibilities for studying certain complex genetic phenomena. In particular, studies with Paramecium have led to new concepts of the relationships between nucleus and cytoplasm in inheritance.

Before considering reproduction in Paramecium, one of the structural complexities of this organism should be understood. It has two kinds of nuclei, as do other ciliates. *Paramecium aurelia* has one very large *macronucleus*, which is the source of phenotypic control. It also has two much smaller *micronuclei*, which act as germinal nuclei in fertilization processes and serve to form new macronuclei. In structure, micronuclei are typical of nuclei as you already know them. The structure of the macronucleus is imperfectly understood, but it is a compound that includes a large number of complete sets of chromosomes.

Binary Fission. The life cycle of Paramecium is not well defined in the sense that each individual goes through the same rigid sequence of reproductive events. Depending upon circumstances, a variety of nuclear processes of reproductive significance may occur. However, there is only one way by which paramecia give rise to more paramecia. This is through *binary fission*, which is essentially a division of one individual into two individuals that are alike in genetic constitution.

At fission, each micronucleus divides by mitosis, and the macronucleus appears to elongate, constrict in the middle, and then separate into two pieces. One of the two products of each of these nuclear divisions goes into the anterior half of the animal, the other to the posterior portion. The animal constricts around the middle and finally separates into two daughter animals, each of which contains a nuclear constitution like that of the original animal.

This kind of reproduction in Paramecium, then, is fundamentally like the proliferation of body cells in multicellular plants and animals. In either instance, single cells through mitosis and cell division give rise to daughter cells that are genetically identical with their progenitors. Paramecium cells, however, separate entirely from one another, and, although complex, they do not show the visible differentiation occurring among some cells of higher organisms.

Conjugation. Under appropriate conditions, paramecia of different mating types pair closely and carry out reciprocal fertilization. Within paired animals several important events occur before nuclear fusions take place. In each animal the macronucleus breaks down into fragments that finally disintegrate, although disintegration may not be complete until after the conjugating cells have moved apart and undergone a number of fissions. Similarly, in each animal each of the two micronuclei undergoes meiosis. Since the consequence of the two divisions of a single meiotic sequence is four haploid nuclei, and since in each conjugant two nuclei undergo meiosis, each conjugant at one stage contains eight haploid nuclei. Seven of the eight disintegrate. In each animal the remaining nucleus divides mitotically, so that each conjugant contains two genetically identical haploid nuclei, a condition that permits reciprocal fertilization. From each conjugant one of the nuclei migrates into the opposite member and fuses with the nucleus that has remained behind. Thus the diploid condition is restored. Because the nuclei that fuse in each of the pair members are alike, the diploid nuclei resulting from fertilization are alike for members of a conjugating pair. The fertilization nuclei then undergo two mitotic divisions. The paired animals move apart, each containing four nuclei. Two of the four become macronuclei, two micronuclei. At the first fission following conjugation the two macronuclei are distributed into the two daughter cells. The micronuclei divide as already described for fission, and the daughter cells end up with the normal complement of two micronuclei and a macronucleus. The genetic result of this process of sexual reproduction is that members of a pair of exconjugants, or any cells derived from them by fission, are of identical genotype. Figure 4-10 summarizes in a simplified way conjugation in Paramecium.

Cytoplasmic Transfer. Members of conjugant pairs of paramecia usually separate after a relatively short period of time. In these normal conjugations, nuclei are exchanged, but little if any cytoplasm is transferred. Under special conditions, which are at least partly under experimental control, a connecting

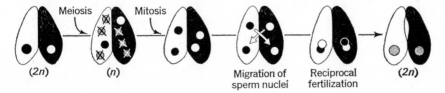

Figure 4-10. *When Paramecia conjugate, reciprocal fertilization takes place. (After Sonneborn; Baitsell*, Science in Progress, Seventh Series, *Yale University Press, 1951, p. 175.)*

strand persists between conjugants for thirty minutes or more. When this occurs, considerable cytoplasm is exchanged. The effects on the conjugants may be profound. As will be discussed in a later chapter, T. M. Sonneborn and co-workers have demonstrated the transmission of nonnuclear, heritable entities by means of the cytoplasmic bridges sometimes formed in Paramecium.

Macronuclear Regeneration. Sonneborn has shown that in conjugating animals subjected to high temperatures the precursors of the new macronuclei fail to develop. Under these circumstances a fragment of the old macronucleus may regenerate into a complete new one.

Autogamy. One of the peculiarities in the life history of Paramecium is a particular kind of nuclear reorganization called *autogamy*, a process that occurs in single, unpaired animals. The initial events are the same as occur in individuals taking part in conjugation. Finally, however, animals undergoing autogamy achieve internal self-fertilization. In more detail, autogamy proceeds as follows. The two micronuclei divide meiotically. Seven of the eight resulting nuclei disintegrate. The remaining nucleus divides once by mitosis, giving rise to a pair of identical haploid nuclei. Here the resemblance to conjugation ends. With no paired animal at hand, the nuclear migration and reciprocal fertilization characteristic of conjugation cannot take place. Instead, there is fusion between members of the pairs of haploid nuclei occurring in the single, unpaired animals. You should perceive that this results in homozygosity within each of the newly formed diploid nuclei. Subsequent events are like those described for an exconjugant, following conjugation. For purposes of simplification we have ignored the macronucleus in this description of autogamy. In fact, the macronucleus during autogamy behaves as it does in conjugation, disappearing and being restored in the same ways. Figure 4-11 summarizes autogamy in principle, although not in detail.

Genetic Applications. Much of what we have to say about the genetics of Paramecium comes better in other contexts, particularly in "Extrachromosomal

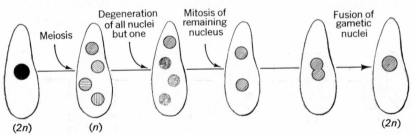

Figure 4-11. *Autogamy in Paramecium is self-fertilization within a single cell. See also Figure 4-12. (After Sonneborn; Baitsell,* Science in Progress, Seventh Series, *Yale University Press, 1951, p. 177.)*

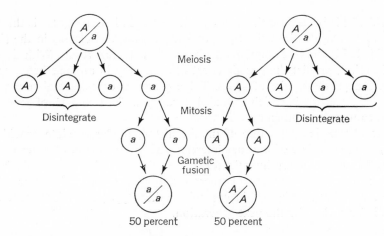

Figure 4-12. *Autogamy in heterozygous Paramecia leads to homozygosity. Relate this figure to Figure 4-11. (After Sonneborn,* Advances in Genetics, **1**:290, *1947.)*

and Epigenetic Systems," Chapter 11. But certain points are worth emphasis at this time. First, the occurrence of autogamy in Paramecium provides a useful and rather unusual tool for genetic investigation. The most significant feature of autogamy is the fusion of sister haploid nuclei, which in a single step brings about homozygosis for all genes.

Since autogamy takes place rather regularly in most strains, and indeed is inducible by experimental means, recessive genes cannot long remain masked by dominants. The occurrence of autogamy following mating in Paramecium greatly facilitates the investigator's attempts to resolve the genetic constitution

Table 4-1. GENETIC RESULTS OF AUTOGAMY IN PARAMECIA HETEROZYGOUS FOR TWO INDEPENDENT ALLELIC PAIRS, g^{60}/g^{90} AND d^{60}/d^{90}.

Genotypes of Progeny	Numbers
$g^{60}g^{60}d^{90}d^{90}$	18
$g^{90}g^{90}d^{90}d^{90}$	20
$g^{90}g^{90}d^{60}d^{60}$	19
$g^{60}g^{60}d^{60}d^{60}$	17
Dead	118

Source: Beale, *Genetics*, **37**:67, 1952.
Note: The deviations from an expected 1:1:1:1 ratio are small, particularly in view of the many dead animals. The allelic pairs under consideration determine antigenic traits to be discussed in Chapter 11.

of hybrids. This is because the result of autogamy in heterozygotes is that each allelic pair segregates into the two possible homozygous classes in the simple ratio of 1:1. Figure 4-12 diagrams the basis of the typical result. With the foregoing in mind, predict the genetic effect of autogamy in animals heterozygous for two independent pairs of genes. Table 4-1 gives experimental results of a study by G. H. Beale in which animals heterozygous for two pairs of genes were induced to undergo autogamy.

A second genetic application of the life history of Paramecium provides an experimental situation of unusual interest. The phenomenon of macronuclear regeneration permits an investigator to obtain exceptional animals having micronuclei of one genotype and macronuclei of another. In these exceptional animals, the macronucleus is in control of phenotype; the micronuclei carry genetic information for the next generation.

Reproduction in Bacteria

Bacteria typically occur as single cells. All are small, but the range of size among the approximately 1,500 known species is great. Some bacteria are about as large as a paramecium. On the other hand certain rod-shaped bacteria have dimensions of only 0.2 to 2.0 microns wide and 1 to 15 microns long. In a rough sort of way, bacteria fall into three classes with reference to shape—rods (bacilli), spheres (cocci), and spirals. A bacterial cell has definite structure, but difficulties of observation leave the precise nature of certain structural aspects in doubt. But a bacterial cell does have a nucleus, a cytoplasm with various components such as ribosomes, a cytoplasmic membrane, and a wall. Outside the wall one sometimes finds a capsule, or protective sheath, whose chemical composition is a genetic attribute. Certain bacteria, those with flagella, are motile.

The statement that bacteria have nuclei needs qualification. Bacteria have internal structures stainable by the Feulgen technique for revealing DNA. We know also that bacteria have genes composed of DNA and that the genes are organized into a system that has the genetic properties of a chromosome (Fig. 4-13). These properties of bacterial nuclei are generally consistent with the attributes of nuclei of higher organisms. But the bacterial nucleus does not have a membrane that separates it from the cytoplasm. Neither does one find a spindle mechanism, nor a condensation cycle of the DNA. The bacterial "nucleus," then, is the functional homologue of the nucleus found in higher organisms, but it represents a different evolutionary stage in the organization of hereditary materials. The attributes of bacterial nuclei are similar to those of the nuclei of the blue-green algae. Organisms having nuclei of this kind are called *prokaryotes*, to distinguish them from *eukaryotes*, organisms with the more familiar and more highly organized kind of nucleus described in Chapter 3.

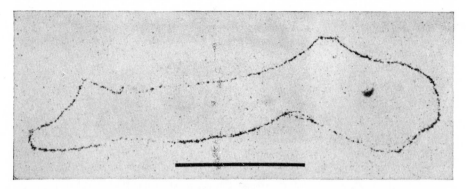

Figure 4-13. *After cells of* Escherichia coli *K12 are grown on a medium containing a radioactively labeled constituent of DNA, one can detect an autoradiographic image that shows the DNA in the form of a circle. The scale designates 100 microns.* (*Courtesy of J. Cairns*, Endeavour, **22**:*144, 1963.*)

Fission. Bacteria multiply by fission. In principle, the process is the same as described for Paramecium. A cell divides to give rise to a pair of daughter cells that are genetically like each other and like the original. In detail, the process differs from fission in eukaryotes. Since neither spindles nor condensed chromosomes are found in bacteria, mitosis cannot be said to be taking place. And since division of the nuclear material is not synchronous with cellular division, cells with more than one nucleus can occur. These may be seen in Figure 4-18, on page 120.

Bacteria can multiply with enormous speed. The colon bacillus *Escherichia coli* is able to double in number with each twenty minute interval. After 48 hours of this kind of growth a single cell would give rise to a progeny whose mass would be many times the mass of the earth. Such growth would occur, however, only under continuingly favorable conditions.

Dealing with Phenotypes in Bacteria. The rapid reproduction and small size of bacteria are favorable attributes from the point of view of one who wishes to do research in genetics. Enormous populations of individuals can be produced in a short time and with great economy. At the same time, these attributes require that certain procedures be followed and certain limitations be recognized that are not characteristic, for example, of maize genetics.

Let us consider first what phenotypes in bacteria are favorable for genetic research. Many of the examples of genetically determined phenotype that have come to your attention are visually observable characteristics of size, shape, color, or the like. The relative simplicity of an individual bacterium as compared with a corn plant or a human being reduces considerably the possibility for observable variation in morphology. In any case, the small size of bacteria makes examination of morphological traits tedious and difficult. If bacteria did

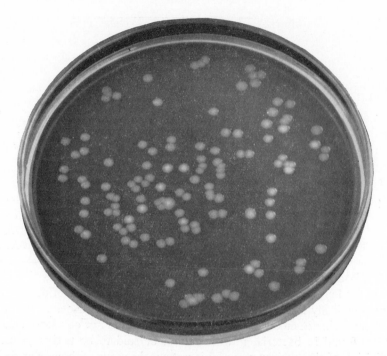

Figure 4-14. *When bacteria are introduced onto an agar medium in a Petri dish, individual cells grow into macroscopic colonies.* (*From G. Stent*, Molecular Biology of Bacterial Viruses, *W. H. Freeman and Co., 1963, p. 31.*)

have noses it would be difficult to see them. Even when individual bacteria grow into macroscopic colonies the potential for visibly definable variation is small (Fig. 4-14). Fortunately, the biochemical and physiological attributes of bacteria provide genetic variables that can be determined with ease and precision. As preliminary examples we can think of sensitivity or resistance to antibiotics such as penicillin, ability or inability to grow in the absence of certain vitamins, the ability or inability to produce particular antigens or other chemical substances, and the capacity or lack of it to be stained by particular dyes.

If you have been thinking while reading, you may have anticipated what is to be discussed next. This is the matter of counting bacteria. Genetics is for the most part a quantitative science. The typical genetic experiment involves not only a discrimination among phenotypes but their enumeration. How does one count huge populations of individuals that can't be seen unless a high-powered microscope is used?

Actually a variety of ingenious ways for counting bacteria have been developed. Here we will deal only with one of the simplest and most widely used in genetic experiments. Suppose that we have a culture of *Escherichia coli* growing in a liquid medium in a culture tube. Such a culture might have as many as twenty million live cells per milliliter of medium. Or there might be many fewer. At any rate, one can find out quite simply. A known volume, say 0.1 ml, is withdrawn from the culture of bacteria. This can be diluted 100 times, for example, by adding it to 9.9 ml of appropriate sterilized fluid that will not harm the bacteria. Withdrawing 0.1 ml from the dilution mixture and adding it to another 9.9 ml of aseptic fluid will give an additional dilution of 100 times, or 1 in 10,000 as compared to the original culture. A series of such dilutions provides material from which samples of known size can be withdrawn for inoculating onto solidified medium in Petri dishes. If the solidified medium is appropriate for growth, each live bacterium inoculated into the Petri dish will reproduce by fission and ultimately will give rise to a visible colony, composed of millions of cells. When platings are made from a dilution series, the least dilute inocula may give too many colonies to count; the most dilute inocula may give no colonies. But, from somewhere in the series, plates will be obtained having a readily countable number that can be accurately determined. A simple multiplication, based on the extent of dilution, gives an estimate of the number of viable cells in the original culture.

Discrimination of phenotype is usually done on colonies rather than on individual cells. This is a valid procedure genetically, since each colony is a *clone*, or the vegetatively reproduced derivatives of a single cell. Thus, in principle at least, all the cells of the colony have the same genotype as the cell from which the colony derives. Colonies may have observable phenotypes, where individual cells do not, or they may be subjected to chemical tests that discriminate phenotype.

Conjugation. The development of genetics as a science has been marked by a number of surprises for biologists. On the whole these surprises have turned out to fit in well with the previously established structure of the science and to confirm that genetic mechanisms, wherever found, have an underlying unity. One of the most startling discoveries, by Joshua Lederberg and E. L. Tatum in 1947, was that a mixture of bacteria of different genotypes may produce genetic recombinants. In bacteria of a particular strain, if two cultures marked with genes *abCD* and *ABcd* are put together, recombinant types such as *ABCD* can be recovered. It is as though bacteria of the parental types had mated and produced recombinant progeny, as happens in sexual reproduction in higher organisms.

The original finding of bacterial recombination was in the K12 strain of *Escherichia coli*. Attempts to demonstrate the same phenomenon in other bacteria have had limited success. But vigorous work in a number of laboratories throughout the world has given us a remarkable picture of mating in *E. coli* and established this bacterium as one of the most fruitful objects of genetic investigation.

The main outlines of bacterial conjugation are clear. As in Paramecium and

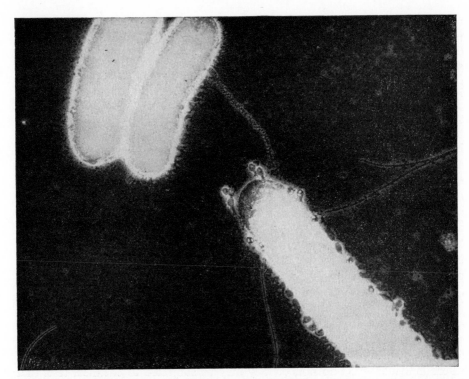

Figure 4-15. *F pili are slender appendages that grow from the surface of those cells of* Escherichia coli *that in a mating pair act as donors of genetic material. Distinguishable from other kinds of pili in that they specifically adsorb M12 phages, they appear essential for the transfer of the bacterial chromosome at conjugation and very likely mediate this transfer. The male-female mating pair shown in the photograph is connected by an F pilus labeled with RNA male phage (M12). The female cell has T6 phage adsorbed; the male shows no adsorbed phage because of resistance to it. (Courtesy C. C. Brinton, Jr.)*

Neurospora, matings occur only between genetically differentiated strains. When cells conjugate, continuity is established by bridges that mediate the injection of genetic material from one cell into another (Fig. 4-15). The kind of cell that has been designated F⁻ in a mating pair is the recipient of the injected material; the F⁺ cell is the donor. How does one know that the transfer of genetic material goes only one way? The answer is provided by pedigree studies of cells that have been isolated following conjugation. Vegetative progeny of the *donor* exconjugants never show recombination of the genes that marked the conjugants. Cells derived by fission of the *recipient* (sometimes called "female") exconjugants do show recombinant types for marker genes of the original conjugants.

One observation, which will be further developed in Chapter 6, should be added here. The injection of genetic material when mating occurs is not always complete. When cells conjugate, the integration of genetic material into the recipient cell is progressive and sequential, and need not proceed to its ultimate conclusion. The phenomenon can be studied experimentally by interrupting conjugating pairs at different times during mating, for example by putting them in a Waring blendor. When mating is interrupted, partial but operative zygotes are formed. The use of marker genes enables analysis of the composition of the zygotes.

Following conjugation, zygotes or partial zygotes remain in the diploid state for a few generations of vegetative reproduction of the cells. Eventually, segregation and reduction to haploidy occur, but the process seems not to be as regular, and certainly is not as rapid, as is meiosis in higher forms.

Bacteria are able to exchange genetic material by exotic means as well as by this relatively straightforward conjugation process. They will be considered in later chapters, when *transformation* and *transduction* are described.

Reproduction in Bacteriophage

Bacteriophage, literally, means "eater of bacteria." Bacteriophages, or phages as they are often called, do not actually eat bacteria but are viruses that infect bacteria and interact with them in ways that are worth our close attention. Among objects of genetic study, they are the simplest and smallest that we will consider. The T4 phage of *E. coli* has overall dimensions of about 65 × 95 millimicrons (millionths of a millimeter). Too small to be seen with the best of light microscopes, bacteriophages are nevertheless observable. Some of them have the shape of a tadpole. Figure 4-16 shows bacteriophage in a picture obtained by use of an electron microscope. The accompanying diagram interprets the structure of the phage. A phage particle has a protein coat, with the core of the head being DNA.

Phages are able to reproduce only within bacterial cells that they infect. When infection occurs, the phages attach themselves by their tails to a suitable host bacterium. This adsorption, which takes place at particular sites on the surface of the bacterium, is followed by injection of the DNA core into the bacterial cell. The protein coat of the virus remains outside as a ghost. The DNA is the principal component that enters the bacterial cell, is replicated there as new phage particles are reproduced, and is the genetic material of the phage. Incidentally, although all viruses include nucleic acids in their chemical makeup, some viruses include RNA rather than DNA.

Dealing with Phenotypes in Phage. Phages are characterized by small size, relative morphological simplicity, and rapid reproduction. Furthermore, they may be acquired in enormous numbers. Therefore, for the purposes of genetic

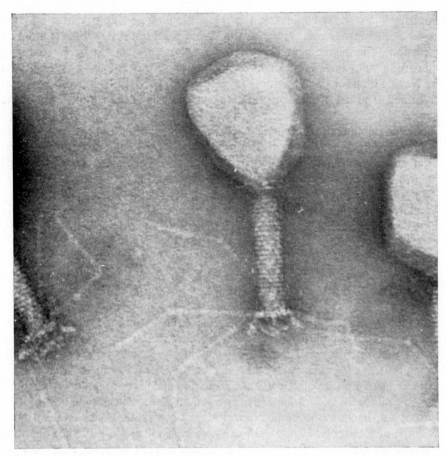

Figure 4-16. *Electronmicrograph and interpretative diagram (facing page) of T4 bacterio-phage. In the micrograph, the phage is enlarged about 300,000 diameters. In the diagram, a complete phage particle is shown at the upper left; below is a particle attached to a bacterium, with its hollow core penetrating the bacterial cell wall. At the right of the diagram various components of the phage are shown separately. (Courtesy Dr. M. F. Moody of The Rockefeller University.)*

study, phages pose particular problems with reference to enumeration and the designation of phenotypes.

Multiplication of phage within a bacterial cell is followed by dissolution, or *lysis*, of the cell. As a result, phage particles are released, ready to infect other cells. If a dilute suspension of phage is introduced to a thin layer of susceptible bacteria growing on an agar plate, each phage particle adsorbs to the bacterium with which it comes in contact. After the phage has multiplied, the bacterium lyses, releasing more phage around it. Adjacent cells become

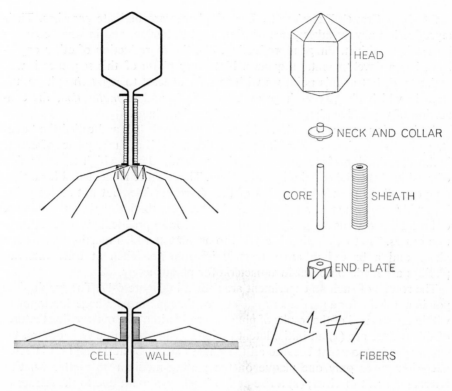

HEAD

NECK AND COLLAR

CORE | SHEATH

END PLATE

CELL WALL

FIBERS

Figure 4-16. (*Continued.*)

infected. Repetition of these events results in an area of the plate becoming free of bacterial growth. This area appears as a clear zone on the plate, and is known as a *plaque*. (For a picture of plaques, see Fig. 6-11 on page 170.) A given kind of virus produces plaques that are remarkably uniform in appearance. Different kinds of phage, however, may produce characteristically different plaques, the variation being, for example, in size or in sharpness of the margin. Plaque type, then, is a genetic attribute, and is one of the phenotypes that is useful in genetic studies of bacteriophage. Another distinctive, easily determined phenotype in phage is host range. Phages differ genetically in their ability to infect bacteria of particular strains.

If you think about it, you will see that a plaque represents a colony of phage, and corresponds to a colony of bacteria. Plaques can be used in counting bacteriophage just as colony counts are used in the enumeration of bacteria. Serial dilutions of phage particles introduced onto plates layered with sensitive bacteria provide a series of plaque counts that are the basis for calculating the numbers of phage in a population. As with bacteria, the calculation is based on colony (plaque) count, dilution, and volume of phage plated.

One-Step Growth Experiments. Bacteriophages are obligate parasites. Their reproduction always takes place within a host bacterium and involves complex interactions with it. Phage reproduction, like the reproduction of other organisms, has a developmental sequence. Not every phase of this sequence is well understood, but at this point we wish to refer at least to *vegetative* phage, the form in which phage (or phage genes) multiply, and to *maturation*, the conversion of vegetative phage into mature, infective phage.

An ingenious procedure, devised by Ellis and Delbrück and called the "one-step growth experiment," permits quantitative analysis of the sequence of events in the reproduction of phage. Bacteriophage are mixed with a high concentration of sensitive bacterial cells in liquid medium. The mixture is incubated for a time long enough to permit adsorption of the phage particles but not so long that lysis will have occurred. At the end of this brief incubation period, the mixture of bacteria and virus is greatly diluted, so that for all practical purposes adsorption ceases. That is, dilution is carried to a point at which meetings between a phage and a bacterium are extremely improbable. Then at brief intervals platings are made onto sensitive bacteria for plaque assay.

The results of such an experiment are shown in Figure 4-17. The graph gives you in numbers, for a particular cycle of growth, a picture of phage development within the cell and of the results of phage reproduction. The outstanding features of this picture are (1) an initial latent period, during which no lysis occurs, the count remains constant because all the phages are in bacteria, and an infected bacterium gives only one plaque on the plating medium no matter what its internal content of virus; (2) a rise period in plaque count due to the release of

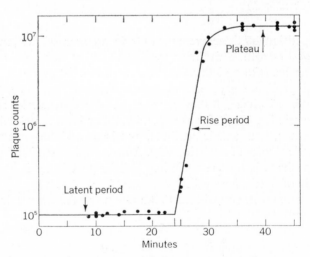

Figure 4-17. *Results of a one-step growth experiment with bacteriophage T-4. See text for details.* (*From Stent,* Molecular Biology of Bacterial Viruses, *W. H. Freeman and Co., 1963, p. 73.*)

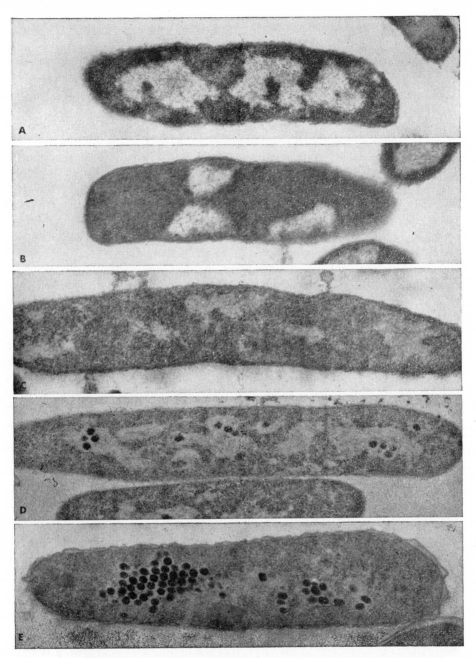

Figure 4-18. *Electron micrographs of phage growth and development and changes in an infected cell.* A: *At time of infection, nuclei of* E. coli *appear as transparent areas.* B: *At 2 to 4 minutes the nuclei are altered.* C: *After 10 minutes vacuoles containing fibrillar phage appear.* D: *After 14 minutes condensates of phage DNA appear.* E: *At 40 minutes phage heads are present along with DNA condensates.* (*By Kellenberger, from* Sci. Am., **204***:93, June 1961*.)

free phage—free to reinfect and lyse many bacteria; (3) a final high plateau, accounted for by failure of newly released viruses to find unlysed host cells that would permit them to reproduce further.

Life Cycle of Bacteriophage. We have now given enough information about phage that you can begin to appreciate Figure 4-18, which includes a series of electron micrographs picturing various stages of the growth of phage within a bacterial cell. Over the period spanned by the photographs phage DNA reorganizes bacterial metabolism so that new phage particles can be produced. This period of production of new phage particles is called the *vegetative phase*. It includes replication of the genetic material (DNA), synthesis of the phage protein, and the morphological elaboration of the phage particles. The vegetative phase ends when phage particles mature to an infective form.

So far, however, we have considered but one kind of life history for a bacterial virus, that of a *virulent phage*, whose host cells are always destroyed through lysis. In addition, we need to construct a life history for phages that behave in a different way. These are *temperate phages*, so called because under certain circumstances they infect without destroying their hosts. The host bacteria *not only survive but transmit genetically the ability to produce phage particles like those involved in the original infection.* Occasionally, however, such a temperate phage ceases to be benign. It produces infective particles within the host cell, lyses the cell, and acts like a virulent virus. This occasional virulence of a temperate phage permits us to recognize its presence in a strain of bacteria.

Ingenious experimentation has told a good deal about temperate phages. When a phage infects an appropriate bacterial host, a decision is made—how is not clearly known—as to whether the phage will act like the virulent viruses we have already considered in some detail or whether it will integrate itself in the bacterium and behave in benign fashion. If the second of the alternatives comes about, the DNA of the phage is incorporated into the bacterial chromosome. The integrated phage is called *prophage* and the bacterium into whose chromosome it is integrated is said to be *lysogenic* (having the heritable property of ability to lyse). The prophage replicates in coordination with the replication of the genetic material of the host, but does not direct the synthesis of other phage materials nor organize mature infective particles. A prophage is like a gene that exercises only a small part of its potential for phenotypic expression. The prophage does make its presence felt in at least one way, however: lysogenic bacteria are immune to further infection by phages like the one they carry. More specifically, the presence of prophage seems to prevent the multiplication of closely related phages that infect a lysogenic cell. This property of immunity is so regularly expressed that it can be used as a means of identifying lysogenic cells. Of course, lysogenic bacterial cells can also be detected when the temperate phage turns intemperate. But such cells are already dead by the time they are detected, and are of no use for further experimentation.

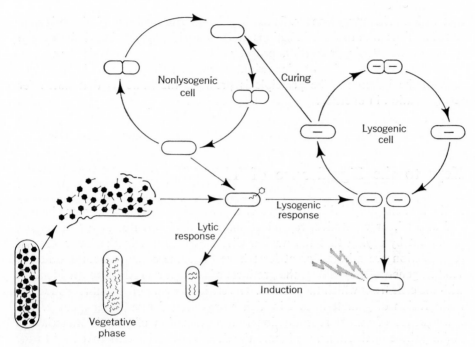

Figure 4-19. *The life cycle of a temperate phage. (After Lwoff, Bacteriol. Rev., 17:269, 1953.)*

In certain lysogenic strains, phage production can be *induced* simultaneously in all cells (by treatment with ultraviolet light or X-rays, for example). Induction has permitted the investigation of phage production by essentially the techniques of the one-step growth experiment. The results tell us that reproduction of the temperate phages is similar to that of the virulent phages. In short, both systems are characterized by an initial latent period followed by the production of a crop of phage particles. Occasionally, lysogenic cells lose their prophage (are cured). The life cycle for a temperate phage is shown in Figure 4-19.

Genetic Applications. Bacteriophage has many advantages for genetic research. Its simplicity of composition permits a fairly direct attack on problems of the nature of genetic material. The enormous populations that can be obtained, and then readily manipulated, permit the detection of extremely rare genetic events. The particular relationships of phages to their hosts are the basis for ingenious procedures not possible for most of the objects of genetic study. These advantages of phage will emerge in various contexts throughout this book.

In this place only one thing further—but an extremely important one—needs to be said about phage. A phage has a genetic system basically like that of a higher organism. In the first place, the genetic material of phage mutates, so that variant phenotypes occur. Secondly, the genetic material of phage is able to recombine: when mixed infections are made with phages that differ by two or more phenotypic properties, progeny phage emerging at lysis may show recombination of the phenotypic attributes characterizing the two kinds of virus used in the infection. The principle is precisely the same as that noted for recombination in bacteria.

Keys to the Significance of This Chapter

In the first chapter of this book, we distinguished between sexual reproduction and asexual, or vegetative, reproduction. This distinction can now be refined somewhat by saying that the nuclear divisions involved in the latter kind of reproduction are mitotic, or at least have the same consequences as mitosis. That is, genetically speaking, the products of asexual reproduction are identical with the cells from which they derive. The critical feature of sexual reproduction is the fusion of gamete nuclei, in higher organisms formed after meiosis. The consequence of sexual reproduction is recombination of genetic material. In making these distinctions, we ignore the developmental events that lead to the production of a mature individual and that we have detailed for a variety of kinds of organisms in our consideration of life cycles.

In fact, this present chapter is intended to show that the principal facets of reproduction have underlying unity while the details are enormously varied. Sexual reproduction occurs generally in higher plants and animals and in microorganisms such as Paramecium and Neurospora. As we have treated matters, both autogamy and conjugation in Paramecium are forms of sexual reproduction. The fact that sexually differentiated parents are not involved in autogamy is interesting but irrelevant.

Both bacteria and bacteriophages are able to reproduce in ways that involve the recombination of genetic materials. If sexual reproduction is defined only in terms of its genetic consequences, then bacteria and phages belong in the same category as man and maize, and may be said to have sexual reproduction. If we wish to emphasize differences, we can point out that the recombinative reproduction of phage and bacteria does not include a regular meiosis and is therefore significantly distinct from sexual reproduction as it occurs in Neurospora, Paramecium, and higher organisms.

Geneticists must be familiar with the details of the life cycles and reproductive processes of organisms with which they work. Both the interpretation of genetic results and the development of techniques of investigation are necessarily related to the particular pattern of reproduction of the plant or animal being studied.

The characteristics of a life cycle are a fundamental biological heritage of the species, and define many of the attributes of that species.

REFERENCES

Allen, C. E., "The Genetics of Bryophytes." *Botan. Rev.*, **1**:269–291, 1935. (A general review of genetic studies in a group of plants with favorable possibilities for investigation.)

Allen, C. E., "Haploid and Diploid Generations." *Am. Naturalist*, **71**:193–205, 1937. (Discussion of the alternation of gamete-producing and spore-producing generations at various levels in the plant kingdom.)

Beadle, G. W., "Genetics and Metabolism in Neurospora." *Physiol. Revs.*, **25**:643–663, 1945. (Pages 643 to 651 are the most pertinent to the purposes of this chapter.)

Demerec, M., and Kaufman, B. P., *Drosophila Guide*. Washington: Carnegie Institution of Washington, 1950. (Simple, accurate guide to the life history of Drosophila and to the handling of Drosophila for genetic work.)

Fincham, J. R. S., and Day, P. R., *Fungal Genetics*. Philadelphia: F. A. Davis Co., 1963. (Generally worth reading. Chap. 2 covers the life cycles and breeding systems of a good many fungi of genetic interest.)

Jacob, F., and Wollman, E. L., *Sexuality and the Genetics of Bacteria*. New York: Academic Press, 1961. (A good source of information about bacterial genetic systems and methods of analysis for bacterial genetics.)

Kiesselbach, T. A., "The Structure and Reproduction of Corn." *Univ. Nebr. Agr. Exp. Sta. Res. Bull. No. 161*, 1949. (Detailed, abundantly illustrated, and with an extensive bibliography.)

Pontecorvo, G., "The Genetics of *Aspergillus nidulans*." *Advances in Genetics*, **5**:141–238, 1953. (A good presentation of the life cycle and the methods of analysis for an important object of genetic investigation.)

Randolph, L. F., "Developmental Morphology of the Caryopsis in Maize." *J. Agr. Research*, **53**:881–916, 1936. (Definitive article on the development of the corn kernel.)

Sharp, L. W., *Fundamentals of Cytology*. New York: McGraw-Hill Book Co., 1943. (The chapters on the cytology of reproduction in animals (Chap. 9), in angiosperms (Chap. 10), and in plants other than angiosperms (Chap. 11) are directly pertinent to the consideration of life cycles.)

Sonneborn, T. N., "Recent Advances in the Genetics of Paramecium and Euplotes." *Advances in Genetics*, **1**:263–358, 1947. (See especially pp. 265–295 in this detailed and authoritative review.)

Stent, G. S., *Molecular Biology of Bacterial Viruses*. San Francisco: W. H. Freeman and Co., 1963. (A scholarly and readable text.)

QUESTIONS AND PROBLEMS

4-1. What do the following terms signify?

aleurone	embryo sac	microspore
ascospore	endosperm	monoecious
autogamy	first-division segregation	pericarp
fission	gametophyte	plaque
chromosome cycle	life cycle	prophage
clone	lysogenic	second-division segregation
complementation	macronucleus	self-pollination
conidia	megaspore	sporocyte
cross-pollination	micronucleus	sporophyte

4-2. List particular advantages of the following organisms for genetic research: corn, Drosophila, Neurospora, Paramecium, bacteriophage.

4-3. What are some of the advantages, for genetic analysis, of haploid organisms as compared with diploids?

4-4. What genetic phenomena important in diploids cannot ordinarily be investigated in haploid organisms?

4-5. In corn the haploid number of chromosomes is 10. How many chromosomes would you normally expect to find in (a) the nucleus of an aleurone cell, (b) the tube nucleus, (c) a microsporocyte nucleus, (d) a root tip nucleus, (e) a polar nucleus, (f) an egg nucleus?

In corn, three of the allelic pairs concerned in aleurone pigmentation are *A-a*, *I-i*, and *Pr-pr*. For pigment to develop, at least one *A* allele must be present and no *I* allele may be present. Aleurone pigment is purple when *Pr* is present but is red with a genotype homozygous for *pr*.

4-6. Suppose that, in an isolated experimental corn plot, seeds of the genotype *AaprprII* are planted in even-numbered rows and seeds of the genotype *aaPrprii* are planted in odd-numbered rows. When the plants grow up, natural pollination is permitted. What aleurone colors might you expect to find in ears of corn formed on plants in the even-numbered rows? In the odd-numbered rows?

4-7. With reference to the preceding question, would you expect the aleurone colors to fall into Mendelian ratios? Justify your answer.

4-8. Designate all the possible genotypes of aleurone that might occur within ears formed on plants of the odd- and even-numbered rows of corn in Question 4-6. (Remember that aleurone is part of the endosperm and therefore is triploid.)

4-9. Starting with the two kinds of seed described in Question 4-6, outline in detail the steps you would take to develop, in the most efficient way possible, a strain of corn that would breed true for purple aleurone color.

4-10. Which do each of the following ascospore arrangements in Neurospora indicate, first-division segregation or second-division segregation?

$$
\begin{array}{cccccccc}
+ & + & c & c & + & + & c & c \\
c & c & + & + & + & + & c & c \\
c & c & c & c & + & + & + & + \\
+ & + & c & c & c & c & + & + \\
\end{array}
$$

4-11. Suppose that you dissected an ascus of Neurospora, and after germinating the spores found the following spore arrangement to be indicated:

$$+ \quad + \quad c \quad + \quad c \quad + \quad c \quad c$$

What would be anomalous about such an arrangement?

4-12. What possibilities might explain the result indicated in the preceding question?

4-13. For various reasons, ascospores of Neurospora sometimes fail to germinate. Suppose that you have made a cross between a strain with wild-type pigmentation $(+)$ and an albino strain (al). You wish to determine the frequency with which the al gene segregates at the second division of meiosis. What is the minimal number of spores out of a set of eight that must germinate to tell whether first- or second-division segregation has occurred within an ascus?

4-14. Suppose that you are still working with a cross where the allelic pair $+/al$ is segregating. Assume equal viability of wild-type and albino ascospores. If only one spore from an ascus germinates, what is the probability that it will produce a wild-type culture? If two spores from an ascus germinate, what is the probability that both will produce wild-type cultures? That you will obtain a wild-type and an albino culture?

4-15. In a laboratory experiment by a student at Cornell University, an albino strain of Neurospora was crossed with a wild-type strain. After isolations of ascospores *at random*, 2018 albino cultures and 1996 salmon-pink (wild-type color) cultures were obtained. Apply the chi-square test to determine whether the result is consistent with expectation.

4-16. The gene *peak* in *Neurospora crassa* affects morphology in such a way that the mycelium assumes the form of a small cone. The inheritance pattern of *pk* is normal, a cross of wild × *peak* giving 4 *pk* and 4 + spores in each ascus. The cross + × + gives, of course, only +, and *pk* × *pk* gives only *peak*. If one examines the asci themselves, however, an unexpected phenomenon emerges. Asci formed after the crosses + × + or + × *pk* are the tubular structures you have learned to expect, containing eight spores in linear arrangement. But the cross *pk* × *pk* produces an ascus in which the spores are found in a double row of four or in some other nonlinear arrangement. Give a genetic interpretation of the results.

Aspergillus nidulans, like Neurospora, is an ascomycete. The conidia of Aspergillus are uninucleate. They occur in chains, and the nuclei of the conidia of a given chain are genetically identical, having arisen by mitotic division of a single original nucleus. The chains occur in clusters, or heads, which are produced from a multinucleate vesicle. The minimal medium for the wild type of Aspergillus consists only of salts and sugar.

4-17. G. Pontecorvo describes an experiment in which a yellow, thiamine-requiring strain of Aspergillus and a green, adenine-requiring strain were inoculated together at a point on the edge of a Petri dish containing minimal medium. Growing mycelium appeared from the inoculum and began to spread over the medium. Give a preliminary interpretation of the occurrence of growth.

4-18. The growing mycelium in 4-17 produced conidia in such fashion that there was a thorough mixture of yellow heads, green heads, and mixed heads consisting of yellow and green chains. Interpret these results.

4-19. Some mycelial sectors of entirely yellow conidia were observed. How might these be accounted for?

4-20. When growth covered about half the Petri dish, thiamine was added to the medium. Subsequent growth produced, almost entirely, yellow conidial heads. Interpret this result. Does this interpretation modify or amplify in any way your answer to 4-19? How?

4-21. Bacteriophage particles show no metabolic activity Are bacteriophages alive? Discuss.

CHAPTER 5

The Chemical Basis of Heredity

As you have seen, genes were first detected by following their patterns of transmission through sequences of generations. The analysis of patterns of heredity permits us to ascribe a number of properties to genes and to make certain verifiable predictions about their behavior and interrelationships. But to understand heredity as a phenomenon of life, we need to know how genes act in determining the characteristics they control and how genes replicate themselves to permit transmission for indefinite numbers of cellular or organismal generations. Such an understanding of heredity must rest primarily on a knowledge of its chemical and physical basis.

A priori, we may deduce that genes must be complex chemical entities; no simple compound could determine in such precise ways all of the many hereditary properties of an organism. If we are to seek candidates for the genic substance, we need consider only complex chemical structures. Knowing that chromosomes are bearers of genes, can we learn something of the chemical nature of genes by considering the chemical nature of chromosomes? Two classes of highly complex chemical compounds, proteins and nucleic acids, are found in chromosomes, but since both are present in the cytoplasm as well as in the nucleus, we cannot identify the genic substance simply by localizing the complex chemical material. More subtle approaches are required. In this chapter we will first discuss the chemical nature of proteins and nucleic acids, and then turn to observations and experiments that point to one of these as the genic substance. Finally we will describe in some detail the chemical structure of the gene, as presently understood, and the implications of this chemical structure for genetic systems.

Proteins. Proteins are composed of molecular subunits called polypeptides. The polypeptides are in turn made up of smaller chemical compounds called

Name	Abbreviation	Formula
Alanine	Ala	CH_3—CH—COOH | NH_2
Arginine	Arg	NH_2—C—NH—$(CH_2)_3$—CH—COOH || | NH NH_2
Asparagine	Asp—NH_2	H_2N—C—CH_2—CH—COOH || | O NH_2
Aspartic Acid	Asp	HOOC—CH_2—CH—COOH | NH_2
Cysteine	Cy—SH	HS—CH_2—CH—COOH | NH_2
Glutamic Acid	Glu	HOOC—$(CH_2)_2$—CH—COOH | NH_2
Glutamine	Glu—NH_2	H_2N—CO—$(CH_2)_2$—CH—COOH | NH_2
Glycine	Gly	CH_2—COOH | NH_2
Histidine	His	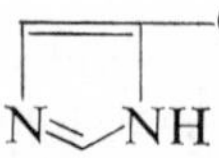 —CH_2—CH—COOH N≈NH NH_2
Isoleucine	Ileu	CH_3—CH_2—CH—CH—COOH | | CH_3 NH_2
Leucine	Leu	CH_3—CH—CH_2—CH—COOH | | CH_3 NH_2
Lysine	Lys	H_2N—$(CH_2)_4$—CH—COOH | NH_2
Methionine	Met	CH_3—S—CH_2—CH_2—CH—COOH | NH_2
Phenylalanine	Phe	 —CH_2—CH—COOH | NH_2

Figure 5-1. *The chemical formulas of the twenty amino acids that occur in proteins. The common groups involved in peptide bonds are shaded.*

Name	Abbreviation	Formula
Proline	Pro	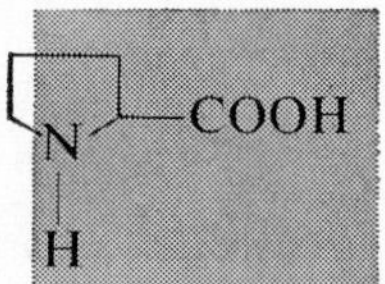
Serine	Ser	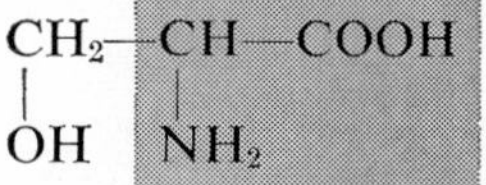 CH_2—CH—$COOH$ with OH and NH_2
Threonine	Thr	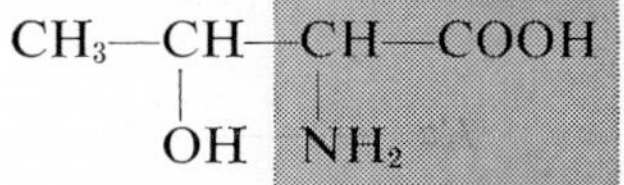 CH_3—CH—CH—$COOH$ with OH and NH_2
Tryptophan	Try	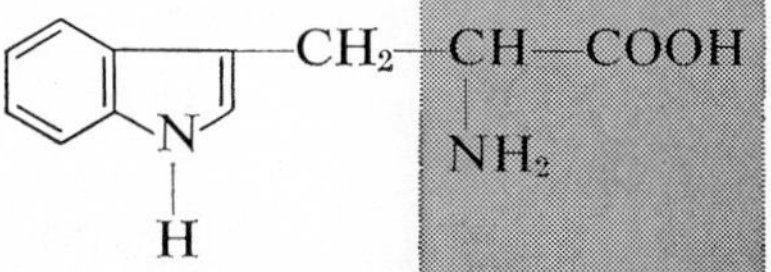 (indole ring)—CH_2—CH—$COOH$ with NH_2
Tyrosine	Tyr	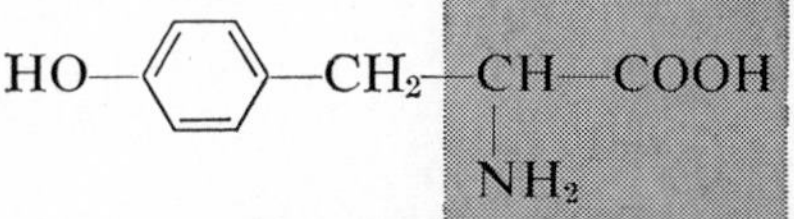 HO—(benzene ring)—CH_2—CH—$COOH$ with NH_2
Valine	Val	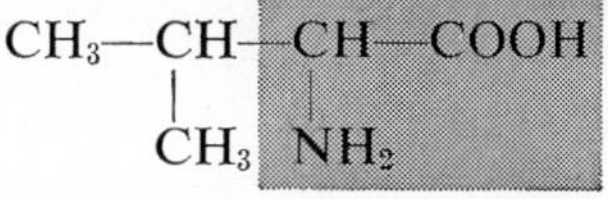 CH_3—CH—CH—$COOH$ with CH_3 and NH_2

Figure 5-1. (*Continued.*)

amino acids linked together in long chains. The polypeptide chains of a protein may contain 200 or more amino acid molecules. However, only twenty different *kinds* of amino acids are known to be present in proteins. Their chemical structures are shown in Figure 5-1. The composition of a given protein is uniform: each molecule is composed of the same number and kind of polypeptide chains and each of these polypeptide chains is in turn composed of the same number and kind of amino acids linked together in precisely the same sequence.

A protein can be considered to have three levels of organized structure. Its *primary structure* is the sequence of amino acids in the polypeptide chain. The *secondary structure* is the spatial orientation of amino acids relative to one another. In most proteins the polypeptide chain is for large portions of its length coiled in what is called an *alpha helix*. The *tertiary structure* is the overall structure of the protein—the way in which the polypeptide chains are held together and the whole structure is folded and superfolded. As an example we can consider hemoglobin. The hemoglobin molecule in the normal adult person is composed of four polypeptide chains, two identical chains called *alpha* chains and two other identical chains called *beta* chains. Large portions of each chain are in an alpha-helix configuration, but each chain has many kinks and the whole structure has a very complex shape. Proteins then, are complex molecules,

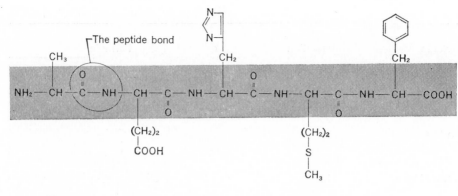

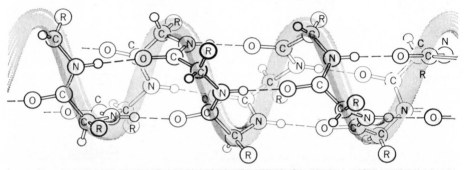

Ala —————— Glu —————— His —————— Met —————— Phe

Primary Structure. Amino Acid Sequence.

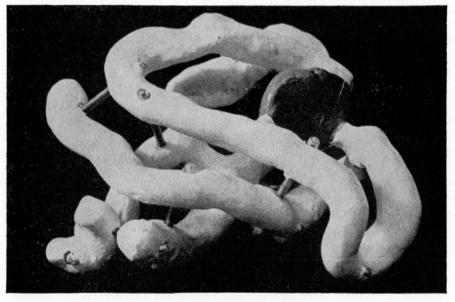

Secondary Structure. Alpha Helix.

Tertiary Structure.

both in shape and composition, and the kinds of different proteins that could exist are as numerous as the stars in the sky. Proteins appear to qualify as a possible genic-substance candidate. See Fig. 5-2 and Color Plate II.

Nucleic Acids. Two kinds of nucleic acids are known. One is called deoxyribonucleic acid (DNA for short) and the other ribonucleic acid (RNA). These compounds, like proteins, are long molecules made of similar subunits. In the nucleic acids, however, the subunits are not amino acids but substances called *nucleotides*. Each nucleotide is composed of three simpler compounds linked together, phosphoric acid, a sugar, and a nitrogen-containing ring compound. In DNA the sugar component is deoxyribose; in RNA it is ribose. A nucleotide includes one of four different nitrogenous ring compounds, two of a class called purines and two of a class called pyrimidines. The purines in DNA are adenine and guanine, the pyrimidines cytosine and thymine. (We will subsequently use the notation A for adenine, G for guanine, C for cytosine, and T for thymine.) RNA, in addition to having ribose rather than deoxyribose, differs in composition from DNA in having the pyrimidine uracil (U) in place of thymine. A nucleotide, then, is a compound containing phosphoric acid, ribose (or deoxyribose), and either a purine or a pyrimidine.

In the nucleic acid, the nucleotides are hooked together end-to-end making the linkage sugar-phosphate-sugar-phosphate, etc., with the purines and pyrimidines attached as side groups to the sugar molecules. Nucleic acids can be very long molecules. Some molecular chains of DNA (from the bacterial virus T4, for example) contain approximately 200,000 nucleotides! Nevertheless, only four different kinds of subunits make up a nucleic acid molecule as compared to the 20 different kinds of subunits (amino acids) found in some proteins. Thus we might think that nucleic acids do not have sufficient complexity to account for the complexity of genes. Indeed, for some time it was generally thought that DNA had a simple repeating structure (A-G-C-T-A-G-C-T, for example). As we will see, however, DNA molecules are more complex than we might think at first, and we will find that evidence supports the candidacy of DNA rather than protein as the substance of which genes are made.

Figure 5-2. *(Facing page). The structure of proteins. A sequence of amino acids linked together by peptide bonds illustrates the primary structure. The relative configuration of the amino acids to each other in an α helix is shown to indicate secondary structure. The orientation of the complete polypeptide chain of myoglobin is shown as an example of tertiary structure. The structure of myoglobin is shown in much more detail in Color Plate II (between pages 36 and 37).*
(The myoglobin model is from Kendrew, Sci. Amer., Dec. 1961.)

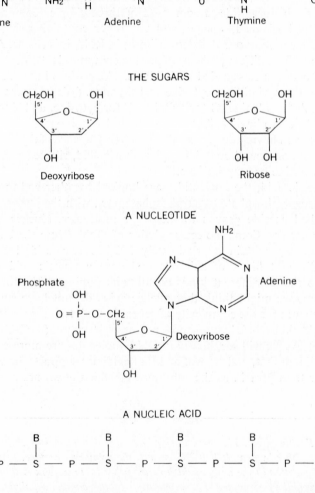

Figure 5-3. *The chemical composition of nucleic acids. The bases (or nitrogen-containing ring compounds) are linked to a sugar and a phosphate to form a nucleotide. Nucleotides are linked together to form a nucleic acid or polynucleotide chain.*

Evidence for DNA as the Genic Substance. The first clue supporting the notion that genes are composed of DNA is that DNA is largely restricted to the chromosomes while protein and RNA are most abundant in the cytoplasm. This observation is of long standing. A much used cytological stain for chromosomes is the Feulgen reagent, which can be used as a fairly specific chemical test for DNA in living systems. Quantitative measurements of the amount of DNA in cells show that there is a correlation between the amount of DNA and the number of chromosome sets. For example, the amount of DNA in a haploid cell is half the amount present in a diploid cell of the same species. Furthermore, the composition of the DNA (the relative amounts of the four nucleotides) is constant for a given organism in all types of cells, but varies from one species to another. These two properties, which we would expect the genic substance to have, do not hold for the proteins or the RNA of the nucleus. To continue our disclosure of the genic substance, let us now turn to some experiments with microorganisms that give us more direct evidence as to the genetic role of nucleic acids.

Direct Evidence for Nucleic Acids as Genetic Substance. In 1928 Griffith found that mice injected with cells of an avirulent strain of the bacterium *Diplococcus pneumoniae* (also called pneumococcus), together with heat-killed cells from a virulent strain, frequently succumbed to infection. From these infected animals living bacteria of the virulent type could be isolated. It thus appeared that some active principle in the heat-killed virulent cells was capable of "transforming" avirulent cells to virulent ones. It was later found that this "transforming principle" could in fact transform certain pneumococcus strains for any genetic trait of the cells that donated the transforming principle. In 1944 Avery, MacLeod, and McCarty demonstrated beyond question that this transforming principle was DNA. The "principle" of the donor cells could be chemically purified and retain its transforming activity. If then treated with DNase, an enzyme that destroys DNA, it rapidly lost its transforming activity but was unharmed by treatment with RNase (an enzyme that destroys RNA) or proteases (enzymes that destroy proteins).

In recent years transforming systems in other bacterial species (notably, *Bacillus subtilis*) have been developed and studied in some detail. It is now clear that the phenomenon of transformation is really a special form of recombination involving the "naked" genes (DNA) of the donor and the gene complement (genome) of the recipient. The phenomenon is considered more fully in Chapter 6. For our discussion here, transformation provides impressive proof that DNA is genetic material.

We have already described (in Chapter 4) the mode of growth of the bacterial virus T4. The following experiment shows that DNA is the component of the virus particle that directs the synthesis of new virus particles in an infected cell. First, virus is grown in bacterial cells in a medium containing radioactive phosphorus. Since DNA contains phosphorus, but protein does not, only the DNA in the multiplied virus is made radioactive. The virus is then allowed to infect nonradioactive bacteria. Shortly after the attachment of the virus, the bacteria are swirled in a food blender. The turbulence does not harm the bacteria but does knock the virus particles off the surface of the bacteria. The mixture can then be centrifuged at a speed at which the bacteria settle to the bottom but the phage particles stay in suspension, thus separating the bacteria from the virus particles. The bacteria nevertheless remain infected and produce a crop of phage. Appropriate tests show that nearly all of the radioactivity is now associated with the bacteria and their new crop of virus, and not with the virus coats stripped off the cell surface and remaining in the supernatant after centrifugation. If this experiment is repeated, but with radioactive sulfur rather than phosphorus, thus labeling the virus protein rather than its DNA, we find that most of the radioactive sulfur is not associated with the bacteria but is associated with the virus coats that remain suspended after the centrifugation. The interpretation of these experiments is that after a virus injects its DNA into a cell, the empty virus coat, composed of protein, remains outside the cell and can be removed without affecting the further growth of the virus. (This has been verified by other studies, particularly with the electron microscope.) It seems that the purpose of the protein coat is to serve as a vehicle for getting the viral DNA from one cell to another, and the purpose of the DNA is to carry the necessary information to direct the production of a new generation of virus. (A small amount of protein is injected into the cell along with the DNA, but it constitutes less than five percent of the total injected material, the rest being DNA.)

An even more elegant experiment further demonstrates the genetic role of nucleic acids in viruses. The tobacco mosaic virus (TMV for short) is composed of protein and RNA. (All viruses contain protein and nucleic acid. Some, like TMV, have RNA rather than DNA as their nucleic acid component.) The protein and RNA can be chemically separated, and the virus can be reconstituted by reassociating the components in a test tube. Hybrid viruses can be produced by combining the RNA from one strain of TMV with the protein from another. When tobacco plants are infected with such a hybrid virus, we find that the new virus produced in the plant is identical with virus of the strain that provided the RNA of the synthetic hybrids; virus of the strain that provided the protein does not appear. Again we see that nucleic acids carry the genetic specificity of the virus.

In recent years it has been found that appropriate host cells can be infected with pure nucleic acid isolated from a large number of different viruses either of the RNA- or DNA-containing kind. The host cells produce new virus particles identical to the parental virus from which the nucleic acid was extracted.

Many facts point to nucleic acids as the bearers of genetic specificity. If we put aside for the moment the exceptional case of the RNA-containing viruses,

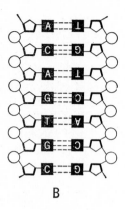

B

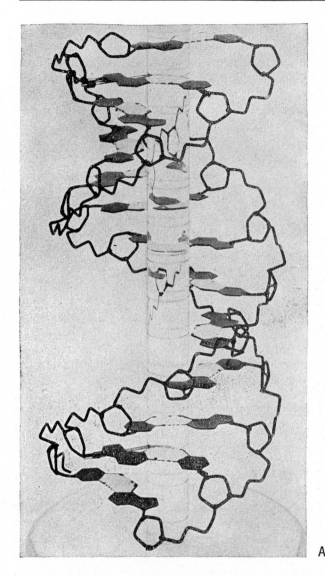

Figure 5-4. *The physical structure of DNA. A: A photo of a model of DNA showing its double helical nature. B: The structure in diagrammatic form. The circles represent the phosphates; the pentagons, deoxyribose. The interrupted lines linking the bases (squares) indicate hydrogen bonds, three for G-C pairs, two for A-T pairs. (5-4A from Crick,* Sci. Amer., *Oct. 1954.)*

A

DNA appears to be the stuff of which genes are made. As you proceed in this book you will find further facts and experiments to support this conclusion. For example, some of the most potent chemical agents for causing gene mutations, that is, heritable changes in genes, are compounds that specifically interact with nucleic acids.

The Nature of the DNA Molecule. What features of the DNA molecule make it appropriate for its genetic role? The breakthrough in our discovery of how DNA works as the hereditary blueprint came with the detailed elucidation of the physical structure of the DNA molecule by Watson, Crick, and Wilkins. Although the relative quantities of the four nucleotides in DNA vary from species to species, one rule always holds: the amount of adenine equals the amount of thymine and the amount of cytosine equals the amount of guanine. Thus what varies from species to species is the ratio A + T/G + C. This suggests that A-T and G-C always occur in pairs in the DNA molecule. From studies of DNA using the diffraction of X-rays, Watson and Crick proposed a model for the structure of DNA. According to their model, the molecule is composed of two polynucleotide chains, antiparallel to one another (if you examine Fig. 5-4 you will see that a DNA strand has polarity) and coiled together in a plectonemic (intertwined) spiral or helix. The two strands are held together partly by weak chemical bonds (hydrogen bonds) between pairs of bases on the opposite chains. Adenine is always opposite from and paired with thymine, guanine with cytosine (thus giving the observed ratios mentioned above). The phosphate-sugar backbone forms the outside of the helix and the bases are tucked inside it. This model, whose accuracy has been supported by a number of critical experimental tests, has very interesting features and implications.

First, we should note that the patterns of pairing make one of the two strands redundant. That is, given the sequence of bases on one strand, we can describe the sequence of bases on the other strand. In a sense then, the structure has built-in complementarity. Watson and Crick were led to propose that this built-in complementarity is the basis for the mode of replication of DNA. (Faithful replication, of course, is a feature of the gene.) They proposed that the two DNA strands could come apart and each strand could then serve as a template for the synthesis of a new complementary strand (Fig. 5-5). Each base in a strand would "attract" into correct position the particular base that is its partner. The result would be two identical, double-stranded, DNA molecules.

Now we can go back to a question raised earlier. With only four different bases as structural subunits, does DNA have sufficient potential for complexity to provide for the many genetic specificities known to exist? We now have good reason to believe that the sequence of bases in DNA is not a simple repeated sequence, but that sequences are complex and differ in molecules of different origins. We can then suppose that a gene contains its information in a code, the symbols of which are the four bases (A, C, G, and T). We know that the English language, with its 26 letters that form thousands of words, can be expressed by a code composed of only two symbols—the Morse code with its dot and dash. The genes can surely carry genetic information in a code composed of four symbols.

Logical extensions of the Watson-Crick structure for DNA, then, are that the specificity of the gene is carried in the *sequence* of nucleotides within the DNA molecule and that the complementary-strand structure of the molecule

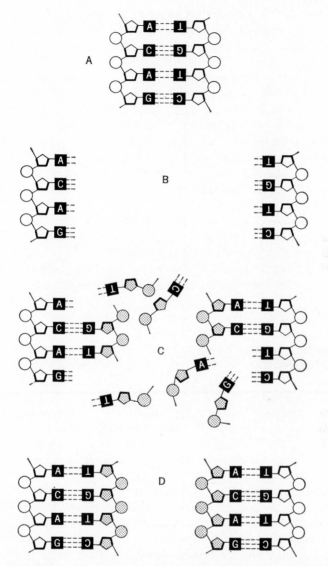

Figure 5-5. *A schematic representation of DNA replication. The two strands of the DNA molecule* (A) *separate in the region undergoing replication* (B). *Free nucleotides pair with their appropriate partners and are linked together* (C). *The whole process proceeds in a zipper-like fashion until two complete DNA molecules are finally formed* (D). *The newly formed polynucleotide chains are indicated by shading.* (*After Crick*, Sci. Amer., *Sept. 1957.*)

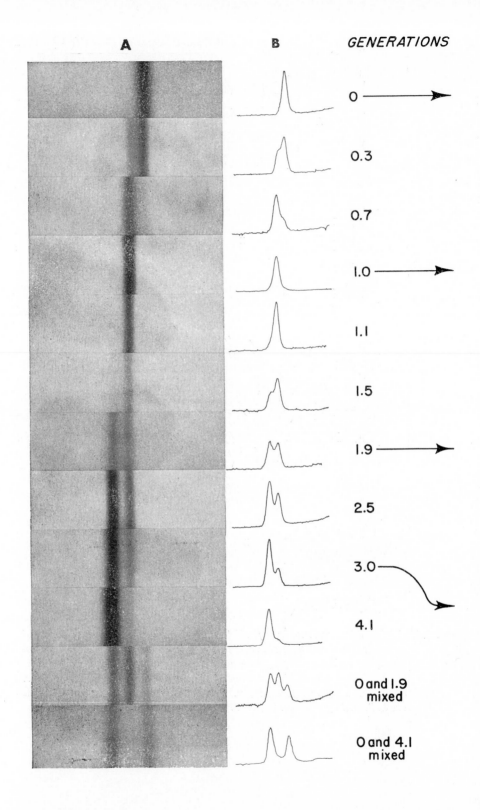

A **B** *GENERATIONS*

0 ⟶

0.3

0.7

1.0 ⟶

1.1

1.5

1.9 ⟶

2.5

3.0 ⟶

4.1

0 and 1.9 mixed

0 and 4.1 mixed

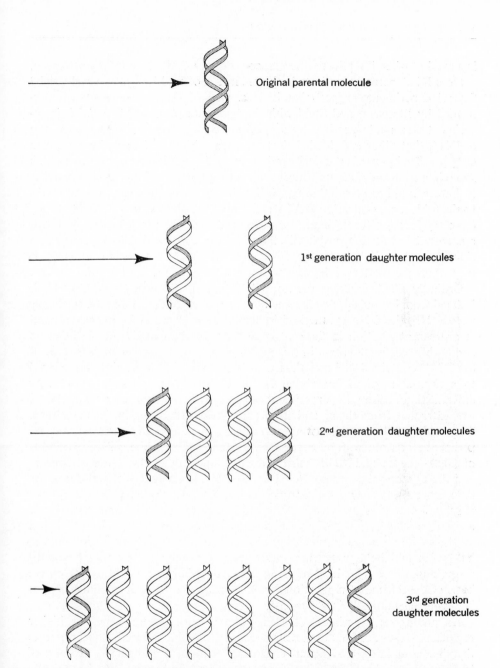

Figure 5-6. (*Facing page.*) *The replication of* E. coli *DNA.* A: *Photographs show the DNA in the centrifuge cells.* B: *Density tracings of these photos indicate the amount of DNA at different positions in the centrifuge cells. The interpretations of these observations are shown above. Parental polynucleotide chains are shown as shaded while newly formed strands are unshaded.* (*After Meselson and Stahl,* Proc. Natl. Acad. Sci. **44**:*671, 1958.*)

Original parental molecule

1st generation daughter molecules

2nd generation daughter molecules

3rd generation daughter molecules

insures that when it replicates the sequence of nucleotides is faithfully preserved.

Does DNA actually replicate in the manner proposed by Watson and Crick? Several experiments suggest that it does. One of the most elegant was performed by Meselson and Stahl with the bacterium *Escherichia coli*. Their experiment was made possible by a special technique for separating molecules of DNA that differ *only* in their isotope composition. This technique is called buoyant density-gradient centrifugation. DNA is centrifuged at extremely high speed in a solution of cesium chloride for a long time. (Cesium chloride happens to have a number of suitable properties, but other solutes can also be used.) In the tube, a concentration gradient of the cesium chloride develops due to the opposing forces of centrifugal force and diffusion. The DNA eventually forms a narrow band at that position in the tube at which the density of the cesium chloride and the DNA are the same. Since the molecular weight of DNA is very great, the molecules diffuse very little and the DNA band is quite narrow.

Ordinary DNA contains the normal isotope of nitrogen (N^{14}). If DNA is isolated from bacteria grown in a medium containing, in place of N^{14}, the heavy isotope N^{15}, the DNA synthesized is slightly more dense and will form a band at a different position in the centrifuge tube. In the experiment of Meselson and Stahl, cells were grown in this medium for a time long enough that all of the DNA contained N^{15} and thus was dense. The cells were then transferred to a normal medium containing N^{14} and allowed to continue growth. At different times during the experiment samples of the bacteria were taken, DNA was extracted from them, and its density was determined in the centrifuge. The results of the experiment are given in Figure 5-6. As you can see, the experiment starts out with all dense DNA. After a short while we see two kinds of DNA—dense and "half-dense." Slightly later, all of the dense disappears and there is only half-dense. Still later we have both light and half-dense. As time proceeds the relative proportions of light and half-dense change but the absolute amount of half-dense DNA stays constant. Even though the bacteria are not growing in a synchronous manner, we can see that after one cell division all of the DNA is half-dense, after two divisions there are equal amounts of half-dense and light, after three divisions three times as much light as half-dense, and so on. This kind of result is exactly what one would expect on the basis of the scheme of replication for DNA proposed by Watson and Crick, as is shown schematically in Figure 5-6.

An analogous experiment (see Figure 5-7) with chromosomes of higher organisms adds confirmation. Chromosomes can be induced to incorporate thymine containing a radioactive isotope (tritium, a radioactive isotope of hydrogen) into the DNA. If cells are allowed to incorporate the isotopically labeled thymine for a number of generations all the chromosomes will be identically labeled. If chromosomes so labeled are spread and stained for cytological examination, and then covered with a thin photographic film for a long time, a "photograph" of the chromosomes is made as the film is exposed to the radio-

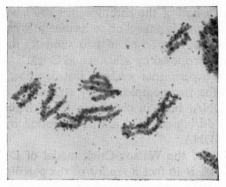

1st generation after labeling.   All chromatids labeled.

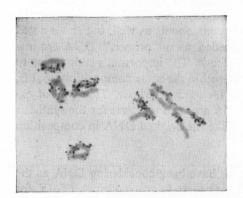 

2nd generation after labeling.   Only one chromatid labeled.

Figure 5-7. *Chromosome replication. The chromosomes have been stained and are seen as gray. The black dots are exposures of the photographic emulsion from the decay of radioactive particles of tritium. The drawing of the second-generation photo shows those chromosome parts that contain tritium (dark) and those that do not (light). For any one segment there is segregation of label from nonlabel, however, in some chromosome pairs (four out of the six) both chromosomes have labeled segments due to recombination (sister-strand crossing over).* (After Taylor, Woods and Hughes, Proc. Natl. Acad. Sci. **43**:*122, 1957.*)

active decay of the tritium isotope. This "photograph" (called a radioautograph) can then be compared to a normal picture of the stained preparation to determine the distribution of the isotope in the chromosomes. If cells are allowed to multiply in the absence of the radioactive isotope after first having been grown in its presence for a number of generations, the following results are obtained: after one cell division all the chromosomes are still labeled, but after two divisions we find that only half of the chromosomes are labeled. This shows that the DNA of a chromosome is passed on to its daughter chromosomes and that the chromosome must be composed of two subunits that segregate at each division, in the manner proposed by Watson and Crick.

Neither of the two experiments we have just described prove that the two subunits are in fact the two complementary strands of the DNA. However, they do show that DNA, both in chromosomes of higher organisms and in bacteria, is disposed during cell division as suggested by Watson and Crick for the replication of DNA.

Another implication of the Watson-Crick model of DNA structure is that newly synthesized DNA is in fact a *replica* of the parental molecule since the old strands act as templates for the synthesis of the new ones. Do we have support for this notion?

Kornberg and his associates have been able to synthesize DNA in a test tube. To do this they use the DNA subunits (the nucleotides), a specific enzyme, and some DNA. (The system needs other components as well, but those noted are the most relevant.) The DNA is needed as a "primer." DNA from a different source can be used for each synthesis. The important result for us to consider is that the newly made DNA always has the same composition (that is, proportion of each nucleotide) as the DNA used as primer. This type of experiment supports our contention that DNA acts as a template for the synthesis of new DNA, since the new DNA is just like the original DNA in composition and can be formed only in its presence.

RNA As Genetic Material. Although we have been considering DNA as the genetic material, we have indicated that at least in some viruses (TMV for example) RNA seems to have a genetic role. How does RNA fit into the general picture? As we shall see later, a certain kind of RNA is the "partner of the gene." It plays an important role in gene action, functioning as an intermediary in the expression of the gene. In the exceptional case of the RNA viruses (which lack DNA) the genetic system appears to be "short circuited," so to speak, with RNA playing both the role of the gene and of its intermediate. In any case, it now appears probable that in all organisms, except certain viruses, DNA is the basic hereditary material.

Keys to the Significance of This Chapter

There is good evidence that the genetic material is DNA. A number of special features of the chemical structure of the DNA molecule help us to understand some of the unique properties of genes. The complementary nature of the two strands of DNA suggests the mode by which genes are faithfully reproduced during cell division. The great specificity and complexity of genes can be accounted for by the sequence of nucleotides within DNA molecules. The specific messages carried by genes are the sequences of nucleotides in the DNA molecules of which genes are composed.

REFERENCES

Asimov, I., *The Genetic Code*. New York: Signet Science Library, 1963. (A popular account of the biochemistry of heredity, from the essentials of organic chemistry to the "cracking" of the code.)

Avery, O. T., MacLeod, C. M., and McCarty, M., "Studies on the Chemical Nature of Substance Inducing Transformation of Pneumococcal Types. Induction of Transformation by a Desoxyribonucleic Acid Fraction Isolated from Pneumococcus Type III." *J. Exptl. Med.*, **79**:137–158, 1944. (This paper first described the evidence that the transforming principle is DNA.)

Chargaff, E., and Davidson, J. N., *The Nucleic Acids*. New York: Academic Press, 1955. (A series of volumes containing many advanced articles.)

Crick, F. H. C., "Nucleic Acids." *Scientific American*, September 1957. Available as Offprint 54 from W. H. Freeman and Co., San Francisco. (A general treatment of the structure of nucleic acids.)

Davidson, J. N., *The Biochemistry of Nucleic Acids*. London: Methuen and Co., 1957. (A general treatment of nucleic acid chemistry.)

Kendrew, J. C., "The Three-Dimensional Structure of a Protein Molecule." *Scientific American*, December, 1961. Available as Offprint 121 from W. H. Freeman and Co., San Francisco. (Describes the work that led to the uncovering of the structure of myoglobin.)

Kornberg, A., *Enzymatic Synthesis of DNA*. New York: John Wiley and Sons, 1961. (This small book is an elegant account of the work on the *in vitro* synthesis of DNA.)

Stein, W. H., and Moore, S., "The Chemical Structure of Proteins." *Scientific American*, February 1961. Available as Offprint 80 from W. H. Freeman and Co., San Francisco. (A good general treatment of protein structure.)

Watson, J. D., and Crick, F. H. C., "A Structure for Desoxyribose Nucleic Acids." *Nature* (London), **171**:737–738, 1953. (The first paper in which the model of DNA structure was described.)

Watson, J. D., and Crick, F. H. C., "Genetic Implications of the Structure of Desoxy-ribonucleic Acid." *Nature* (London), **171**:964–969, 1953. (These authors make some predictions derived from their model.)

Potter, Van R., DNA Model Kit. Minneapolis: Burgess Publishing Co., Copyright 1959. (A paper model of DNA, easy to build.)

QUESTIONS AND PROBLEMS

5-1. What do the following terms signify?

polypeptide	TMV
nucleotide	RNA
protein	radioactive isotope
nucleic acid	template

If DNA is heated above 100°C the two strands of the double helix come apart. If the DNA is then cooled slowly, the single strands find their complements again and reform double-helical DNA.

5-2. If dense DNA (DNA labeled with N^{15}) from the virus T4 is mixed with light DNA (containing the normal isotope of N) and the mixture is centrifuged in a cesium chloride density gradient, two bands of DNA are formed. If the mixture is first heated to 100°C and then cooled slowly prior to centrifugation, a new intermediate band appears, in addition to the two bands corresponding to those formed by the unheated material. How do you account for its presence?

5-3. If labeled T4 DNA and unlabeled DNA from another virus, T5, are heated together and slowly cooled, only two bands are formed in the density gradient. How do you account for the absence of the intermediate band in this experiment?

5-4. What do the results of the experiments described in 5-2 and 5-3 tell you about the nature and specificity of the base pairing and base sequence in DNA?

5-5. The replication scheme for DNA proposed by Watson and Crick has been called a "semiconservative" replication scheme since one of the two parental strands is passed on to each daughter molecule. Two other modes of replication could be envisaged: in "conservative replication" one of the two daughter molecules would contain both parental strands, the other neither strand; in "dispersive replication" small fragments of the parental molecules would be passed randomly to the daughter molecules. What results would Meselson and Stahl have found if the first of these other models for DNA replication were correct? If the second model were correct?

5-6. High-frequency sound chops DNA into small fragments that are still double stranded and that have the same density as the unbroken molecules. How many bands should be formed in the cesium chloride density gradient from hybrid DNA (half dense) treated with high-frequency sound?

5-7. If we define a gene as a segment of DNA with a unique sequence of base-pairs, how many different genes could there be if genes consisted of DNA sequences containing

a) one base pair?
b) three base pairs?
c) one thousand base pairs?
d) n base pairs?

(Note that in DNA there exist four different base pairs A-T, T-A, G-C, and C-G.)

CHAPTER 6

Linkage, Crossing Over, and Chromosome Mapping

The principle of independent assortment of members of different allelic pairs at the time of germ-cell formation is one of the cornerstones on which an understanding of genetic systems has been built. You have met with and utilized this principle repeatedly, both in reading and in problem solving. Now you need to learn that independent assortment is not a rule without exceptions. When certain different allelic pairs are involved in crosses, deviations from independent assortment regularly occur. Study of the basis of these deviations has added much to our present detailed knowledge of the structure of germ plasm.

Linkage

Deviation from Independent Assortment Seen in a Testcross Ratio. For an initial example of systematic deviation from independent assortment we may turn to the frizzle fowl. In an experiment conducted by F. B. Hutt, colored frizzle females were crossed with a White Leghorn male. The male was homozygous for gene *I*, a dominant that acts as an inhibitor for melanin pigmentation, and for the normal, recessive allele of frizzle. The hens were *iiFF;* that is, they did not have the dominant white gene but were homozygous frizzles. F₁ females obtained from the cross of colored, frizzle females × white, normal males were testcrossed. The crosses and their results are tabulated as follows:

P: *iiFF* (colored, frizzle) ♀ ♀ × *IIff* (white, normal) ♂
F₁: *IiFf* (white, frizzle) ♀ ♀ and ♂♂
testcross: *IiFf* ♀ ♀ × *iiff* ♂

white, frizzle	18
colored, frizzle	63
white, normal	63
colored, normal	13
	157

(Data from F. B. Hutt, *Genetics*, **18**:84, 1933.)

You see that the results are markedly different from the 1:1:1:1 ratio we expect from testcrosses of dihybrid individuals. If frizzle and normal feathering are considered alone, however, the proportion of 81 frizzles to 76 normals fits closely the expected 1:1 ratio. Likewise, the count of 76 colored to 81 white shows no greater deviation from the expected than might reasonably be accounted for by chance. The real deviation from the expected is therefore not in the behavior of either allelic pair alone, but in their behavior with respect to each other.

At this point the value of the testcross is conspicuous. Since the male parent in the testcross was homozygous recessive, the respective frequencies of the different kinds of egg cells formed by the dihybrid females can be determined directly by inspection of the phenotypes of the testcross progeny. Clearly, the F₁ females gave rise to egg cells in the following proportions: 63*If*, 63*iF*, 18*IF*, and 13*if*. The two classes of gametes found in unexpectedly great numbers have the same allelic combinations as were present in the gametes formed by the two parents of the F₁ dihybrids. The new combinations (*IF* and *if*) make up only 19.7 percent ($^{31}/_{157}$) of the total.

The Meaning of Linkage. The tendency of parental combinations to remain together, which is expressed in the relative infrequency of new combinations, is the phenomenon of linkage. Genes show linkage because they are in the same chromosome. New combinations of linked genes are called *recombinations*.

For some of you who have been thinking independently about various aspects of segregation, an introduction to linkage will suggest the beginning of an answer to a troublesome question. We have said earlier that independent assortment of members of different allelic pairs depends upon independent assortment of members of homologous chromosome pairs at meiosis. But there are many more different genes than there are chromosome pairs in an organism, so that at least some chromosomes must carry many genes. How, then, you may have asked, can all genes segregate independently of one another? The answer is that they do not. You have just seen one illustration of this fact. The rest of the chapter will be an elaboration of the phenomenon of linkage, in the course of which you will find ample evidence that linked genes are indeed genes in the same chromosome.

The Method of Notation for Linked Genes. The crosses summarized on page 147 can now be shown in more meaningful fashion:

$$\text{P: } \frac{i\,F}{i\,F}\text{ (colored, frizzle) } ♀♀ \times \frac{I\,f}{I\,f}\text{ (white, normal) } ♂$$

$$\text{eggs: } i\,F \qquad\qquad \text{sperm: } I\,f$$

$$\text{F}_1\text{: } \frac{i\,F}{I\,f} ♀♀ \quad \text{and} \quad ♂♂$$

$$\text{testcross: } \frac{i\,F}{I\,f} ♀♀ \times \frac{i\,f}{i\,f} ♂$$

sperm:

eggs:		$i\,f$	
	$I\,F$	white, frizzle	(18)
	$i\,F$	colored, frizzle	(63)
	$I\,f$	white, normal	(63)
	$i\,f$	colored, normal	(13)

Pay particular attention to the way linked genes are represented. For instance, the genotype of colored, frizzle females in the parental generation is written as $\frac{i\,F}{i\,F}$. This is a simplification of $\dfrac{i\,F}{i\,F}$, where the two horizontal lines represent the homologous chromosomes in which genes i and F are situated. The designation of F_1 individuals as $\frac{i\,F}{I\,f}$, then, indicates that genes i and F are in one homologue and genes I and f are in the other.

Examination of this new summary of the crosses will make it clear to you that the colored frizzles and the white normals in the testcross progeny are determined by egg cells having the same allele-in-chromosome combinations as found in the different gametes of the P generation, and also in the chromosomes of the F_1 females used as parents in the testcross. Egg cells giving rise to colored normals or to white frizzles in the testcross progeny show recombination of the original parental arrangement of alleles.

Results of Another Testcross Involving Frizzle and Dominant White. Your understanding of the effects of linkage may be amplified if we consider another cross carried out in the course of Hutt's investigations. In this instance also, females heterozygous for the F and I genes were testcrossed to a double-recessive male. The results showed:

white, frizzle	15
colored, frizzle	2
white, normal	4
colored, normal	12
	33

(Data from F. B. Hutt, *Genetics*, **18**:85, 1933.)

Superficially, these results appear to represent an exact reversal of the tendency noted in the earlier cross. Here, colored frizzles and white normals make up the least frequent classes, whereas previously they were most frequent. The explanation of the apparent discrepancy is simple, however, and helps to substantiate what we have said about the chromosomal basis of linkage. In the experiment presently under consideration, the dihybrids being testcrossed were of a genetic constitution $\frac{I\,F}{i\,f}$ instead of $\frac{i\,F}{I\,f}$. The greater frequency of $I\,F$ and $i\,f$ gametes in the second testcross is accounted for by the fact that they represent the arrangement of alleles in the dihybrid parent.

Observe that the 2 colored frizzles and the 4 white normals of the second cross make 6 recombinations in the total of 33 progeny, or 18.2 percent recombinations. This value for recombinations is very near the 19.7 percent nonparental combinations found in the first experiment. The conclusion that the tendency of linked genes to recombine is equally strong whether the arrangement of alleles is both dominants in one chromosome and both recessives in the other, or a dominant and a recessive in each homologue, would seem warranted. These different kinds of allelic arrangements in a double heterozygote are designated by the terms *coupling*, for the arrangement $\frac{AB}{ab}$, and *repulsion*, for the arrangement $\frac{Ab}{aB}$.

Linkage and F_2 Ratios. The typical Mendelian dihybrid ratio in F_2 results from the formation by F_1 individuals (*AaBb*) of equal numbers of gametes of the four possible genotypic combinations (*AB, Ab, aB, ab*). Linkage of the segregating genes of a dihybrid leads to deviations from a 9:3:3:1 ratio of phenotypes just as it leads to deviations from the testcross ratio of 1:1:1:1. The magnitude of such deviations depends upon the frequency with which recombinations occur. Another way of expressing the same idea is to say that the magnitude of such deviations depends upon the strength of the linkage.

For particular pairs of linked genes, recombination frequencies are relatively stable. Once established, they may be used as a basis of prediction for genotypic and phenotypic ratios to be obtained from future crosses involving those gene pairs. Let us consider an example of how a recombination frequency derived from the results of a testcross can be made the basis for forecasting an F_2 ratio.

In tomatoes, gene O, for round fruit, is dominant over its allele o, for fruit of elongate shapes. Gene S, a gene for simple inflorescence, is dominant over s, which determines compound inflorescence. A cross was made between Yellow Pear tomatoes, characterized by elongate fruit and simple inflorescence (*ooSS*), and Grape Cluster tomatoes, having round fruit and compound inflorescence

(*OOss*). The F_1 consisted of plants having nearly round fruits and simple inflorescences. A testcross of the F_1 gave the results shown below.

$$\frac{Os}{oS} \times \frac{os}{os}$$

round, simple	(*OS*)	23
long, simple	(*oS*)	83
round, compound	(*Os*)	85
long, compound	(*os*)	19

(Data from J. W. MacArthur, *Genetics*, **13**:414, 1928.)

Linkage is obviously indicated by the data. The two recombination types, round, simple and long, compound, make up only 42 out of 210, or 20 percent of the total progeny. This means that of gametes formed by the dihybrid $\frac{Os}{oS}$, 80 percent are types *Os* and *oS*, while 20 percent are *OS* and *os*. Since the two recombination types are formed in about equal numbers, and since the two nonrecombination (parental) types seem also to be formed in equal numbers, we are justified in supposing that of 100 gametes formed by the hybrid, the following average frequency of genotypes will be maintained:

$$\left.\begin{matrix} oS & 40 \\ Os & 40 \end{matrix}\right\} \text{80 parental types}$$

$$\left.\begin{matrix} OS & 10 \\ os & 10 \end{matrix}\right\} \text{20 recombination types}$$

With these values in mind, we can make predictions about an F_2 generation arising from the cross of $\frac{Os}{oS}$ individuals. The necessary calculation requires no difficult alteration in the familiar methods for computing Mendelian ratios. In fact, a checkerboard square can be utilized again, with the qualification that we now take into account that the gametic types are not equally frequent. Weighting the gametic frequencies is simple for the example at hand, since the gametes *OS*, *Os*, *oS*, and *os* are expected to occur in a proportion of 10:40:40:10. If the terms of this proportion are converted into decimal fractions (0.1, 0.4, 0.4, and 0.1), the probabilities for given gametic unions may be calculated as shown in Figure 6-1. A summation of phenotypes in this checkerboard square is:

round, simple	(*OS*)	0.51
long, simple	(*oS*)	0.24
round, compound	(*Os*)	0.24
long, compound	(*os*)	0.01
		1.00 = 100%

If these particular phenotypes were determined by genes that were inherited independently, the corresponding values would be approximately 0.56, 0.19, 0.19, and 0.06 (a 9:3:3:1 ratio).

$$\frac{Os}{oS} \times \frac{Os}{oS}$$

Pollen

		0.1 OS	0.4 Os	0.4 oS	0.1 os
		0.1 *OS*	0.4 *Os*	0.4 *oS*	0.1 *os*
	0.1 *OS*	0.01 OS/OS	0.04 OS/Os	0.04 OS/oS	0.01 OS/os
Eggs	0.4 *Os*	0.04 Os/OS	0.16 Os/Os	0.16 Os/oS	0.04 Os/os
	0.4 *oS*	0.04 oS/OS	0.16 oS/Os	0.16 oS/oS	0.04 oS/os
	0.1 *os*	0.01 os/OS	0.04 os/Os	0.04 os/oS	0.01 os/os

Figure 6-1. *Prediction of F$_2$ genotypes when the F$_1$ is $\frac{Os}{oS}$ and the recombination frequency is 20 percent.*

Now let us compare our prediction for an F$_2$ with actual counts of F$_2$ individuals from an experimental cross of this kind. In the actual F$_2$, 259 individuals were classified. From our checkerboard prediction, we would expect 51 percent of these to be round, simple (*OS*), 24 percent long, simple (*oS*), 24 percent round, compound (*Os*), and 1 percent long, compound (*os*). See Table 6-1 for the comparison of real F$_2$ values with those expected on the basis of recombination frequency obtained from the testcross. Examination of these values will convince you that the correspondence of actual segregants to predicted segregants is very good.

Table 6-1. COMPARISON OF ACTUAL F$_2$ SEGREGANTS WITH THOSE PREDICTED ON THE BASIS OF TESTCROSS RESULTS.

Phenotypes		Number of Actual F$_2$ Segregants	Number of Predicted F$_2$ Segregants
round, simple	(*OS*)	126	132
long, simple	(*oS*)	66	62
round, compound	(*Os*)	63	62
long, compound	(*os*)	4	3

Source: J. W. MacArthur, *Genetics*, **13**:414, 1928.

The F_2 progeny from a hybrid ($OoSs$) can fall into quite different proportions of phenotypes, depending on whether the alleles are in a repulsion or a coupling phase. Suppose that instead of F_1 hybrids of the combination $\dfrac{Os}{oS}$, we were to consider individuals of the constitution $\dfrac{os}{OS}$. Utilize again the recombination value of 20 per cent derived from the testcross data, and see whether you can predict the relative frequencies of F_2 phenotypes under this condition of coupling in the F_1. (On the basis of a progeny of 100 the answer to the problem is 66:9:9:16.)

Calculating Strength of Linkage from F_2 Data. It is easier to estimate linkage values on the basis of testcross data, from which the totals of recombination and nonrecombination types of gametes can be read directly, than on the basis of F_2 data like those shown in the first column of numbers in Table 6-1. Calculations of recombination frequency can, in fact, be made from such F_2 data, and sometimes investigators in genetics find it practical to utilize F_2's for establishing the strength of a linkage. The methods of calculation, however, are relatively devious. Since they add little to one's understanding of genetics in general, we will mention them only to say that they exist, and that the procedure for the necessary mathematical manipulations is readily available in the literature of genetics. (See "References" at the end of this chapter.)

Crossing Over

It has been stated that genes show linkage because they are on the same chromosome. Recombination of linked genes is accomplished through a process by which homologous chromosomes exchange parts. This process is called *crossing over*. We shall first list briefly certain points necessary to the explanation of crossing over and then substantiate and amplify them in the subsequent discussion.

1. The genes in chromosomes are in linear order, somewhat like beads on a string.
2. When a gene (A) and its allele (a) are present in different members of a pair of homologous chromosomes, the gene and its allele occupy corresponding places in the homologues.
3. In order to produce recombinations between two different allelic pairs situated in the same chromosome pair, crossing over must occur between the locations (*loci*) of the genes involved.
4. Crossing over characteristically occurs in the first division of meiosis. It is this meiotic crossing over that concerns us most. However, somatic crossing over is also known to occur, and is discussed later in the chapter.

5. Meiotic crossing over takes place during the time in the nuclear reproductive cycle when four chromatids are present for each pair of chromosomes.

With these points in mind, we may initially visualize crossing over as occurring between chromatids in the manner indicated in Figure 6-2.

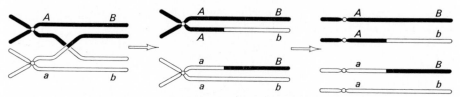

Figure 6-2. *Crossing over is the interchange of corresponding segments between chromatids of homologous chromosomes.*

Cytological Detection of Crossing Over. Before considering crossing over in more detail, let us confirm that recombination actually does result from a physical interchange of parts of homologous chromosome strands. To demonstrate this fact unequivocally is not as easy as might be imagined. The difficulty is that under ordinary circumstances members of a pair of homologous chromosomes are not visibly distinguishable, even by close microscopic examination. Only under unusual circumstances, when the ends of homologous chromosomes are somehow "labeled," can the recombination of linked genes be related to physical exchange of parts between the homologues. Curt Stern has exploited a set of unusual circumstances of this kind to fashion one of the classical experiments in genetics. His proof of a cytological basis for crossing over is summarized in Figure 6-3. Other investigators have utilized corn to demonstrate in equally elegant ways that recombination of linked genes is accompanied by interchange of chromosome material. The clear results obtained both in experiments with Drosophila and corn experiments strengthen impressively the conviction that recombination has a definite physical basis.

Crossing Over in the Four-Strand Stage. If crossing over occurs at some time during the close association of homologous chromosomes at meiosis, it would seem possible that it might occur either when only two strands are present or after the two strands have doubled. A consideration of the life cycle of Neurospora makes it plain that this organism is tailor-made for resolving the question of whether crossing over occurs at a two- or at a four-strand stage. As a matter of fact, in the chapter on "Life Cycles" we have already indicated in a preliminary way how first- and second-division segregations in Neurospora relate to this question. The consideration of chief importance is that all the products of a single meiosis can be recovered, since they are retained for a time within the ascus. With reference to the question posed, it is only necessary to

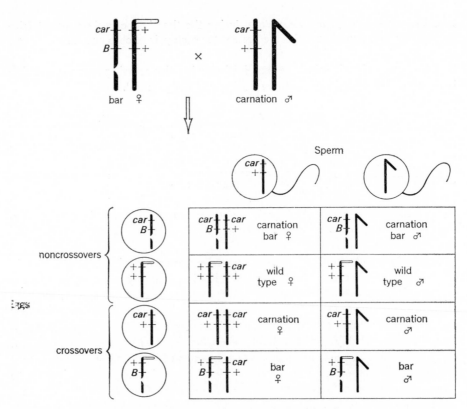

Figure 6-3. *Curt Stern found aberrant X-chromosomes in Drosophila that could be utilized for cytological demonstration of crossing over. One was an X-chromosome to which a portion of a Y-chromosome (shown in outline) was attached. The other was a broken X-chromosome whose acentric portion was attached to chromosome 4. Under a microscope, both kinds of aberrant chromosomes could be easily distinguished from each other and from normal X-chromosomes. Female Drosophila carrying these abnormal chromosomes and heterozygous for* carnation *(a recessive eye-color variant) and for* bar *(dominant) were crossed with carnation males. Examination of the chromosomes of progeny whose phenotypes indicated genic recombination revealed that the genic recombination was accompanied by appropriate exchange of identifiable chromosome segments. (After Stern, Biol. Zentr.,* **51***:586, 1931.)*

determine whether linked-gene pairs may give rise to both parental and recombination types within a single ascus or whether only one type or the other is recovered after a meiotic division.

For an example, we may look at one kind of result obtained after the cross of an albino (*al*) strain, producing white conidia, with a strain (*ag*), unable to synthesize the amino acid arginine. The wild-type allele of the *al* gene conditions orange-pigmented conidia, and that of the *ag* gene permits the biosynthesis of arginine. These genes are linked, and *ag* is known to be nearer the centromere of the chromosome than is *al*. The parental strains in this cross can be represented, then: $\overset{\text{+}\qquad al}{\underline{\quad\text{o}\quad\quad\quad\quad}}$ and $\overset{ag\qquad\text{+}}{\underline{\quad\text{o}\quad\quad\quad\quad}}$, with the small circle in the

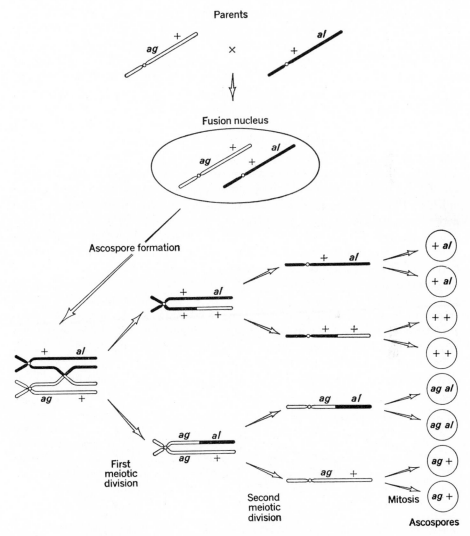

Figure 6-4. *Types of ascospores produced within an ascus of Neurospora in which a crossover has occurred between two linked genes. Gene* ag *determines an inability to synthesize arginine;* al *determines nonpigmented spores. The fact that ascospores are produced that give rise to parental-type mycelia in addition to ones that give double mutant and wild-type mycelia shows that crossing over occurs at a 4-strand stage.*

chromosome representing the centromere. Remember that Neurospora is haploid. Crossover types in the progeny of a mating between the two strains would be ——o———— (wild type) and ——o———— (double mutant). These crossover types are recovered from certain asci dissected in order, but along with them, in the same asci, are parental types. Figure 6-4 shows the analysis of a typical ascus dissected in order. Observe that the allelic pair nearest the centromere shows first-division segregation, while the *al* pair segregates in the second meiotic division. Verify that this ascospore distribution could not be formed unless four discrete chromatids were already present when the exchange occurred. In short, crossing over in Neurospora occurs at a four-strand stage. (Would the same sort of analysis be possible in corn?)

Crossing over in the four-strand stage has been demonstrated in Drosophila equally satisfactorily but in not quite so simple a manner as is possible in Neurospora. One proof in Drosophila depends upon the existence of attached-X female flies that are heterozygous for a sex-linked gene, such as that for vermilion eyes (*v*). Females of this kind are phenotypically wild type for eye color, inasmuch as the vermilion gene is recessive. Since the attached-X complex is inherited as a unit, one might expect these females to give rise only to wild-type daughters. Vermilion daughters do occur, however, in individuals where detachment of the X's has not taken place. This becomes understandable on the basis that crossing over occurs between the locus of *v* and the centromere of its chromosome at a time when four strands are closely associated during meiosis. Figure 6-5 diagrams the effects of crossing over at the four-strand stage in attached-X females. Verify that, if crossing over occurred at a two-strand stage, the effect would be merely to exchange *v* and its allele on the two arms of the attached-X, and that parental-type gametes would still be produced.

Evidence for crossing over at a four-strand stage in other organisms suggests that this aspect of the mechanism of meiosis is widespread among living things

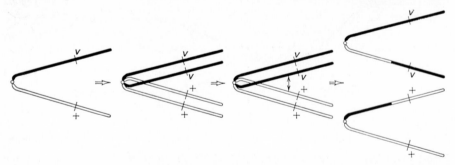

Figure 6-5. *When attached-X females of Drosophila are heterozygous for a sex-linked gene, homozygous egg cells may be produced if crossing over occurs between the gene locus and the centromere. This is another way of demonstrating that crossing over occurs at a 4-strand stage.*

Chromosome Mapping

The Basis for Chromosome Mapping. In Drosophila, the genes for the mutant characters *white eyes, yellow body,* and *cut wings* are all sex-linked. Appropriate matings reveal that the white and yellow genes consistently show about 1 percent crossing over. But white and cut give around 20 percent crossing over. Results of this kind are typical for Drosophila and for other organisms whose linkage patterns have been studied. Crossing over between particular pairs of linked genes occurs at quite characteristic and stable frequencies, but these frequencies may differ rather widely, depending upon the gene pairs involved. This generalization might lead to, or at least it is consistent with, the conclusion that each gene has a particular and well defined locus in its chromosome. Such a conclusion is supported by the fact that crossover values can be utilized to demonstrate a definite serial order for the genes in a chromosome and even as a basis for mapping "distance" relations among linked genes.

Map Distance. The results of crosses involving linked genes provide no basis for estimating the distances between these genes in terms of standard linear measurements. However, chromosomes can be mapped effectively in terms of percentages of crossing over obtained from genetic experiments. In this system, one unit of map distance between linked genes is the space within which 1 percent crossing-over takes place. Thus 5 percent crossing over between genes *A* and *B* establishes that they are situated 5 units of map distance apart in their chromosome.

$$A \longleftarrow 5 \text{ units} \longrightarrow B$$

The Orderly Arrangement of Genes in a Chromosome. The kind of reasoning utilized in chromosome mapping can be illustrated if we now assume that gene *A*, above, and a third linked gene, *C*, give 7 percent crossing over, and that *B* and *C* are shown by crossover data to be 12 units apart. In such an instance, we can deduce that *A* is situated between *C* and *B* in its chromosome.

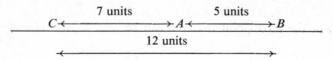

If *B* and *C* give 2 percent crossing over, we would conclude that *B* lies between *A* and *C*.

The situations just presented are hypothetical, but real genetic situations in which the crossover relationships among three or more linked genes attest to a precise linear order of genes in chromosomes have been disclosed repeatedly.

Double Crossing Over. Before proceeding to the analysis of actual crossover data, we need to consider a major pitfall to be avoided in the mapping of chromosomes. This pitfall is revealed by close examination of the relationship between recombinations and crossovers, which, although closely related, are not quite the same things. Recall that *recombinations* are new combinations of linked genes. *Crossing over*, the process that produces recombinations, is the interchange of corresponding segments between chromatids of homologous chromosomes. *Crossovers* are the chromatids resulting from such interchanges. Ordinarily we recognize a crossover by a recombination; testcross data, at the same time that they give us recombination frequencies, provide a measurement of frequency of crossing over. This measurement is not always accurate.

Let us suppose that crossing over occurs more or less at random up and down the length of a chromosome. Relatively greater distance between two genes, then, should mean relatively better chances for simultaneous occurrence of two or more instances of crossing over between these genes. If two genes, *A* and *B*, are rather far apart in their chromosome, two exchanges between a pair of chromatids will have the effect shown in Figure 6-6. You can see that the *double-crossover* chromatids do not show recombination of the marker genes, and, in fact, are indistinguishable from *noncrossover* chromatids. Since the concept of map distance is based on the number of actual physical interchanges, recombination values may give far too low an estimate of the distance between genes, if double crossing over has frequently occurred.

How may this difficulty be obviated? Double crossing over usually occurs only infrequently within distances under 5 map units, or, for certain chromosome segments, within distances up to 15 or 20 units. Therefore, data for chromosome mapping should be taken from linked-gene pairs that are quite close together, so that the recombination values provide an accurate measure of crossing over.

Geneticists have found that an efficient way to obtain good recombination data is to employ so-called *three-point* testcrosses, involving three different genes situated within a relatively short segment of a chromosome. One of the advantages of the three-point testcross can be seen if you refer again to Fig-

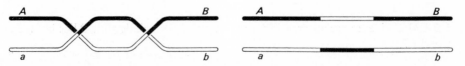

Figure 6-6. *If only two allelic pairs in the same chromosome are available as "markers," a double crossover between two chromatids is not genetically detectable, since the gene combinations produced are the same as in parental-type chromatids.*

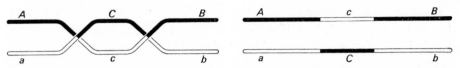

Figure 6-7. *Double crossing over between two allelic pairs is genetically detectable if a third allelic pair is situated in the segments exchanged. Under such circumstances, chromatids with new genic combinations are produced. Compare with Figure 6-6.*

ure 6-6. What if between genes *A* and *B* in this figure a third gene pair *C-c* were segregating? Then double crossing over between *A* and *B* might be detected genetically by the alteration of relationships of the middle allelic pair, *C-c*, as shown in Figure 6-7.

A Three-Point Testcross in Corn. Now, starting with the experimental data shown in Table 6-2, let us apply some of these principles to the specific problem of mapping genes in a chromosome. The data are testcross values from studies by G. W. Beadle, and involve the segregation of three mutant recessives in corn. These are: *v = virescent* seedling; *gl = glossy* seedling (leaves have a particularly shiny appearance); and *va = variable sterile* (characterized by irregular distribution of chromosomes at meiosis). In conformity with the usage of the investigator, the normal allele of each mutant gene is indicated by a "+."

Beadle's testcross data are shown in columns I and II of the table. Column III will be referred to in detail later, but its significance should already be apparent to you.

Table 6-2. A THREE-POINT TESTCROSS IN CORN.

I	II	III		
		Genotype of		
	Number of	*Gamete from*		
Phenotypes of Testcross Progeny	*Individuals*	*Hybrid Parent*		
normal	235	+	+	+
glossy, variable sterile	62	*gl*	*va*	+
variable sterile	40	+	*va*	+
variable sterile, virescent	4	+	*va*	*v*
glossy, variable sterile, virescent	270	*gl*	*va*	*v*
glossy	7	*gl*	+	+
glossy, virescent	48	*gl*	+	*v*
virescent	60	+	+	*v*

Source: Emerson, Beadle, and Fraser, *Cornell Univ. Agr. Exp. Sta. Mem. 180,* p. 59, 1935.

Parental- and Double-Crossover Types. In analyzing the progeny from a three-point testcross, it is well to work from the known toward the unknown. An obvious point of attack on the data is to identify the classes of progeny that represent parental (nonrecombination) type gametes from the hybrid parent. This parent, in the cross being considered, was of the genotype $\dfrac{+ \quad + \quad +}{gl \quad va \quad v}$ having all the dominants in one chromosome and all the recessives in its homologue. Therefore, normal progeny and virescent, glossy, variable-sterile individuals are the two classes among the testcross segregants that represent parental-type gametes. Reference to the table will show you that these classes include by far the largest numbers of individuals. Even if the allelic arrangement in the heterozygous parent were unknown, we might deduce, on the basis of relative frequency of types, that $+ \qquad + \qquad +$ and *gl   va   v* are parental-type gametes. Why?

It should be emphasized that our representation of the sequence of genes in parental and gamete chromosomes is so far purely arbitrary and *may not represent the true gene order at all.*

The double-crossover progeny may also be picked out. Double crossing over is the simultaneous occurrence of two relatively improbable events, and double-crossover gametes are the least frequent type. Glossy plants, represented only seven times, and virescent, variable steriles, of which there are only four, clearly are the double-crossover progeny. Notice that when one member of a type, such as the double crossovers, is established, the other member can immediately be ascertained. If *gl   +   +* (glossy) represents one double-crossover type, its partner must be of the complementary allelic constitution, $+ \qquad va \qquad v.$

Determining Gene Order. Once the parental- and double-crossover type chromosomes have been established, the relative order of genes in the chromosome may be deduced. The reasoning here is based on the consequences of double crossing over as shown in Figure 6-7. Reaffirm that double crossing over in such an instance has the effect of changing the associations of the members of the middle allelic pair.

Therefore, if we note from three-point testcross data which allelic pair is transposed in order to make double crossover from parental-type chromosomes, that allelic pair must be situated between the other two. For ease of reference, we can write down again the parental and double-crossover types indicated by the linkage study in corn.

PARENTAL TYPES			DOUBLE CROSSOVERS		
+	+	+	*gl*	+	+
gl	*va*	*v*	+	*va*	*v*

You can now see that if *gl* and its wild-type allele are transposed in their allelic associations, double crossovers can be obtained from parental types. Therefore, in the actual linear order of genes in the chromosome, *gl* must be situated between *va* and *v*. The heterozygous parent in the testcross can now be designated correctly, both for gene order and association of alleles in the homologous chromosomes, as follows:

$$\frac{+ \quad\quad + \quad\quad +}{v \quad\quad gl \quad\quad va}$$

If we diagram double crossing over between chromatids of the parental type written in correct order, it is readily demonstrated that the gametic types derived correspond to those indicated by the data from the testcross.

$$\frac{+ \quad + \quad +}{v \quad gl \quad va} \; \times\!\times \; \longrightarrow \quad \frac{+ \quad gl \quad +}{v \quad + \quad va}$$

Looking at the correct hybrid parental type shown at the left above, you will observe that a single crossing over between the loci of *v* and *gl* gives rise to chromatids *v*____+____+ and +____*gl*____*va*. This region of the chromosome between *v* and *gl* may be called, for convenience, Region I. Also, a single crossing over between *gl* and *va* gives rise to chromatids that are +____+____*va* and *v*____*gl*____+. These chromatids may be said to result from crossing over in Region II. According to genetic convention, crossover regions are arbitrarily designated by numbers in sequence, with I indicating the interval between the first two genes at the left, as the chromatid or chromosome is represented.

We can now resummarize our three-point testcross data in meaningful fashion. Showing gametic frequencies of different chromosome types, the data are:

$$\frac{+ \quad\quad + \quad\quad +}{v \quad\quad gl \quad\quad va} \quad \begin{matrix} 235 \\ 270 \end{matrix} \Big\} \text{ parental types}$$

$$\frac{+ \quad\quad gl \quad\quad +}{v \quad\quad + \quad\quad va} \quad \begin{matrix} 7 \\ 4 \end{matrix} \Big\} \text{ double-crossover types}$$

$$\frac{v \quad\quad + \quad\quad +}{+ \quad\quad gl \quad\quad va} \quad \begin{matrix} 60 \\ 62 \end{matrix} \Big\} \text{ single crossovers, Region I}$$

$$\frac{+ \quad\quad + \quad\quad va}{v \quad\quad gl \quad\quad +} \quad \begin{matrix} 40 \\ 48 \end{matrix} \Big\} \text{ single crossovers, Region II}$$

$$\text{total} \quad\quad\quad 726$$

Map Distance. Distances in the chromosome map under construction may be computed in the usual way. You will remember that crossover frequencies can be translated directly into units of map distance. In the present instance, crossing over in Region I is represented by gametes of the following chromosome types, which have derived from chromatids of the heterozygous parent.

v	+	+	60
+	*gl*	*va*	62
+	*gl*	+	7
v	+	*va*	4
total			133

Out of 726 chromosomes recovered in gametes, 133 represent interchanges of chromatid material between the loci *v* and *gl;* 133 is 18.3 percent of 726. The map distance between *v* and *gl* is 18.3 units.

Similarly, crossing over between the loci of *gl* and *va* is represented by these types:

+	+	*va*	40
v	*gl*	+	48
+	*gl*	+	7
v	+	*va*	4
total			99

The instances of crossing over in Region II are 99 in 726 chromatids, or 13.6 percent of the total. The map distance between *gl* and *va* is 13.6 units. The segment of chromosome we have charted may be represented as:

v	*gl*	*va*
18.3 units	13.6 units	

The concept of map distance is based on total frequency of crossing over in a region of a chromosome. In calculating map distances from three-point testcross data, do not forget to add the numbers of double crossovers to each of the sets of single-crossover figures. Remember that each instance of double crossing over represents single crossing over in both Regions I and II. If double crossovers are neglected in the computation of map distance, the computed values will indicate smaller distances than are actually shown by the genetic results.

Interference and Coincidence. Once a portion of a chromosome map is established, it provides the geneticist with rather specific probability values. For instance, the map distance of 18.3 units between v and gl means that we can expect 18.3 percent of gametes to represent crossing over between the loci of these genes. The value 13.6 for distance between gl and va has similar significance.

If, in the portion of the corn chromosome we have just mapped, crossing over in Region I is independent of crossing over in Region II, the probability of simultaneous crossing over in the two regions can be arrived at by applying a principle that we first encountered in Chapter 2.

> When two events are independent, the probability of their simultaneous occurence is the product of their separate probabilities.

In other words, if crossing over in the two regions is independent, $0.183 \times 0.136 = 0.025 = 2.5$ percent double crossovers might be expected. Actually, only $\frac{11}{726} = 1.5$ percent double crossovers occurred.

In a single instance, a deficiency in the expected number of double crossovers might not worry us very much, since some deviations from the expected are simply the effects of chance. It is fairly characteristic, however, that double crossovers do not appear as often as the frequency of crossing over in individual regions might lead us to expect. This seems to mean that once crossing over occurs the probability of another crossing over in an adjacent region is reduced. The phenomenon is called *chromosomal* or *chiasma interference*.

Interference may vary from one portion of a chromosome to another, and also among different chromosomes. For a chromosome map to function with greatest effectiveness as a table of probabilities, interference in various portions of the map should be designated. Strengths of interference are commonly summarized as *coefficients of coincidence*. These are no more than ratios between observed and expected frequencies of double crossing over, and are calculated as $\frac{\text{actual frequency of doubles}}{\text{expected frequency of doubles}}$. For the portion of the chromosome map of corn we have been studying, the coincidence can be computed as $\frac{1.5}{2.5} = 0.6$.

Coincidence varies inversely as interference varies, and you may expect coincidence values to vary from 0 to 1. Complete interference gives a coincidence value of 0; a coincidence of 1 indicates no interference whatsoever. In some cases, there appears to be a correlation of neighboring crossovers resulting in a number of double crossovers in excess of the random expectation. Here, the coincidence values are greater than 1. This situation is usually referred to as *negative interference*. This is the general case with bacteriophage recombination as we will shortly describe.

Mapping Functions. As we have pointed out, map distance is not synonymous with recombination frequency. Map distance is based on the total frequency of

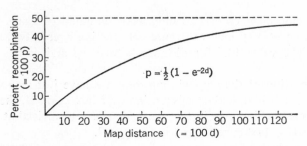

Figure 6-8. *Haldane's mapping function.* e *is the base of natural logarithms;* d, *the frequency of crossing over.*

crossovers; recombination frequency measures only those crossovers that lead to exchange of markers. Thus, double crossovers in a given region affect map length but do not contribute to recombination of genes "bracketing" the region. Recombination frequency *is* a measure of map distance for distances short enough that multiple crossovers constitute an insignificant class, but as distances become large, this proportionality no longer holds.

Often the relationship between map distance and recombination frequency is expressed in terms of a mapping function. A mapping function developed by J. B. S. Haldane is shown in Figure 6-8. This function assumes no interference between crossovers. Clearly, other functions can be, and have been, developed on the basis of particular assumptions regarding interference. The inclusion of positive interference gives curves that approach the limiting value of 50 percent recombination more rapidly than the curve in Figure 6-8; if negative interference is assumed, curves are shallower. Mapping functions show graphically that as distances become longer, recombination frequencies are no longer additive and eventually become independent of map distance. Map distances, however, will be additive if the function employed is applicable.

Mapping by Tetrad Analysis. In many organisms such as certain fungi, bryophytes, and algae, all the products of a single meiotic event may be recovered and analyzed genetically. Such an analysis is called tetrad analysis. In some of these organisms the immediate products of meiosis undergo one or two mitoses, giving rise to eight or sixteen products, but these are exact replicates of the original four. In most organisms with recoverable tetrads, the four products, or derivatives of them, are *unordered*. However, in Neurospora and some species of yeasts, as we have already pointed out in the chapter on life cycles, the products are *ordered* and the order in the ascus is indicative of the pattern of meiotic events. A great deal of information about crossing over can be obtained from tetrad analysis.

Unordered Tetrads. Three segregation patterns are possible if two gene pairs are segregating in a tetrad. Let us consider the segregation of two linked markers in a cross in which the parental combinations are *AB* and *ab*. A tetrad of the type $AB/AB/ab/ab$ is called a parental ditype (PD for short), since there are only two types of products and they are of the parental combinations. Tetrads of the type $AB/Ab/aB/ab$ are called tetratype (TT), and tetrads of the type $Ab/Ab/aB/aB$ are called nonparental ditype (NPD). Assuming no more than two exchanges between the two loci, these three types of tetrad arise in the following manner: parental ditypes arise from meioses in which either no crossover or a two-strand double occurs between A and B; tetratypes arise from meioses in which either a single crossover or a three-strand double occurs between A and B; nonparental ditypes (which are rare for linked markers) arise from meioses in which two crossovers occur between A and B, the two exchanges involving all four strands. (Confirm these statements by considering the markers A and C in Fig. 6-9).

Tetrad analysis is effective for determining whether two markers are linked. As we have just pointed out, if two markers are linked, the NPD's will be rare relative to the other types of tetrads. However, if the two markers are not linked, the NPD's will be just as numerous as the PD's since both types arise by random assortment of the chromosomes. If the markers are not linked, the tetratypes arise as a consequence of crossovers between either A and its centromere or B and its centromere. Hence, irrespective of the frequency of TT's, inequality of the two ditypes is evidence for linkage of the two loci.

To compute recombination frequencies from the frequency of the various tetrad types, we must remember that only one-half of the products of a TT tetrad are recombinant, but that all the products from an NPD are recombinant. Thus, to compare tetrad frequencies with randomly isolated progeny frequencies we make the relation: recombination frequency between A and $B = \dfrac{\frac{1}{2}TT + NPD}{\text{total tetrads}}$.

Ordered Tetrads. More information can be obtained from ordered than unordered tetrads. As was pointed out in the chapter on life cycles, second-division segregation for a particular marker indicates that crossing over has occurred between the marker gene and the centromere of its chromosome. Thus, from an analysis of ordered tetrads, recombination frequencies can be readily obtained not only between different genes but also between genes and their centromeres. (Gene-centromere distances can also be obtained by indirect calculation from data involving unordered tetrads with a suitable array of gene markers. References at the end of the chapter include information on this subject.)

As an example of the analysis of ordered tetrads, let us consider the patterns of segregation of three linked genes. A very large number of types of tetrads are possible, depending upon the number and location of the crossovers that occur. Figure 6-9 shows only three possible types of tetrads that result from crossing over in the two intervals between the three linked markers. Since crossing over can involve either of the two chromatids of each homologous chromosome, three types of double-crossover tetrads are possible. The two exchanges may involve the same two chromatids. Such exchanges are called

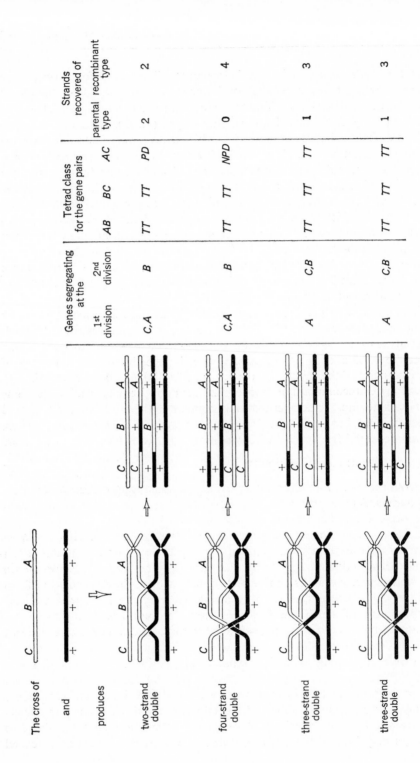

Figure 6-9. *Tetrad analysis. The consequences of double exchanges involving three linked genes.*

two-strand doubles. Each of the two exchanges may involve a different pair of chromatids, resulting in a *four-strand double*. Either possible combination of three strands may participate in the two exchanges, one strand being involved twice. This gives rise to the two types of *three-strand doubles*. As is shown in the figure, these different classes of double exchanges give rise to different segregation patterns in the tetrads. We see that exchanges between a given marker and the centromere result in second-division segregation for that marker. If we classify each tetrad as to whether it is PD, TT, or NPD with regard to each of the three pairs of segregating markers, we can differentiate among the two-, three-, and four-strand doubles. As well as those types of tetrads diagramed, single-exchange types, double-exchange types involving the interval between the centromere and *A* and one of the other intervals, triple-exchange types, and so on may also be obtained from a cross of the type described. Our purpose here is only to indicate the kind of detailed analysis of recombination possible with tetrads.

Chromatid Interference. Chromosomal or chiasma interference, which we have already discussed, has to do with the influence of crossovers upon the proximity of other crossovers along the chromosome. Tetrad analysis reveals that another possible source of interference between crossovers may exist. If the involvement of all four chromatids in crossing over is random, all four types of double-exchange classes would be equally frequent. That is, the ratio of two-:three-:four-strand double exchanges would be 1:2:1. However, if deviations from this expectation occur, they would be due to "chromatid interference;" that is, crossovers influencing the occurrence of other crossovers involving the same chromatid. Thus *negative* chromatid interference would result in an excess of two-strand doubles, *positive* chromatid interference, an excess of four-strand doubles. However, no compelling evidence for the existence of chromatid interference, either positive or negative, has yet been found in any organism.

Half-Tetrads. Studies of recombination in Drosophila involving attached-X chromosomes permit the analysis of two of the four strands from a single meiotic event. Also, after mitotic recombination the recovery of two of the four chromosomal strands that may be involved in exchanges is possible, as described below. Analysis either of attached-X's or of the products of mitotic recombination gives more information about the details of events that occur during crossing over than do studies of randomly obtained products of meiosis.

NONMEIOTIC METHODS OF MAPPING GENES

Various nonmeiotic phenomena have been discovered that facilitate the mapping of genes both in organisms whose reproductive processes do not include meiosis and in organisms that have meiotic stages. Such phenomena are valuable tools for providing information about the genetic structure of organisms without meiotic cycles; they also furnish a comparison of genetic structure determined by two or more independent methods.

Mitotic Crossing Over. In some organisms crossing over can occur during mitosis as well as meiosis. Over twenty-five years ago Curt Stern disclosed convincing evidence of crossing over in somatic tissue of Drosophila. As in meiosis, the exchanges occur at the four-strand stage. However, the chromatids then segregate in the manner expected for mitosis. As Figure 6-10 shows, one consequence of mitotic crossing over is that half of the time chromosome segregation is such that the daughter cells are rendered homozygous for any

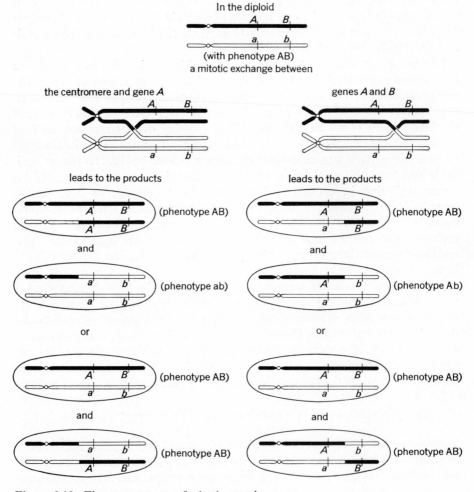

Figure 6-10. *The consequences of mitotic crossing over.*

locus distal to the centromere and the point of the exchange. Thus, any recessives in this region that were masked by dominant alleles in the parent cells become homozygous and may be detected by the appearance of the recessive phenotype in daughter cells.

Mitotic crossing over has been used extensively by G. Pontecorvo and his collaborators for mapping genes in *Aspergillus nidulans*, a fungus with a life cycle very similar to that of Neurospora. Aspergillus differs from Neurospora, however, in that true mating types do not exist; hyphae from any two strains may fuse, form a heterokaryon, and have nuclear fusions followed by regular meiotic cycles. The vegetative spores of Aspergillus are mononucleate, permitting the selective isolation from the heterokaryon of rare diploids. For example, haploid spores produced by an heterokaryon between an adenineless and a leucineless strain will not grow on minimal medium, since they contain either adenineless or leucineless nuclei. However, heterozygous diploid spores will grow on minimal medium, providing that the normal alleles of the biochemical mutant markers are dominant. In these diploid strains, mitotic recombination can be demonstrated. As may be seen from Figure 6-10, the order of linked markers relative to the centromere and to each other is easily determined. In the example shown, two markers are linked in the order centromere-*A*-*B*. An exchange between the centromere and *A* gives homozygosis for both *A* and *B*. An exchange between *A* and *B* permits homozygosis for *B* but not *A*. Homozygosis for *A* but not *B* would require a rare double crossover. (In which regions and involving which strands?)

Since mitotic recombination is rare in the organisms in which it is known to occur, a "signal" marker is usually used for detection of mitotic crossovers. As an example, in Aspergillus a particular mutant gene gives resistance to acriflavine, a compound that kills the wild-type strain. The mutant gene is a semidominant with respect to its wild-type allele. The heterozygote *Acr/acr*[+] is partially resistant to acriflavine; any diploid individuals that become homozygous for *Acr* also become more resistant to the compound. Thus introduction of the drug into the growth medium will select and isolate clones that have become homozygous for *Acr* as the result of mitotic recombination. Such isolates can then be analyzed further for the segregation of other markers. Complete analysis of the selected clones can be accomplished either by making crosses with another diploid (giving a tetraploid meiosis from which some haploid segregants emerge), or by selecting for immediate haploid derivatives as discussed in the next section.

In yeasts, diploids occur as a regular feature of the life cycle. After fusion of cells of opposite mating type, cells with diploid nuclei are formed. These diploid cells may multiply extensively before undergoing meiosis, and mitotic recombination can occur. Analysis of the mitotic recombinants is easier in yeast than in Aspergillus since the yeast diploids can regularly be induced to undergo meiosis without further nuclear fusion, provided that the cells are

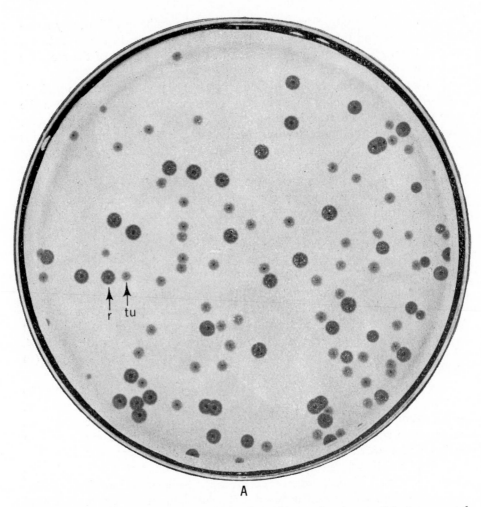

A

Figure 6-11. *A two-factor cross in phage T4. A: a plating of a mixture of the two parental types, r and tu. B (facing page): a plating of the progeny of the cross. The recombinant types rtu and + are present as well as the parental types.*

heterozygous for the mating-type alleles. Since the products of the meiosis are haploid, marker genes can be identified directly and the consequences of earlier mitotic recombination are readily determined.

Haploidization. The diploid strains of Aspergillus occasionally produce haploid segregants. These segregants are apparently derived through progressive chromosome loss. If we analyze the *Acr* derivatives of an *Acr/acr+* diploid, as mentioned earlier, some of the isolates will be diploid, but homozygous for

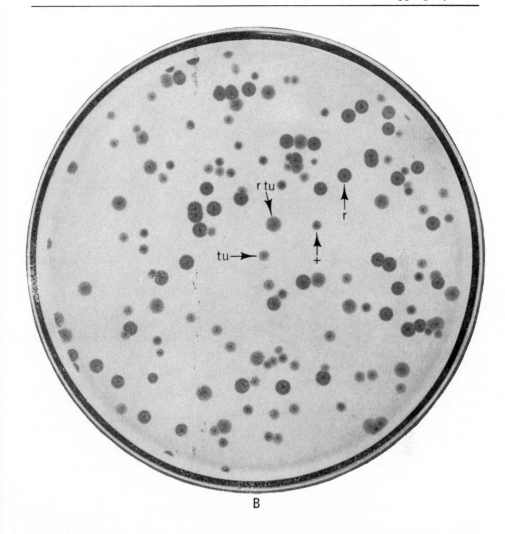

B

Acr as a consequence of mitotic recombination. In addition, however, some of the fully acriflavine-resistant derivatives will be haploids carrying the *Acr* chromosome but not its homologue. During haploidization the chromosomes assort independently of one another. From one diploid a variety of different haploids are obtained due to this independent segregation of the various chromosomes. This chromosomal segregation process is independent of mitotic crossing over; consequently, genes on the same chromosome generally do not segregate from one another during the formation of the haploids. You can see that haploidization provides a means of assigning mutations to chromosomes uncomplicated by the often ambiguous consequences of crossing over.

This general system of diploid formation, mitotic crossing over, and haploidization has been termed, by Pontecorvo, the *parasexual cycle*. It accomplishe⌐ the major feats of the sexual cycle—fusion, recombination, and segregation— without the intervention of a meiotic phase. The system has been used to study the genetics of *Penicillium notatum* and other fungi for which a sexual cycle with meiosis is not known.

Recombination in Bacteriophage. Genetic recombination has been found to occur in a number of different bacteriophages. A "cross" consists of infecting bacteria with a sufficiently large number of phage of two different genotypes (about five of each type per bacterium) that virtually all of the bacteria are infected with phages of both genotypes. Among progeny phage issuing from the infected bacteria, recombinant types are found (see Fig. 6-11). As with crosses of higher organisms the frequency of recombinants depends upon physical relationships between the particular markers employed. Reciprocal recombinants are found in equal numbers, and the same frequency of recombinant types is found whether the mutant markers are in coupling or repulsion. Recombination values give rough additivity. Thus, mapping of genes in bacteriophage is possible in much the same way as for higher organisms.

Here, however, the correspondence of a phage cross with the meiosis of higher organisms ends. From many detailed experimental investigations the process of recombination in phage appears to be as follows. After infection of a bacterium, the phage gene complements, or genomes, (mainly or exclusively DNA) multiply, forming a "pool" of vegetative (or noninfectious) phage. These genomes interact in some way to form recombinants. Phage geneticists refer to this interaction as a "mating." Matings are essentially random with respect to partner (since there are no mating types in phages) and occur repeatedly during the growth cycle. About half-way through the growth cycle the maturation of phage genomes into complete, infectious, phage particles starts. During the maturation period, multiplication and mating of vegetative phage continue, but genomes are continually withdrawn from the pool and matured and these no longer participate in multiplication and mating. Finally the infected cell lyses, liberating the accumulated mature phage particles. Thus, one cycle of growth of phage in actuality consists of many cycles of multiplication and mating of the phage, and a progeny phage particle may have been involved in more than one mating event.

Since matings occur at random with respect to partner, in a cross $AB \times ab$ one-half of the matings would be between genomes of like genotype ($AB \times AB$ and $ab \times ab$). These matings cannot produce detectable recombinants. Thus even if the frequency of recombinants coming from a single mating were 50 percent, the maximum frequency observed in a hypothetical population of phage all of which had experienced one mating could only be 25 percent. Moreover, if the matings are truly random in time, all progeny will not have experienced the same number of matings; in fact, some progeny may not have mated at all.

For these reasons the maximum frequency of recombination observed in phage is generally less than the 50 percent found in higher organisms. In phage T4 it is 43 percent; in phage lambda it is 15 percent. These differences from one phage species to another occur in part because the frequency of matings differs from phage to phage, lambda being less "sexy" than T4.

What constitutes a "mating" in phage is not yet known. Reciprocal recombinants are formed with equal frequency, but it appears that they are not formed in the same event, as happens in meiotic crossing over. Furthermore, although phages appear to be genetically haploid organisms, in some phage species *partial diploids* are found. Such phage are detected among progeny of crosses as being heterozygous only for short regions of the genome. These partial diploids are thought to be intermediates in the formation of recombinants but their precise structure and their role in the recombination process are yet to be clarified.

Many features of phage recombination, then, bear little resemblance to the known details of meiosis in higher organisms and we may wonder whether the underlying mechanisms of crossing over in meiosis and of recombination during phage growth are basically similar. (For a more complete treatment of phage recombination, consult the references.)

Recombination Mechanisms in Bacteria. As indicated in Chapter 4, bacteria have various mechanisms for exchanging genetic material. Geneticists have been able to use a variety of means for studying recombination in bacteria. The consequence is that our understanding of recombination has been considerably enriched.

Conjugation. Cells of *E. coli* K12 show sexual differentiation. "Male" cells (designated F^+) attach to "female" cells (F^-) and a conjugation tube forms between them. Although recombinants are produced as a consequence of crosses between F^+ and F^- cells, their frequency is 10^{-5} or less. However, in 1952 two different highly fertile male strains were found, derived from cultures of "normal" (F^+) males. Crosses of the highly fertile males to F^- cells resulted in the production of a proportion of recombinants as high as 20 percent. Frequently since then, other such fertile males have been isolated from F^+ cultures. These fertile strains have been designated Hfr (for high frequency recombination). Hfr individuals appear to arise rarely but regularly in F^+ populations and the rare recombinants found in $F^+ \times F^-$ crosses probably derive from matings between F^- cells and the rare Hfr cells present in F^+ cultures rather than from $F^+ \times F^-$ matings. F^+ cells can be shown to conjugate with F^- cells very efficiently, but no recombinants are formed as a consequence of the conjugation.

E. L. Wollman and F. Jacob have shown that during conjugation between Hfr and F^- cells the linearly arranged genome of the male passes slowly through the conjugation tube, always with a particular end first, and enters the female cell. However, the conjugation tube is fragile and cells rarely remain conjugated until the entire male genome is transferred into the female. After the cells separate they reproduce, and among the descendants of the female cell, recombinants bearing marker genes from the male may appear. The female exconjugant may be likened to a zygote. However, the actual mechanism by which the recombinants are formed as segregants during multiplication of the female exconjugant is yet to be clarified. Furthermore, in most cases the female is diploid for only a portion of the genome since the whole male genome is rarely passed into the female. Partial diploidy of an F$^-$ exconjugant may persist for a number of generations, with recombinant segregants continuing to arise, probably by some type of process that involves crossing over and the elimination of genome fragments.

As shown by Wollman and Jacob, the genes in the male genome can be mapped on the basis of their time of entry into the female cell. At different times samples of conjugating pairs are taken and are agitated in a food blender for a short interval to separate them. The single cells are then permitted to multiply, and the fraction of females that produce progeny with a given marker derived from the male is determined. As you can see (Fig. 6-12) the transferred markers may be arranged in the order in which they became resistant to the separation of conjugating pairs, or, in other words, in order according to the time they entered the female. Markers can also be mapped by conventional linkage studies, but linkage relationships are distorted by the fact that markers

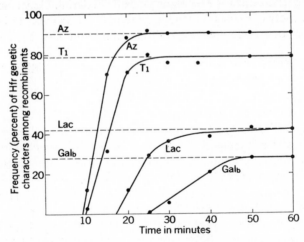

Figure 6-12. *The transfer of Hfr markers into the zygote. At the time indicated on the abscissa, mating pairs were disrupted in a blender and plated. The ordinate indicates the percent of recombinants inheriting a very early marker from the male (leucine) which also inherit various other markers from the male (Az, T1 resistance, etc.). The time scale differs from that of Figure 6-13. (After Jacob and Wollman,* Sexuality and the Genetics of Bacteria, *1961, Fig. 21.)*

Figure 6-13. *A map of* E. coli *K12. The scale of the map is in terms of time of entry. On this scale it takes about 89 minutes for the whole male chromosome to enter the female. The relative locations of only a few markers are shown. The inner circle shows the origins and direction of transfer of several different Hfr strains. The symbol leu represents leucine; azi, azide resistance; T1ᴿ, resistance to phage T1; arg, arginine; lac, lactose fermentation; gal, galactose fermentation; (λ), location of prophage lambda; try, tryptophan; his, histidine; cys, cystine; phe, phenylalanine; lys, lysine; str, streptomycin resistance; rha, rhamnose fermentation; thi, thiamine; met, methionine; pyr, pyrimidine.*

that enter late are present in fewer of the female exconjugants than are "early" markers, even when mating pairs are not forced to separate prematurely.

If recombination experiments are performed with a number of independently isolated Hfr strains, we obtain a most remarkable result. The order of entry into the F⁻ of the markers from the different Hfr's is different in each case! However, all the orders are consistent with one another if we assume that the genome of the F⁺ cell from which the Hfr's were derived is a ring and that in different Hfr's the point of entry of the male genome into the female and the direction of entry can differ. That is, one Hfr may transfer markers in the order and direction abcd . . . xyz, another in the order yzabc . . . uvwx, and a third cbaz . . . gfed. The Hfr has a ring chromosome that breaks at a specific point prior to transfer.

Another feature of this bizarre system is that all of the recombinants produced in crosses of Hfr × F⁻ are themselves F⁻ except for rare individuals that have received the most terminal markers from the male. These rare recombinants are generally Hfr's and transfer their genomes in the same sequence as did their father. It appears that the "mating-type" marker is always the last to enter the female, no matter what the order of entry of all the other markers. The inheritance of the fertility character is itself a strange story that we will treat separately in Chapter 11.

Conjugation systems like the one just described appear to be rare among bacteria. However, there exist two other "sexual" systems in bacteria that are more widespread and that are also useful for genetic studies. These are *transformation*—the transfer of free DNA from one bacterium to another—and *transduction*—the transfer of genes from one bacterium to another mediated by certain temperate bacteriophages.

Transformation. The ultimate in simplicity of genetic transfer is the phenomenon of transformation. Cells of genotype *a* are grown under special conditions and then exposed to DNA extracted from cells of genotype *A*. Among the progeny of the treated cells there may appear up to 10 percent transformed cells, that is, cells of genotype *A*. To date transformation systems are well characterized only in the bacterial species *Diplococcus pneumoneae, Haemophilus influenzae,* and *Bacillus subtilis.*

The mechanism of transformation is not as yet understood in detail. Cells must take up the DNA from the medium. The recipient cells must be grown under special conditions because cells must be in a *competent* state in order to be transformed. Competence apparently is associated with the ability of the cells to take up DNA molecules from the medium. An introduced DNA molecule must then pair with a homologous portion of the bacterial genome and interact in some manner to permit the *integration* of the gene in the DNA molecule into the genome of the recipient. Presumably, integration involves recombination between the bacterial genome and the DNA fragment. The maximum frequency of transformation obtainable is thus dependent on the size of the fraction of the bacterial population that is competent and also on the *efficiency of integration* (recombination probability).

When the DNA of a bacterium is isolated it is in the form of molecules whose molecular weight is about 1 percent of that of the total DNA of the cell nucleus. As a consequence most genes of the organism end up in different molecules and during transformation are taken up and integrated independently of one another.

However, linkage of genes in the same molecule can be demonstrated in two ways. If two genes *A* and *B* are linked, the frequency of cells transformed for *A* and *B* simultaneously will be greater than the product of the frequencies of the single transformations. This apparent linkage can be proven by comparing the frequency of double transformants obtained using DNA from a bacterium of genotype *AB* with the frequency found after using a mixture of DNA obtained from cells of genotype *Ab* and *aB* (in both cases the recipient being *ab*). If the genes are truly linked, the frequency of double transformants (*AB*) will

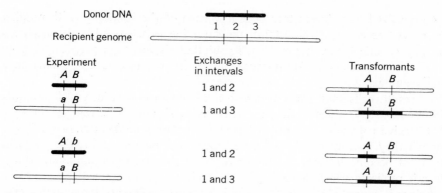

Figure 6-14. *Linkage in transformation. If* AB *DNA is used to transform* aB *cells (upper experiment),* AB *transformants arise by exchanges in intervals 1 and 2, and in intervals 1 and 3 as well. If* Ab *DNA is used to transform* aB *cells (lower experiment),* AB *transformants arise as a consequence of exchanges in intervals 1 and 2 only.*

be greater in the case in which the DNA was obtained from the *AB* strain. Another method is used for mapping closely linked mutations. If strain *Ab* is treated with DNA from strain *aB*, *AB* transformants can arise by recombination between the genome and the fragment. If *A* and *B* are closely linked, the frequency of *AB* transformants should be less than the frequency obtained if we treat *Ab* or *aB* cells with DNA obtained from *AB* cells. The explanation for this result is shown in Figure 6-14. In fact, this principle can be used to construct linkage maps.

Transduction. Transduction and transformation have in common the feature that genetic material is transferred from one bacterial cell to another without the donor and recipient cells ever coming into contact. When a virus is the agent in this transaction, the process is called transduction. As a bacterial virus multiplies within its host cell, fragments of the bacterium's genetic material are incorporated in the newly formed virus particles. After the host cell disintegrates and virus particles are released, these virus may infect other bacterial host cells. Upon infection, the bacterial genetic material that has been carried along within the virus enters the new host bacterium and may become incorporated into the genome of the recipient. It is obvious that only temperate, not virulent, bacteriophages may act as transduction vectors since the host cell must survive the infection in order to be transduced.

Transduction, a widespread phenomenon, has been most extensively studied in the bacteria *E. coli* with the phage vectors P1 and lambda and *Salmonella typhimurium* with its vector P22. Two types of transducing phages exist, those that transduce any bacterial gene (generalized transducers) and those that transduce genes only from a restricted region of the bacterial genome (specialized transducers).

Generalized transducers transduce any gene of the bacterium with roughly the same frequency; about 10^{-5} of the surviving infected cells are transduced for any genetic locus. It has been found that those phage particles that do transduce are unusual in that they contain few if any phage genes. In some cases they may even consist of phage overcoats with a bit of bacterial DNA inside! As a result, cells that are transduced can become lysogenic only if they are simultaneously infected with a normal phage. About 90 percent of the cells that receive a bacterial fragment become *abortive transductants*. That is, although they have received a genome fragment from the donor, the fragment does not become integrated into the recipient genome by recombination; nor does it replicate when the cell genome replicates. The fragment can function but since it cannot replicate it is passed at random to one of the daughter cells at each cell division. An amusing demonstration of this phenomenon was made by B. Stocker using the transduction of a gene that permits the *Salmonella* bacterium to produce a flagellum and thus to swim. If the recipient cell is non-motile, upon receiving the gene for motility from the donor, the recipient manufactures a flagellum and starts swimming across the Petri dish. When the cell divides only one of the daughters receives the gene for motility and is able to develop a flagellum and to swim. The other daughter cell becomes nonmotile and forms a colony on the agar surface. Abortive transduction in this case results in the formation of a trail of colonies, which indicates the wandering path of the cell with the fragment and each place where this cell divided. It is worth noting that an abortively transduced cell is diploid for those genes contained in the fragment and thus in some cases the functional properties of abortive transductants can be used to test for recessiveness and allelism.

About 10 percent of the cells that receive the fragment from the phage achieve stable integration of genes from the fragment into the genome of the recipient. As in transformation only a small portion of the whole genome is contained in any one fragment, so that most genes are unlinked in transduction. However, linkage is detectable. The methods of analysis of transduction are similar to those used in transformation, and the underlying mechanisms of integration are probably similar.

The Mechanism of Crossing Over

To this point the chapter has dealt largely with the consequences of crossing over (or the analogous phenomenon of recombination of linked genes in phage and bacteria) and its use in mapping genes. In spite of the theoretical and practical importance of the phenomenon, little is known about the actual mechanism of crossing over. The following facts are sufficiently well established that they must be taken into account in the construction of any models of crossing over.

1. Crossing over appears to be exact in that it does not often result in deletion or addition of chromosome parts. As we will see in Chapter 9, this precision may be maintained even down to the level of the nucleotide.
2. Crossing over occurs at the four-strand stage of meiosis or mitosis.
3. Four-strand double-crossover tetrads occur, therefore all of the four chromatids may be involved in crossing over; only two, however, are involved in any one exchange.
4. Crossover events are reciprocal, but this fact requires qualification of the sort included in the next section.

It has generally been accepted, although without direct evidence, that crossing over occurs at pachytene, when the chromosomes first pair and are observed to be structurally double. This is reasonable, since crossing over seems to require pairing and since the chromosomes must be double for crossing over to occur at the four-strand stage. By diplotene the homologues separate, but are held together by chiasmata (see Fig. 3-8G). Cytogenetic evidence suggests that chiasmata arise as a consequence of crossing over, indicating that crossing over occurs prior to diplotene. Not all chiasmata are the result of crossing over, however, for they occur in male Drosophila where crossing over is absent. (See Problems 6-10 to 6-14).

Simple theories of crossing over fall into two classes, breakage theories and copy-choice theories. Breakage theories, which were the earliest proposed, explain crossing over as the simultaneous breakage of two chromatids and the joining of the broken ends to form recombinant chromatids. No theory of this type so far proposed suggests a satisfactory mechanism by which two chromatids would break at precisely equivalent points (No. 1 in our list of facts above). Copy-choice theories explain crossing over as due to a "switching" from one parental chromatid to the other by the newly forming strands as they are copied from the parental strands during their synthesis. Theories of this type do not adequately account for the fact that crossing over can involve all four of the

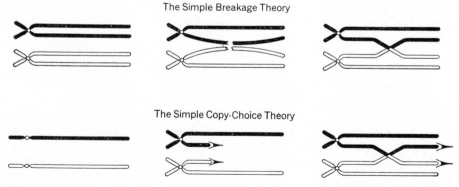

The Simple Breakage Theory

The Simple Copy-Choice Theory

Figure 6-15. *The simple theories of crossing over.*

chromatids, not just the two new chromatids (No. 3). Furthermore, the synthesis of new chromosomal material—at least the DNA—occurs during the interphase preceding meiosis; on such a copy-choice model one must assume that crossing over occurs in interphase rather than in pachytene, the only known time at which most chromosomes regularly pair. You can see that neither type of simple theory is satisfactory.

The preceding discussion did not take into account the fact that phenomena analogous to crossing over occur in bacteria and in phages, which are without visibly organized chromosomes although their hereditary material is linearly organized. A mechanism for these phenomena must be sought as well.

Recombination of Closely Linked Markers. Further insight into recombination phenomena has been obtained as a result of studies designed to explore the "fine structure" of genes (discussed in Chap. 9). With all organisms that have been carefully studied, the rules of crossing over, as previously described, appear to break down when recombination events between very closely linked markers are examined. The following example from the work of Mary Mitchell illustrates a characteristic kind of result and, incidentally, shows the power of tetrad analysis.

Our story concerns the behavior in crosses of two distinguishable mutants of a gene controlling a step in the synthesis of the vitamin pyridoxine in Neurospora. These mutants, which at first seem to be allelic, are designated *pdx* and *pdxp*. The pyridoxine locus is bounded closely on either side by two markers *pyr* (pyrimidine requirement) and *co* (variant with colonial morphology). Mitchell analyzed many hundreds of tetrads from the cross $pyr^+ pdxp co \times pyr pdx co^+$. As expected, the vast majority of these tetrads showed segregation of *pyr* from pyr^+, *pdx* from *pdxp*, and *co* from co^+ in the characteristic 2:2 ratio. In addition, however, she found a few tetrads containing spores of genotype pdx^+. Three such tetrads are shown below.

SPORE PAIR	ASCUS 1	ASCUS 2	ASCUS 3
1	$pyr^+ pdxp co$	$pyr^+ pdxp co$	$pyr^+ + co^+$
2	$pyr^+ + co$	$pyr^+ + co$	$pyr^+ pdxp co$
3	$pyr pdxp co^+$	$pyr pdx co^+$	$pyr pdx co^+$
4	$pyr pdx co^+$	$pyr pdx co^+$	$pyr + co$

(Data from M. B. Mitchell, *Proc. Natl. Acad. Sci.* **41**:215–220, 1955.)

How can we explain the occurrence of these aberrant tetrads? One obvious explanation is that they arose by mutation of *pdx* or *pdxp* to pdx^+. However, such pdx^+ spores were never found in the crosses $pdx \times pdx$ or $pdxp \times pdxp$; they were confined to progeny of the crosses $pdx \times pdxp$. This would seem to rule out mutation as the origin of the pdx^+ spores.

We might then suggest that the *pdx*+ spores arise as a result of crossing over. If this is the case, we must postulate that *pdx* and *pdxp*, while functionally allelic, are not allelic in the structural sense and that in fact we are dealing with mutations at two sites separable by recombination within the *pdx* gene. Thus, we must designate the cross as, *pyr*+ *pdxp pdx*+ *co* × *pyr pdxp*+ *pdx co*+, admitting that we do not yet know the order of *pdx* and *pdxp*. The *pdx*+ spores can then be assumed to arise from crossing-over between the *pdx* and *pdxp* sites.

However, we run into further difficulties. First, the reciprocal product of the crossover, the *pdx pdxp* class, is missing from the tetrads. Furthermore, if we diagram the tetrads on the basis of *pdx* and *pdxp* being at different sites, we immediately see that for one or other of the two allelic pairs (*pdx* and its alternative and *pdxp* and its alternative) the tetrads do not show the 2:2 segregation customarily found for members of allelic pairs, but rather a 3:1 segregation. This phenomenon was originally termed *gene conversion*, because the apparent violation of the 2:2 segregation rule suggested that one gene in the tetrad was "converted" into its allelic form.

In some of the aberrant tetrads, recombination of the markers on either side of the *pdx* locus occurs. Additional information about this association of recombination and gene conversion can be gained from an analysis of randomly collected spores deriving from the cross. (A large number of the rare *pdx*+ spores can easily be selectively germinated on medium not containing pyridoxine.) Listed below are some data Mitchell obtained with regard to the two "outside" markers associated with *pdx*+ spores selectively germinated out of spore progeny.

GENOTYPE		NUMBER
pyr	*co*+	13
pyr	*co*	7
pyr+	*co*+	7
pyr+	*co*	5

We see that with regard to the alleles present at the two bracketing loci all four classes of spores are found. If the origin of the *pdx*+ spores were a simple conversion of one *pdx* allele by the other we would expect the great majority of *pdx*+ spores to have the outside markers in parental combinations because of close linkage between *pyr* and *co*. On the other hand, if crossing over were the mechanism by which the *pdx*+ spores arise, we would expect the *pdx*+ spores to be predominantly of one recombinant type or the other depending upon the order of *pdx* and *pdxp* with respect to *pyr* and *co*. Neither explanation seems very satisfactory because all four possible types of segregants are found in high frequency.

Aberrant tetrads are not restricted to crosses of *pdx* × *pdxp* in Neurospora. In most organisms, unexpected phenomena associated with recombination are the rule rather than the exception when recombinant products are recovered from crosses involving mutants that are apparently allelic.

How are we to explain this peculiar phenomenon of so-called gene conversion? From many studies it is clear that one can construct linear maps of the mutant sites within genes on the basis of the frequency of wild types arising in crosses between a number of different allelic mutants. Thus, the events that give rise to the wild types appear to be recombination events. Furthermore, if suitable crosses are performed, the double-mutant recombinants (reciprocal product) can be shown to occur, although usually not in the same event that gives the wild type. These facts indicate that the various mutations do in fact occupy different sites in the gene and can be recombined by the "conversion" mechanism.

Two types of interpretations of conversion have been proposed. One kind of theory maintains that conversion is a phenomenon separate and distinct from crossing over. As a specific model we might imagine that chromosomes consist of alternating regions of DNA "gene material" and protein "linkers." In this model conventional crossing over occurs by breakage and reunion in the protein regions while gene conversion occurs only within the DNA regions— perhaps by some kind of copy-choice mechanism. An alternative type of theory maintains that this phenomenon of exceptional recombination in small regions is the same phenomenon as crossing over and appears unusual only because we are examining crossing over in such minute detail. For example, R. H. Pritchard suggests that crossing over requires that chromosomes be especially closely paired for very short regions, regions of "effective contact." Within these regions crossing over may occur frequently and need not be reciprocal. Such regions of effective pairing might even occur while the chromosomes are duplicating during interphase and not be detected.

The mechanism of crossing over remains nearly as much a mystery as it was when discovered in 1915. A more detailed knowledge of the structure of chromosomes and of the mechanism of their duplication seems a necessary preliminary to clearing up the mystery of crossing over, one of the fundamental unsolved problems of genetics.

Keys to the Significance of This Chapter

Genes situated in the same chromosome tend to be inherited in blocks rather than to assort independently. This is the phenomenon of *linkage*.

Linkages are broken when homologous chromatids exchange corresponding parts at meiosis, a process called *crossing over*. It is through crossing over that *recombinations* among linked genes occur.

Frequencies of crossing over between particular gene loci can be used as the basis for mapping chromosomes, with these frequencies corresponding to units of map distance between the genes.

The analysis of frequencies of crossing over proves that the different genes within a chromosome are situated in a definite serial order and that their loci are fixed. Thus chromosome mapping reveals a high degree of organization in the germ plasm, which is of extreme importance in our understanding of genetic systems.

Recombination of linked genes can also occur during mitosis and in organisms without either meiosis or mitosis such as bacteria and viruses, and can serve as a basis for gene mapping in these organisms.

REFERENCES

Barratt, R. W., Newmeyer, D., Perkins, D. D., and Garnjobst, L., "Map Construction in *Neurospora crassa*." *Advances in Genetics*, **6**:1–93, 1954. (Describes the uses of tetrad analysis and summarizes the results obtained in Neurospora.)

Bridges, C. B., and Brehme, K. S., "The Mutants of *Drosophila melanogaster*." *Carnegie Inst. Wash. Pub.* 552, 1944. (An extensive compilation of the descriptions of Drosophila mutants. Chromosome maps are found on pp. 238–252.)

Burdette, W. J., editor, *Methodology in Basic Genetics*. San Francisco: Holden-Day, 1963. (See especially chapters on bacterial recombination by W. Hayes, on tetrad analysis by S. Emerson, and on mitotic recombination by R. H. Pritchard.)

Cooper, K. W., "The Cytogenetics of Meiosis in Drosophila. Mitotic and Meiotic Autosomal Chiasmata without Crossing Over in the Male." *J. Morphol.*, **84**:81–122, 1949. (An important paper on the subject of chiasmata and crossing over.)

Creighton, H. B., and McClintock, B., "A Correlation of Cytological and Genetical Crossing Over in *Zea mays*." *Proc. Nat. Acad. Sci.*, **17**:492–497, 1931. (A landmark in experimental genetics.)

Emerson, R. A., Beadle, G. W., and Fraser, A. C., "A Summary of Linkage Studies in Maize." *Cornell Univ. Agr. Exp. Sta. Mem. No. 180*, 1935. (Gives a brief description of many of the known genes in corn, summarizes linkage data to 1935, and presents a chromosome map.)

Fincham, J. R. S., and Day, P. R., *Fungal Genetics*. Philadelphia: F. A. Davis Co., 1963. (See Chap. 4 on chromosome mapping and Chap. 7 on genetic exchange. Centromere mapping from unordered tetrads is described in Chap. 4.)

Hotchkiss, R. D., and Weiss, E., "Transformed Bacteria." *Scientific American*, November 1956. Available as Offprint 18 from W. H. Freeman and Co., San Francisco. (A description of transformation.)

Hutt, F. B., and Lamoreux, W. F., "Genetics of the Fowl: II—A Linkage Map for Six Chromosomes." *J. Heredity*, **31**:231–235, 1940. (Gives an illustrated chromosome map showing approximate locations of 21 genes in six linkage groups of the fowl.)

Immer, F. R., "Formulae and Tables for Calculating Linkage Intensities." *Genetics*, **15**:81–98, 1930. (The calculation of linkage intensities from F_2 data.)

Jacob, F., and Wollman, E. L., *The Sexuality and the Genetics of Bacteria*. New York: Academic Press, 1961. (This is a lucid account of the studies of the conjugation system in *E. coli*.)

MacArthur, J. W., "Linkage Groups in the Tomato." *J. Genetics*, **29**:123–133, 1934. (Presents a tentative chromosome map for the tomato, giving approximate locations for twenty genes in ten of the twelve pairs of chromosomes.)

Mather, K., *The Measurement of Linkage in Heredity*. New York: Chem. Pub. Co. of N.Y., 1938. (A general treatment of problems of measuring linkage.)

Morgan, T. H., *The Theory of the Gene*. New Haven: Yale University Press, 1926. (A classic by a geneticist who received a Nobel prize for his work. The first chapter is pertinent here. Note the "theory of the gene" formulated on p. 25.)

Owen, A. R. G., "The Theory of Genetical Recombination." *Advances in Genetics*, **3**:117–157, 1950. (Suitable for the student who is mathematically inclined.)

Pontecorvo, G., "The Genetics of *Aspergillus nidulans*." *Advances in Genetics*, **5**:142–238, 1953. (A treatise on the analysis of the genetics of Aspergillus.)

Pontecorvo, G., and Käfer, E., "Genetic Analysis Based on Mitotic Recombination." *Advances in Genetics*, **10**:71–104, 1958. (Details the uses of mitotic crossing over and haploidization in the analysis of genome structure.)

Stent, G., *Molecular Biology of Bacterial Viruses*. San Francisco: W. H. Freeman and Co., 1963. (Contains several excellent chapters on the genetics of bacteriophage.)

Stern, C., "Somatic Crossing Over and Segregation in *Drosophila melanogaster*." *Genetics*, **21**:625–730, 1936. (A very extensive analysis of mosaic formation in Drosophila resulting from somatic crossing over.)

Whittinghill, M., "Consequences of Crossing Over in Oögonial Cells." *Genetics*, **35**:38–43, 1950 (The possible effects of premeiotic crossing over in gonial cells upon recombination phenotypes. Discussion accompanied by a useful diagram.)

Wollman, E. L., and Jacob, F., "Sexuality in Bacteria." *Scientific American*, July 1956. Available as Offprint 50 from W. H. Freeman and Co., San Francisco. (A description of the early work on the conjugation system in *E. coli*.)

Zinder, N. D., "Transduction in Bacteria." *Scientific American*, November 1958. Available as Offprint 106 from W. H. Freeman and Co., San Francisco. (A description of transduction.)

QUESTIONS AND PROBLEMS

6-1. What do the following terms signify?

chromosome map	interference
coincidence	linkage
coupling	locus
crossing over	map distance
crossovers	parental-type gamete
double crossing over	recombinations
double crossover chromatid	repulsion
independent assortment	three-point testcross

6-2. In your study of genetics thus far, what evidence has been presented that genes are situated in chromosomes?

6-3. How do undetected double crossovers affect estimates of map distance?

6-4. What are the cytological and genetic conditions necessary for a straightforward demonstration that the recombination of linked genes results from a physical interchange of appropriate parts of homologous chromosomes?

6-5. In tomatoes, round fruit shape (O) is dominant over elongate (o), and smooth fruit skin (P) is dominant over peach (p). Testcrosses of F_1 individuals heterozygous for these pairs of alleles gave the following results, reported by MacArthur:

smooth, round	smooth, long	peach, round	peach, long
12	123	133	12

In the F_1, were the two pairs of alleles linked in the coupling or the repulsion phase? Calculate the percentage of recombination.

6-6. In Neurospora, the frequency of second-division segregations of a gene is a function of the distance between the gene and the centromere of its chromosome. Explain this statement.

6-7. In order to calculate *map distance* between gene and centromere in Neurospora it is necessary to divide the percentage of second-division segregations by two, if the map units are to be comparable to those utilized, for example, in chromosome maps of Drosophila. Why is the division by two necessary?

6-8. A summary of data revealing segregation of the mating type alleles, *A* and *a*, in Neurospora has been made by Barratt and Garnjobst. Out of 755 asci analyzed, 117 second-division segregations have been observed. Calculate the percentage of second-division segregations and the distance of the mating-type locus from the centromere of its chromosome.

6-9. Two mutant genes (*ag* and *thi*) in Neurospora are known to interfere, respectively, with the syntheses of the amino acid arginine and the vitamin thiamine. After a cross in which these genes were segregating, the following spore arrangements were found in the frequencies noted. (Only one member of each pair of spores is indicated.)

PAIR 1	PAIR 2	PAIR 3	PAIR 4	
ag thi	ag thi	+ +	+ +	42
+ thi	+ thi	ag +	ag +	40
+ +	+ +	ag thi	ag thi	39
ag +	ag +	+ thi	+ thi	42

How are the *ag* and *thi* genes located in the chromosomes with respect to their centromeres and with respect to each other?

6-10. Bridges and Morgan reported a series of crosses in Drosophila beginning with a mating between a dachs ♂ and a black ♀. In F_2 there were obtained 186 wild type, 71 dachs, 93 black, and 0 dachs, black. Do these results indicate that the loci of genes *d* and *b* are very closely linked? Explain your answer. Remember that there is no crossing over in male Drosophila.

6-11. Curled wings (*cu*) and spineless bristles (*ss*) are autosomal recessive characters in Drosophila. The genes giving rise to these characters are both in chromosome 3. Starting with a wild-type ♀ and a curled, spineless ♂, prepare a diagram showing parents and progenies through an F_2 generation.

6-12. Suppose that after the crosses indicated in the preceding question, the following were obtained in F_2:

wild	292
curled	9
curled, spineless	92
spineless	7
	400

Remembering again the special circumstance of no crossing over in male Drosophila, is there some reasonably straightforward way to arrive at an estimate of the map distance between *cu* and *ss*? Explain your method and calculate the distance.

6-13. Assume in Drosophila three pairs of alleles, $+/x$, $+/y$, and $+/z$. As shown by the symbols, each mutant gene is recessive to its wild-type allele. A cross between ♀♀ heterozygous at these three loci and wild-type ♂♂ gives the following results:

♀♀:	$+$ $+$ $+$	1010	
♂♂:	$+$ $+$ $+$	30	
	$+$ $+$ z	32	
	$+$ y $+$	441	
	$+$ y z	1	
	x $+$ $+$	0	
	x $+$ z	430	
	x y $+$	27	
	x y z	39	

How were members of the allelic pairs distributed in the members of the appropriate chromosome pair of the heterozygous ♀♀? What is the sequence of these linked genes in their chromosome? Calculate the map distances between the genes and the coefficient of coincidence. In what chromosome of Drosophila are these genes carried?

6-14. In Drosophila, assume three pairs of alleles, $+/n$, $+/o$, and $+/p$. Genes n, o, and p are all recessives and sex-linked. They occur in the order n-o-p in the X-chromosome, with n being 12 map units from o, and o being 10 units from p. The coefficient of coincidence for this region of the X-chromosome is 0.5. From a cross between ♀♀ of the genotype $\dfrac{+\ \ +\ \ p}{n\ \ o\ \ +}$ and wild-type ♂♂, predict the kinds and frequencies of phenotypes that would be expected to occur in a progeny of 2000 individuals.

6-15. In corn, the following allelic pairs have been identified in chromosome 3:

$+/b$ = plant-color booster *vs* nonbooster
$+/lg$ = liguled *vs* liguleless
$+/v$ = green plant *vs* virescent

A testcross involving triple recessives and F_1 plants heterozygous for the three gene pairs gave in the progeny the following phenotypes:

$+$	v	lg	305
b	$+$	lg	128
b	v	lg	18
$+$	$+$	lg	74
b	v	$+$	66
$+$	$+$	$+$	22
$+$	v	$+$	112
b	$+$	$+$	275
			1000

Give the gene sequence, the map distances between genes, and the coefficient of coincidence.

6-16. In corn, the genes *an* (anther ear), *br* (brachytic), and *f* (fine stripe) are all in chromosome 1. From the data of R. A. Emerson, summarized below, determine the sequence of genes in their chromosome, the map distances, and the genotypes of the homozygous parents used to make the heterozygote.

TESTCROSS PROGENY

+	+	+	88
+	+	*f*	21
+	*br*	+	2
+	*br*	*f*	339
an	+	+	355
an	+	*f*	2
an	*br*	+	17
an	*br*	*f*	55
			879

6-17. Curved wing and brown eye are recessive characters in Drosophila. The genes for these characters are located about 30 units apart in chromosome 2. Suppose that you wish to obtain a double-mutant stock that is curved, brown. You have available for breeding purposes a single-mutant stock having brown eyes, and a double-mutant stock homozygous for curved and also true-breeding for *yellow body*, a sex-linked recessive.

Outline the steps you would take in obtaining as efficiently as possible a true-breeding stock for curved wings and brown eyes.

For each step show the genotypes of the parents to be used in the cross, the genotypes of progeny that will be useful in succeeding steps, and the probability of obtaining such progeny.

6-18. In Figure 3-10 (p. 83), it is suggested that, of the four sperm produced by a primary spermatocyte in a trihybrid, two are alike and of one kind, and the other two are alike and of another kind. Show how a single exchange between the locus of *a* and the centromere, for example, would result in the formation of four kinds of sperm from a single primary spermatocyte.

6-19. What do the following terms signify?

abortive transduction	haploidization
four-strand double exchange	Hfr
gene conversion	transformation
half-tetrad	

6-20. Suppose that in an organism with unordered tetrads the following tetrads were recovered from the cross $xyz \times +++$

$xyz/xyz/+++/+++$	94
$xy+/xy+/++z/++z$	86
$xyz/x+z/+y+/+++$	8
$xy+/x++/+yz/++z$	12

Compute the relative frequencies of PD, TT, and NPD tetrads for each pair of markers (xy, xz, and yz). From the ratio PD/NPD what can you conclude about the linkage of these markers? Remembering that tetratypes for unlinked markers arise from crossing over between the centromere and one of the markers, what can you conclude about the linkage of x, y, and z to their respective centromeres?

6-21. From the diploid Acr/acr^+, w^+/w, pro^+/pro in *Aspergillus nidulans*, Pontecorvo and Käfer isolated a number of segregants that were resistant to high concentrations of acriflavine. Of these, 31 turned out to be haploids: 19 *Acr w pro*, and 12 *Acr w pro^+*. A total of 363 diploids were also recovered. Of these 315 were *Acr/Acr, w/w, pro^+/pro*, while 48 were *Acr/Acr, w^+/w, pro^+/pro*. On the basis of these results assign the markers to linkage groups and indicate the relative order of linked genes to centromeres.

CHAPTER 7

Variations in Genome Structure

Among organisms with chromosomes, each species has a characteristic set of genes, or *genome*. In diploids a genome is found in each normal gamete. It consists of a full set of the different kinds of chromosomes, each of which is of a particular and typical size and morphology. Since a genome is a coded blueprint for the various genetic attributes of an organism, much effort by many geneticists has gone to the analysis of genomes.

On viewing chromosomes for the first time, the student frequently remains unimpressed. This is because chromosomes are usually seen as inert things—fixed, artificially colored, and laid out dead on a microscope slide. To think of chromosomes correctly, it is necessary to realize that living chromosomes are the quintessence of biological dynamism. As carriers of genes, chromosomes are the seat of control of the varied biochemical activity that is the basis of life. Within a given species, the numerical and structural constancy of chromosomes is impressive. Their precise movements at mitosis and meiosis and their astonishing powers of replication are other evidences of extreme vitality. In some ways, however, we get an even more striking picture of the liveliness of chromosomes if we consider them when they do not behave according to rule. Their irregularities have important genetic results that help us to appreciate better the meaning and consequences of their usual, predictable action.

Variations in Chromosome Structure

Chromosomes are structures with definite organization. They are not unchangeable, however. Through various means they may be broken and their normal structure disrupted. X-rays, atomic radiations, and various chemicals are among the agents that can cause breaks in chromosomes. Breaks also sometimes occur under natural conditions, where in most instances the reason for breakage is undefinable. As an introduction to the exceptional behavior of chromosomes, let us look at an instance that is experimentally analyzable. Note particularly how an initially single deviation from the normal can give rise to a whole series of unusual cytological events.

Breakage-Fusion-Bridge Cycles. In the gametophyte and endosperm of corn, ends of chromosomes that have recently been broken behave as though they were "sticky," as is shown by their tendency to adhere to one another. Extensive studies of broken chromosomes in corn have been made by Barbara McClintock. She finds that following reduplication of a broken chromosome the two sister chromatids may adhere at the point of previous breakage. This situation and some of its possible consequences are diagramed in Figure 7-1. You will note

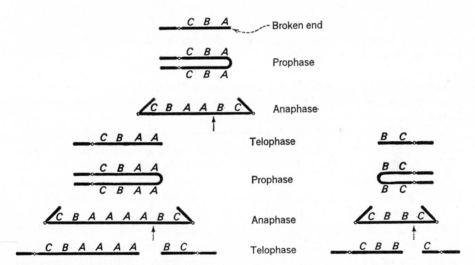

Figure 7-1. *A breakage-fusion-bridge cycle in corn. The chromosome at the top has been recently broken, near gene* A. *(From McClintock,* Genetics, **26**:*235, 1941.)*

that the fused sister chromatids are unable to separate readily. In effect, they constitute a single chromatid with two centromeres, a *dicentric* chromatid. As the centromeres move to opposite poles at anaphase, the dicentric chromatid stretches out, forming a chromatin bridge from one pole toward the other. This bridge eventually breaks, but the break does not always occur at the point of previous fusion. Therefore, chromosomes may be formed that show duplications or deficiencies if compared with an original type.

Thus, if the original chromosome is

$$\frac{C \qquad B \qquad A,}{\text{---o---}}$$

the type

$$\frac{C \qquad B \qquad A \qquad A}{\text{---o---}}$$

is a *duplication* type, since region A is twice represented. The type

$$\frac{C \qquad B}{\text{---o---}}$$

is a *deficiency*, since region A is absent.

When a chromosome bridge such as we have described breaks, perhaps as the result of tension caused by the movements of the centromeres of the dicentric chromatid, two new broken ends are provided. Each of these has the same qualities of adhesiveness that gave rise to the original fusion. This situation permits repetition, in cyclic series, of more events like those diagramed in Figure 7-1. Spontaneous production of chromosome aberrations through breakage-fusion-bridge cycles may occur in this manner for some time. But when a broken chromosome is introduced into the sporophytic generation such cycles cease, for in the zygote the broken ends heal.

The Visual Detection of Chromosomal Aberrations. The particular chromosomal complement of an individual or of a related group of individuals is its *karyotype* and is defined both by the number and morphology of the chromosomes. The karyotype of corn is exceptionally favorable cytological material. Each of the ten chromosomes making up a genome in maize is morphologically distinguishable. Especially at pachytene in meiosis, when the chromosomes exist as long paired threads, the distinguishing features of the chromosomes can be clearly observed. These features include knobs, chromomeres, and the location of the centromere. Each stands out as a fixed landmark for the cytologist. For particular chromosomes, the relative sizes and spacing of these landmarks provide constant patterns in linear arrangement. Absence or duplication of landmarks, or their unusual arrangement as compared with the typical topography of a chromosome, is visible evidence of aberration. Probably the chromosomes of all organisms possess longitudinal differentiation in the sense that we have described it for the chromosomes of corn. But not all chromosomes are well suited for detailed microscopic examination. In some organisms, different members of the genome are so similar in gross morphology that they are not readily distinguishable. Other organisms have notably small chromosomes. In still other instances, effective cytological work awaits the development of appropriate techniques. Study of the chromosomes of man was once thought to be a quite hopeless project. Today, karyotype analysis in man is a lively area of research.

Salivary-Gland Chromosomes. Chromosomes at meiosis in Drosophila happen to be small and difficult to work with. Despite this fact, Drosophila provides extraordinary advantages for cytological study. Nuclei of cells in the salivary glands of larvae of Drosophila and other Dipterans have remarkably large chromosomes with distinct longitudinal differentiation. Salivary-gland chromosomes may be up to 200 times the size of corresponding chromosomes at meiosis or in the nuclei of ordinary somatic cells.

A photograph of the salivary-gland chromosomes of Drosophila is reproduced in Figure 7-2. Note particularly the appearance of the *cross bands*. These

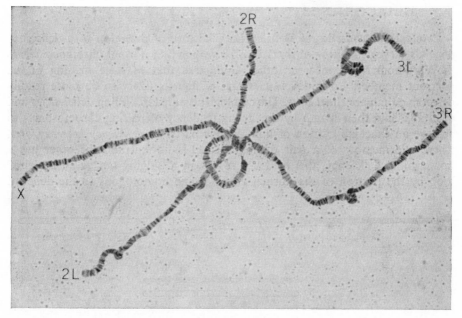

Figure 7-2. *The salivary-gland chromosomes of Drosophila show distinct longitudinal differentiation, so that particular chromosomes and particular parts of chromosomes are readily identified. (Courtesy of E. Vann.)*

are striking when stained with basic fuchsin or other appropriate dyes, but are also obvious in unstained living nuclei. The salivary-gland chromosomes are thought to be *polytene* units, composed of many reduplicated chromonemata in close longitudinal association. A widely accepted interpretation of the cross bands is that they represent a fusion, or at least a juxtaposition, of corresponding chromomeres of the multiplied chromonemata. In any case, for a given chromosome the various cross bands show constancy of relative size and spatial arrangement.

A further advantage of the salivary-gland chromosomes for cytological study is that they appear constantly to be in a prophaselike state. For practical purposes, the chromosomes are at all times in a condition appropriate for effective staining and detailed observation. Moreover, homologous salivary-gland chromosomes show the kind of close pairing you have learned to associate with the zygotene stage of meiosis. Differences between the homologous chromosomes, therefore, become relatively easy to detect. The size of salivary-gland chromosomes together with the fact that they show *somatic pairing* have facilitated the explanation of many puzzling phenomena characteristic of aberrant chromosomes at meiosis.

DEFICIENCIES

Cytological Properties of Deficiencies. The ease of detection of a deficiency depends largely on how obvious in a corresponding normal chromosome are the chromomeres, knobs, or other landmarks that may be missing in the aberrant member. Long deficiencies are, of course, likely to be more readily detectable than are short ones. The cytologist is aided in finding deficiencies and in establishing their length by the strong tendency of pairing chromosomes to achieve an exact apposition of homologous parts. In *deficiency heterozygotes*, where one member of a pair of homologues is normal and the other has a nonterminal deficiency, this tendency results in a bulging out as an unpaired loop by the portion of the normal chromosome homologous to the deficient

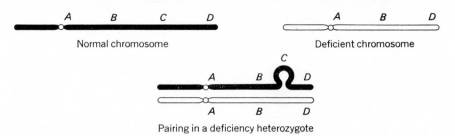

Pairing in a deficiency heterozygote

Figure 7-3. *In this deficiency heterozygote, when the homologues synapse region C of the normal chromosome bulges out in an unpaired loop because it has no corresponding part in the deficient chromosome.*

segment. This situation is diagramed in Figure 7-3. An unpaired loop is characteristic of deficiency heterozygotes, either in meiotic cells at synapsis or in somatic cells if chromosome pairing occurs.

Genetic Effects of Deficiencies. When a segment of a chromosome is absent, the genes normally situated in the missing segment also are absent. The severity of consequences for the organism depends on the physiological importance of the gene-controlled processes involved. Usually, if a chromosome is deficient in any considerable number of gene loci, lethality results. Deficiency homozygotes are even less likely to be viable than are deficiency heterozygotes. This is understandable since in the homozygotes important functions of absent genes are entirely lost to the organism.

The same principles hold in general for the viability of deficiency gametes and gametophytes of plants. For a particular chromosomal deficiency in a higher plant, the effects on viability of the male and female gametophytes may

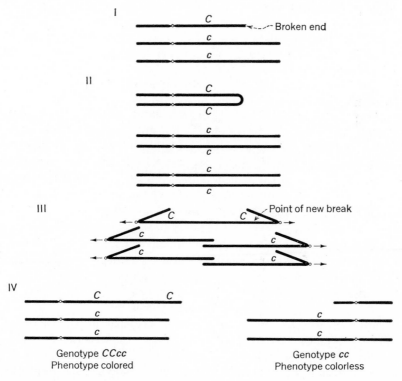

Figure 7-4. *Variegated aleurone color in corn may result from the operation of breakage-fusion-bridge cycles. Gene C is essential for the development of aleurone color; its allele, c, when homozygous, determines colorless aleurone. Aleurone nuclei are triploid (3n). I: A nucleus of genotype Ccc; the chromosome carrying the C allele has a recently broken end near the C locus. II: Prophase in this nucleus, showing fusion of the chromatids in which the C allele is found. III: Anaphase, showing a bridge formed by the dicentric chromatid; the other chromatids separate in regular fashion. IV: Sister telophase nuclei; the bridge has broken in such fashion that one nucleus has a duplication CC, and the other is deficient for the C locus. The former gives rise to a colored phenotype, the latter to colorless. Breakage-fusion-bridge cycles may continue throughout the development of the aleurone tissue.*

be quite different. In such instances, female gametophytes are the more likely to survive. In animals, gametes with sizable chromosomal deficiencies are frequently able to function.

Nonlethal deficiencies may give rise to unusual phenotypic effects. In an organism heterozygous for any allelic pair (*Aa*), loss of a chromosome segment carrying the dominant allele (*A*) permits the recessive allele (*a*) to express itself phenotypically. For example, corn aleurone of the genotype *Ccc* is colored, provided that other genes necessary for the development of pigmentation are present in addition to the dominant *C*. If a chromosome carrying *C* acquires a broken end, so that it goes through breakage-fusion-bridge cycles, the chromosomal segment in which *C* is situated may be lost from certain cells. The phenotypic result is a variegated aleurone—that is, an aleurone having patches of both colored and colorless tissue. Figure 7-4 shows how such variegation may come about. This kind of unexpected expression of a recessive trait, caused by the absence of a dominant allele, is called *pseudodominance*.

In corn, a series of very small, nonlethal deficiencies has been studied by McClintock. These deficiencies, when homozygous, produce phenotypic effects similar to those obtained after "gene mutations" that presumably did not involve chromosomal loss. Particular deficiencies give rise, for example, to the characters *white seedling*, *pale yellow seedling*, and *brown midrib*. In the case of brown midrib, the phenotypic effect produced by the appropriate deficiency resembles in every way the effect of a mutant gene, bm_1. The locus of this gene is known to be in the very chromosomal segment that was deficient in McClintock's material. Similarly, the mutant character *yellow body* is produced in Drosophila that are homozygous deficient for a particular, very small portion of the X-chromosome. The segment of the chromosome involved is that which has been established as the locus of gene *y*.

Utilization of Deficiencies in the Physical Mapping of Chromosomes. We have already emphasized in Chapter 6 that linkage-map distances between genes are not necessarily proportional to physical linear measurements. Special cytological techniques must be used to determine the physical location of a gene in a chromosome. Localization is accomplished by identifying a gene locus with relation to some visible landmark such as a chromomere or cross band. In this process of *cytological mapping*, deficiencies have been particularly useful.

In an originally heterozygous individual, recessive genes may show pseudodominance as a consequence of deficiencies involving the loci of their dominant alleles. Suitable chromosomal material permits the positions of deficiencies to be visually identified. The geneticist therefore has means not only for the genetic detection of loci involved in a deficiency but also for the cytological delimitation of deficiencies of appreciable size. If, for a particular deficiency, certain genes can be shown to be absent, and a particular chromosomal segment is also seen to be missing, a reasonable inference is that the loci of the absent genes are in the missing segment. In order to locate genes as accurately as possible, it is necessary in cytological mapping either to study short deficiencies, involving as few cross bands and gene loci as are feasible to detect, or to study a series of more extended deficiencies that overlap one another in varying degree.

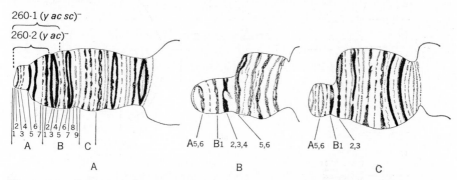

Figure 7-5. *Deficiencies can be utilized to localize genes in the salivary-gland chromosomes of Drosophila. The left-hand tip of the X-chromosome is shown in A. The pointer lines A 1, 2, 3, etc., and B 1, 2, 3, etc., serve to identify cross bands in the chromosome and are used as points of reference. The bracketed region 260-1 designates that portion of the X-chromosome absent in the deficiency heterozygote shown in drawing B. Similarly, 260-2 defines the limits of the deficiency in the heterozygote shown in C. The larger deficiency includes the loci for* yellow, achaete, *and* scute, *while the smaller includes only the loci for* yellow *and* achaete. *Therefore, the scute locus must be in the narrow band of chromosome material that differentiates the two deficiencies. (After Demerec and Hoover,* J. Heredity, **27:**206, 1936.)*

Figure 7-5 illustrates the use of deficiencies to localize genes in the salivary-gland chromosomes of Drosophila. Diagram A shows the nondeficient left-hand tip of the X-chromosome as it usually appears in the salivary glands. Because of exceedingly close pairing, what is actually a pair of homologous chromosomes appears to be single. Diagrams B and C show different deficiency heterozygotes, also with strong pairing. The complete complement of cross bands in the nondeficient member may be used as an index for identifying the missing bands in the deficient homologue. Since both deficiencies shown are terminal deficiencies, no unpaired loop is formed by the nondeficient chromosome. Instead, the tip of the deficient chromosome looks as if it were chipped off. Genetic tests show the deficiency chromosome represented in B to be lacking in the loci for *y* (*yellow*), *ac* (*achaete*), and *sc* (*scute*). The deficiency chromosome in diagram C lacks the loci for *y* and *ac*. Since the *scute* locus is absent in the first deficiency chromosome but is present in the second, it must be situated in the narrow chromosomal segment whose presence differentiates the second deficient chromosome from the first.

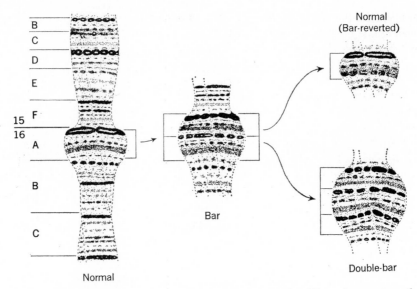

Figure 7-6. *Portions of X-chromosomes from salivary glands of* D. melanogaster, *showing the* bar *region. Notice the segment that is duplicated in bar and triplicated in double-bar. (From Bridges,* Science, **83**:*210, 1936.)*

DUPLICATIONS

Occasionally a nucleus is found to be aberrant in having material additional to that found in the normal chromosomal complement. Special terms, to be introduced later, are applied to instances of extra whole chromosomes or extra sets of chromosomes. Extra parts of chromosomes are called *duplications.* Various kinds of duplications have been observed. Some exist attached to the chromosome whose segments are "repeated"; some are attached to different chromosomes; others may exist as independent fragments.

One reason duplications are interesting is that they make it possible to investigate the effects of unusual dosages of the genes whose loci are involved. Ordinarily, a given allele may be represented either once or twice in a nucleus, although, as pointed out in Chapter 4, in the endosperm an allele may occur one, two, or three times within a nucleus. With duplications, an allele may be present three or more times; hence duplications may be utilized in studying the effects of various quantitative ratios between members of an allelic set.

You have already seen that breakage-fusion-bridge cycles produce duplications as well as deficiencies. When the *C* locus, for aleurone color in corn, is involved, various numbers of repeats of gene *C* may be accumulated. Variegated endosperm arising out of circumstances like those indicated in Figure 7-4 may show colored patches that vary from quite intense to extremely light coloration. Presumably the color intensity depends upon the number of dominant alleles present. This situation correlates well with the fact that in the normal aleurone one dose of *C* gives lighter color than two *C* alleles, whereas triple dosage (*CCC*) conditions the darkest coloring of all.

Position Effect on Bar Eye in Drosophila. Close cytological examination of the appropriate region in salivary-gland chromosomes of *D. melanogaster* discloses that the variant *bar eye* is associated with duplication of a chromosomal segment (Fig. 7-6). Homozygous stocks of bar flies do not breed quite true, since about 1 in 1,600 offspring is a reversion to wild type. In the same pro- portion, a new mutant type, called *double bar*, also arises out of homozygous bar strains. The effect of double bar is to reduce the eye even more than does bar. If a double-bar chromosome is examined cytologically the cross bands that are duplicated in bar are found now to exist in triplicate. Figure 7-7 indicates how homozygous bar females might produce normal and double-bar gametes if synapsis were oblique and crossover products unequal.

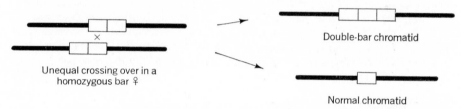

Double-bar chromatid

Unequal crossing over in a homozygous bar ♀

Normal chromatid

Figure 7-7. *The normal and double-bar individuals that appear in bar stocks of Drosophila can be accounted for on the basis of oblique synapsis and unequal crossing over. For simplification, only the two strands involved in the crossing over process are shown.*

The existence of normal, bar, and double-bar chromosomes provides an opportunity for study of the different effects of the bar segments in various numerical and positional combinations. Keeping in mind that a particular chromosomal segment is represented once in a normal chromosome, twice in bar, and three times in double bar, you can see that both in homozygous bar females and in females heterozygous for normal and double-bar there is a total of four of these segments per chromosome complement. In terms of quantity of basic hereditary material these flies are equivalent for the loci we have been con- sidering. However, the two situations prove not to be equivalent physiologically. As shown in Figure 7-8, eyes of the heterozygous female are smaller than those of the comparable homozygote. It is clear that segments lying side by side in a single chromosome somehow reinforce each other's action to a greater degree than do the same segments when situated in separate chromosomes. The analysis of these *position effects*, made by Sturtevant in 1925, is significant as early evidence that rearrangements of pieces of chromosomes may affect the func- tional expression of genetic material.

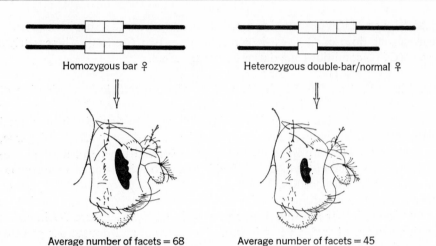

Figure 7-8. *In homozygous bar females, a particular segment of the X-chromosome is represented four times, twice in each homologue. In females heterozygous for double-bar and normal this same segment is also represented four times, once in the normal X-chromosome and three times in the double-bar chromosome. Eye size in the two kinds of females is appreciably different, providing an instance of position effect.*

TRANSLOCATIONS

The kind of aberration in which a fragment of one chromosome becomes attached to a nonhomologous chromosome is called a *translocation*. Non-homologous chromosomes may interchange parts in *reciprocal translocation*. If two normal nonhomologous chromosomes have a sequence of regions designated ABCDE and LMNO, reciprocal translocation might give rise to the aberrant types ABCNO and LMDE. The segments involved in reciprocal translocation may or may not be similar in size.

Although the translocation process itself is imperfectly understood, the general situation resulting in translocation can be said to be one where chromosome breaks are followed by new kinds of unions among the broken ends of chromosomes. X-ray treatment of nuclei, which induces chromosome breakage, is known to increase translocation frequency.

The Cytology of Translocations. The cytology of translocations is a large and complicated subject in itself, and we can deal only with its simplest manifestations. In the following discussion, we will confine ourselves to *reciprocal translocations* since these seem to be the most frequent and important in genetic systems. You should keep in mind from the outset three basic chromosomal types: (1) the *standard*, untranslocated types, (2) translocation homozygotes, (3) translocation heterozygotes. All these are illustrated in Figure 7-9.

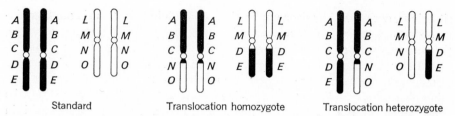

Figure 7-9. *If the chromosome pairs at the left have the standard sequence of chromatin material for a species, those in the center represent an arrangement in nuclei homozygous for a reciprocal translocation. The pairs at the right are heterozygous for the translocation.*

Translocation homozygotes may have no obvious cytological peculiarities. Their pairing at meiosis is regular, and the transmission of chromosomes from one nuclear generation to another may be as uncomplicated as in the original untranslocated types. In fact, when translocated pieces are about the same size, cytological observation of translocation homozygotes may not reveal the aberration at all, although suitable genetic experiments will. Naturally, if outstanding chromosomal landmarks are involved, translocation is disclosed by the change in their physical relationships with one another.

In translocation heterozygotes, the chromosomes get into various awkward

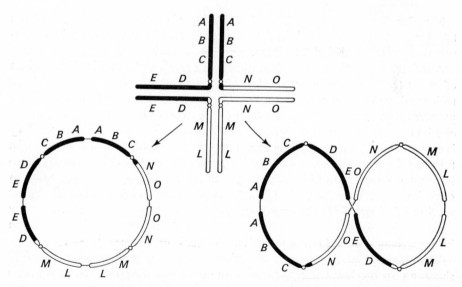

Figure 7-10. *Chromosome pairs heterozygous for a reciprocal translocation typically conjugate in a crosslike configuration at pachytene (center). Later the configuration may open out into a ring (left) or a figure-eight (right). For simplicity, the figures are not shown with doubled strands.*

situations, occasioned by the complications of attaining close pairing of homologous parts at meiosis. To achieve conjugation in a translocation heterozygote, the chromosomes characteristically assume a crosslike configuration at pachytene. Later, as terminalization (see p. 77) of chiasmata in the arms of the cross nears completion, the figure may open out into a four-membered ring. Or, as a fairly frequent variation, the cross-like configuration of pachytene may be transformed into a "zig-zag." The basic configurations are diagramed in Figure 7-10.

In the "zig-zag" arrangement, alternate members of the configuration go to the same pole at anaphase. When this is the case, the resulting nuclei have complete chromosomal complements, with half of them carrying reciprocal-translocation type chromosomes. Looking at the plain ring, you will see that distribution of adjacent members of the configuration to the same pole results in nuclei with duplications and deficiencies.

The cytology of translocation heterozygotes is complicated considerably when three or more chromosomes in a nucleus are involved in translocation. In *Oenothera lamarckiana*, which has been the object of several classical studies in cytogenetics, multiple reciprocal translocation results in all but 2 of the 14 chromosomes forming a single ring at meiosis. More or less complex ring formation has also been observed in onions, in the Jimson weed, in Paeonia, and in a large number of other kinds of plants.

The Genetics of Translocations. You would, of course, expect translocation to alter the linkage associations for genes contained in the exchanged chromosomal segments. This indeed occurs.

Another genetic effect of translocation is the *semisterility* that is characteristic of many translocation heterozygotes. If, in a plant heterozygous for a single reciprocal translocation, chromosomes from the meiotic translocation figure pass two by two at random to opposite poles, two-thirds of the resulting spores can be expected to be defective because of duplications and deficiencies. This is shown in Figure 7-11. Actually, random separation by twos, of members of a translocation ring, appears to be relatively rare. The centromeres of the chromosomes usually separate disjunctionally, with homologous centromeres going to opposite poles. Therefore, of the six types of gametes diagramed in the upper portion of Figure 7-11, the two at the left are less frequently found. In many plant species, about half rather than two-thirds of the gametes of translocation heterozygotes are produced by duplication-deficiency spores.

The phenotypic consequences of semisterility are easily observed. Figure 7-12 shows an ear produced after the pollination of semisterile corn. The gaps in this unfilled ear are due to abortion of about half the ovules.

If pollen of semisterile corn is observed under the microscope, about half the grains are seen to be abnormally small. Staining tests reveal that they are

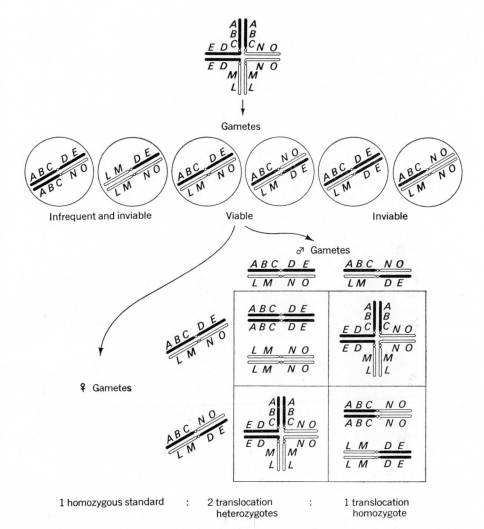

Figure 7-11. *Gamete formation and the results of self-fertilization of a plant heterozygous for a translocation.*

deficient in starch content. The physiological defects of these aberrant pollen grains result from their carrying chromosomal duplications and deficiencies.

INVERSIONS

The validity not only of chromosome maps but also of many of our broad concepts of genetics depends upon the truth of the assumption that the linear

Figure 7-12. *Phenotypic consequences of semisterility in corn. At the right is a normal ear. At the left is an ear produced on a plant heterozygous for a reciprocal translocation between chromosomes 1 and 7. Can you think of a reason why the kernels on the ear of semisterile corn are generally larger and of more irregular shape? (Courtesy of R. A. Brink.)*

order of the genes in a chromosome is well fixed. This assumption has survived many tests, and in general must be taken to be true. However, a given portion of a chromosome occasionally provides genetic results indicating a gene order just the reverse of one already found for the same kind of plant or animal. If visible landmarks exist in the portion of the chromosome in question, these too can be seen to occur in inverted order. The term *inversion* is given to this kind of aberration in which a chromosomal segment exists in reverse relationship to the rest of its chromosome.

Just as with translocations, organisms may be homozygous for an inversion,

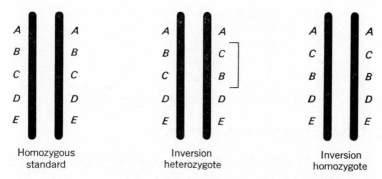

Figure 7-13. *Comparison of a pair of chromosomes having the standard sequence of chromatin material with a pair of chromsomes heterozygous for an inversion and with a pair of homologues homozygous for the inversion.*

heterozygous for an inversion, or they may be homozygous for the standard order of parts in the chromosome. (See Fig. 7-13). The *standard* chromosome order is the generally established order within the group of organisms. So far as cytological activities go, inversion homozygotes may behave perfectly normally. If in the course of time an inversion type gains numerical ascendancy in a population, it has a better claim to the title of "standard" than does the original order.

How do inversions arise? Probably in a variety of ways. Very likely some inversions originate in the manner diagramed in Figure 7-14. That is, a chromosome may form a loop, with breakages occurring at the point where the chromosome intersects itself. A single unit may then be constituted because of the tendency of sticky ends to adhere. When this happens the recently broken ends need not unite in the original combination. If they find new partners, inversion results.

Pairing in Inversion Heterozygotes. Inversion heterozygotes have no special difficulties in mitosis. At meiosis, however, inversion heterozygotes cannot achieve a true synapsis by simple linear pairing. Where direct conjugation must fail, however, circuitous means accomplish very nearly the same purpose. The

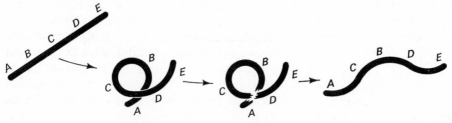

Figure 7-14. *A possible manner of origin for inversions.*

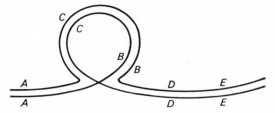

Figure 7-15. *Pairing in an inversion heterozygote.*

characteristic way in which inversion heterozygotes synapse is by forming a looped configuration such as is shown in Figure 7-15.

In salivary-gland chromosomes and other favorable material, an inversion loop makes it possible to determine with precision the extent of the aberration. This is illustrated in Figure 7-16. However, loop formation does not occur in every inversion heterozygote. Sometimes an inverted segment and the portion of a normal chromosome homologous to it simply fail to pair.

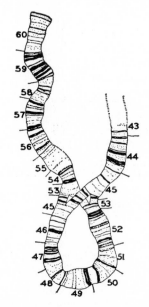

Figure 7-16. *An inversion loop in the salivary-gland chromosomes of an inversion heterozygote of* Drosophila azteca. *In this favorable material for cytological observation, it is possible to determine that the inversion involves the segments labeled 45 to 53. (From Dobzhansky and Socolov,* J. Heredity, **30**:*9, 1939.)*

Inversions as Crossover Suppressors. In 1921, part of the research activities of A. H. Sturtevant were being devoted to the study of corresponding genes in the related species *Drosophila melanogaster* and *D. simulans*. A comparison of maps for chromosome 3 in each of these species disclosed a most interesting fact. The corresponding genes for *scarlet*, *peach*, and *delta* had a gene order *st-p-Δ* in *D. melanogaster* but existed in the sequence *st-Δ-p* in *D. simulans*. Sturtevant's discovery was significant in several ways. First, the mere demonstration of an inversion added an important and previously unknown fact to our knowledge of genetic systems. Second, the study of comparative gene arrangements in related organisms was the forerunner of a series of investigations that have significantly enlarged our understanding of evolutionary processes. Third, finding the inversion suggested to Sturtevant an explanation for a phenomenon that had been puzzling geneticists for some time. Almost as soon as linkage began to be studied in Drosophila, investigators uncovered occasional instances of *crossover sup‹*

pressors. Their effect, in the heterozygous state, was to reduce markedly crossing over as it is measured by recombination.

The nature and mode of action of crossover suppressors was at the outset entirely obscure. With his discovery of an inversion, however, Sturtevant perceived that synapsis along an inverted segment would necessarily be abnormal in inversion heterozygotes, perhaps even absent. He predicted that, as a result of abnormal conjugation, crossing over would probably be considerably reduced. Subsequent experimentation saw the prediction fulfilled.

Why should recombination be suppressed by an inversion? You will find it enlightening to diagram what happens when a single exchange occurs *within the inversion loop* of an inversion heterozygote. We have done this for you in Figure 7-17, but for practice try it yourself. Remember that crossing over occurs at a four-strand stage. Notice that the results of crossing over differ, depending upon whether the inverted segment includes the centromere of the chromosome.

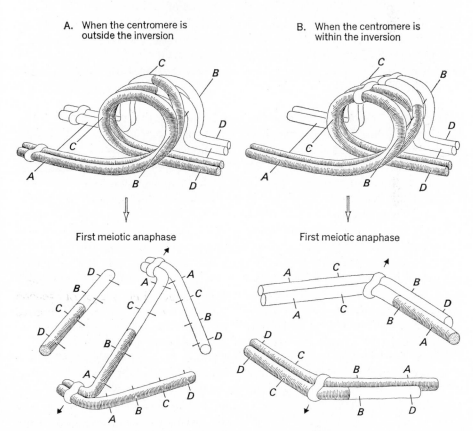

A. When the centromere is outside the inversion

B. When the centromere is within the inversion

First meiotic anaphase

First meiotic anaphase

Figure 7-17. *Crossing over within the inversion loop of an inversion heterozygote results in aberrant chromatids with duplications or deficiencies.*

But in either case, the end effects are the same: Aberrant chromatids with duplications and deficiencies are produced. If the centromere is outside the inverted segment, a *dicentric* chromatid and an *acentric* (lacking a centromere) fragment are formed. The dicentric chromatid forms a chromatin bridge, which may break. The acentric fragment has an uncertain future, but it usually is lost rather quickly because it has no means for systematic orientation at the time of nuclear division. On the other hand, if the centromere is included in the inverted segment, a single exchange gives rise directly to duplication and deficiency in each of the chromatids involved.

Inversions, then, do not necessarily suppress cytological crossing over. However, genetic recombinations in the inversion segment are very effectively suppressed, because single crossover chromatids carrying such recombinations are of a duplication-deficiency type, and either fail to function in fertilization or fail to produce viable zygotes. But this is not a complete accounting for the suppression of recombinations, since the presence of an inversion often reduces recombination immediately outside the inverted segment as well as within it. We may say, then, that the "suppression of crossovers" probably involves both nonappearance of crossover types and actual reduction in the frequency of the process of crossing over, the latter perhaps because of unsatisfactory pairing in the neighborhood of the inversion.

Inversions, Translocations, and Position Effects. Both inversion and translocation have the effect of rearranging segments of chromosomes, and, therefore, of putting genetic material into new associations. In such situations, position effects are occasionally observed, some of them resulting in apparent instability of gene action. Among such position effects in Drosophila is the effect of chromosomal rearrangements involving the locus for white eyes. Heterozygotes having the *w* gene on a normal chromosome and its "+" allele on a rearrangement chromosome may show mosaicism for eye color. The variegated eyes may have scattered red or pink patches on a light background, light patches on a red background or any of several variations on these color schemes, depending upon the modifying influences of genotype and environment. Most, if not all, of the variegation phenomena related to position effect in Drosophila seem to occur, when genes normally situated in euchromatin are transferred near heterochromatin. Position effects that determine petal-color mosaicism in Oenothera resemble closely the variegated-type position effects in Drosophila.

The various interpretations of position effect are as yet highly speculative. One set of hypotheses assumes that reversible alterations in the physical structure of genes may be involved in position effects. Another approach is based on the idea that chemical reactions among substances produced through genic activity may be affected by the proximity of particular genes. It is interesting in this regard that certain position effects are long-range effects—long in terms of chromosomal segments—that affect different loci to different degrees.

PREFERENTIAL SEGREGATION

In the 1930's, A. E. Longley examined the chromosomal morphology of a large number of strains of maize cultivated by various Indian tribes. Among many interesting variants he discovered an unusual chromosome 10 deriving from corn grown in Latin America and the Southwest of the United States. Abnormal 10, as it is called, differs from normal chromosome 10 by a variation in chromomere pattern near one end of the chromosome and by the presence of a large extra segment attached to that end. The additional material includes euchromatin and a prominent knob. No evidence exists to indicate that the extra material of abnormal 10 is homologous with any chromosomal regions found in the standard genome of corn.

Abnormal 10 has an interesting and significant attribute. M. M. Rhoades has found that in plants heterozygous for normal and abnormal 10 the two kinds of chromosomes do not have equal chances of being in the surviving member of the quartet of megaspores derived from a meiosis. In fact, about 70 percent of the time abnormal 10 is preferentially segregated into the basal megaspore, the only one of the quartet that functions in reproduction. (In this context, you may find it useful to review the life cycle of maize, in Chapter 4.)

Genes closely linked with the distinguishing segment of the abnormal 10 chromosome also show preferential segregation. This is readily seen if members of the allelic pair *R-r* are used as markers. *R* determines colored aleurone, the recessive allele colorless. The *R* locus is about one map unit from the knob in abnormal 10. In one of his experiments Rhoades bred a heterozygous stock in which the abnormal 10 chromosomes carried gene *r* and normal 10 *R*. Plants of this kind were pollinated with *r* pollen. The result was 1,881 colored seeds and 4,441 colorless. By contrast, control crosses, employing the same markers but all normal chromosomes, give the expected 1:1 backcross ratio. The actual results were 53,740 colored and 53,492 colorless seeds.

Abnormal 10 has another curious property. It is responsible for the formation of *neocentromeres* at the two divisions of meiosis. In other words, in the presence of abnormal 10, secondary sites of centromere activity occur in other regions than that of the true centromere. The neocentromeres occur precociously in knob-bearing chromosomes. Chromosomes having neocentromeres move more rapidly to the poles than do chromosomes without them.

Rhoades has utilized the foregoing observations to construct a hypothesis accounting for preferential segregation involving abnormal 10. He suggests that in plants heterozygous for abnormal 10 crossing over between the centromere and the knob gives heteromorphic dyads, one knobbed and the other knobless. As the result of neocentric activity the knobbed chromatid of a dyad at anaphase I lies closer to the pole; the knobless chromatid faces the spindle plate. Assuming this orientation to be maintained until the second metaphase of meiosis, the knobbed monads would tend to migrate to the terminal poles, the knobless to the inner poles. Since in maize the basal cell becomes the functional mega-spore, the knobbed chromosome would be recovered preferentially.

Whether or not this hypothesis proves to be entirely correct, the phenomenon of preferential segregation stands as a significant exception to the standard genetic mechanism. Moreover, the fact that abnormal 10 is found fairly frequently in strains of corn having wide geographic distribution poses a puzzle. When the solution is obtained we are likely to have additional understanding of genetics in relation to evolution. Incidentally, preferential segregation has been observed and studied in Drosophila and other organisms as well as in maize (see Chapter 13).

Variations in Chromosome Number

You are doubtless aware that there are different kinds of cultivated wheat, each suited to particular purposes and growing conditions. But you may not realize the full extent of diversity within the genus Triticum, to which wheats belong. To indicate this diversity we can mention three examples. Plants of *einkorn* wheat are small, yielding comparatively little grain, which threshes out with the glumes attached. Today, einkorn is planted chiefly for experimental purposes and is of minor agricultural importance. *Durum* wheat, grown largely in the Dakotas and Minnesota, has thick heads and large, hard grains. Because of its high gluten content, durum flour is used in macaroni and other pastes. Probably most familiar to you is *common* wheat, or *bread* wheat, which occurs in many varieties and is more widely grown than any of the other wheats. If einkorn is examined cytologically, the chromosome number in the nuclei of somatic cells is found to be 14, whereas durum has 28 chromosomes, and common wheats have 42. These wheats represent the three groups into which different species of the genus Triticum can be naturally divided on the basis of their chromosome numbers. A grouping of Triticum species is summarized in Table 7-1.

Table 7-1. GROUPING OF TRITICUM (WHEAT) SPECIES ACCORDING TO CHROMOSOME NUMBER.

14 Chromosomes	28 Chromosomes	42 Chromosomes
T. monococcum (einkorn)	*T. dicoccum* (emmer)	*T. vulgare* (common wheat)
	T. durum (durum wheat)	*T. compactum* (club wheat)
	T. polonicum (Polish wheat)	*T. spelta* (spelt)
	T. turgidum (poulard wheat)	
	T. dicoccoides (wild emmer)	

In oats, also, 14-, 28-, and 42-chromosome species can be distinguished. Different species of the genus Chrysanthemum are found to have chromosome numbers of 18, 36, 54, 72, and 90. Note that in wheat and oats the chromosome numbers exist as different multiples of the number 7, and in Chrysanthemum of the number 9. In a great many groups of related plants, similar progressions of chromosome number have been observed. What is their significance?

To answer this question, we can return to the situation in wheat. Each *genome* in einkorn consists of 7 members; in other words, the basic chromosome number (*n*) is 7. Somatic cells of einkorn have the diploid (2*n*) number, 14. It is an easily drawn inference that the higher multiples of *n* among wheats may have arisen through accumulation of genomes. Thus, durum, with 28 chromosomes, may be thought of as a *tetraploid* (4*n*) and bread wheat, with 42, as a *hexaploid* (6*n*).

As early as 1917, a Danish geneticist, Winge, observed the frequent occurrence of different multiples of a single basic chromosome number among related groups of higher plants. He suggested that reduplication of chromosome complements may occur rather frequently in the evolution of taxonomic series. More recently, variations in the number of genomes per nucleus have been found to occur occasionally among members of the same species. Figure 7-18 pictures mitoses in epidermal cells of the tail tips of salamanders, *Triturus*

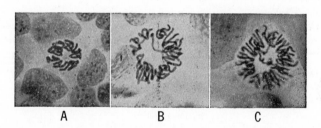

A   B   C

Figure 7-18. *Nuclei of haploid* (A), *diploid* (B), *and triploid* (C) *salamanders seen at mitotic metaphase.* (*Courtesy of R. B. Griffiths,* Genetics, **26:**76, 1941.)

viridescens, that differ from one another in having 1, 2, or 3 genomes per nucleus.

At about the time that Winge's publications were focusing attention on problems of genome reduplication, A. F. Blakeslee and J. Belling, in the United States, were concerned with another aspect of the relationship of variation in chromosome number to genetic diversity. They found that Jimson weeds (*Datura stramonium*) do not invariably have their normal diploid chromosome number of 24. Occasionally, plants have 25 chromosomes, with one of the 12 kinds represented in triplicate, the others in duplicate. This type of aberrant condition is called *trisomic*. Over a period of years, each of the 12 different chromosomes, which make up a genome in Datura, was observed one or more times in the trisomic condition. All of these trisomics are characterized by separate and recognizable phenotypic effects. For example, the so-called *Globe* phenotype is typical of plants trisomic for the particular chromosome that Datura geneticists have labeled 21·22. Globe plants differ from normal in that their leaves are broader and less indented and their seed capsules are more globular and have stout spines.

The Terminology of Ploidy. We have just considered, in an introductory way, two sorts of exception to the rule that somatic nuclei contain two of each kind of chromosome characteristic of the species, and germ-cell nuclei contain one. An organism may have an unusual number of full chromosome complements (as in tetraploidy) or an unusual number of chromosomes for only part of its genome (as in trisomics). Distinguishing these two as different kinds of situations is in a way arbitrary. From one point of view all variations in chromosome number are matters of duplication and deficiency. Variations in the presence or absence of *parts of chromosomes*, or of *whole chromosomes*, or of *whole sets of chromosomes* form an almost continuous series of situations having to do with gene dosage. The number of different possibilities for duplication and deficiency in this larger sense is so vast that there is real need for some sort of working classification. Our aim is to introduce you to a reasonable minimum of this necessary terminology.

1. *Aneuploidy* is a general term referring to nuclei containing chromosomes whose numbers are not true multiples of the basic number in the genome or genomes involved. In other words, *aneuploid* plants or animals are characterized by incomplete genomes. An organism lacking one chromosome of a diploid complement is called a *monosomic*. A *trisomic* has two complete genomes plus a single extra chromosome, as previously illustrated in Datura. *Tetrasomics* carry a chromosome in quadruplicate; the remaining chromosomes are present twice. If there are two extra chromosomes that are different members of the genome, the organism is called *double trisomic*. There are other types of variations of aneuploidy described by a nomenclature similar to that just indicated.

2. *Euploidy* is a term covering situations in which the total chromosome number involves complete genomes. Among *euploids*, a *monoploid* organism has just one genome per nucleus. A *triploid* has 3 genomes per nucleus, a *tetraploid* 4, a *hexaploid* 6, an *octaploid* 8, and so on. Multiple genomes including three or more sets of chromosomes per nucleus (in other words triploids and above) are frequently designated by the term *polyploid*. Polyploids are sometimes differentiated as *autopolyploids* and *allopolyploids*.

Autopolyploids are those in which the multiple genomes are identical, or very nearly so. Genome reduplication within a normally diploid species gives rise to autopolyploidy.

Allopolyploids are those in which the genomes making up a multiple set are not alike. You will see later that in particular instances it may be difficult to tell whether a polyploid organism should be called auto- or allopolyploid.

A guide to the terminology of ploidy is summarized in Table 7-2.

Table 7-2. EXAMPLES OF BASIC TYPES OF CHROMOSOME COMPLEMENTS.

Name of Type	Shorthand Formula	Chromosome Complement in Which C, B, A, and S Are Nonhomologous Chromosomes
I. IN ANEUPLOIDY†		
monosomic	$2n - 1$	(CBAS) (CBA)
trisomic	$2n + 1$	(CBAS) (CBAS) (C)
tetrasomic	$2n + 2$	(CBAS) (CBAS) (C) (C)
double trisomic	$2n + 1 + 1$	(CBAS) (CBAS) (CA)
II. IN EUPLOIDY		
monoploid	n	(CBAS)
diploid	$2n$	(CBAS) (CBAS)
triploid	$3n$	(CBAS) (CBAS) (CBAS)
autotetraploid	$4n$	(CBAS) (CBAS) (CBAS) (CBAS)
allotetraploid	$4n$	(CBAS) (CBAS) (C'B'A'S') (C'B'A'S')

† Examples of aneuploidy in this table are based on a diploid complement. Chromosome complements that are polyploid may show analogous aneuploid variations, as (CBAS) (CBAS) (CBAS) (C), for example.

ANEUPLOIDY

Monosomics. Monosomics, as an aberrant type of chromosomal complement, have a good deal in common with deficiencies. You will recall that deficiencies of any considerable length usually result in lethality. It is not surprising, then, that monosomics are rather rarely found. However, viable monosomics have been studied in Drosophila, tobacco, and other organisms.

Monosomics behave at meiosis as you might expect. Being of a chromosome constitution $2n - 1$, they produce two kinds of gametes, n and $n - 1$. The odd chromosome, which has no pairing partner, tends to pass at random to either pole at meiosis. Frequently, however, it acts as a laggard at anaphase and is not included in either of the daughter nuclei. For this reason, n gametes occur less frequently than the $n - 1$ type. In plants, nuclei with a missing chromosome seldom survive the gametophytic generation, presumably because the complete absence of certain genes means that fundamental biochemical functions cannot be carried on. This fits well with what we learned about deficiencies.

Perhaps the best known monosomic is the *haplo-IV* type of *D. melanogaster.* Flies of this kind lack one of the fourth chromosomes, to be remembered as the smallest by far in the *D. melanogaster* chromosome complement. C. B. Bridges described haplo-IV flies as differing from normals in having pale body

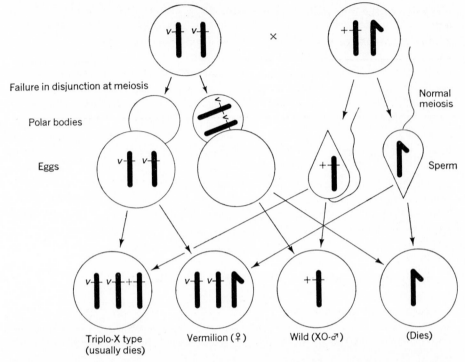

Failure in disjunction at meiosis

Polar bodies

Eggs

Normal meiosis

Sperm

Triplo-X type
(usually dies)

Vermilion (♀)

Wild (XO-♂)

(Dies)

Figure 7-19. *The origin of triplo-X and XO Drosophila through failure of normal disjunction of the X-chromosomes at meiosis. The usual two meiotic divisions are assumed but not shown in the figure.* (*After Bridges,* Genetics, **1**:*10, 1916.*)

color, shortened wings, roughish eyes, slender bristles, reduced fertility, and high mortality.

Haplo-III and haplo-II types have not been found in Drosophila. Apparently the absence of a second or a third chromosome results in the drastic physiological difficulties we expect to follow from any sizable deficiency. *XO* flies do occur, however, and they turn out to be scarcely distinguishable from normal males (*XY*), except that *XO* flies are sterile.

Trisomics. Trisomics are a complementary type to monosomics. The latter lack one chromosome; the former have one extra chromosome. These types of aneuploidy may be of similar origin. Sometimes at the first division of meiosis in a diploid, homologous chromosomes fail to pair properly or to orient themselves in regular fashion at metaphase. If both members of a pair of homologues go to the same pole (*nondisjunction*), half of the eventual products of meiosis will have one chromosome too many, and the other half will be one chromosome short of a full complement. Union of the first type of abnormal meiotic product with a normal gamete of the opposite sex produces a trisomic. Meiotic products with a missing chromosome can give rise to monosomics. A possible origin for triplo-X and *XO* Drosophila is given in Figure 7-19.

If disjunction in trisomics were such that two chromosomes always went to one pole, and the third chromosome went to the opposite pole, and if the distribution of individual chromosomes under this condition were random, then predictions as to phenotypic ratios might be made for crosses involving organisms of known genotype. Take, for example, a triplo-IV Drosophila having the genotype $++ey$ for the recessive character *eyeless*. Assume absence of crossing over between the locus of eyeless and the centromere. As noted in Figure 7-20, random segregation would give the following gametes in the designated frequencies: $2 + ey : 2 + : 1 ++ : 1 ey$. In a cross to ordinary eyeless flies (*ey ey*), the resulting phenotypic ratio would be 5 normal to 1 eyeless, a proportion closely approached when actual crosses of this kind are made. In tomatoes, Datura, corn, and other plants, however, analogous crosses do not give these results. And different types of trisomic matings (e.g., $AAa \times AAa$) often fail also to yield results conforming to expectation based on the conditions that account for the 5:1 ratio just described.

There are various reasons for these deviations from the expected. First of all, the assumption of no crossing over between the locus of the segregating gene and the centromere of its chromosome may not hold. Moreover, although a trisomic is expected to give rise to equal numbers of $n + 1$ and n gametes, $n + 1$ types may not be formed as frequently; and even if they are, $n + 1$ pollen may not function as effectively in fertilization. For instance, Buchholz and Blakeslee found in trisomics of Datura that $n + 1$ pollen grows considerably more slowly down the style. In a study involving the trisomic type *cocklebur*,

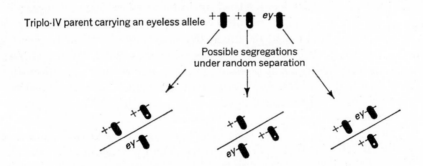

Triplo-IV parent carrying an eyeless allele

Possible segregations under random separation

Summary of gametic types: $1++, 2+, 2+ey, 1ey$

Figure 7-20. *The gametic types produced by a triplo-IV Drosophila carrying an* eyeless *gene. The dot in one of the chromosomes is utilized to distinguish it from its homologue which also carries a wild-type allele.*

where the 11·12 chromosome is present in triplicate, $n + 1$ pollen grew at an average rate of 1.9 mm per hour; n pollen at 2.6 mm per hour.

When a chromosome exists in triplicate, regular meiotic segregation of two homologues to one pole and one to the other depends on all three homologues pairing in a single *trivalent* configuration. Such a configuration is diagramed in Figure 7-21, where you will note that at a given place pairing is between two

Figure 7-21. *The trivalent configurations at pachytene. Note that at any given point along the chromosomes, pairing occurs between only two of the homologues.*

chromosomal segments only. If two of the homologues form a *bivalent* configuration and the third exists as a *univalent*, the unpaired chromosome may be lost during meiosis, as happens in monosomics. Trisomics of corn frequently fail to transmit their extra chromosome. In the female gametophyte, failure of transmission is not due to poor viability of the $n + 1$ gametes. Rather, the extra chromosome is lost at meiosis. Work by J. Einset has shown that the transmission of $n + 1$ gametes through the egg is variable. If we compare trisomics for different members of the genome, longer chromosomes are found more likely to be transmitted than are shorter ones. With the shorter chromosomes, meiotic configurations that were originally trivalent frequently "desynapse" to form a bivalent and a univalent. The univalents lag at division and often are not incorporated into the daughter nuclei.

The study of trisomics has revealed an instance of preferential segregation. As stated earlier, segregation in triplo-IV Drosophila usually gives two chromosomes at one pole, and one at the other. If we designate these chromosomes A, B, and C, the expected patterns of segregation are AB/C, AC/B, BC/A. If the chromosomes segregate at random the three patterns of segregation should be found in equal numbers. A. H. Sturtevant, utilizing fourth chromosomes of different origin, tested randomness of segregation in triplo-IV Drosophila and found that the three segregation types seldom occurred with equal frequency. The chromosomes he studied showed such consistent patterns of nonrandom segregational behavior that preference orders could be established.

More recently, instances of trisomy and other forms of aneuploidy have been found in man. For example, trisomy for chromosome 21 is associated with, and indeed appears to account for, Down's Syndrome, sometimes called Mongoloid imbecility. Such instances will be considered in detail in Chapter 15.

Tetrasomics. Aneuploid nuclei may have chromosomes in quadruplicate or even further reduplicated. Phenotypic expression of the tetrasomic condition may take the form of accentuation of the effects found for the corresponding trisomic. Or, sometimes, the extra chromosomes of a tetrasomic give rise to no obvious changes in phenotype.

The four homologues of a tetrasomic often tend to form a quadrivalent figure at meiosis. If disjunction by twos is regular, a fairly stable genetic system can be maintained. But quadrilvalents are not always formed, and disjunction is not always regular. On the whole, however, tetrasomics and other even-numbered extra-chromosome types behave in more stable and regular fashion at meiosis than do odd-numbered types.

EUPLOIDY

Monoploidy. In a monoploid organism, each kind of chromosome is represented only once in a nucleus. This is the typical condition in eggs, sperm, and in the gametophytic generation of plants; but to describe nuclei having a *gametic* chromosome number the term *haploid* is generally used. There is a valid reason for this distinction in terms, since polyploid organisms may regularly produce gametes having two or more genomes. Such gametes are sometimes called *polyhaploid*. The gametic chromosome number of a polyploid may be *haploid* with reference to the chromosome number characteristic of the adult organism but not *monoploid* in terms of number of genomes.

Monoploids are usually smaller and less vigorous than their diploid prototypes. Nevertheless, monoploids hold a certain amount of interest for plant breeders. This is because doubling the chromosomes of a monoploid can give rise to diploid individuals homozygous for all the gene pairs in the organism. The advantages in having such genetically "pure lines" for the accurate reproduction of superior germ plasm are obvious.

Characteristically, monoploid plants are sterile. The reason is that the chromosomes have no regular pairing partners during meiosis. At the first division, some of the chromosomes may go to one pole and some to the other. The nuclei that are formed are deficient in one or more chromosomes, leading to the production of large numbers of nonfunctional spores or gametes. Occasionally, by chance, all the members of the genome pass to one pole at the first meiotic division. Then a normal spore or gamete is produced. Union of two "complete" gametes of this kind gives rise to a diploid organism.

If the distribution of chromosomes to the poles is random in the meiosis of a monoploid, we can predict the probability of a daughter nucleus receiving the entire genome. The probability of any chromosome going to a given pole is ½ under conditions of random distribution. In a monoploid tomato (12 chromosomes), the probability of all the chromosomes passing to a given pole at meiosis would be $(\frac{1}{2})^{12}$, or one in 4,096 times. In fact, the probability is lower than that calculated here, since unpaired chromosomes at meiosis often lag and fail to be included in a nucleus. Small wonder that monoploids show low fertility.

Triploidy. Triploids have three complete genomes per nucleus. Our discussion here will deal with *autotriploids*, where the genomes have been derived by reduplication within one species of organism and, cytologically at least, are essentially identical.

In general, triploids originate by the fusion of a monoploid gamete with a diploid gamete, a situation diagramed in Figure 7-22. Diploid gametes occur

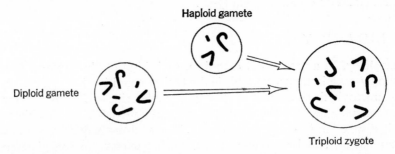

Figure 7-22. *The origin of a triploid, showing two genomes coming from one gamete and one genome from the other.*

sporadically as *unreduced* germ cells in a diploid organism. They are also produced by meiosis in tetraploid organisms or in sectors of otherwise diploid organisms where doubling of the somatic chromosome number has taken place. *Somatic doubling* is sometimes spontaneous, but it may also be induced experimentally.

If you think of a triploid nucleus as one having a complete set of trisomics, you will immediately realize that the triploid state is highly unstable when it comes to sexual reproduction. The difficulties of triploids, like those of trisomics, arise from the fact that at meiosis the centromeres of three homologous chromosomes have no way to orient themselves to give equivalents at the two poles of the spindle apparatus. As a result, if out of every three homologues two go to one pole and one to the other, a meiotic product may contain either n or $2n$ chromosomes, or any number in between. This is generally true whether three homologues originally align themselves as a trivalent or whether they exist as a bivalent and a univalent. Therefore, many of the gametes arising from a triploid have unbalanced genomes. In fact, the probability of getting gametes that are complete in being either n or $2n$ is relatively small. To calculate this probability, on the basis that segregation of each set of three chromosomes is independent of the other sets, you may use a procedure similar to that employed in computing the probability for obtaining a complete genome after meiosis in a monoploid.

In other words, the probability of a nucleus obtaining a complete genome and no more is $(\frac{1}{2})^n$. (What is the probability that a nucleus will obtain two complete genomes?)

Examples of triploidy in animals are rare but cover a range of forms including Drosophila, salamanders, and the land isopod Trichoniscus. Triploidy is somewhat more common and widespread in plants; it has been found in grasses, forest trees, garden flowers, crop plants, and elsewhere. On the whole, however, triploids as a group have been rather unsuccessful in establishing themselves in natural populations, no doubt because a usual consequence of triploidy is sterility.

Experimental Techniques for Inducing Polyploidy. In a low frequency among cell divisions, chromosomes reduplicate without an accompanying nuclear division. The result is a *restitution nucleus* having a doubled chromosome number. Techniques for inducing polyploidy increase the frequency of such events by creating disturbances in the normal synchronization of the processes of nuclear division. The same general effects may be achieved in a variety of ways. For example, if tomato plants are decapitated, some of the new shoots that arise may turn out to be tetraploid. Also, shoots arising near the point of union in certain graft combinations are frequently of a doubled chromosome constitution. L. F. Randolph has induced chromosome doubling in corn by surrounding the ear-shoot regions of growing plants with a heating pad. Cold shock has been used with similar effect in Drosophila.

A number of chemical agents induce polyploidy. Among them are chloral hydrate, acenaphthene, sulfanilamide, ethyl-mercury-chloride, hexachlorocyclohexane, and colchicine. Of these, colchicine, an alkaloid obtained from the autumn crocus Colchicum, has been most widely and effectively used. Colchicine is water soluble, and has the added advantage of being almost nontoxic at concentrations that are effective in producing a high proportion of polyploid nuclei.

Certain aspects of the action of colchicine on dividing nuclei, especially in plants, are well understood. At critical concentrations of colchicine, spindle fibers do not form, and the normal process of mitosis is modified to a sequence of events called *C-mitosis*. In the absence of a functioning spindle, chromosomes fail to move into an equatorial plate but remain scattered in the cytoplasm (a stage called C-metaphase). However, the chromosomes eventually separate at the centromeres, and a C-anaphase is initiated. (See Fig. 7-23.) Following this, the distributed and reduplicated chromosomes go through regular telophasic transformations, with a membrane finally developing around a nucleus that has the doubled chromosome number. If the effects of the colchicine are dissipated, this new polyploid cell may regenerate a bipolar spindle and produce daughter nuclei that are polyploid like their immediate progenitor. But if critical concentrations of colchicine remain, additional C-mitoses may follow, resulting in

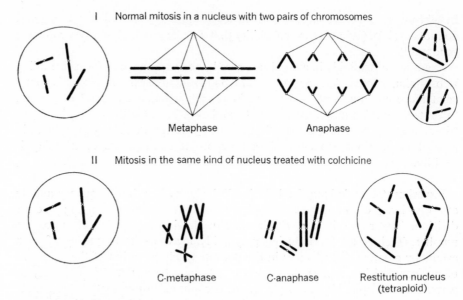

I   Normal mitosis in a nucleus with two pairs of chromosomes

Metaphase   Anaphase

II   Mitosis in the same kind of nucleus treated with colchicine

C-metaphase   C-anaphase   Restitution nucleus
                                          (tetraploid)

Figure 7-23. *Normal mitosis compared with mitosis in a nucleus treated with colchicine (C-mitosis).*

further reduplications of the chromosome complement. More than 1,000 chromosomes have been found in onion root-tip nuclei left in colchicine for four days. However, when the reduplication of genomes exceeds octoploidy, the regeneration of a regular spindle mechanism does not readily take place.

Autotetraploidy. Autotetraploids have four similar genomes per nucleus. We have just considered how tetraploidy arises through somatic doubling, but the simplest way in which tetraploidy can occur and be maintained is through a fusion of diploid gametes.

At meiosis in autotetraploids, pairing usually takes place in quadrivalent groups, as illustrated in Figure 7-24. Sometimes, however, two pairs of bivalents

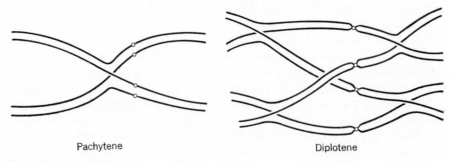

Pachytene   Diplotene

Figure 7-24. *Quadrivalent association in early stages of meiosis in an autopolyploid.*

or a trivalent and a univalent occur. In these cases, the products of meiosis are irregularly formed. Incomplete genomes find their way into gametes, and sterility is a frequent result.

Where quadrivalents are regularly formed, and where disjunction from the quadrivalents is two by two, the chromosomal basis for a stable system of sexual reproduction is at hand. Even then the genetics of tetraploids is far more complex than the genetics of diploids. For a given pair of alleles, *A-a*, in an autotetraploid, five genotypes are possible—the homozygotes *AAAA* and *aaaa*, and the heterozygotes *AAAa*, *AAaa*, and *Aaaa*. The complexities of genotype in an autotetraploid, however, are small in comparison to the complexities of what happens when segregation occurs.

Let us consider briefly an autotetraploid plant of the genotype *AAaa*. If the locus of gene *A* is near the centromere of its chromosome, the assortment of genes into gametes can be predicted by a summation of random separation of the four chromosomes by twos. The prediction says that gametes can be expected in a proportion of 1*AA*:4*Aa*:1*aa*. If, on the other hand, the *A* locus is sufficiently distant from the centromere that crossing over permits random assortment of these points on the chromosome, then the prediction becomes quite different. We need to take into account 8 chromatids, four carrying gene *A* and four carrying its allele *a*, assorting by twos. In this instance the gametes are predicted in a proportion of 3*AA*:8*Aa*:3*aa*. Many genes, of course, are neither very close to the centromere nor far enough from the centromere to be independent of it.

The foregoing paragraph is no doubt sufficient to convince you that the genetics of autotetraploids is indeed complicated, although undoubtedly interesting. We will leave further treatment to more specialized works in the literature of genetics, having said only enough to be able to make the following statements. Given our first prediction of the proportion of gametes derived from genotype *AAaa*, self-fertilization of such a plant would give a progeny in a phenotypic ratio of 35*A*:1*a*, assuming complete dominance of allele *A*. Given our second prediction, self-fertilization would give a progeny in a phenotypic ratio of about 21*A*:1*a*. The important point is that under any prediction the recessive phenotype appears relatively infrequently as compared to the 3:1 relationship found in F_2 progeny of a diploid hybrid.

In dihybrids an analogous relationship would hold. Take the dihybrid autotetraploid *AAaa BBbb*, and assume that the *A* and *B* loci are very near the centromeres of nonhomologous chromosomes. If the dihybrid is selfed we expect each pair of characters to be represented in the progeny in a proportion of 35:1, as stated above. In other words, phenotype *a* is expected $\frac{1}{36}$ of the time, the same being true for phenotype *b*. Since the *a* and *b* loci are on separate chromosomes, segregations at these loci are independent events. The probability of recovering the double recessive phenotype *ab* is therefore $\frac{1}{36} \times \frac{1}{36} = \frac{1}{1296}$.

Phenotypic Consequences of Autopolyploidy. The accumulation of genomes beyond the diploid level is usually detectable in one or more phenotypic attributes. As you shall see, however, generalizations as to the phenotypic effects of polyploidy are dangerous.

Morphological Effects. Polyploids are sometimes huskier and more vigorous than corresponding diploids. In fact, the phenotypic differentiation of polyploids often has led to their first being recognized as unusual types worthy of study. Triploid aspen trees owe their discovery to exceptionally large leaves, which attracted the attention of investigators. Subsequently, triploid aspens were produced experimentally by crossing diploids with tetraploids. Analysis of the triploids revealed the striking "gigantism" of their leaves to be due to appreciably larger cell size. Pollen grains, stomatal cells, and wood cells of xylem also were found to be larger in the triploids than in the diploids.

G. Fankhauser, has found that for all tissues examined the nuclei and cells of triploid salamanders are consistently larger than those of the diploids. The number of cells in most body organs of the triploids is somewhat reduced, however. Through this kind of compensation, the size of organs and overall body size of many polyploids remain approximately the same as the corresponding diploids.

The effects of chromosome doubling may differ for various genotypes within a single species. In the grass *Stipa lepida* certain autotetraploids have broader leaves than the diploids from which they derive, but other diploids produce tetraploids that have narrower leaves.

The fact that polyploidy is sometimes accompanied by increases in size and vigor has been used to advantage by plant breeders, as you can see by comparing the Easter lilies shown in Figure 7-25. Among cultivated plants, even triploids occupy a fairly important position, despite their difficulties in reproducing sexually. Since many plants can be produced vegetatively, the sterility of triploids can be bypassed, and advantage can be taken of their sometimes superior qualities. A number of important varieties of fruits are triploids, like the Gravenstein and Baldwin apples. The Keizerskroon tulip, which has unusually large flowers, is a triploid, as are a good many other ornamental plants.

Physiological Effects. The ascorbic acid content has been reported to be higher in tetraploid cabbages and tomatoes than in the corresponding diploids. Tetraploid corn, besides being occasionally more vigorous than comparable diploid maize, gives cornmeal containing something like 40 percent more vitamin A than cornmeal from the diploid. On the other hand, an investigation of the riboflavin and pantothenic acid in grapes showed diploids and tetraploids to have about the same content of these vitamins.

Effects on Fertility. One of the most important effects of autopolyploidy is that it often reduces fertility, occasionally to an extreme degree. This effect is of consequence when polyploidy is considered either in relation to evolution or to plant improvement. A great deal of evidence indicates that sterility in autopolyploids is only partly due to irregular chromosome distribution following

Figure 7-25. *The smaller Easter lily was produced on a diploid plant, the larger on a tetraploid that was obtained by means of chromosome doubling induced by colchicine treatment. (Courtesy of S. L. Emsweller, Bureau of Plant Industry, U.S. Dept. of Agriculture.* Science in Farming, *Yearbook of Agriculture, 1943–1947, ff. p. 80.)*

multivalent associations at meiosis. The addition of genomes seems frequently to result in disturbances of gene-controlled physiological processes, and thus to alter a variety of the conditions essential to normal fertility.

Origin of Allopolyploids: The Genesis of an Amphidiploid. When certain crosses are made between members of distinct taxonomic groups, the first-generation hybrid is highly sterile, because the genomes are so different that each kind of chromosome lacks a homologue to act as its pairing partner at meiosis. Because of sterility, such hybrids face extinction unless they are able to reproduce vegetatively. Even with unsystematic distribution of the chromosomes at meiosis, however, some unreduced gametes can be expected to occur merely by chance. The situation is comparable to the formation of unreduced gametes in monoploids. Or, somatic doubling in the hybrid may result in the regular formation of gametes containing one complete genome of each of the parental types. A fusion of such gametes gives rise to a fertilized egg cell containing the complete diploid complement of each original parent. Double

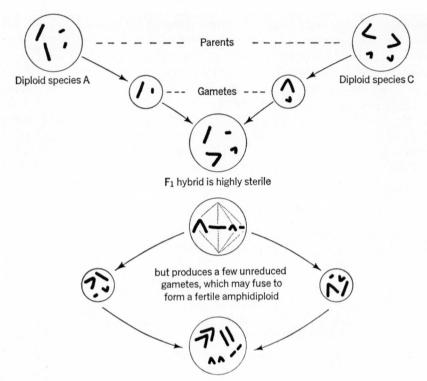

Figure 7-26. *A sequence of events leading to amphidiploidy.*

diploids of this kind—called *amphidiploids*—are sometimes fertile, since each chromosome now has a pairing partner. In well adjusted amphidiploids, cytological behavior at meiosis may be like that of a diploid with a relatively large number of chromosomes. Figure 7-26 outlines the sequence of events just described as one path to amphidiploidy.

An almost diagrammatic example of the genesis of an amphidiploid has been given us by the Russian geneticist Karpechenko, who made intergeneric crosses involving radishes and cabbages as parents, and ultimately obtained a fertile and generally true-breeding hybrid. Karpechenko's experimental results are summarized in Figure 7-27.

The cabbage, *Brassica oleracea*, and the radish, *Raphanus sativus*, both have a diploid chromosome number of 18. F_1 plants obtained from a cross between the two also have 18 chromosomes, 9 from each parent. These individuals differ morphologically from both parental types, and in many ways are a kind of compromise between them. Examination of the fruit structures of the different kinds of plants, parents and progeny, shows this particularly well. The F_1 plants are highly sterile because of failures in pairing at meiosis. They do produce a few seeds, however, and some of the plants arising from these seeds are fertile.

Cytological examination reveals 36 chromosomes in the nuclei of somatic cells. Chromosomal behavior is regular, even at meiosis, where pairing takes place in such fashion that 18 bivalents are formed. It is apparent that the fertile F_2 plants have arisen from a fusion of unreduced gametes from F_1 individuals, and that these F_2 plants contain the full diploid complements of the original cabbage and radish progenitors. Morphologically, the amphidiploids are a good deal like the F_1 hybrids. But the fruits of the amphidiploids are characteristically somewhat larger, with the plants as a whole possessing a good deal of the robustness that sometimes is a consequence of reduplication of genomes. This stable new form of plant may be considered a distinct taxonomic entity. In recognition of its hybrid origin, it has been given the name *Raphanobrassica*.

Diverse Forms of Allopolyploidy. The Raphanobrassica was chosen to introduce you to allopolyploidy because it offers a simple and straightforward example. Among naturally and experimentally produced allopolyploids there

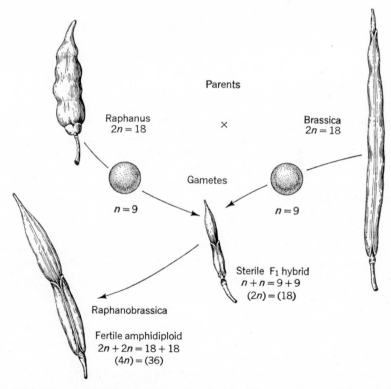

Figure 7-27. *The origin of Raphanobrassica, showing chromosome numbers and charac teristic fruit structures for the different plants involved. (After Karpechenko Z. Indukt. Abst. Vererb., **48**:27, 1928.)*

are many variations on the theme we have introduced. Some of these variations can be seen among plant types obtained following crosses of members of the genus Rubus. The European raspberry, *R. idaeus*, has 14 chromosomes, and we can represent its two genomes by the symbols I and I. The dewberry, *R. caesius*, is a tetraploid having 28 chromosomes; its genomes can be designated $C_1C_1C_2C_2$. Hybridization of the European raspberry and the dewberry produces sterile F_1 individuals (IC_1C_2) that are triploids with 21 chromosomes. Chromosome doubling in the F_1's has given rise to hexaploid amphiploids. Backcrosses to the parents designated above have given rise respectively to tetraploids and pentaploids. Apparently only unreduced gametes of the triploid F_1 function in these crosses. These results can be summarized:

	GENOMES	CHROMOSOME NUMBER
F_1 is	IC_1C_2	21
backcross to *R. idaeus* gives	IIC_1C_2	28
backcross to *R. caesius* gives	$IC_1C_1C_2C_2$	35
amphiploid (*R. maximus*) is	$IIC_1C_1C_2C_2$	42

Source: After Clausen, Keck, and Hiesey, "Experimental Studies on the Nature of Species," *Carnegie Inst. Wash. Pub.* 564, p. 111, 1945.

Difficulties of Distinguishing between Auto- and Allopolyploidy. Our preliminary way of distinguishing between auto- and allopolyploids was on the basis that the former have similar and the latter have dissimilar genomes. A more sophisticated and more useful definition has been suggested by G. L. Stebbins. He proposes that "an autopolyploid is a polyploid of which the corresponding diploid is a fertile species, while an allopolyploid is a polyploid containing the doubled genome of a more or less sterile hybrid." (Try to outline the general argument on which such a definition is based.)

Whatever definitions we use, in practice the distinction between auto- and allopolyploidy is difficult, more often than not. Within a species, genic constitution and the structural arrangement of chromosomes may vary considerably among individuals. On the other hand, the parents involved in the widest possible crosses may have many genes in common and may have many homologous chromosome segments. In any case, a given individual may combine characteristics of auto- and allopolyploidy. A simple example is provided by any redoubled amphidiploid.

Attempts have been made to distinguish kinds of polyploidy on the basis of chromosome homologies as indicated by pairing affinities at meiosis. The principle here is that autopolyploids should show multivalent associations at meiosis and allopolyploids should not. This criterion breaks down, however, in many instances. A well known example involves members of the Primrose family. *Primula kewensis* arose at Kew, in England, as a result of somatic doubling in a sterile hybrid plant obtained from a cross between *P. floribunda*, a native of Afghanistan, and *P. verticillata*, an Arabian species. Two facts about chromosome pairing in *P. kewensis* are worth noting. First, pairing in the diploid F_1 is almost regular (an indication that not all hybrid sterility is due to failures in pairing). Second, some quadrivalent formation occurs in the tetraploid. Thus, if *P. kewensis* were not known to have arisen as a species hybrid, criteria of chromosome pairing might lead to its classification as an autopolyploid. Conversely, in some undoubted autopolyploids, of which there are examples in the tomato, conjugation may occur largely as bivalents. Here classification on the basis of chromosome pairing would probably lead to an autoploid being designated as an alloploid. As a final complication, over a period of time auto-polyploids may shift from a multivalent to a bivalent type of synapsis. Cyto-logical examination has been made of autotetraploid corn that originated following heat treatment in 1937 and has been propagated every year since. Plants grown from seed of the 1937 harvest showed more quadrivalents and fewer bivalents at meiosis than did plants grown from seed of the 1947 harvest.

Aneuploidy in Polyploids. You have learned that a genome, under a variety of conditions, may suffer loss of component chromosomal material. Or, chromo-somes may be added, as in trisomy or tetrasomy. Ask yourself whether a diploid or a polyploid would be better able to withstand the frequently deleterious effects of aneuploidy. You are probably going to answer that its reduplicated genomes should provide the polyploid with a greater buffering capacity against the effects of addition and loss of chromosomal material. In general, such an answer, based on such a line of reasoning, appears to be correct. At least, we can say that remarkable series of aneuploids are obtainable from polyploid plants. Elegant examples are provided by the work of E. R. Sears.

Common wheat (*Triticum vulgare*) has 42 chromosomes, and is a hexaploid of complex origin whose chromosomes characteristically pair by twos at meiosis. Over a period of years, Sears has obtained from one variety of common wheat, Chinese Spring, all 21 possible monosomics and trisomics. From the trisomics he has been able to derive a complete series of tetrasomics, and from selfed monosomics a complete series of *nullisomics*, in which a single homologous pair of chromosomes is entirely lost. In short, Sears' various lines of wheat comprise a 0 to 4 dosage series for each of the 21 chromosomes (haploid number).

Grown under favorable conditions, most of the monosomics are scarcely distinguishable from standard Chinese Spring wheat. In general, the nullisomics are considerably reduced in vigor and fertility, but all survive to maturity and one of the nullisomics shows only slight reduction in vigor and height. This remarkable ability to withstand chromosome loss is comprehensible only in terms of the polyploid origin of common wheat.

Sears has found that particular tetrasomes will compensate for particular nullisomes. Detailed study of the compensation effect reveals that the 21 chromosomes of wheat fall neatly into seven groups of three. A group has the property that any of the three chromosomes within it is able to compensate for the nullisomic condition of either of the other two. For example, (the chromosomes of wheat being numbered from I to XXI) group 1 is composed of I, XIV, and XVII. Compensation for nulli-I is provided by tetrasomics XIV or XVII but not by tetrasomics of other chromosomes of wheat. The basis of this systematic compensation, of course, is the fact that the haploid chromosome complement of wheat is composed of three genomes of seven chromosomes each. These genomes have a good many homologies and similarities with each other, even though pairing ordinarily does not occur between corresponding members of different genomes. Any nullisomic plant, by reduction in fertility and vigor and in other ways, shows the effects of a change in genic balance. But survival is possible because vital genic functions are retained in other partially duplicate genomes. The compensation effect of tetrasomics, which restores something close to normalcy, identifies those chromosomes in other genomes that are basically duplicates for those lost by nullisomy.

Polyploidy and Evolution. Some authorities estimate that a third or more of the species of angiosperms are polyploids. Among them are many crop plants (see Table 7-3). And polyploidy is common, although not everywhere found, in

Table 7-3. CHROMOSOME NUMBERS INDICATING POLYPLOIDY IN GENERA THAT INCLUDE REPRESENTATIVE AGRICULTURAL PLANTS.

Genus and Species		Common Name	n
Sorghum	*versicolor*	————	5
	vulgare	Sorghum	10
	halepense	Johnson Grass	20
Gossypium	*arboreum*	Asiatic treecotton	13
	hirsutum	Upland cotton	26
Trifolium	*hybridum*	Alsike clover	8
	repens	White clover	16
	medium	Zigzag clover	40 (?)†
Medicago	*hispida*	Bur clover	8
	sativa	Alfalfa	16
Avena	*brevis*	Short oat	7
	Abyssinica	Abyssinian oat	14
	sativa	Common cultivated oat	21

† Different chromosome numbers have been reported for this species.

the rest of the plant kingdom. This argues strongly for the evolutionary significance of polyploidy.

To leap to immediate judgment of the importance of polyploidy in evolution may lead to precarious conclusions, however. Autoploidy as such adds no new genes to the chromosome complex. The phenotypic effects of autoploidy are frequently seen largely as exaggerations of characteristics already present in the diploid. Seldom do significantly divergent characters appear. Allopolyploids are sometimes strikingly different from their progenitor types; but also they may be nearly indistinguishable from one of the parental species; and they are most often intermediates. Allopolyploidy is probably more involved in the production of new combinations of characters, than in the origin of characters themselves.

What, then, is the role of polyploidy in evolution? A few concepts that have emerged from the voluminous work of many investigators appear in the following paragraphs.

Polyploidy and the Sudden Emergence of Species. In many of its aspects evolution moves slowly and involves gradual processes by which many differences in organisms arise and are accumulated. In contrast to these time-consuming processes, at least one mechanism operates by which species may originate in "cataclysmic" fashion. This mechanism is polyploidy. You have seen in the case of the Raphanobrassica how a new taxonomic entity can be produced in a relatively few simple steps. Similar steps have led to the natural formation of many amphiploids.

This kind of evolution in action was more or less directly observed in *Spartina townsendii*, called "probably the only amphiploid which is known to have arisen spontaneously in historic times." *Spartina townsendii*, a kind of cord grass, was first collected in England in 1870. By 1907 it had spread over thousands of acres on the south coast of England, and now it is established on the coast of France as well. The fascinating history of *S. townsendii* is summarized in a paper by Huskins cited among the references at the end of this chapter.

There is a good deal of evidence for the amphiploid origin of *S. townsendii*. First, it combines several distinctive characteristics of *S. stricta*, a European species, and *S. alterniflora*, a species thought to be American in origin and transferred to Europe via shipping. Natural populations of *S. stricta* and *S. alterniflora* meet in precisely the locality in which *S. townsendii* appeared. Huskins' cytological studies showed that *S. townsendii* has 63 pairs of chromosomes, the sum of the chromosome numbers of its putative parents. *S. alterniflora* has 35 pairs of chromosomes, and *S. stricta* 28 pairs.

Polyploidy as a Conservative Force in Evolution. A diploid that can give rise to polyploids is flexible in the evolutionary sense because of its potential for rapid production of new types. This is especially important in the event of sudden environmental changes unfavorable to previously existing members of the group.

Polyploidy, once developed, however, makes for a certain kind of inflexibility. Mutations to the recessive form of a gene have a reduced chance of expressing themselves phenotypically in polyploids. The brief discussion of autotetraploid ratios contrasted with diploid ratios should have given you an indication of this.

Allopolyploidy has much the same general effect on the phenotypic expression of mutant recessive genes. At least the effect is the same for loci in common among the diverse genomes of an allopolyploid. Let us take any amphidiploid and designate its different genomes as 1 and 2. If a gene A occurs in both genomes, the genotype for a homozygous dominant individual might be written $A^1A^1A^2A^2$. Suppose now that gene A^1 mutates to a recessive form a^1. Reproduction of the mutant allele followed by recombination may produce individuals of the genotype $a^1a^1A^2A^2$. But dominant alleles are still present, and the dominant phenotype will be preserved.

Assuming the amphidiploid to behave cytogenetically like a diploid, no normal recombination can give rise to completely homozygous-recessive individuals under the conditions we have outlined. If such homozygous recessives are to occur, independent mutation of the A gene must take place in genome 2. On the assumption that mutations are random events, the probability of simultaneous mutations in the two different genomes would be something like the square of the probability for one mutation. Of course, a recessive mutation occurring in one of the genomes might be reproduced at length until a corresponding mutation of the wild-type allele took place in the other genome. But this possibility scarcely argues for the genetic flexibility of polyploids. Some experimental results obtained by L. J. Stadler are pertinent. In studies of the frequencies of visible mutations induced by X-rays, he found that significantly fewer mutations appeared in tetraploid wheats than in diploids. In a hexaploid wheat, he observed no mutations at all.

The visible effects of recessive-gene mutation tend, then, to be suppressed by polyploidy. The result is that polyploids have largely lost one of the important sources of variability available to diploids. Since inherited variability is essential for evolution, one of the evolutionary disadvantages of polyploidy becomes apparent.

But if in some ways polyploidy imposes a reduction in phenotypic variability, in other ways it provides for flexibility. Consider once again gene A, which occurs in two different genomes of a polyploid, where it can be designated as A^1 and A^2. Even though A has a vital function, either A^1 or A^2 may be free to mutate to an allele with a new function. Because of duplication of genetic material in polyploids, new genetic functions may arise without accompanying loss of old ones. By contrast, in a diploid the loss of a genetic function often is lethal.

Polyploidy in Relation to Asexual Reproduction. The frequency of occurrence of polyploidy within plant groups has been found to be positively correlated with the presence of means for asexual reproduction, which takes many forms in the plant kindgom. Both the reasons for this association and its consequences have been objects of active investigation accompanied by fruitful debate. We cannot detail this debate here. Clearly, however, polyploidy can become established more readily where means for asexual reproduction are at hand. This is because one of the frequent immediate consequences of polyploidy is reduction in sexual fertility.

Continued asexual reproduction has the consequence that the possibility for genetic recombination is lost. Here, then, is another aspect of the genetic inflexibility often characteristic of polyploidy.

Polyploidy and Geographical Distribution. When polyploids show physiological and morphological differences from their diploid prototypes, they can be expected to show adaptation to different environmental conditions. And in fact, polyploids frequently have different geographical distributions from those of their diploid relatives.

For an example we can turn to the genus Crepis, which belongs to the family of plants called the composites. Races of Crepis that occupy the Mount Hamilton range in California appear to be allopolyploids. The climate of their habitat combines a hot, dry summer and a mild winter. This is not a climate typical of the regions inhabited by any of the diploid species of Crepis. Indications are, however, that the allopolyploids of the Mount Hamilton region arose from diploid progenitors, one of which is adapted to a cold, dry climate, and the other to rather mild, relatively moist conditions. The implication is that the tolerance of dryness on the one hand and tolerance of a mild winter on the other comes about in the allopolyploids as a result of the combination of different genomes.

We can generalize to say that polyploids, through the accumulation of diverse genomes, have the ability to acquire wide ranges of environmental tolerance. This, plus the fact that different polyploid forms may be *rapidly* evolved, makes polyploids expert colonizers, especially when environmental change opens new areas to plant invasion. For any given instance, however, the success or failure of polyploidy as an evolutionary mechanism depends on a variety of factors. In the higher northern latitudes, the proportion of polyploids is particularly great. Stebbins has emphasized that the "success" of polyploidy in these regions is partly attributable to the prevailing growth habit among arctic plants. Arctic flora include a great many perennials with means for asexual reproduction, a condition that favors the establishment of polyploidy.

Keys to the Significance of This Chapter

Ordinarily, chromosomes behave regularly and predictably at mitosis and meiosis. The genes carried in chromosomes reproduce themselves with accuracy at definite times in the nuclear cycle. A genome is basically stable in the structure and number of its chromosomal components. Genetic systems, therefore, are characterized by an underlying conservatism. If they were not, we would not be commonly confronted with the familiar genotypic and phenotypic ratios that attest so impressively to regular patterns of inheritance and that facilitate our study and understanding of genetics. More important, life would be quite different than we now know it.

But suppose that genes always reproduced other genes exactly like the originals. Suppose that genome structure were perfectly stable. Again life would be very different than we find it. But, in fact, variation is characteristic of life and is the basis of the evolution of living matter. One source of heritable variation is gene mutation, which will be considered at length in the following chapter. Another continuing source of heritable variation is alteration in genome structure.

Genetic systems, then, are merely conservative; they are not inflexible. The standard structure of a genome is subject to a bewildering array of realizable variations. Individual chromosomes may be altered by translocation, inversion, deficiency, duplication. Whole chromosomes may be lost or may be replicated to an unusual degree. Entire sets of chromosomes may be gained, or they may disappear. One aberrant circumstance may set in motion whole cycles of extraordinary chromosomal behavior.

Alterations in genome structure affect fertility and such vital components in the genetic system as synapsis and crossing over. Gene action is influenced by position effects, altered dominance relations, and changes in gene dosage. The physiology of the organism, therefore, is not independent of genome structure; structural changes account for part of the variety of life.

Organisms with unbalanced or unstable genomes are usually at a disadvantage, sometimes in viability, frequently in fertility. But not every alteration of a genome is accompanied by imbalance or instability. Reduplication of complete genomes need not unbalance an organism at all. And some polyploids are able to behave quite regularly at meiosis and to undertake sexual reproduction without paying the penalty of sterility.

By means of various forms of auto- and allopolyploidy, taxonomic groups within the plant kingdom have elaborated a variety of new forms whose fitness for survival is tested by their environments. The relative rapidity with which this process of elaboration can take place is particularly significant under circumstances of sudden change in environment. Many plant groups constantly participate in unconscious experimentation to provide types better adapted to altering conditions and to the different ecological circumstances occurring at the boundaries of the original locale of the group and within it. Thus polyploids, and also other deviants from standard genome structure, are participants in evolution.

REFERENCES

Avery, A. G., Satina, S., and Rietsema, J., *Blakeslee: The Genus Datura.* New York: Ronald Press Co., 1959. (Chaps. 5 and 6 are particularly pertinent. The book includes a short biography of the geneticist A. F. Blakeslee.)

Blakeslee, A. F., "New Jimson Weeds from Old Chromosomes." *J. Heredity,* **25**:80–108, 1934. (A readable summary of extensive investigations of trisomics. The article includes many useful diagrams and photographic illustrations.)

Clausen, J., Keck, D. D., and Hiesey, W. M., "Experimental Studies on the Nature of Species. II. Plant Evolution through Amphiploidy and Autoploidy, with Examples from the Madiinae." *Carnegie Inst. Wash.* Pub. 564, 1945. (Includes brief summaries of our knowledge of the origin of many allopolyploids. For example, the origin and history of the loganberry, pp. 112–115, is likely to be of interest to many students.)

Cleland, R. E., "The Cytogenetics of Oenothera." *Advances in Genetics,* **11**:147–237, 1962. (Major review of an exotic cytogenetic system.)

Darlington, C. D., *Chromosome Botany.* London: Allen & Unwin, 1956. (Brief and readable. Considers chromosomes in a variety of contexts.)

Huskins, C. L., "The Origin of *Spartina townsendii.*" *Genetica,* **12**:531–538, 1930. (Describes the natural origin of a new amphidiploid species.)

Little, T. M., "Gene Segregation in Autotetraploids. II. *Botan. Rev.,* **24**:318–339, 1958. (This, together with an earlier article written for the same journal by Little, summarizes autotetraploid ratios and discusses their basis.)

McClintock, B., "The Stability of Broken Ends of Chromosomes in *Zea mays.*" *Genetics,* **25**:234–282, 1941. (An admirable paper in experimental cytogenetics. Gives a detailed discussion of breakage-fusion-bridge cycles.)

McClintock, B., "The Association of Mutants with Homozygous Deficiencies in *Zea mays.*" *Genetics,* **26**:542–571, 1941. (Description of the production of homozygous minute deficiencies through the aberrant behavior of ring chromosomes and of the phenotypic effects of these deficiencies.)

Painter, T. S., "Salivary Chromosomes and the Attack on the Gene." *J. Heredity,* **25**:464–476, 1934. (A well illustrated, general discussion of the use of salivary-gland chromosomes in localizing genes in Drosophila.)

Rhoades, M. M., "Preferential Segregation in Maize." In *Heterosis,* 66–80. Ames: Iowa State College Press, 1952. (Clear treatment of a difficult subject.)

Rhoades, M. M., "The Cytogenetics of Maize." In *Corn and Corn Improvement*, 123–219. New York: Academic Press, 1955. (A definitive review.)

Rick, C. M., "Cytogenetics of the Tomato." *Advances in Genetics*, **8**:267–382, 1956. (A comprehensive review that includes lists of genes, a chromosome map, and other information on the tomato in addition to what is implied by the title.)

Sears, E. R., "The Aneuploids of Common Wheat." *Univ. Missouri Agr. Exp. Sta. Res. Bull. No. 572*, 1954. (Summary of massive and precise work on aneuploidy in an important crop plant.)

Stebbins, G. L., *Variation and Evolution in Plants*. New York: Columbia University Press, 1950. (Chaps. 8, 9, and 10 deal with aspects of polyploidy important in evolution. Scholarly, vigorous treatment.)

Sturtevant, A. H., and Beadle, G. W., *An Introduction to Genetics*. Philadelphia: W. B. Saunders, 1939. Available as a paperback reprint, New York: Dover Publications, 1962. (Chaps. 8, 9, and 11 give an excellent introduction to the basic cytogenetics of chromosomal aberration.)

Sturtevant, A. H., and Randolph, L. F., "Iris Genetics." *Am. Iris Soc. Bull., No. 99:* 52–66, 1945. (Includes an elementary discussion of autotetraploid ratios and the practical problem of recovering recessive types from autotetraploids.)

Swanson, C. P., *Cytology and Cytogenetics*. Englewood Cliffs: Prentice-Hall, 1957. (An excellent text. Detailed and comprehensive. Valuable as a reference to students of genetics who wish to go more deeply into cytological matters.)

QUESTIONS AND PROBLEMS

7-1. What do the following terms and phrases signify?

acentric	inversion loop
amphidiploid	physical mapping of chromosomes
aneuploid	preferential segregation
chromatid assortment	reciprocal translocation
chromosomal ring	semisterility
chromosome bridge	somatic doubling
cross band	somatic pairing
crossover suppressor	standard chromosome type
deficiency	sticky chromosome
deletion	translocation homozygote
dicentric	translocation heterozygote
euploid	trivalent

7-2. Certain mice execute bizarre steps—in contrast to the normal gait for mice—and are called "waltzers." The difference between normals and waltzers is genetic, with waltzing being a recessive characteristic. W. H. Gates crossed waltzers with homozygous normals and found among several hundred normal progeny a single waltzing mouse, a ♀. When crossed to a waltzing ♂, she produced all waltzing offspring. Crossed to a homozygous normal ♂, she produced all normal progeny. Some ♂♂ and ♀♀ of this normal progeny were intercrossed, and there were no waltzing offspring among their progeny. Painter examined the chromosomes of waltzing mice that were derived from some of Gates' crosses and that showed a breeding behavior similar to that of the original, unusual waltzing ♀. He found that these individuals had 40 chromosomes, just as in normal mice or the usual waltzing mice. In the unusual waltzers, however, one member of a chromosome pair was abnormally short. Interpret these observations, both genetic and cytological, as completely as possible.

7-3. Why is it easier to localize the genes carried in the sex chromosomes than it is to localize the genes carried in particular autosomes?

7-4. Give reasons why it is generally easier in Drosophila than in mice to localize genes in particular regions of particular chromosomes.

7-5. C. R. Burnham, working with a semisterile line of corn, found that crosses with normal gave progeny in a ratio of 1 normal to 1 semisterile. Further genetic tests showed *semisterility* giving 4.1 percent recombinations with gene *an* (*anther ear*) and 1.1 percent recombination with *ra* (*ramosa tassel*). From linkage studies carried out with standard chromosomal lines of corn, it is known that *an* is in chromosome 1, and *ra* in chromosome 7. How can you account for this failure of semisterility to recombine at random with members of two separate linkage groups?

7-6. Do you think it likely that the semisterility in Burnham's line of corn is due to gene mutation? Elaborate.

7-7. What unusual cytological configuration might you expect to see if chromosomes of the semisterile line of Question 7-5 were examined at pachytene?

7-8. In corn, a plant *pr/pr*, having standard chromosomes, was crossed with a plant homozygous for a reciprocal translocation between chromosomes 2 and 5, and for the *Pr* allele. The F_1 was semisterile and phenotypically *Pr* for aleurone color. A backcross to the parent with the standard chromosomes gave: 764 semisterile, *Pr;* 145 semisterile, *pr;* 186 normal, *Pr;* and 727 normal, *pr.* What can you say about the location of gene *pr* with reference to the translocation point?

7-9. Summarize the crosses outlined in the preceding question with a series of diagrams, showing both the genes and the chromosomes involved for all parents and progenies. When homologous chromosome pairs are shown at meiosis, represent them at a four-strand stage. To make the diagrams clearer, you might draw chromosomes 2 with colored pencil and chromosomes 5 with ordinary pencil. (The locus of *pr* is in chromosome 5.)

7-10. In the *standard* arrangement of genes in chromosome 3 of *D. melanogaster*, *sr* (*stripe*), *e* (*ebony*), and *ro* (*rough*) occur in the order listed. One strain of Drosophila has been found to have a gene sequence *sr—ro—e*. Name the type of aberration shown in this strain. How, besides showing different gene order indicated by linkage tests, might the aberration be demonstrated?

7-11. Single crossovers within the inversion loop of inversion heterozygotes give rise to chromatids with duplications and deficiencies. Diagram what happens when, within the inversion loop, there is a *double exchange* involving the same two strands and when the centromere is outside the inversion loop. (Remember that crossing-over occurs at a four-strand stage.) Are aberrant chromatids produced?

7-12. In corn, how many chromosomes would be typical of the root-tip nuclei in (a) a monosomic, (b) a trisomic, (c) a monoploid, (d) a triploid, (e) a tetrasomic, (f) an autotetraploid?

7-13. In corn, R is a gene for red aleurone; its recessive allele r determines colorless. McClintock and Hill made a cross between a diploid rr and a trisomic Rrr, using the diploid as the female parent. There were 1282 red kernels and 2451 colorless in the progeny. Confirm that this approximation of a 1:2 ratio is expected, if the $n + 1$ pollen grains are nonfunctional.

7-14. On the basis of the result noted in the preceding question, what ratio of colored to colorless would be expected from the cross rr ♀ $\times$ RRr ♂? (Actually McClintock and Hill obtained 646 red kernels and 355 colorless after making such a cross.)

7-15. Assume that gene x is a new mutant in corn. An xx plant is crossed with a triplo-10 individual (trisomic for chromosome 10) carrying only dominant alleles at the x locus. Trisomic progeny are recovered and crossed back to xx ♀ ♀. What ratio of dominant to recessive phenotypes is expected if the x locus is not in chromosome 10? If it is in chromosome 10?

7-16. Assuming that you had a complete set of trisomics in maize, how could you utilize the type of experiment indicated in the preceding question to determine which of the chromosomes of corn was the site of any new gene mutation you happened to discover?

7-17. What advantages would this trisomic method for determining the linkage group of a new mutant have over the familiar method of utilizing testcrosses involving the new gene and known members of already established linkage groups?

7-18. A gene A is situated very near the centromere of its chromosome. What phenotypic ratio of A to a is expected if $AAaa$ plants are testcrossed against $aaaa$ individuals? (Assume that the dominant phenotype is expressed whenever at least one A allele is present.)

7-19. What would be the answer to Question 7-18 if the locus of A were far enough from the centromere to permit "random assortment of chromatids?"

7-20. The American cultivated cotton *Gossypium hirsutum* has 26 pairs of chromosomes. The Asiatic cotton *G. arboreum* has 13 pairs of chromosomes, as does an American cotton, *G. thurberi*. When *hirsutum* and *arboreum* are crossed, the resulting triploids have 13 pairs of chromosomes and 13 single chromosomes at meiosis. Thirteen pairs and 13 single chromosomes are also observed in triploids derived from the cross *hirsutum* × *thurberi*. When *thurberi* and *arboreum* are crossed, the F₁ individuals are highly sterile, with chromosome pairing being very irregular. What do these observations suggest about the possible origin of American cultivated cotton?

7-21. What steps might an investigator take to produce experimentally from diploid species a cotton chromosomally equivalent to American cultivated cotton?

7-22. How may amphidiploids arise other than through chromosome doubling of a sterile F₁ interspecific hybrid?

CHAPTER 8

Mutation

Mutations constitute the principal raw material with which nature works to bring about evolution. They also provide the principal tools of the geneticist, for only through a comparison of alternative states of genes is the geneticist able to gain insight into the structure and function of genes. In fact, an immutable genetic locus would be an undetectable locus.

General Properties of Mutations

Mutations are sudden, heritable changes in the structure of the genetic material. Plant and animal breeders for centuries have observed occasional *sports*—plants or animals with new phenotypes that sometimes occur and which, after their occurrence, pass on the new traits to their descendants. Not all such sports, of course, are due to *de novo* changes in the genetic material of the sport, for a recessive mutation can be carried for many generations as a hidden impurity and show up only when two heterozygotes mate.

The study of dominant mutations requires no special techniques because such mutations express themselves even in a heterozygote (Fig. 8-1). If a mutation proves, by its behavior in subsequent generations, to be a dominant, then it must have arisen in the generation in which it first appeared.

On the other hand, the detection of recessive autosomal mutations requires special breeding projects, because such mutations express themselves only in the homozygous state, and only in inbred strains can they be easily distinguished from recessives already present in a heterozygous state. Sex-linked recessive mutations, however, can be detected directly in organisms such as man and

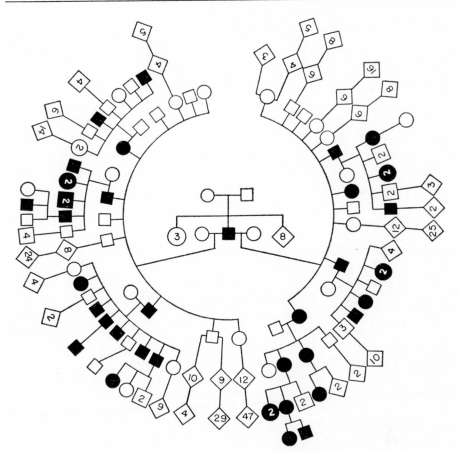

Figure 8-1. *A human pedigree of a dominant mutation, chondrodystrophic dwarfism. A female is indicated by a circle, a male by a square. Individuals whose sex was not known to the investigator are indicated by a diamond. Individuals showing the trait in question are generally designated by a black symbol. A number enclosed by a large symbol indicates the number of siblings (that is, brothers or sisters) who are not listed separately. The mutant individual (center black square) had normal parents, three normal sisters, and eight other normal siblings of unidentified sex. He had two normal wives, twenty-two children, seven of whom were affected. Only affected children had affected children. (From Stern,* Principles of Human Genetics, *W. H. Freeman and Co., 1960, p. 448.)*

Drosophila, where the males are hemizygous for all or part of the X-chromosome. Figure 8-2 gives a pedigree involving Queen Victoria of England. It is probable that she was the source of a sex-linked recessive mutation for hemophilia, a defect of the blood-clotting mechanism, since none of her male ancestors displayed the trait.

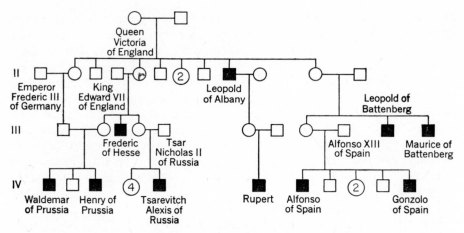

Figure 8-2. *Pedigree of hemophilia, a sex-linked mutation, in the royal families of Europe. All of Queen Victoria's children are entered. Later generations comprise many more individuals than are indicated here. (From Stern,* Principles of Human Genetics, *W. H. Freeman and Co., 1960, p. 449.)*

Mutations occur in somatic tissues as well as in the reproductive cells of organisms. In such cases, the mutant organism is genotypically, and in many instances phenotypically, a *mosaic* of normal and mutant tissue. However, in higher animals and in certain plants if the reproductive cells are not affected, the mutant trait is not passed on to future generations and genetic tests cannot be performed to verify that what is detected as a phenotypic change is of a mutational nature. In many other organisms, particularly microorganisms, a distinction between somatic and reproductive cells is largely irrelevant since whole organisms can be grown from somatic cells (for example, from asexual spores or bits of hyphae of the mold Neurospora). Mutations occurring in asexual stages can pass later through the sexual stages and be subjected to genetic tests.

What kinds of mutations are available for study? We can study only mutations that result in readily identifiable phenotypic changes. Some mutations are detected by an obvious change produced in the appearance or behavior of an organism, for example, white eyes or curly wings in Drosophila, shrunken kernels in corn, or the waltzing trait in mice. Other mutations alter the phenotype of the organism so slightly that they can be detected only by special techniques. Mutant genes that give these slightly modified phenotypes are called *isoalleles*. At the other extreme are mutations that result in the death of the organism. Such lethal mutations are relatively frequent. This should not be surprising since most changes in genes might be expected to be deleterious to an organism already well adapted to its environment.

Dominant lethals, of course, cannot be directly studied genetically unless death occurs only after the reproductive stage of the organism. Recessive lethal mutations, however, may be studied genetically since they can be transferred from generation to generation in a heterozygous condition. If a recessive lethal is located on the X-chromosome, it survives as a heterozygote only in the sex having two X-chromosomes. In Drosophila, for example, the inheritance of a sex-linked lethal (*l*) follows the pattern illustrated in Figure 8-3. The lethal is carried only by the female, from generation to generation; and the sex ratio observed in crosses with heterozygous females, instead of being 1:1, is 1♂:2♀.

The effect of sex-linked lethals on the sex ratio makes them relatively easy to detect. For this reason, much of the work on mutation in Drosophila has utilized this particular kind of mutation. Special techniques have been worked out to make the detection of new sex-linked lethal mutations easy and objective; we will describe one of these methods in a later section.

In haploid organisms recessive lethals cannot ordinarily be maintained from generation to generation, although in Neurospora, for example, they can be carried in heterokaryons that include wild-type nuclei.

A special class of lethals, *conditional lethals*, can be passed on and studied, even in haploid organisms. Mutations affecting the nutritional properties of the organism may be considered conditional lethals when the mutant organism fails to propagate or long survive on minimal medium but grows on appropriately supplemented medium. (As a specific example, in Neurospora mutations that result in the loss of ability to synthesize tryptophan are lethal in an environment lacking tryptophan but not lethal if tryptophan is supplied.) Temperature-sensitive lethal mutations form another category of conditional lethals. These are mutations that render essential genes inactive at high temperature but not at low temperature, so that a strain carrying such a lethal can be

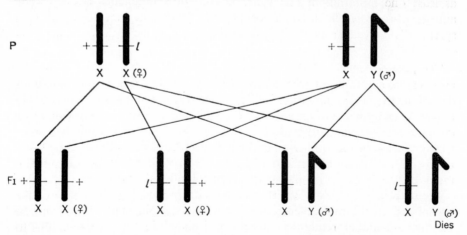

Figure 8-3. *The progeny of a female heterozygous for a sex-linked lethal show a 2:1 sex ratio.*

propagated at low temperature. In other instances, mutant genes behave normally at high but not at low temperatures. Since lethals, as a class, represent the sum of all kinds of mutations that cause the premature death of the organism, they are often used in studies of mutation rates because they offer the opportunity of studying the summed frequency of mutation of many genes.

How frequently do mutations occur? First let us point out that the same mutation can recur. Some genes appear to mutate at much higher rates than other genes. Back-mutation can also occur. That is, a mutant gene may apparently "revert" to the normal or wild-type state. The frequency with which this occurs varies from gene to gene and also varies for different mutations of the same gene.

The essential characteristics of spontaneous mutation so far discussed can be summarized by means of a simple diagram:

$$A \underset{v}{\overset{u}{\rightleftharpoons}} a$$

in which A and a are alleles and u and v represent mutation rates in the two directions. The diagram implies that:

1. Genes spontaneously change to allelic forms, at rates generally constant for any particular gene but varying from gene to gene.
2. Mutation is often reversible; the rates in the two directions are often different.

We have so far considered only mutation of particular genes. Higher organisms have tens or hundreds of thousands of different genes. Even if mutations were relatively rare, we might expect a high proportion of gametes to contain at least one mutation in one gene. Studies of spontaneous sex-linked and autosomal-recessive lethals in Drosophila indicate that over 1 percent of all sperm cells carry at least one newly occurring lethal in one of their chromosomes.

The Selection and Isolation of Mutants. Mutation of a specific gene is, in most cases, a rare event whose careful study requires special methods for its detection. In general, mutation studies demand that the investigator be able to handle large numbers of organisms from which he can isolate and identify rare deviants from a standard type. For this reason much of the recent work on gene mutation has used microorganisms, which are easily cultured in enormous numbers. Moreover, most microorganisms are haploid, a condition that permits mutations to be expressed phenotypically soon after they occur.

Let us consider a few of the techniques used in studying mutation in microorganisms. In the mold Neurospora, strains containing mutations that cause various nutritional requirements can be isolated in the following manner. Asexual spores are collected and then incubated in minimal liquid medium for a time just long enough to allow spores to germinate. The suspension of germinating spores is then poured through fine gauze. The spores that have germinated are trapped because their mycelial web catches in the gauze. Ungerminated spores, having no hyphae, pass through the gauze. The ungerminated spores are then plated on a nutritionally enriched medium and the cultures that germinate are tested for their nutritional requirements. This filtration method utilizes the phenomenon of conditional lethality to separate mutants from nonmutants. Specifically, it selects and preserves mutants with nutritional requirements unsatisfied by the original incubation medium. In the filtration step nonmutants are removed from the population. The environment is then so adjusted that the conditional lethals grow and can be collected. By this "enrichment" technique, a population originally containing many nonmutants and few mutants is converted into a population consisting largely of mutants.

A similar method is often used with bacteria. Penicillin is an antibiotic that kills only growing bacteria. If a population of bacteria is placed in a minimal medium containing penicillin, those cells that can grow are killed. Since growth is not possible for mutants having nutritional requirements not supplied by the medium, such mutants do not grow and are not killed. After the penicillin is washed away, the population is transferred to a nutritionally supplemented medium and the cells that grow are later tested for nutritional requirements.

Neither of these "enrichment" methods is highly efficient for the quantitiative recovery of mutants and therefore they are not good ways of measuring the frequency of mutation. However, once such nutritional mutants are obtained, the frequency of reversion to prototrophy (ability to grow without a specific growth supplement) is easily determined by plating populations of cells (in the case of bacteria) or spores (in the case of molds) on a minimal medium not containing the nutrient specifically required by the mutant under study. Only those cells or spores that are genetically and phenotypically revertant can grow to form colonies.

Other types of mutants as well as nutritional revertants are easily recovered by selective techniques. For example, in bacterial populations the number of mutants resistant to an antibiotic like streptomycin or resistant to a particular phage can easily be measured by plating a large population of cells onto plates containing the antibiotic or phage and then determining the fraction of the total cells plated that are able to grow in the presence of the selective agent.

This type of selective experiment is *not*, however, a measure of the *mutation rate* to, for example, streptomycin resistance. To illustrate why the experiment is not a true measure of the mutation *rate* let us consider an example. Suppose that a particular gene mutates with a probability x per cell per generation, and that mutant cells multiply at the same rate as nonmutant ones. In every generation, as the population grows, another fraction, x, of mutants is added to the fraction already present in the population; the frequency of mutants present in a population is not constant but increases with time owing to the continued accumulation of new mutants. The frequency of mutants in a population at any

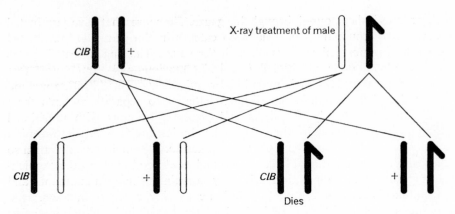

Figure 8-4. *The first step in the* ClB *method for detecting sex-linked lethals is to mate the treated males with females heterozygous for the* ClB *chromosome.*

one time is related, but not identical, to the rate at which new mutants are formed. Most mutation rates are low, and the rate at which mutants accumulate in a population is in some cases negligible. Nevertheless, various considerations, including differential reproduction of mutants relative to nonmutants, make determination of actual mutation rates a tricky business. This is true of higher organisms as well as of microorganisms.

Induced Mutations

Our discussion to this point has been limited to new alleles that originate spontaneously. Geneticists have long hoped that the process of mutation could be brought under experimental control, so that new hereditary qualities could be induced at will in an experimental population. This hope has been in part realized, but only to the extent that spontaneous mutation rates can be enormously "speeded up" by a variety of external agents. The increase in mutation rate is a general one, and it is still impossible to single out particular genes and mutate them at will.

The *ClB* Method for Detecting Sex-Linked Lethals. The first clear evidence for the successful induction of mutations was reported in 1927 by H. J. Muller. His demonstration that X-rays induce mutations was made possible through the development of an efficient technique for the easy identification of new lethal mutations in the X-chromosome of Drosophila. The technique is called the *ClB* technique, because of the three factors essential to its operation:

C stands for a long inversion of the X-chromosome that acts as a crossover suppressor (recall Chap. 7) and therefore maintains the integrity of the chromosome from generation to generation.

l stands for a known recessive lethal on the X-chromosome carrying the inversion.

B stands for bar eye, which serves as a phenotypic marker for this chromosome.

The technique is simple in operation: Male Drosophila are treated with the agent to be tested for its ability to induce mutation. Muller used X-rays. These males are then mated with *ClB* females—that is, with females heterozygous for the *ClB* chromosome (Fig. 8-4).

From the F_1 progeny, the wide-bar females (carrying the *ClB* chromosome from their mothers and a "treated X-chromosome" from their fathers' sperm) are selected and mated, each in a separate bottle, with their normal brothers (almost any male Drosophila can be used in this mating). Figure 8-5 shows the result of this cross.

The advantage of the *ClB* method is that it depends for its interpretation only on noting whether or not males appear in F_2 of any mating. The investigator can simply look through the side of the glass bottle in which the flies are grown; if he sees males, he can conclude that no lethal was induced in the treated X. If he sees no males, he can check more carefully and, if they are really absent, he can conclude that a lethal was induced in the treated X.

The summed spontaneous mutation rate for all lethals on the X-chromosome is about two per thousand X-chromosomes per generation (0.2 percent). Observed frequencies that exceed this figure significantly indicate that the agent tested has increased the mutation rate.

The *ClB* method has contributed greatly to genetic knowledge. But it has

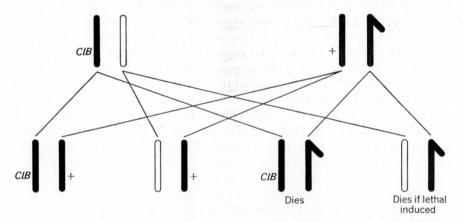

Figure 8-5. *The bar-eyed female progeny of the cross in Figure 8-4 are heterozygous for the ClB chromosome (from their mothers) and the "treated X-chromosome" (from their fathers). If a lethal was induced on the treated X, these females, mated to normal males, will have no sons. If they have sons, no lethal was induced.*

been found to have certain limitations, chiefly because the inversion of the X-chromosome in the *ClB* stock does not completely suppress crossing over, and misleading or inaccurate observations occasionally result. For this reason, the method has been improved by relatively minor modifications also devised by Muller.

Induction of Mutations by Ionizing Radiations. X-ray is representative of *ionizing radiations*, a group that also includes the α, β, and γ radiation of radioactive substances, protons, and neutrons, in addition to X-rays. The atomic and hydrogen bombs are now familiar and terrible sources of ionizing radiation.

As almost everyone knows nowadays, an atom is made up of a positively charged atomic nucleus and a surrounding constellation of negatively charged electrons, the charges so balanced that normal atoms are electrically neutral. When ionizing radiations pass through matter, they dissipate their energy in part through the ejection of electrons from the outer shells of atoms, and the loss of these balancing, negative charges leaves atoms that are no longer neutral but are positively charged. Such charged particles are called *ions*. When an atom becomes ionized, the molecule of which it is a part probably undergoes chemical change; when this changed molecule is a gene or a part of a gene, and when the modified gene duplicates its new pattern, the result of the change is a mutation. Probably X-rays induce mutations in other ways as well.

The amount of X-ray treatment is measured in *Roentgen units* (*r* units), which are calibrated in terms of the number of ionizations per unit volume of air under standard conditions. It is important to note that *r* units do not in themselves involve a time-scale for irradiation; a given dosage in *r* units may be obtained by low intensities over long periods of time or by high intensities over shorter periods. The frequency of lethals induced by X-rays seems, for moderate dosages, to be simply proportional to the dosage of X-rays. Roughly, 1,000 *r* units increase the frequency of sex-linked lethals in Drosophila from the spontaneous level of about 0.2 percent to about 3 percent; 2,000 units to about 6 percent; 4,000 units to about 12 percent, and so on. In a graph, this relationship between dosage and frequency appears to be a straight line, over a considerable range of dosages (Fig. 8-6). It drops away from a straight line markedly at high X-ray dosages; this may be mainly because the *ClB* and similar techniques really measure the frequency of lethal-bearing chromosomes rather than of the lethals themselves. At high dosages the chance is good that a given chromosome will carry two or more induced lethals. This chromosome would be counted as only a single induction by the *ClB* technique. There is, however, debate on the question of whether the shape of the curve may have some further significance beyond this simple relationship.

If the essentially straight-line relationship between dosage and effect can be accepted as fact, at least two interesting conclusions are indicated. The first is that a specific effect is produced by a single "hit." If two or more independent hits were required to produce a lethal mutation, then the relationship between density of ionization and percentage of lethals would plot as a curve, not as a straight line. It is with this question that some of the debate over the true shape of the curve is concerned. Some of the effects of X-radiation on chromosome structure clearly require two hits. The question is whether the X-ray-induced lethal mutations may involve these same kinds of changes in chromosomes, or whether they are mostly specific, single-gene changes involving perhaps chemical changes in the genes themselves, such as might be induced by single hits. We will consider the chromosomal changes later in this chapter.

Is radiation cumulative in its induction of genetic changes? This is an important question from a practical point of view since the background radiation to

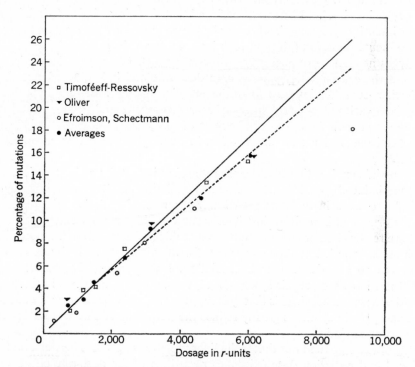

Figure 8-6. *As the dosage of X-ray (in* r *units) increases, the percentage of mutations rises. Points on the graph represent the observations of investigators as shown with regard to the induction of sex-linked lethals by X-rays. For low dosages, the relationship is approximately linear; the solid line is calculated from this linear relationship at low dosages. At higher dosages, the observed data fall away from this line. The dotted line is based on a calculation recognizing the probability that more than one lethal may be induced on any particular chromosome. (After Sturtevant and Beadle,* An Introduction to Genetics, *W. B. Saunders Co., 1940, p. 211. From Timoféeff, Zimmer, and Delbrück.)*

which we are all exposed, although still quite low, has been significantly increased as a consequence of the testing of nuclear weapons. It has been estimated that in a generation (30 years) we each receive a dose of about 3.0r to our gonads from natural background radiation and an additional 0.1r from the nuclear-weapons tests already performed. Are these doses in a thirty-year period as effective in inducing mutations as the same dose would be if administered in a short time?

The frequency of lethal mutations induced in Drosophila sperm by X-rays is not a function of the rate at which the radiation is applied. Low intensities over long periods of time, high intensities over short periods of time, or intermittent radiation, all produce the same percentage of lethals from the same total dosage in r units. However, this does not appear to be true of spermatogonia or oöcytes in mice. W. L. Russell has found that chronic radiation (administered at low intensity for a long time) results in fewer induced mutations than the same dose of acute radiation (given at high intensity for a short time). Thus in some cell types at least there is clearly a dose-rate effect. More work is required to find out the exact relationship between dose rate and induction of mutations.

Further complications are that the effectiveness of radiation in inducing mutations varies at different stages of spermatogenesis and oögenesis and that variations in such factors as the amount of oxygen present can also influence the effectiveness of a dose of radiation in causing mutations or chromosome breaks.

A possible explanation of the dose-rate effect is that for a short period after they are induced genetic lesions caused by radiation can be repaired by physiological processes in the cells. The lower the rate of radiation, the fewer the lesions produced in a given time interval and the higher the fraction of lesions that are repaired before they become irreparable (perhaps through duplication). Such an explanation could also account for the different sensitivity of different cell types to genetic damage by radiation. For example, Drosophila sperm might not have such repair systems, and, thus, all genetic lesions in these cells would remain unhealed. This theory is supported by evidence that active metabolism and protein synthesis greatly reduce the frequency of chromosomal breaks induced by radiation.

It should be pointed out that if the linear relationship holds between mutants induced and radiation dose (Fig. 8-6) regardless of the dose-rate effect noted above, even the small amounts of radiation already produced by fallout have induced many mutations, in man as well as in plants and animals. How many have been induced cannot be determined with any degree of certainty since the dose-rate effects are not accurately known, nor is the intrinsic sensitivity of human germ cells to induction of mutations by radiation. Clearly, these matters are of great practical importance and ones about which much more knowledge is needed.

Induction of Mutations by Ultraviolet Light. Radiations of the same sort as visible light, but too short in wavelength to be visible to the human eye, fall within that range of the spectrum called the *ultraviolet*. Many chemical compounds absorb ultraviolet in rather narrow wavelength bands. That is, they behave as if they were "colored," but their "color" does not affect our senses directly because it involves a part of the spectrum that our eyes do not detect.

Ultraviolet is not an ionizing radiation. When compounds absorb sufficient energy in the ultraviolet, however, certain of their electrons are raised to higher energy levels, a state known as *excitation*. This is less drastic than the complete ejection of electrons, but it does result in greatly increased chemical reactivity on the part of the affected molecules. This may be the direct effect of ultraviolet involved in its mutation-producing (*mutagenic*) activity.

In contrast to X-ray, ultraviolet penetrates tissues only very slightly. It is difficult, for example, to irradiate a male Drosophila sufficiently with ultraviolet to affect the sperm in his body. For this reason, most of the work with ultraviolet has been done with other forms—bacteria, molds, and the pollen of higher plants—although the first work with ultraviolet was done by irradiating Drosophila eggs. When male flies are compressed between quartz plates, to bring their testes close to the surface, and their ventral sides are irradiated, some ultraviolet can penetrate to the sperm. Properly used, ultraviolet is a very effective mutagenic agent. The most effective wavelength of ultraviolet for inducing mutations is about 2,600 Å. This is the wavelength that is best absorbed by DNA and a wavelength at which proteins absorb little energy. This ultraviolet-wavelength correlation of maximum mutagenicity and of maximum energy absorption by DNA can be construed as evidence that, mutagenically, ultraviolet acts directly on DNA.

Chemical Mutagenesis. Many chemical substances have been shown to be mutagenic in various organisms, but in most instances their precise mode of action in inducing mutations is not known. However, in recent years a number of substances that interact with DNA in ways that are at least partly understood have been found to be highly mutagenic in microorganisms. Work with these mutagens has revealed a great deal about mutation.

The mutagens whose chemistry is somewhat understood are known either to interact with DNA directly, or to be effective only during DNA synthesis. Nitrous acid and alkylating agents such as nitrogen mustard react chemically with intact DNA and modify its structure. The so-called *base analogues*, such as 5-bromouracil and 2-aminopurine, are mutagenic because of their incorporation during DNA synthesis in place of the normal DNA nitrogen base constituents, which they closely resemble. We will return to a fuller discussion of the mode of action of these mutagens.

Of practical concern is the possible mutagenic nature of many of the substances, natural and artificially produced, to which man is exposed. For example, in western society, the average man in a lifetime ingests enormous amounts of caffeine (in coffee, tea, and some soft drinks), a substance that has been proven to be mutagenic in bacteria! It remains to be determined, however, whether ingested caffeine is mutagenic in mammals. Recently, certain chemicals have been found to be highly mutagenic at doses that are only slightly toxic to the treated cells. One example is N-methyl-N-nitro-N-nitrosoguanidine. Compounds of this sort may be particularly hazardous because the hazard is not immediately obvious.

Directed Mutation. A practical goal of geneticists for many years has been to find mutagens that act only on specific genes. In some organisms *mutator* genes exist that appear to induce mutations in other specific genes. An example is the gene *dotted* in corn, which causes heritable changes at another locus (the *A* locus). However, as will be described in Chapter 11, examples of this kind are interpreted other than as directed mutation. Nonspecific mutator genes exist that cause a high mutation rate of all genes. These mutator genes are most likely to be explained as causing the production within the organism of chemical mutagens, such as base analogues.

Before the era of modern genetics it was believed that acquired characteristics could be inherited. This notion of adaptive directed mutation is unsubstantiated and is now largely discredited. More recently, the appearance of antibiotic-resistant mutants in bacteria seemed perhaps to be an instance of directed mutation. The case is worth examining because it can be used to illustrate methods for making the important discrimination between mutagenic agents and agents that simply select preexisting mutants from mixed populations. As was pointed out earlier, the frequency of mutants resistant to a particular antibiotic like penicillin can be determined for a bacterial population by plating the population on plates containing penicillin. Only those cells resistant to the drug are able to grow and form colonies. Since in such plating experiments penicillin is always used to detect cells that are resistant to it, one can at the outset question whether exposure to the antibiotic in fact induces the mutants in addition to permitting their detection. How can we determine whether the drug induces changes to penicillin resistance in cells not resistant before exposure to the drug or only acts as a selective agent permitting preexisting mutants to grow?

This apparent dilemma may be attacked by a variety of techniques, one of the most elegant being "replica plating," which was devised by J. Lederberg (Fig. 8-7). A large population of bacteria is introduced onto plates not containing the drug. All viable bacteria grow and form colonies. Then a piece of velvet is stretched over a block of wood and pressed gently on a plate containing discrete colonies of the bacteria. Next the velvet is pressed on a fresh plate containing antibiotic. Some bacteria from the first plate will have been caught in the nap of the velvet and will serve as inocula for the second plate. The second plate is thus a replica of the first. Of course, only cells from resistant colonies on the first plate will produce new colonies on the second, antibiotic-containing plate; cells from sensitive colonies will not. By comparing the position of colonies on the original plate and on its replica, sister colonies may be identified. Samples can then be taken from colonies on the first plate, which have never been exposed to the drug, and determinations of sensitivity or resistance can be made. When such experiments are done, it is found that resistance arises in the absence of exposure to penicillin. Therefore, the antibiotic does not induce mutations; it merely detects and selects preexisting mutants by imposing differential survival on a mixed population of resistant and sensitive cells.

So we see that most apparent instances of directed mutation have other explanations. The most important examples of apparent directed mutation yet to be fully explained involve cytoplasmically rather than chromosomally determined characteristics such as chloroplasts. These systems are described in more detail in Chapter 11.

The Structural Basis of Mutation

Up to now we have only considered mutations in the simplest sense, as heritable changes in phenotype. We can imagine a variety of changes in the genetic

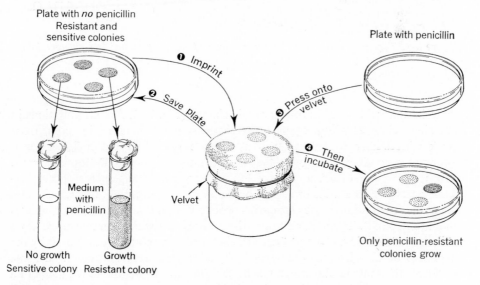

Figure 8-7. *The method of indirect selection of mutants by replica plating.*

away, and already the doomsday warnings are arriving, the foreboding accounts of a Russian horde that will come sweeping out of the East like Attila and his Nuns.
—*Red Smith in the Boston Globe.*

Substitution

"I can speak just as good nglish as you," Gorbulove corrected in a merry voice.
—*Seattle Times.*

Deletion

"I have no fears that Mr. Khrushchev can contaminate the American people," he said. "We can take in stride the best brain-washington he can offer."—*Hartford Courant.*

Insertion

He charged the bus door opened into a snowbank, causing him to slip as he stepped out and ran beneath the bus, which fell over him.—*St. Paul Pioneer Press.*

Inversion

Tomorrow: "Give Baby Time to Learn to Swallow Solid Food."
etaoin-oshrdlucmfwypvbgkq
—*Youngstown (Ohio) Vindicator.*

Nonsense

Figure 8-8. *Typographical errors as illustrations of types of mutational changes in the genetic code. (From Benzer, "Genetic Fine Structure,"* Harvey Lectures 56, *1961.)*

material that will produce a phenotypic change (Fig. 8-8). Are all mutations in fact single changes within a gene? The best answers to this question come from work with higher organisms such as Drosophila, where cytological investigations are possible. It is found that many mutations, especially those induced with ionizing radiations, are associated with gross rearrangements of chromosome parts (see Chap. 7). In Drosophila, many, though not all, of such changes are lethal when homozygous (and thus probably would not survive in organisms that are haploid). The apparent mutations associated with such rearrangements are probably in most cases owing to the breakage or loss of genes caused by the breakage of the chromosomes and their subsequent rearrangement. In some instances the mutant phenotype has been shown not to be due to a change in a gene or to a loss of a gene, but rather to the fact that a gene may have an altered expression when placed in a new location in the genome by translocation or inversion—the position-effect phenomenon (see page 208). The frequency of rearrangements such as inversions and translocations induced by X-rays, in contrast to the frequency of gene mutations, is not directly proportional to the dose of X-rays administered. Instead, such frequencies are roughly proportional to the square of the dose. This is what we might expect, since for a translocation or inversion to take place, two or more chromosome breaks must be present in the cell at the same time.

Many mutations, both spontaneous and induced, are not, however, associated with gross chromosomal rearrangements and do appear to be changes *within* genes. In such cases, as you might expect, a number of different mutational possibilities exist for any particular gene. When, for example, the *R* gene, a compound locus in corn, mutates, it may change to any one of a large number of mutant alleles. Alleles at this locus may be classified into four main groups, depending on how they affect aleurone color and certain pigments (not including chlorophyll) of the plant:

R^r = colored aleurone, pigmented plant—red or purple.
R^g = colored aleurone, unpigmented (therefore, green) plant.
r^r = colorless aleurone, pigmented plant—red or purple.
r^g = colorless aleurone, unpigmented (green) plant.

These groups may again be subdivided; there are, for example, at least 22 different alleles that fall within the R^r class but which differ among themselves in such characteristics as the amount and distribution of pigment they produce.

The various mutational possibilities for a particular gene are not equally likely to occur. For example, when an allele of the R^r class mutates, it is much more likely to change to one of the R^g or of the r^r class than to an r^g allele.

Similarly, the mutant alleles can in turn give rise by mutation to new alleles, or they can presumably back-mutate to R^r. Here again, the different possible mutations are not equally likely; some of the changes that you might expect to occur have in fact never been observed.

Analyses of such *multiple allelic series* indicate that mutation can modify a gene in a great variety of ways. However, to learn more precisely what kinds of changes mutations cause in genes we must turn to microorganisms, where highly sensitive techniques are available for the study of gene structure. We will consider mainly work on the *rII* genes of phage T4. This system has produced our most detailed knowledge of the nature of gene mutations. Work on other microorganisms supports the general picture derived from the phage.

In the next chapter we will describe in more detail our knowledge of the structure of the gene and specifically of the *rII* genes. For our purposes here we may note that the two *rII* genes of phage T4 consist of linear sequences of three hundred or more mutable elements that are separable by recombination. The resolving power of recombination analysis in this system is such that recombination events occurring between adjacent nucleotides of the DNA molecule are detectable.

S. Benzer found that about 20 percent of the spontaneously occurring *rII* mutations are deletions of parts of the gene. These range in size from very small deletions involving only a few base pairs, up to complete deletions of both *rII* genes, involving a region thousands of nucleotide pairs in length. The rest of the spontaneously occurring mutations behave as point changes in the gene (presumably changes of one nucleotide pair). These point changes can occur at hundreds of different locations within the gene. We can see that the overall frequency of mutation from *rII*+ to *rII* has a limited meaning; this frequency measures the sum of mutation rates at many different sites within the gene and also includes a miscellaneous collection of deletions of various sizes. You can get an idea of the relative mutation rates at different sites from Figure 9-6 on page 274. At some sites mutations have been observed repeatedly (note particularly the two very mutable sites; these Benzer terms "hot spots"); at other sites only one mutation has been found. We conclude that the rates of mutation from *r*+ to *r* at different sites within the gene vary enormously, and that the overall mutation rate, in these specific cases at least, depends primarily on a few very mutable sites.

Enormous differences in back mutation or reversion rates (*r* to *r*+) are also observed. Such reversion frequencies can be measured quantitatively and vary from 10 percent (such mutations are "unmappable" since the revertants are as abundant as recombinants in many crosses) to nondetectable (fewer than one *r*+ in $10^{10}r$ phage). Even different mutations apparently located at the same site can vary greatly in their revertibility; in fact, this is one criterion by which they may be distinguished.

Thus we see that from one viewpoint the mutability of a gene has little meaning, for genes are composed of a large number of subunits that may differ greatly in both forward and back mutation rates.

Back Mutation. Reverse mutations, by phenotypic tests, restore the original phenotype. Do reverse mutations in fact restore the original genetic structure? This is a difficult question to answer categorically. In many cases apparent back mutations are demonstrably not reversions at the site of the initial change, but are secondary changes at a different location in the genome that restore the original phenotype (see Fig. 8-9). Mutations of this kind are called *suppressor mutations*. The action of at least some such suppressors is to alter the metabolism of the organism so as to compensate for the original defect. Suppressor mutations can also occur *within* the same gene. These presumably act by making secondary modifications of the structure of the gene so that it can function in a way similar to, although usually not identical with, the wild-type gene function. Often, these suppressor mutations themselves determine a mutant phenotype. Such intragenic changes are difficult to distinguish from true reversions. Although

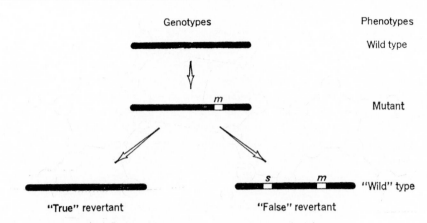

Figure 8-9. *Possible genetic structures of revertants.*

we can in some instances demonstrate that a particular back mutation is *not* a true back mutation but is owing to a suppressor, it is very difficult to prove the reverse, namely that a given back mutant is identical in genotype to the original wild type. These considerations re-emphasize that what we regard as measurements of back-mutation rates may in fact be the sums of rates for many different kinds of mutations—true reversions and various types of suppressor mutations.

The Chemical Basis of Mutation

With the identification of DNA as the genic material we may now hopefully inquire into the precise chemical nature of mutational events. In chemical terms, what are the changes in genes that we detect as mutations?

Following their description of the structure of DNA, Watson and Crick suggested a mechanism for the origin of spontaneous mutations. They proposed that the faithful reproduction of DNA depends on the specificity of hydrogen bonding between the paired bases of the DNA molecule, A always pairing with T, and G always pairing with C. As described in Chapter 5, this pairing specificity results in replication of the DNA, with each parental strand acting as a template for the polymerization of the new complementary strand. Watson and Crick suggested that occasionally the bases can undergo what is called a *tautomeric shift;* that is, as shown in the example given in Figure 8-10, occasionally in adenine a hydrogen atom involved in bonding with the complementary strand can shift from the amino group to the nitrogen in the ring. The consequence is that adenine can pair, during replication, with cytosine rather than with its normal partner, thymine. This situation is exceptional, so upon the next replication adenine is expected again to pair with thymine. The cytosine introduced

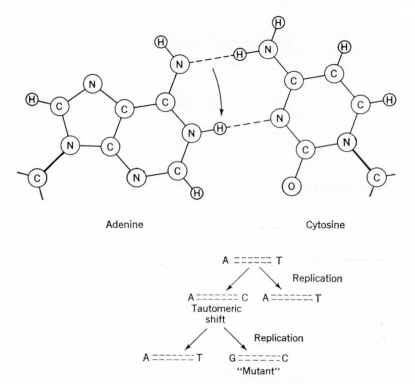

Figure 8-10. *A possible chemical basis for spontaneous mutations.*

into the strand in place of thymine would pair with guanine. Thus the descendants of the strand with the cytosine would now have a G-C pair in place of an A-T pair. At this point in the gene the code would be altered, and in many instances would presumably result in a recognizable mutation.

The suggestion is, then, that spontaneous mutations arise as *copy errors* during duplication of DNA. While some spontaneous mutations may well be attributable to this proposed mechanism, it is clear that spontaneous mutations can occur in the absence of DNA synthesis, as, for example, in stored Drosophila sperm or in nondividing bacteria. If we consider the mode of action of various chemical mutagens we see that some mutagenic agents act directly to induce mutations in nonduplicating DNA while others induce mutations only during DNA synthesis. Apparently both copy errors and direct gene changes can be induced to occur, and both presumably occur spontaneously.

A good example of a direct-acting mutagen is provided by nitrous acid. Phage can be treated with nitrous acid, removed from the nitrous acid, and then allowed to multiply in bacteria. Many of the treated phage are found to be mutant. Careful examination of the progeny of the mutants reveals that most of the induced mutants are in fact only *half-mutant*. That is, half their progeny consist of mutant and half of nonmutant phage. This result is expected if a chemical change is induced in only one of the two DNA strands. Upon replication the modified strand produces mutant progeny; the unmodified strand produces normal phage.

The frequency of mutants induced is directly proportional to the length of treatment with nitrous acid. This result suggests that a single chemical event is sufficient to induce a mutation. The probable mechanism of the mutagenic event is shown in Figure 8-11. If this scheme is correct, nitrous acid, by deamination, converts adenine to hypoxanthine and cytosine to uracil. Since it is thought that hypoxanthine should behave like guanine rather than adenine during replication, and uracil like thymine rather than cytosine, the net result of these changes should be, after a cycle of DNA synthesis, the conversion of A-T pairs to G-C pairs (adenine to hypoxanthine to guanine) and also of G-C pairs to A-T pairs (cytosine to uracil to thymine). Such chemical changes, in which one purine replaces another purine and pyrimidines are similarly switched have been termed *transitions;* mutations due to such chemical changes are called *transition mutations.*

Alkylating agents, such as nitrogen mustard or ethyl ethanesulfonate, form

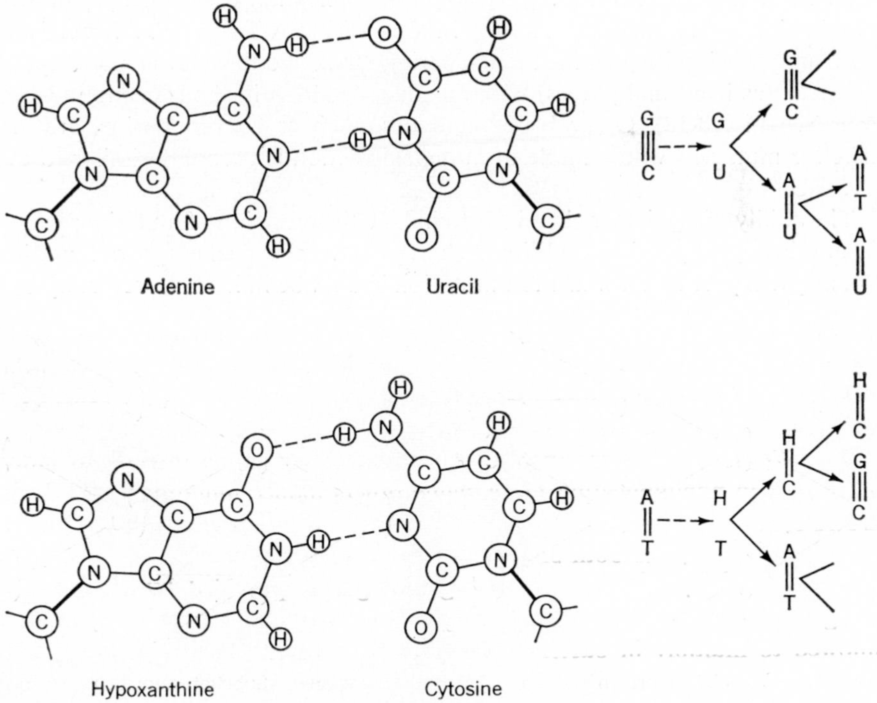

Figure 8-11. *The probable mechanism of nitrous acid mutagenesis.*

another class of mutagens that acts upon nonreplicating DNA. Their mode of action differs from that of nitrous acid in producing many delayed mutations, both in phage and in higher organisms. In Drosophila, progeny flies from crosses in which the sperm have been treated with nitrogen mustard are often *mosaic*, indicating that the mutation probably occurred after several divisions of the treated chromosomes in the zygote. Similarly, many of the mutations induced in phage by alkylating agents arise only after a number of duplications. This delay is thought to be owing to the occurrence of copy-error mutations several generations after the DNA has been treated. One of the chemical reactions of alkylating agents is the addition of an ethyl group to guanine. A possible explanation of the delayed mutagenic action of alkylating agents is that ethylated guanine eventually is broken off from the DNA leaving a gap, which can be filled, in some cases, by adenine.

In contrast to the alkylating agents, base-analogue mutagens are mutagenic only if present during DNA synthesis. One of these, 5-bromouracil (5-BU), is an analogue of thymine and can quantitatively substitute for it in DNA. However, 5-BU is thought to pair occasionally with guanine. Since mutations are induced either during the incorporation of 5-BU in place of thymine or during incorporation of thymine in place of 5-BU, it would appear that 5-BU that enters DNA in place of thymine can during subsequent duplication occasionally pair with guanine instead of adenine, resulting in a transition A-T to G-C. Alternatively, 5-BU might pair with guanine upon entering DNA and then at the next duplication pair (as it normally does) with adenine, resulting in the transition G-C to A-T (see Fig. 8-12). Other base analogues, such as 2-aminopurine, are also effective mutagens whose mode of mutagenic action is probably similar to that of 5-BU.

The ability of particular chemicals to induce the reversion of mutations caused by another agent provides additional clues as to the chemical nature of mutation. We might expect that a transition mutation should be induced to revert by base

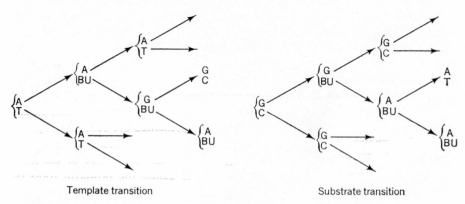

Template transition   Substrate transition

Figure 8-12. *The probable mechanism of 5-bromouracil mutagenesis.*

analogues since we postulate that base analogues cause *both* types of transitions. The results of appropriate studies involving the *rII* mutants of T4 are summarized in Table 8-1. With the exception of proflavine, the chemical mutagens used

Table 8-1. *rII* STUDIES OF CHEMICAL MUTAGENS AND BASE ANALOGUES.

rII Mutants Induced by	*Percent Mutants Revertible by* AP *and/or* BU	*Percent Mutants Not Revertible by* AP *or* BU	*Percent Mutants of Possible Spontaneous Origin*	*Number of Mutants Tested*
2-aminopurine (AP)	98	2	2	98
5-bromouracil (BU)	95	5	2	64
nitrous acid	87	13	15	47
ethyl ethanesulfonate	70	30	10	47
proflavine	2	98	—	55
spontaneous	14	86	—	110

Source: Adapted from E. Freese, "Molecular Mechanism of Mutation." In *Molecular Genetics*, Part I, edited by J. H. Taylor.

appear to cause transition mutations since in all cases the mutants induced are revertible by base analogues.

On the other hand, most spontaneous and virtually all proflavine-induced mutations in the *rII* region of phage T4 appear not to be transition mutations since they are not revertible by base analogues. Since most spontaneous and proflavine-induced mutants are caused to revert by proflavine, they most likely involve the same kind of change.

Two types of suggestions have been made as possible explanations for such nontransition mutants. They might be base substitutions involving a substitution of a purine for a pyrimidine or vice versa, so-called *transversion mutations.* Alternatively the mutations might involve the addition or deletion of single bases. A considerable amount of indirect evidence suggests that the spontaneous and proflavine-induced mutations in phage are of the addition or deletion kind. The existence of transversion mutants has not yet been demonstrated. It should be pointed out that the addition-deletion type mutations may well occur in only some organisms since most spontaneous mutations in bacteria, for example, are not of this type but are transition mutations.

Our understanding of the mechanism of mutation as yet is admittedly small, and many of the interpretations that have been presented are somewhat speculative. Nevertheless, problems of mutation can now be defined on a chemical-physical basis. With increased knowledge of the chemical nature of the gene and of the chemistry of agents that act to alter genes, the powerful tools of physical science can be applied to explain the previously mysterious phenomenon of mutation.

Keys to the Significance of This Chapter

Sudden, heritable changes occur in living things; these changes are called mutations. They may involve gross changes in chromosome quantity or structure, or they may, in other instances, be specific chemical changes in genes.

Two primary characteristics of spontaneous gene mutations are that they are generally recurrent and often reversible. The mutation rate typically differs in the two directions, $A \rightleftharpoons a$, and varies from one gene to another. These rates actually measure the sum of a variety of different changes within genes. In general the forward rate is a measure of the sum of mutation rates at many different sites within a gene, while the reverse rate is, with the exception of suppressor mutations, a measure of the rate of reverse mutation at one site.

The rate of mutation can be greatly accelerated by the use of mutagens. These include ionizing radiations, ultraviolet light, and a variety of chemicals, some of which interact in specific ways with DNA. These mutagens are not selective for particular genes. Many of the chemical mutagens do show specificity in their mutagenic behavior, however, by causing within genes specific chemical changes such as transition of one base pair to another or additions or deletions of base pairs.

REFERENCES

Adelberg, E. A., editor, *Papers on Bacterial Genetics*. Boston: Little, Brown, and Co., 1960. (A selection of original papers in bacterial genetics; contains a number of papers on mutation and a concise summary of the findings in the field.)

Auerbach, C., "Chemical Mutagenesis." *Biol. Rev.* **24**:355–391, 1949. (A competent and extensive review of the early work on chemical mutagenesis.)

Burdette, W. J., editor, *Methodology in Basic Genetics*. San Francisco: Holden-Day, 1963. (Contains chapters on mutation in phage, bacteria, and Drosophila.)

Freese, E., "Molecular Mechanism of Mutation." In *Molecular Biology*, edited by J. H. Taylor, New York: Academic Press, 1963. (A good summary of the chemical basis of mutation.)

Hollaender, A., *Radiation Biology*. New York: McGraw-Hill, 1954. (A comprehensive source of information concerning all aspects of radiation biology including mutation.)

National Academy of Sciences, *The Biological Effects of Atomic Radiation*. Summary reports, 1960. Washington: National Academy of Sciences-National Research Council. (The present consensus of scientific opinion.)

Russell, W. L., "Genetic Hazards of Radiation." *Proc. Amer. Phil. Soc.*, **107**:11–17, 1963. (A summary of studies of induced mutation in the mouse.)

Scientific American, September 1959. (A special issue devoted to the biological effects of radiation.)

Stent, G. S., *Molecular Biology of Bacterial Viruses*. San Francisco: W. H. Freeman and Co., 1963. (Chap. 10 contains a very lucid discussion of the molecular basis of mutation as analyzed in phage T4.)

Stern, C., *Principles of Human Genetics*. San Francisco: W. H. Freeman and Co., 2nd ed., 1960. (Chaps. 21 and 22 are excellent summaries of the subject of mutation, with particular regard to human genetics.)

QUESTIONS AND PROBLEMS

8-1. What do the following terms signify?

base analogue	mutation frequency
ClB	mutation rate
dominant lethal	replica plating
gene mutation	reversion
induced mutation	suppressor
ionizing radiation	somatic mutation
isoallele	spontaneous mutation
conditional lethal	transition mutation
mutagen	

8-2. In experimental plants and animals most of the mutant genes known are recessive to the wild type. In man, however, most of the known simple autosomal abnormalities depend on dominant alleles, and only among sex-linked characteristics are recessive mutations common. Can you offer explanations for this apparent difference between human and experimental populations?

8-3. When corn seed subjected to atomic radiation in atom-bomb tests was planted, the seedlings' first leaves showed many fine light stripes among normal green tissues. The amount of striping was related to the distance the seed had been from "ground zero" for the bomb. Can you explain these facts? (Note that a well formed embryo, complete with several leaf primordia, is present in the kernel.)

8-4. A dwarf bull calf is born to apparently normal parents in a herd of cattle. How might you decide whether the dwarf is the immediate result of a mutation, or a segregant resulting from the chance mating of "carriers" of a recessive dwarfism, or the result of a nongenetic (environmental) modification?

8-5. In certain systematic inbreeding programs, dairy bulls are regularly bred to their own daughters. Suppose that a bull is heterozygous for a recessive lethal, and that the cows to which he is originally bred are all homozygous for the normal allele of this lethal. What will be the outcome of sire-daughter matings in the next generation with regard to this particular lethal?

In mice, the following are members of an autosomal, multiple-allelic series:

$$A^Y = \text{lethal yellow}$$
$$A \;= \text{agouti (wild type)}$$
$$a \;= \text{nonagouti (black)}$$

The agouti/nonagouti alternative was utilized in Problems 1-16 to 1-23. The A^Y allele when heterozygous produces a clear yellow coat color; it is dominant to both A and a. Embryos homozygous for A^Y die. (There are other alleles in this series not listed here.)

8-6. What phenotypes, and in what proportions, would result from the following matings?

a. $A^Y a$ (yellow) $\times$ $A^Y a$ (yellow)
b. $A^Y a$ (yellow) $\times$ $A^Y A$ (yellow)
c. $A^Y a$ (yellow) $\times$ aa (black)
d. $A^Y a$ (yellow) $\times$ AA (wild type)
e. $A^Y a$ (yellow) $\times$ Aa (wild type)

Example: In (a), zygotes will be $A^Y A^Y$, $A^Y a$, $A^Y a$, and aa in equal numbers. The first class will die as embryos, so the phenotypic ratio will be 2 yellow:1 black.

8-7. Assuming an average litter size of eight mice born per litter for many matings of type (e) above, what would you expect to be the average litter size for many matings of type (b) under the same conditions?

8-8. You would never expect both wild-type agouti and nonagouti mice among the progeny of a single lethal yellow male mated to nonagouti females. Why?

8-9. The attached-X ($\widehat{XX}$) Drosophila stock described in Chapter 2 is useful for the detection of new, visible sex-linked recessive mutations. Assume that in studying mutation in the progeny of an X-rayed adult male you are dealing mainly with mutations induced in the sperm the male has formed at the time of treatment. If you contrast the result of mating such a male to an $\widehat{XX}$ female with the result of using a normal female, you will see why the $\widehat{XX}$ technique offers advantages in detecting sex-linked recessive visibles. (Ignore for the present the few $\widehat{XX}X$ progeny that survive in $\widehat{XX}$ matings.)

8-10. What would be the effect of a sex-linked recessive lethal induced in a treated male on the progeny of his mating to an $\widehat{XX}$ female? Contrast the *ClB* technique and the $\widehat{XX}$ technique with respect to the types of mutations they are best adapted to detect.

8-11. How would you distinguish between an autosomal dominant and a sex-linked recessive visible mutation in the $\widehat{XX}$ technique?

8-12. If you assume that sex-linked dominant lethals are induced rather frequently in the sperm of X-rayed males, how would the sex ratio in the progeny of treated males mated to $\widehat{XX}$ females be affected? If the males were mated to *ClB* females? Wild-type females?

8-13. In the *ClB* technique, assume a recessive lethal induced on the X-chromosome of a sperm that has fertilized a *ClB* egg. Now this zygote has a recessive lethal on both X-chromosomes. Why does it not die?

8-14. How might a particular somatic mutation be established in a large population of cultivated plants? Is this likely in animals? Why?

In pigeons, a sex-linked, multiple allelic series includes:

$$B^A = \text{ash-red}, \ B = \text{wild type (blue)}, \ b = \text{chocolate}.$$

Dominance may be considered complete, in the order listed. Remember that in birds the male is XX, the female XY.

8-15. Male pigeons of genotype B^Ab are ash-red but sometimes have "flecks" of chocolate areas in certain of their feathers. Offer two explanations (chromosomal and genic, respectively) of this phenomenon.

8-16. When flecks occur in female pigeons of genotype $B^A(-)$, the flecks are usually chocolate, but blue flecks have also been observed in such birds. How might this bear on the respective likelihoods of your two explanations in Problem 8-16?

Assume the following information about a particular bacterial strain: (1) the mutation rate *a* to *A* is 10^{-5} per cell per generation as is the rate of mutation *A* to *a*, and (2) the growth rates of *A* and *a* cells are the same.

8-17. Starting with a population containing 10^7 *a* cells and 10^2 *A* cells, what will be the frequency of *A* cells in the population after one generation (i.e., when there are 2×10^7 cells in the population)? After two generations? After an infinite number of generations? What do these calculations suggest about the effect of the mutation rates in both directions on the equilibrium ratio A/a (the ratio of the two cell types after a very long time)?

Hydroxylamine is a chemical mutagen that is thought to induce the transition G-C to A-T, but not A-T to G-C.

8-18. J. W. Drake found that about one-half of the *rII* mutations of phage T4 induced with ultraviolet light were induced to revert with base analogues and of these nearly all were found not revertible with hydroxylamine. What is most likely the chemical target of ultraviolet light in the phage DNA for the induction of these transition mutations?

8-19.   A cross is performed between a revertant obtained from a particular *rII* mutant (*rIIx*) and wild type. About 1 percent of the cross progeny are found to be genotypically *rII*. One half of these *rII* segregants, when crossed to *rIIx*, yield no r^+ segregants; the other half yield about 0.5 percent r^+ segregants. What is the probable genetic structure of the revertant? Diagram the various crosses and indicate the genotypes and phenotypes of the various phages involved.

CHAPTER 9

Gene Structure

Around the turn of the century the word "gene" was coined to describe the Mendelian particle, the basic unit of inheritance. In the last fifteen years significant advances have been made in our knowledge of the gene. Along with this knowledge has come the realization that the gene is complex, even subdivisible, and that early definitions of the gene are no longer sufficiently meaningful or accurate. In this chapter we will describe the "classical" gene, as conceived, for example, by T. H. Morgan, and then show how the classical conception has been revised. Our treatment of these matters will necessarily involve the introduction of some new terminology.

The Classical Gene

With standard genetic techniques, the gene can be studied by observing recombination, phenotype, and mutation. The classical conception of a gene assumed it to be a unitary particle by criteria involving each of the three kinds of observation.

1. A gene is a unit of chromosomal structure not subdivisible by chromosomal breakage or crossing over.
2. A gene is a unit of physiological function or expression.
3. A gene is a unit of mutation.

To see what these criteria mean, consider a hypothetical example. Suppose that we isolate two independent mutants, $a1$ and $a2$, of some diploid organism. The two mutants have similar phenotypes, and we would like to know whether the two different mutations have affected the same gene. If *a*1 and *a*2 are each recessive to wild type, an enlightening test is to construct the heterozygous diploid *a*1/*a*2 and determine its phenotype. If the phenotype of the heterozygote is wild type, the implication is that the mutations are in different genes. We deduce that wild-type genes are present in the diploid and we justifiably denote the genotype of the heterozygote as $a1a2^+/a1^+a2$. We deduce that the mutations are not allelic in the functional sense because the mutants complement one another. If, on the other hand, the phenotype of the heterozygote were mutant, the same kind of reasoning process would lead to the conclusion that the mutations are functionally allelic because the mutants do not complement one another. We would imagine, on the basis of the genetic test, that we were dealing with two different mutations of the same gene. This is the standard test for functional allelism.

Still following the classical conception of the gene: if the two mutations are functionally allelic, they must also be allelic in a structural sense. That is, if we examine the meiotic segregants from the heterozygous diploid *a*1/*a*2, for example by a testcross, we should find no chromosomes of genotype $a1^+a2^+$ or *a*1*a*2 (Fig. 9-1). This result would indicate that the two mutations are indeed structurally allelic. If such recombinant segregants did emerge at meiosis, we would be dealing with nonalleles. This is the test for structural allelism. Note, however, that a satisfactory structural test may be difficult to apply in instances where genes are closely linked or where recombination is suppressed, as by an inversion.

From the rebirth of genetics in 1900 until fairly recently, few exceptions to the rule of identity of structural and functional alleles were reported. Most evidence

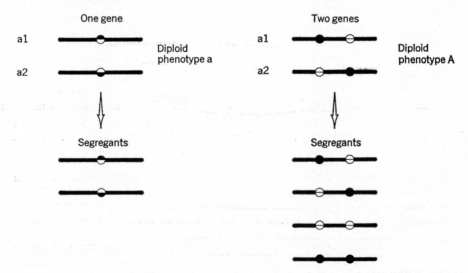

Figure 9-1. *The predicted segregation of allelic and of nonallelic mutations.*

supported the notion that the gene is an indivisible particle, and that the chromosome is a string of genes that may be broken only by crossing over between genes.

At the same time, however, this "indivisible particle," the gene, was seen to be clearly complex from a mutational point of view. Many cases of "multiple" allelism were known and described. For example, a large number of alleles of the so-called white gene, or white locus, in Drosophila had been found and studied. A variety of mutations of this gene were found to result in eye colors varying from white to red, almost like wild type. The mutants that give intermediate phenotypes are distinguishable one from the other and have been given appropriate colorful names such as coral, apricot, cherry, eosin, and so on. The existence of such a range of multiple alleles indicates that a gene can exist in many forms, which must have corresponding differences in function to account for the various distinct phenotypes observed. The classical view considers such mutations to be qualitative changes in the unitary gene particles.

Pseudoalleles. By the late 1940's, a number of studies had been made which showed that many mutations, allelic by the functional criterion, could be separated by recombination and thus were not structurally allelic. An example may be cited from the white locus just described. Two mutations w and w^a are functionally allelic in that the heterozygote w/w^a has very light-colored eyes rather than red (Color Plate I). If many progeny of such heterozygotes are examined, w^+ and ww^a recombinants are found, proving that w and w^a are *not* allelic in the classical structural sense. The recombinants are indeed rare (less than one in a thousand), indicating that the sites of the mutations are closely linked. Nevertheless, the fact of recombination dictates that we write the genotypic formula of the heterozygote as $\dfrac{w^+ w^a}{w\ w^+}$. Now from the recombinants another form of the heterozygote can be constructed, $\dfrac{w\ w^a}{w^+ w^+}$. Such a heterozygote has wild-type phenotype. The phenotype of the heterozygous diploid depends upon the configuration of the mutations! If the wild-type alleles are together in the same chromosome, the *cis* position, the organism is wild type. If the wild-type alleles are on opposite chromosomes, the *trans* position, the organism is mutant. The term *pseudoallelism* has been used to describe this phenomenon, *pseudo* because mutant genes that behave in this way are allelic in a functional but not in a structural sense. Since cases of pseudoallelism are numerous, how do we define a gene? E. B. Lewis suggested that we use structural allelism as our criterion. He pointed out that the difficulty with functional allelism as a criterion in reference to instances of pseudoallelism might be resolved on the basis of a simple assumption. Assume that certain neighboring genes, one having arisen from the other through duplication, have common but slightly different functions that must be carried out in sequence at the site of the

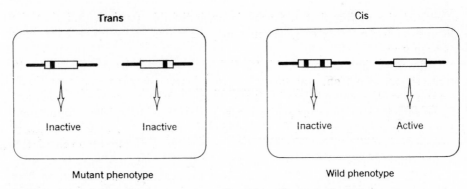

Figure 9-2. *Pontecorvo's interpretation of position-effect pseudoallelism.*

chromosome. Heterozygotes in the *trans* arrangement would be unable, then, to carry out the total functions. G. Pontecorvo suggested an alternative interpretation and proposed that the gene be defined by the functional criterion. He considered pseudoallelism as a demonstration that the gene was subdivisible into parts. As you will see, the Pontecorvo interpretation has turned out to be more generally applicable. But some instances of pseudoallelism in all probability are due to gene interaction as Lewis suggested rather than to subdivision of the gene. As will be discussed in Chapter 10, certain groups of genes do act in groups. However, let us defer further interpretation until we have turned to microorganisms for more detailed analyses of gene structure. The importance of the early work with pseudoallelism is to show that, whatever the interpretation of the phenomenon in a particular instance, the gene cannot be defined as a unit *both* of structure and function.

As geneticists began to look for them, many instances of pseudoallelism were found in diverse organisms in the 1940's and 50's. But even such convenient objects of genetic study as Drosophila and maize are relatively unsatisfactory for work that seeks the detection of extremely infrequent recombinants. The most detailed work in this area has come from microorganisms (especially fungi, bacteria, and phage), of which the research worker can easily examine enormous populations and for which selective techniques have been developed for the isolation of rare mutants or recombinants out of large populations. In this chapter we will examine only a few "case histories" to develop our concept of the gene. These cases will be typical of the kind of result found for numerous genes in a variety of organisms.

The Phage Gene

Much of the impetus to analysis of gene structure comes from the elegantly detailed work of S. Benzer on the so-called *rII* region of bacteriophage T4.

Wild-type T4 makes small fuzzy plaques on plates of *E. coli*, strain B. Among the many types of plaque-morphology mutants of this phage, *r*, or rapid lysis, mutants are particularly easy to recognize because they form large sharp plaques on plates of strain B (see Fig. 6-11). Earlier genetic studies had shown that the *r* mutations can occur at several locations within the genome of the phage. Benzer discovered that *r* mutations that map to one particular region of the genome (the *rII* region) are distinguishable from those that occur in other regions of the genome. These *rII* mutants, for a reason not yet understood, are unable to multiply in *E. coli*, strain K(λ), although they can infect and kill it. Wild-type phage and the other *r* mutants grow perfectly well in strain K(λ).

Benzer isolated over 3,000 independent *rII* mutants and studied them in detail. We will consider several aspects of this study, as well as the general techniques for working with the *rII* system, because they illustrate selective methods widely applicable to study of the genetics of microorganisms.

The Phage Gene—Phenotype. As we have pointed out, K(λ) bacteria do not propagate *rII* mutants. If, however, K(λ) bacteria are infected simultaneously with both *rII* and *r+* phage, phage of both genotypes are reproduced in the bacteria. The *r+* phage performs the function that the *rII* phage cannot, with the result that both phages can grow. With reference to this function, *rII* is recessive to *r+*. In other words, when one mixedly infects K(λ) bacteria with phage of two appropriate genotypes, one has a test for recessiveness and allelism that is analogous to the diploid test of heterozygotes. If various combinations of *rII* mutants are used to infect K(λ) bacteria in this test, the results are straightforward. The *rII* mutants fall into two groups, which Benzer has designated *A* and *B*. Any combination of an *rIIA* and *rIIB* mutant results in the production of normal numbers of phage in K(λ)—phages of both genotypes being produced. In other words, the two mutant types complement one another to permit growth. Any combination of *rIIA* with *rIIA* or of *rIIB* with *rIIB* results in the production of few, if any, phage (that is, the different mutants of the same type give no complementation). By analogy with the phenotypic tests of diploid organisms already described, the functional criterion designates two *rII* genes, the *A* gene and the *B* gene. All *A* mutant genes are functional alleles and all *B* mutant genes are functional alleles, but the *A* and *B* functions are independent of one another. Rather than calling these functional groups genes, Benzer has designated them *cistrons*. Mutations that are functionally allelic he terms *members of the same cistron*. The term "cistron" is derived from the *cis-trans* test already described.

The Phage Gene—Recombination. All *rII* mutations are closely linked. When *rII* mutants are crossed, we do not find, however, that *r+* recombinants arise only in *rIIA* $\times$ *rIIB* crosses. They are formed in *A* $\times$ *A* and *B* $\times$ *B* crosses as well. Crossing depends on using two different *rII* mutants to infect bacteria in which the *rII* mutants can grow (*E. coli*, strain B, for instance). Because *rII*

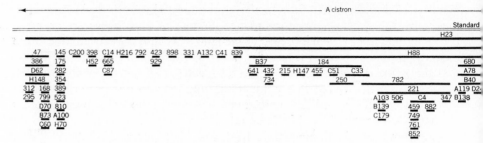

Figure 9-3. *A map of deletions in the rII region of phage T4. The extent of each mutation is indicated by a bar under the name of the mutant. Some deletions are very large (H23), others very small (C200). (From Benzer,* Proc. Natl. Acad. Sci. **45**:*1607, 1959.)*

mutants cannot produce plaques on K(λ) whereas r^+ phage can, r^+ recombinant progeny occurring at frequencies as low as 0.00001 percent (one recombinant in 10 million) can be selected out of a large population of r phage by plating a suitable dilution of the population on K(λ). The total population size of a progeny can be determined by plating on *E. coli*, strain B. Since the double-mutant recombinants have been demonstrated to occur with the same frequency as the reciprocal r^+ recombinants, for any particular pair of r mutants the frequency of recombination can be assumed to be twice the frequency of r^+ phage among the progeny.

If many different r mutants are mapped in the way described, a linear map emerges. All the A mutations are located in one section of the map and all the B mutations are located in an adjacent section. The positions of A and B mutations do not overlap. The A group and the B group are each functional units, but the functional unit of the genome is larger than the recombinational unit. Many mutations that act as functional alleles are separable by recombination. Rather than defining one indivisible "particle," a group of functional alleles appears to define a discrete segment of the genome that is divisible by recombination.

Deletion Mapping. After showing that the *rII* functional regions can be subdivided by recombination, Benzer continued mapping his 3,000-odd mutants in order to see to what extent the *rII* region can be subdivided. Even in phage, where perhaps fifty crosses can be done every day, to map 3,000 mutants presents a formidable task. In fact, about 5 million crosses would be required to do all possible two-point tests [$n(n - 1)/2$, where n is the number of mutants]. Benzer devised some tricks to make things easier. He first developed a "tester" system of deletion mapping to localize unknown mutations.

Among the spontaneous *rII* mutants isolated, a number give no recombination with members of a series of mutants that give recombination among themselves. They behave in recombination as though they contain deletions for small

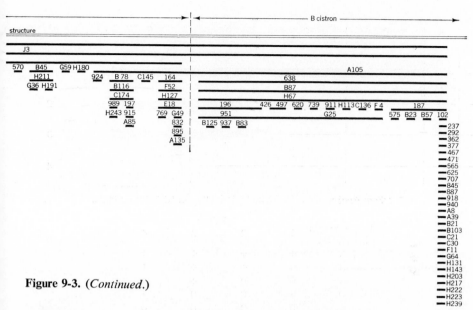

Figure 9-3. (*Continued.*)

segments of the *rII* region. Furthermore, although most mutants are capable of reversion, these mutants, as one would expect of deletion mutants, are incapable of reversion to wild type. A wide variety of deletion types have been isolated, ranging from mutants with very tiny deletions to ones in which both *A* and *B* cistrons are absent. Deletion types constitute about 20 percent of the spontaneous *rII* mutants. A few of the deletions are exceptions to the complementation pattern we stated as designating the *A* and *B* cistrons. The exceptional deletion mutants complement with neither *A* nor *B* mutants. In every such instance, however, the deletion has been shown by genetic mapping to cover at least part of the *A* and part of the *B* cistron and thus the mutant is at least partly deficient for both cistrons. Such mutants are "exceptions that prove the rule."

An elegant proof of the linear structure of the gene comes from a study of the recombination properties of the deletion mutants. A deletion mutant will only give *r+* recombinants in crosses with another deletion mutant if the two do *not* lack a common segment. If a number of different deletion mutants are tested in pairwise combinations for the production of *r+* recombinants, the qualitative results permit the construction of a deletion map (see Fig. 9-3), which is not only a topological representation of the results but also a representation of the locations of the various deletions relative to each other. If the gene is linear in structure and the mutants are simple, continuous deletions of segments of the gene, it is clear that certain types of recombination patterns cannot be observed. Suppose, for example, that given four deletions *A*, *B*, *C*, *D*, no combinations give recombinants except *A* with *C* and *B* with *D*. Such a result is not consistent with linearity but requires a two-dimensional map, or for instance, a circle.

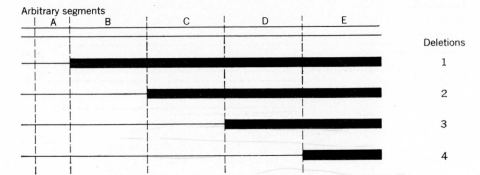

Results of crosses of point mutant with "tester" deletion mutants.

Point mutants in segment	Deletions 1	2	3	4
A	+	+	+	+
B	0	+	+	+
C	0	0	+	+
D	0	0	0	+
E	0	0	0	0

Figure 9-4. *Deletion mapping. The selection of four deletions with different ends permits the recognition of five segments. Point mutants that are located in the different segments will behave differently from each other when spot tested against the four deletions. 0 means no recombination; + means some recombination.*

Benzer analyzed the recombination properties of over 145 deletion mutants and found that with one exception the results were in complete accord with the concept of a linear gene structure. Indeed, the exception, on further analysis, proved to be a double mutant! These results give elegant support, then, to the linear topology of the gene.

A new mutation can be assigned to a given portion of the genome by the pattern of recombination of the mutant with members of a "tester" set of deletion mutants (Fig. 9-4). The test is qualitative rather than quantitative. If the mutant site of the strain being tested lies within a segment absent in a deletion tester, no recombinants are formed in the progeny of the cross. If the mutation lies outside the region deleted in the tester, then recombinants can occur. Since the tests are only qualitative, the crosses can be done in a simple way. The mutant to be located is spread on plates seeded with a mixture of many K(λ) cells and a few B cells. Then drops of suspensions of the different deletion phages are placed on marked areas of the plate. Within the region of a given drop, the B bacteria will be infected with both the unknown r mutant and the tester. If the mutation and deletion are nonoverlapping, r^+ recombinants will be formed in the infected B bacteria. The r^+ recombinants grow in the K(λ) bacteria and produce clearing in the spot when the bacteria lyse. If the mutation and deletion overlap, no r^+

recombinants form, and the spot remains turbid due to the presence of un-infected K(λ) bacteria (Fig. 9-5). In this way, a given mutation can be readily assigned to a region of the genome defined by the system of deletions used. After a number of mutants have been sorted into arbitrary regions by the deletion mapping tests, further mapping determines the number of sites represented in the mutants that are being localized. This more refined localization also can be

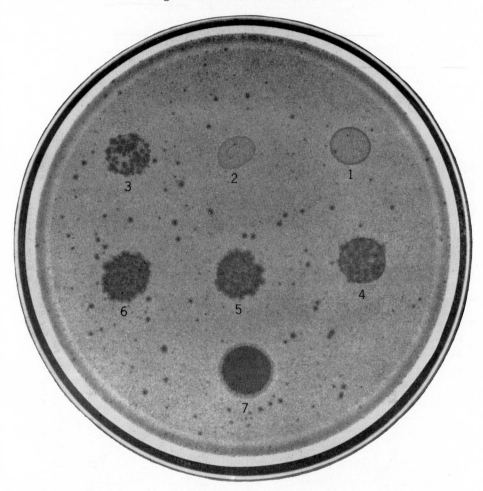

Figure 9-5. *A photograph of a deletion spot-test plate. About 10⁸ particles of a point mutant are plated on a Petri dish with a mixture of strains B and K(λ). The small scattered plaques are due to revertants. Various deletions are then spotted on the plate. The point mutant gives no recombinants with deletions 1 and 2, some with 3, 4, 5, and 6. Deletion 7 is in a different cistron from the point mutant, thus full complementation takes place and the spot clears completely.*

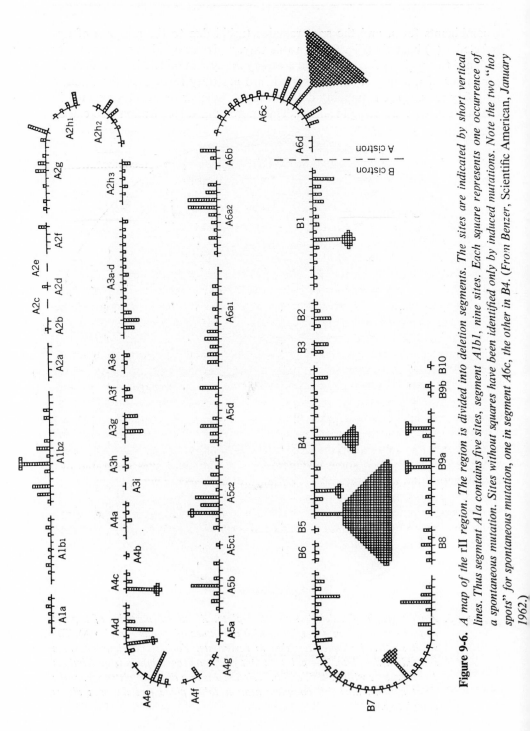

Figure 9-6. A map of the rII region. The region is divided into segments. The sites are indicated by short vertical lines. Thus segment A1a contains five sites, segment A1b1, nine sites. Each square represents one occurrence of a spontaneous mutation. Sites without squares have been identified only by induced mutations. Note the two "hot spots" for spontaneous mutation, one in segment A6c, the other in B4. (From Benzer, Scientific American, January 1962.)

accomplished by "spot testing" the mutants against one another in a manner similar to that used for the deletion mapping.

Using these methods to study more than 3,000 mutants, Benzer has shown that the *rII* region can be subdivided into more than 300 recombinable sites (Fig. 9-6). What does this mean in physical terms? The *rII* region constitutes less than 1 percent of the total map length of the phage. Since the DNA of this phage contains about 200,000 nucleotide pairs, the *rII* region probably consists of less than 2,000 base pairs. This means that "sites" have already been identified, which, on the average, are less than seven base pairs apart. Recombination, therefore, permits very fine dissection of the phage genome, dissection down to the structural level of the constituent nucleotides of the DNA. At least when considering phage, we are led to the conclusion that the gene as a functional unit must be equated with the cistron and that the recombinable elements of which it is composed are probably the nucleotides.

The Phage Gene—Mutation. The large collection of mutations of the *rII* region includes many mutations of independent origin that were not separable by recombination. We may ask: do these mutations differ in more ways than just their independence of origin, or do they merely represent recurrences of an identical mutational event? From Benzer's study it appeared that mutations at the same site, in some cases, differ with regard to their phenotype, frequency of reversion to wild type, or on the basis of response to suppressor mutations. Recent studies by I. Tessman show that many of these heterogeneous sites can be broken up by very sensitive recombination tests into several sites that are homogeneous with regard to the mutant properties. It appears that certain mutations at neighboring sites recombine so poorly that recombination is detected only by very sensitive tests. Tessman's study shows that what we presume is a site may in some instances consist of several sites separable by recombination but at frequencies lower than we can detect. Thus in many cases the basis of the heterogeneity of mutations at what is apparently only one site cannot be determined. One would expect, for example, that transition mutations would give rise to only one class of mutant at a site since only two forms of the base pair are possible, one form giving a mutant and the other a wild-type phenotype.

The Neurospora Gene

In Neurospora, intensive studies have been made on the structure of a variety of genes involved in the biochemical pathways of intermediary metabolism (see Chap. 10). We will take as an example the *pan*-2 gene, studied by N. H. Giles and M. Case. Mutations in this gene result in the inability of the mutant strain to convert keto-valine into keto-pantoic acid. In order to grow, the mutants require pantothenic acid, which is a vitamin and the end product of this particular pathway of synthesis. The missing step is probably controlled by a single enzyme, which is defective or lacking in the *pan*-2 mutants. Some 75 of these mutants have been isolated and mapped to one small region of chromosome 6.

Functional tests. Tests for functional allelism in Neurospora are performed in heterokaryons, diploid tests not being practicable. (You will remember from Chap. 4 that a heterokaryon consists of mycelium containing a mixture of two types of nuclei.) In this case, heterokaryons are established that contain two different kinds of nuclei to be tested with respect to mutations at the *pan*-2 locus. The heterokaryons are then tested to see whether they can grow in the absence of pantothenic acid. Since heterokaryons between any *pan*-2 mutant and wild type are able to grow in the absence of pantothenic acid, the *pan*-2 mutants can be considered recessive to wild type. If different *pan*-2 mutants are combined by pairs in heterokaryons, most combinations are unable to grow in the absence of pantothenic acid. The result indicates that the mutant genes are functional alleles. However, some combinations of mutants are able to grow in the absence of growth supplement, although more slowly than the wild type. The mutants, therefore, fall into two classes with regard to their growth requirements in heterokaryons with other mutants. Members of the larger group of mutants show no growth on unsupplemented medium when in combination with other mutants. These mutants are designated *noncomplementing* mutants, because they do not complement the growth defect of any of the other mutants. Mutants of the second class, termed *complementing*, are able to grow in combination with particular other mutants of this class.

Among the complementing mutants, the pattern of complementation is complex. Results of complementation tests are shown in Figure 9-7. We can derive a pattern out of these results by drawing a linear functional map for

Complementation results					Complementation map
	1	2	3	4	5	
NC's	0	0	0	0	0	1 2 3
5	+	0	0	0		4
4	0	0	+			5
3	+	+				NC's
2	+					

Figure 9-7. *Complementation mapping. On the left the results of pairwise complementation tests with a group of hypothetical mutants are shown. 0 indicates noncomplementation; + indicates some complementation. The results can be represented in the form of the map shown at right. NC indicates noncomplementers. Numbers indicate different mutant types.*

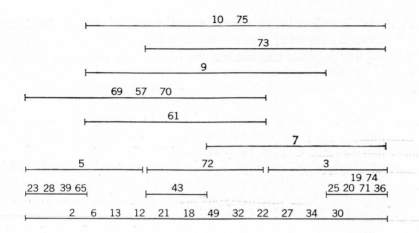

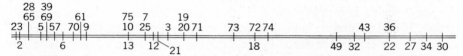

Figure 9-8. *A map of the* Pan-2 *locus in* Neurospora crassa. *The genetic map is plotted along the double line at the bottom. Mutations indicated below the double line are noncomplementing; those above the double line are complementing. The complementation map is given above the genetic map. Mutations on the same line (for example 10 and 75) give identical complementation patterns. (Adapted from Case and Giles,* Proc. Natl. Acad. Sci. **46:**659, 1960.)

this region on the basis that pairs of mutants that do not complement each other are assumed to overlap in function. This "functional" map is *analogous* to the map constructed for the set of *rII* deletions. However, we must stress that the complementation map is only a formal representation of the complementation results and does not necessarily imply anything about the genetic location or structure of the mutations.

For the most part, the complementing *pan*-2 mutants can be arranged into a linear "complementation map," where each of the mutants can be represented, as in Figure 9-8, by an unbroken line, which overlaps only the lines representing mutants with which it does not complement. However, "exceptional" mutants, which do not fit the linear complementation map, have been found. The main point to consider here is that the *pan*-2 region cannot be divided into discrete functional segments such as the *A* and *B* cistrons of the *rII* region of phage T4. In many instances a given mutant does not complement either of two other mutants that do complement with each other. For example, in

A   C
_____ _____

B
_____

clearly *A* and *C* are functionally related, even though indirectly. In fact in most cases the noncomplementary mutants are the most abundant. Furthermore, we should note that even in cases where *pan*-2 mutants complement one another, the rate of growth of the heterokaryon never approaches that of wild type, or even that of a heterokaryon between either of these mutants and wild type. In other words, two mutants, while complementing to some extent, nevertheless display a partially mutant phenotype in the *trans* configuration. All the *pan*-2 mutants, then, even those that do complement, must have some degree of functional relationship. These considerations lead us to think of the *pan*-2 region as a single, albeit complex, functional region—and by the functional criterion of complementation, a gene. The complementation patterns can be thought of as reflecting processes of *intragenic complementation.* Their functional basis will be discussed in Chapter 10.

The Neurospora Gene—Recombination. Like the phage *rII* genes, however, the *pan*-2 region is certainly not a single element by the criterion of recombination. If crosses are performed between two *pan*-2 mutants, the frequency of recombinants can be measured by determining the fraction of the total ascospores from the cross that are *pan*-2$^+$ (that is, those that can grow in the absence of pantothenic acid). Most pairwise crosses produce recombinants, although in very low frequency (about 10^{-3} to 10^{-5}). On the basis of such crosses a linear map can be constructed giving the relative locations of the various *pan*-2 mutations. The 75 mutations are found to be located at a minimum of 23 different sites within a small region of the map. The length of this region in recombination units is about 0.4 percent. The noncomplementing mutations are distributed throughout the genetic map shown in Figure 9-8.

A Gene in Maize

With most higher organisms it is difficult to determine whether the functional unit is subdivisible by recombination. The difficulty is at least partly owing to the general difficulty with a higher organism of collecting and screening sufficiently large numbers of progeny to detect rare recombinational events. One gene in corn has been studied by O. Nelson with techniques whose sensitivity approaches that of the techniques commonly used with microorganisms. In corn, the *waxy* locus is involved in the synthesis of a form of starch called amylose. Standard-type corn plants (*Wx*) have two forms of starch, amylose and amylopectin, in the kernels and also in the pollen. Waxy strains (*wx*) produce amylopectin but no amylose. Nelson has shown that waxy strains lack a particular enzyme required for the formation of amylose. As discussed in Chapter 4, the starch composition of the haploid pollen grains is determined by the genotype of the pollen grain itself. After staining with iodine, *Wx* pollen grains turn blue whereas *wx* pollen grains stain light red. In other words, in this instance the microspores of maize can be observed directly for a phenotypic attribute determined by a particular gene.

A number of different naturally occurring *wx* strains exist. Nelson has crossed a number of these different strains in all possible pairwise combinations. Where *a* and *b* stand for two different *wx* alleles that enter a cross, the endosperm of the resulting kernels is of either genotype $wx^a/wx^b/wx^b$ or $wx^b/wx^a/wx^a$ (review the life cycle of maize, Chap. 4). The kernels contain little or no amylose, indicating that the different alleles do not complement one another. Great quantities of pollen produced by the heterozygous plants can easily be screened under the microscope for the presence of rare, dark-staining pollen grains which are *Wx* in genotype, and which in fact occur.

Wx pollen grains might arise either by back mutation or by recombination. The frequency of back-mutations can be ascertained, however, from the frequency of dark pollen grains arising in self crosses ($a \times a$ and $b \times b$). The mutation rates observed in such crosses are not sufficiently high to account for the results obtained when different *wx* strains are intercrossed. Nelson's result seems to show, then, that different *wx* sites in the *Wx* gene are separable by recombination. The maximum frequency of recombination between *wx* sites is low, under one in a thousand. The number of separable elements so far demonstrated is five, with some of the characters behaving as though determined by small rearrangements or deletions. Although the data collected are not so extensive as those obtainable in studies of microorganisms, they do serve to illustrate that in corn as in microorganisms the functional unit is composed of subunits, which are separable by recombination.

The Drosophila Gene

Here we will just note briefly that for every Drosophila gene locus that has been critically studied, the region of integrated function, defined by complementation tests, is subdivisible by recombination into a number of elements. The white gene mentioned earlier consists of at least four elements separable by recombination. The frequency of recombination between these elements is approximately one in ten thousand.

General Conclusions

The foregoing examples represent only a small sample of the many studies on gene structure in different organisms including yeast, *E. coli*, Salmonella, and especially in Aspergillus, where pioneering work was done by G. Pontecorvo. However, the examples given should be sufficient to indicate the kinds of studies performed and the general picture of the gene that has emerged: it appears that the gene is composed of a large number of mutable elements, which are linearly arranged and are separable by recombination. Each such element may exist in different states, which may be transformed from one to another by mutation. The functional interactions of mutant genes are in some instances (for example, the *pan*-2 gene of Neurospora) complicated by *intragenic complementation*. Such partial nonallelism, however, does *not* permit us to subdivide the gene into discrete, autonomously functioning elements.

This picture of the gene derived from formal genetic analyses agrees well with modern ideas and facts concerning the chemical nature of the gene and its mode of action. The mechanism of gene action will be discussed in more detail in the next chapter, but at this point we can make a brief comparison between the current genetical picture of the gene and our notions of the chemical structure of the gene.

The gene is now generally conceived as a segment of DNA that serves as a code for a particular protein. Thus the gene must be linear in structure and composed of many subelements, the nucleotides. In Chapter 8 we showed that the nucleotide is the fundamental unit of mutation. If recombination occurs within DNA molecules, neighboring nucleotides should be separable by recombination. We are led to equate the element of recombination and mutation with the nucleotide. Since each gene is conceived as acting as a "transcription" unit in the synthesis of its product (presumably protein), the functional allelism of genes containing mutations separable by recombination means that the formation of a normal gene product can be prevented by alterations of the DNA code at many different sites within a gene. The intragenic complementation observed in some systems is explained on the basis of interactions between "mutant" gene products. This topic is re-examined in the next chapter.

Although the above account of gene structure is consistent with findings in microorganisms, it may not be sufficient to explain all cases of pseudoallelism, especially some of those examined in Drosophila. The uncertainty arises from possible instances of functional interactions *between* mutations in different genes, interactions that may mimic the behavior of mutations within the same gene. Well analyzed examples of this type are described in the next chapter in the section on the *operon*.

Terminology

As soon as the classical definition of the gene was found no longer satisfactory, many geneticists began to seek new terms for describing the new findings or to redefine old words in ways that would take into account the new findings. The key word, *gene*, was the first to lose its original meaning. We have chosen in this book to retain use of the word gene for the genetic unit of function. What we term gene, some workers prefer to call *locus* or *functional unit*. At this point we can operationally define the gene only in terms of complementation tests. For cases where biochemical knowledge of gene function is lacking, Benzer coined the word *cistron* as the operational equivalent of gene; it is used where complementation is the only functional test employed. As we have pointed out, the phenomenon of intragenic complementation makes the application of the operational tests for the cistron or gene difficult but not impossible. In the next chapter biochemical knowledge is brought to bear on the clarification of the gene concept.

Some workers prefer to restrict the word gene to describe only the basic indivisible elements of recombination. We have referred to these elements as *sites;* Benzer terms them *recons*.

Another classical term whose meaning has inevitably become fuzzy is *allele*. Should this word refer to elements that are functional alternatives (that is, mutant genes), or to those elements that are both structural and functional alternatives (that is, alternatives at the same site)? Since in the original definition, alleles were conceived as different states of the gene, the former definition would seem most appropriate. To help clarify the situation, Roman has coined the terms *homoalleles* (genes differing at the same site) and *heteroalleles* (genes having mutations at different sites).

Keys to the Significance of This Chapter

The gene was originally conceived as the elementary unit of inheritance, indivisible functionally or structurally. Recent studies on gene structure in a variety of organisms demonstrate that the unit of function is subdivisible into many linearly arranged structural elements. These findings are in accord with the notion that gene products (proteins) are coded in DNA as a linear sequence of nucleotides separable by recombination.

REFERENCES

Benzer, S., "Genetic Fine Structure." In *Harvey Lectures 56*. New York: Academic Press, 1961. (A beautiful account of the genetic structure of the *rII* genes of phage T4.)

————, "The Fine Structure of the Gene." *Scientific American*, January 1962. Available as Offprint 120 from W. H. Freeman and Co., San Francisco. (A good summary of the *rII* story.)

Carlson, E. A., "Comparative Genetics of Complex Loci." *Quart. Rev. Biol.*, **34**:33–67, 1959. (A review of complex loci emphasizing the studies on Drosophila gene structure.)

Morgan, T. H., *The Theory of the Gene*. New Haven: Yale University Press, 1926. (A summing up of the classical view of the gene.)

Pontecorvo, G., *Trends in Genetic Analysis*. New York: Columbia University Press, 1958. (Pontecorvo develops his views concerning the subdivisibility of the gene.)

QUESTIONS AND PROBLEMS

9-1. Define or specifically identify the following:

pseudoallele	site
cistron	complementation map
deletion mapping	heteroallele

9-2. Seven *rII* deletion mutants of bacteriophage T4 were spot tested in all pairwise combinations to test for their ability to give r^+ recombinants. The results are given in the table below. A "+" indicates that recombinants are formed; a "0" indicates that no recombinants are formed. Draw a topological representation of these mutations.

	1	2	3	4	5	6	7
1	0	0	0	0	0	0	0
2		0	+	+	+	+	0
3			0	0	+	0	+
4				0	0	0	+
5					0	0	0
6						0	0
7							0

9-3. Five point mutants were tested against the deletions described in Problem 9-2 with the results given below. What is the relative order of the point mutants?

		DELETION MUTANTS						
		1	2	3	4	5	6	7
	a	0	+	+	+	0	0	0
POINT MUTANTS	*b*	0	0	+	+	+	+	0
	c	0	+	+	0	0	0	+
	d	0	+	0	0	+	0	+
	e	0	+	+	+	+	0	0

CHAPTER 10

Gene Function

Ordinarily we detect the presence of a gene by some phenotypic characteristic of the organism far removed from the primary activity of the gene. For example, people differ as to whether they have hair on the mid-digital segments of their fingers (excluding thumbs). The evidence available indicates that presence of mid-digital hair may be due to a single dominant gene. Obviously, the gene involved somehow accomplishes its effect in terms of the chemistry of the organism. It is almost equally apparent that numerous and complex events intervene between the original chemical process initiated by the gene and the eventual manifestation of the character. In fact, the trail of events leading from gene to mid-digital hair appears to be so tenuous that no one knows as yet how to back-track from phenotype to gene in order to find out just what is the primary chemical action of the gene.

Many heritable characteristics studied by geneticists offer the same kind of difficulties that mid-digital hair does. The phenotypic attributes that geneticists are able to detect and to work with frequently give few clues as to the nature of gene action. However, in a rapidly increasing number of instances, particularly in microorganisms, the activities of genes have been traced to the molecular level.

Gene Control of Metabolism in Man

In 1909, an English physician named Garrod wrote a book called *Inborn Errors of Metabolism*. The title might be paraphrased as "heritable defects in body chemistry." Garrod, at that early date in the history of genetics, was already interested in the control of specific chemical reactions by genes, a concept often thought to be very modern. Garrod, like many of the pioneers of science, was little appreciated by his contemporaries for the profundity of his thought. Relatively recently, geneticists have come to realize the importance of his contribution to our understanding of the kind of problems raised in this chapter.

As a physician, Garrod was particularly concerned with genetic anomalies of the human organism. Among other heritable diseases dealt with in his writings, he considered at length *alcaptonuria*, a disorder characterized by a hardening and blackening of the affected person's cartilage and—more strikingly—by the fact that the urine turns black upon exposure to the air. Alcaptonuria is inherited as an autosomal recessive trait.

The molecular basis of alcaptonuria has been known for a long time, largely because in this disease the presence of a colored substance offers a focal point for chemical analysis. In alcaptonurics, the blackness of the urine is due to the presence of an unusual component, *homogentisic acid*. The abnormality of the situation lies in the fact that homogentisic acid accumulates in the urine of alcaptonurics, whereas in normal people it is broken down into simpler substances. This reaction is accomplished under the influence of an enzyme, which is present in the liver of normal persons but seems to be absent in alcaptonurics. We may summarize the situation as follows:

gene *A*
(enzyme)

NORMAL: homogentisic acid ⎯⎯⎯⎯⎯⎯⎯⎯⟶ maleylacetoacetic acid

gene *a*
(no enzyme)

ALCAPTONURIC: homogentisic acid ⎯⎯⎯ ⎯⎯⎯⟶ homogentisic acid
(no reaction)

STEPWISE METABOLISM UNDER GENE CONTROL

So far we have been considering single genes in relation to single biochemical reactions. If our purpose is to obtain a meaningful picture of life processes, this is a great oversimplification. Neither genes nor chemical reactions occur in isolation in the cell. As a first step toward amplifying the picture, we need to ask questions such as the following: Where does homogentisic acid come from? What happens to the maleylacetoacetic acid that is formed from homogentisic acid as the result of enzyme action?

The Sources of Homogentisic Acid. The answer to the first question is particularly enlightening. Homogentisic acid is derived from phenylalanine and tyrosine, two of the amino acids that serve as building blocks for proteins. When alcaptonurics are fed increased quantities of phenylalanine and tyrosine, there is a corresponding increase of homogentisic acid excreted in the urine.

Increased consumption of phenylalanine and tyrosine by normal individuals is not followed by such an accumulation of homogentisic acid in the urine. These results are easily explained if we assume that homogentisic acid is only an intermediate product in the normal metabolic breakdown of phenylalanine and tyrosine. It represents a way station in a series of transformations leading to the final degradation of the original amino acids. In alcaptonurics, this way station becomes a final stopping place. In normal persons, maleylacetoacetic acid, the normal conversion product of homogentisic acid, is another and later way station. Maleylacetoacetic acid is ultimately transformed into the very simple substances carbon dioxide and water.

Phenylketonuria. Other way stations in the degradative metabolism of phenylalanine and tyrosine are known. Some of these have been recognized on the same basis as homogentisic acid; that is, gene mutation resulting in a block in metabolism has led to the accumulation of an intermediate substance that normally would be subject to further transformation. This is clearly true in the disease *phenylketonuria*, which is inherited as a simple recessive. The major symptom of phenylketonuria is a type of extreme mental defect; affected individuals are called "phenylpyruvic idiots." This symptom is accompanied by the excretion of abnormally large amounts of phenylalanine and phenylpyruvic acid in the urine. Persons with a normal gene at the locus for phenylketonuria are able to convert phenylalanine into tyrosine. The latter substance is one of the precursors of homogentisic acid.

Overall Picture of Phenylalanine-Tyrosine Metabolism in Man. Based on the type of information we have just presented, there has emerged a picture of the degradative metabolism of phenylalanine and tyrosine in man, which is summarized in Figure 10-1. You will note that albinism, which is usually inherited as a simple recessive condition in man, finds a place in this scheme. The *melanin* compounds that are the basis of much of our pigmentation are derivatives of phenylalanine and tyrosine. Albinism represents a genetically determined inability to convert precursor substances into melanin pigments.

A fourth probable instance of genetic block in the sequence of transformations pictured in the figure is represented by the disease *tyrosinosis*. Here the anomaly is inability to oxidize *p*-hydroxyphenylpyruvic acid. Ingestion of precursors of this compound, including tyrosine, leads to its accumulation in the urine, in the same manner as the accumulation of phenylpyruvic acid in phenylketonurics and of homogentisic acid in alcaptonurics. Since only a single individual with tyrosinosis has been studied, nothing is known about the heritability of this disease. A reasonable speculation is that tyrosinosis is the consequence of gene mutation leading to a particular block in metabolism.

Similarly, goiterous cretinism appears to be another instance of genetic inability to carry out one of the normal biochemical transformations of

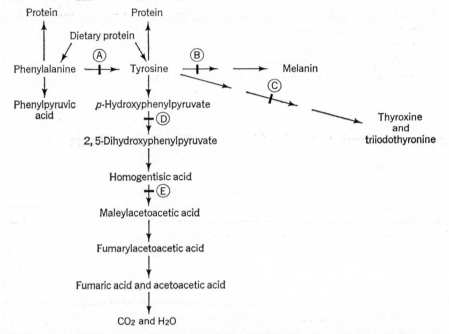

Figure 10-1. *Scheme for the metabolism of phenylalanine and tyrosine in man. The circled letters indicate the metabolic blocks due to hereditary defects described in the text. A: phenylketonuria; B: albinism; C: goiterous cretinism; D: tyrosinosis; E: alcaptonuria. (Adapted from Harris,* Human Biochemical Genetics, *Cambridge University Press, 1959.)*

phenylalanine and tyrosine. This condition, one of extreme mental and physical retardation, is associated with enlarged goiter and failure to produce the thyroid hormones, thyroxine and triiodothyronine.

The scheme presented in Figure 10-1 is partly tentative. Its main features are almost certainly correct, but it could be made more precise and accurate if the effects of mutation of other genes involved in that area of metabolism could be studied. We will have to wait for such mutations to turn up naturally, since there are good reasons why persons should not be bombarded with X-rays in order to create mutations for the geneticist or biochemist to study. Amplification of the scheme may therefore come relatively slowly. Nevertheless, even as it stands, the scheme represents an important contribution to our understanding of the biochemistry of man. More important than the reactions themselves are the concepts involved. Briefly, these are:

1. Metabolism occurs in stepwise fashion, with compounds being converted into other compounds in orderly sequences of transformation.
2. Specific unit processes in the chains of chemical events that make up metabolism are under the control of particular genes.

3. Mutation of genes governing such unit processes may lead to blocks at various points in the pathways of metabolism.

4. The primary consequences of such blocks may be (a) inability to produce certain compounds that are normal metabolites, and (b) accumulation of precursors normally converted into other compounds in a reaction sequence.

Nutrition and Genetic Capacity for Biosynthesis

The concepts just summarized have such far-reaching implications that it is important to affirm their validity and to discover how widely they apply. In general, the study of metabolism through genetic differences in natural populations has severe limitations. Most important of these limitations are the relatively low rate of spontaneous mutation and the fact that mutations of many genes that control vital functions are lethal. The first difficulty can be resolved by inducing mutations with any of various mutagens. The more difficult problem of identifying and preserving mutations that may ordinarily be lethal was met by G. W. Beadle and E. L. Tatum in a method of experimental approach directly related to the fundamental nutritional attributes of Neurospora. In principle, it applies to organisms in general, and has been widely used, especially in microorganisms.

Utilization of Minimal and Complete Media in Detecting and Preserving Biochemical Mutant Strains. Wild-type Neurospora has simple nutritional requirements. If these needs are reduced to the fewest and simplest substances on which the mold will grow normally, they are found to include the following: (1) certain inorganic salts; (2) some suitable carbohydrate, such as the sugar sucrose; and (3) one fairly complex organic compound, the vitamin biotin. These substances constitute the "minimal" medium of Neurospora. From the minimal medium, wild-type Neurospora can synthesize all the other components of its living substance, including amino acids, vitamins, purines, and pyrimidines; the mold incubated on minimal medium under aseptic conditions contains all these kinds of compounds in its mycelium.

Beadle and Tatum assumed that Neurospora transforms the components of minimal medium into vitamins, amino acids, and the like through orderly processes of stepwise metabolism under gene control. The basis of this assumption was the picture of phenylalanine-tyrosine metabolism in man, and other similar findings. It was thought that mutations of genes concerned in the biosynthesis of compounds essential for growth would result in new growth factor requirements for the strains in which the mutations had occurred.

Suppose, for example, that mutation should inactivate some gene that normally promotes a step in the biosynthesis of thiamine from appropriate

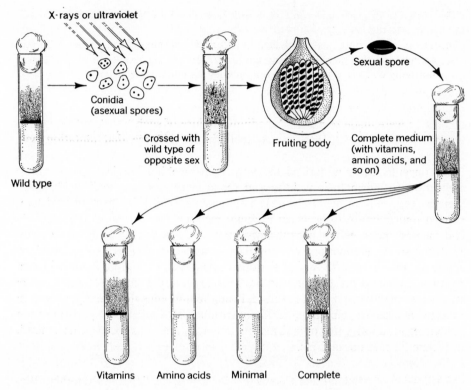

Figure 10-2. *Outline of procedure for producing, detecting, and classifying biochemical mutant strains in Neurospora. The illustration indicates production of a mutant deficient in the synthesis of some vitamin. The particular vitamin involved could be determined by inoculating the mutant strain into culture tubes containing individual vitamins such as thiamine. (After Beadle. In Baitsell,* Science in Progress, Fifth Series. *Yale University Press, 1947, p. 176.)*

substances in the minimal medium. Since thiamine is an essential growth substance, with many important functions in metabolism, a strain of Neurospora carrying the mutant gene would presumably be unable to grow on the minimal medium. In other words, on the minimal medium such a mutant gene would be a lethal. We learned in Chapter 8, however, that lethality may be conditional and that the "lethality" of mutant genes may sometimes be circumvented. For example, a mutant strain of Neurospora unable to synthesize thiamine might be preserved if thiamine were provided as a nutritional supplement to the minimal medium.

With these ideas in mind, Beadle and Tatum irradiated wild-type Neurospora with the expectancy of producing mutant strains with various newly acquired nutritional needs. Anticipating that mutations of this kind would be lethal for the mold cultured on minimal medium, the investigators grew their isolates from irradiated Neurospora on so-called "complete" media. The components of complete media included yeast and malt extract, hydrolyzed casein, and other materials rich in a variety of vitamins, amino acids, and all kinds of substances that might be expected to be essential metabolites and whose biosynthesis might be interfered with by mutation. From individual cultures of this kind, transfers of conidia were made to minimal medium. Failure of such transfers to grow on the minimal medium was taken as preliminary evidence that mutation had occurred.

Out of a series of thousands of isolates from irradiated wild type, many different strains were found to be unable to grow on minimal medium although they could grow on complete media. Crosses of these strains back to wild type proved that inability to grow on minimal medium was due to gene mutation, and in most instances the growth deficiency was assigned to the locus of some single gene. Systematic investigation of the nutritional attributes of a mutant strain almost always revealed that some single substance added to the minimal medium satisfied the growth requirement. As a working hypothesis, such substances can be thought of as representing metabolites in Neurospora whose biosynthesis has been blocked in the mutant strain. The general procedure for producing, detecting, and classifying biochemical mutant strains is summarized in Figure 10-2. A number of different kinds of biochemical mutant strains have been produced in Neurospora. Some of these are included in the following listing, in which the different mutants are identified by the compounds they require for growth.

VITAMINS	AMINO ACIDS	OTHER COMPOUNDS
thiamine	arginine	adenine
pyridoxine	leucine	pyrimidine
p-aminobenzoic acid	lysine	succinic acid
pantothenic acid	methionine	sulfonamide
inositol	phenylalanine	
nicotinic acid	proline	
choline	threonine	
riboflavin	tryptophan	
	valine	
	serine	
	histidine	

(Information from Houlahan, Beadle, and Calhoun, *Genetics*, **34**:493-495, 1949.)

The existence of these diverse mutant strains is substantial confirmation of the validity of the basic assumptions of Beadle and Tatum and of the soundness of their reasoning.

Utilization of Biochemical Mutant Strains in Establishing Pathways of Biosynthesis. The procedure by which wild-type Neurospora synthesizes a metabolite from the constituents of minimal medium may often be expected to include a number, sometimes a large number, of steps. If each of these steps is under the control of a different particular gene, mutation at any one of several different loci might conceivably give rise to the same growth-factor requirement. Assume that essential metabolite A is normally synthesized via a series of precursor substances, B, C, D. Assume also that the transformations of precursors, which eventually lead to A, are each under gene control. As shown below, mutation of either gene *H* or gene *F* might block the production of A and give rise to a growth-factor requirement for A.

$$\xrightarrow{\text{gene } E} \text{substance D} \xrightarrow{\text{gene } F} C \xrightarrow{\text{gene } G} B \underset{h}{\dashrightarrow} (A)$$

$$\xrightarrow{\text{gene } E} \text{substance D} \underset{f}{\dashrightarrow} (C) \xrightarrow{\text{gene } G} (B) \xrightarrow{\text{gene } H} (A)$$

Although the growth-factor requirement after either mutation might be satisfied by a medium containing substance A, mutation at the two different loci would produce quite different sorts of biochemical situations. In the first instance (mutation of gene *H*), the requirement might be specifically for substance A. In the second instance (where genes *G* and *H* are left intact, and where C can be converted to B and then into A if only the organism can procure some C), either C or B should be able to replace A as a growth-factor supplement for the mutant strain. Many actual situations corresponding in principle to this hypothetical example have been found.

Such situations lend themselves to the experimental investigation of pathways in biosynthesis. The basis of such investigations is a study of substitutions of dietary supplements that can be made for a growth-factor requirement common to a series of genetically different biochemical mutant strains. This can be illustrated from a study of mutant strains of Neurospora unable to synthesize the amino acid arginine.

Genetic Control of the Biosynthesis of Arginine in Neurospora. Treatment of wild-type Neurospora with mutagens has produced several genetically different mutant strains each of which is characterized by a requirement for arginine. Biochemically these strains may be classified on the basis of possible substitutions that may be made for arginine in satisfying the growth-factor requirement. One strain has a specific requirement for arginine; other strains grow normally if either arginine or citrulline is added to the minimal medium; still different strains are able to grow on a supplement of arginine, citrulline, or ornithine. These relationships are summarized in Table 10-1.

Table 10-1. GROWTH OF MUTANT STRAINS OF NEUROSPORA.

Mutant Strain No.	Arginine	Citrulline	Ornithine	Unsupplemented Minimal Medium
21502	37.2	37.6	29.2	0.9
27947	20.9	18.7	10.5	0.0
34105	33.2	30.0	25.5	1.1
30300	37.6	34.1	0.8	1.0
33442	35.0	42.7	2.5	2.3
36703	20.4	0.0	0.0	0.0

Source: Data from Srb and Horowitz, *J. Biol. Chem.*, **154**:133, 1944.
Note: The values represent the increase in dry weight in mg after 5 days on liquid minimal medium supplemented with 0.005 mM of arginine, ornithine, or citrulline, or without supplement.

A logical interpretation is that the mutant genes characteristic of the different strains represent blocks at points in a metabolic pathway where their normal alleles control a biosynthetic sequence that is $\longrightarrow$ ornithine $\longrightarrow$ citrulline $\longrightarrow$ arginine. This is illustrated in Figure 10-3. Notice how the wild-type genes control simple chemical additions to precursor molecules until arginine is finally built up. An interesting point is that a search among different lactic acid fermenting bacteria has revealed the existence of types whose growth-factor requirements are the same as in the different biochemical mutant types of

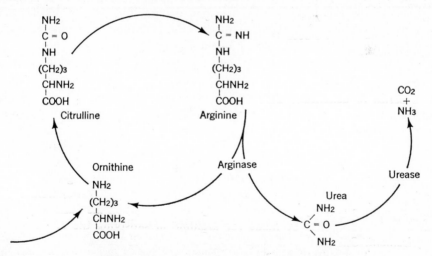

Figure 10-3. *The ornithine cycle in Neurospora. All steps in this sequence of reactions, except those involving urease and arginase, are known to be gene controlled. (After Srb and Horowitz, J. Biol. Chem., 154:137, 1944.)*

Table 10-2. COMPARATIVE EFFECTS OF ARGININE AND RELATED COMPOUNDS
ON GROWTH OF LACTOBACILLI.

Organism	Additions to Arginine-free Media			
	None	Ornithine	Citrulline	Arginine
L. fermenti	0	+	+	+
L. casei	0	0	+	+
L. mesenteroides	0	0	0	+

Source: After Volcani and Snell, *J. Biol. Chem.*, **174**:895, 1948.
Note: 0 = no growth; + = good growth, assuming an appropriate concentration of the supplement.

Neurospora. This finding, shown in Table 10-2, emphasizes that the laboratory production of biochemical mutations in Neurospora is only an acceleration of processes that occur spontaneously in nature. In fact, loss in synthetic capacity due to gene mutation is doubtless one of the major causes of the diversity in nutritional requirements among various kinds of organisms.

Accumulation of Precursors as the Result of Genetic Blocks in Biosynthetic Reactions. Where the reaction chains leading to the synthesis of growth factors in wild-type Neurospora are unbroken, precursor substances are seldom found in sufficient quantity to be easily detected. But when gene mutation breaks the reaction chain of a biosynthesis, precursors may accumulate behind the genetic block. This kind of situation provides remarkable opportunities for isolating and identifying intermediate substances in metabolism. A particularly good example is found in an investigation, made by N. H. Horowitz, of mutant strains of Neurospora unable to synthesize the amino acid methionine.

On the basis of substitutions that could be made for the methionine require-

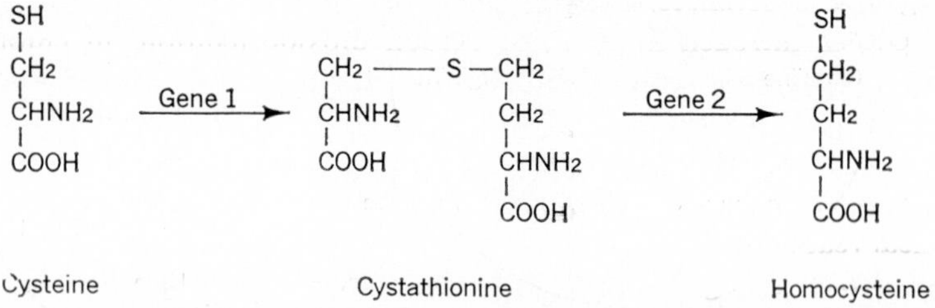

Figure 10-4. *In wild-type Neurospora, the conversion of cysteine to homocysteine is a stepwise process under genic control. A mutant strain in which gene 2 has mutated accumulates cystathionine in its culture medium. When this substance is added to minimal medium, it permits growth of a mutant strain in which gene 1 has mutated. (Based on work of Horowitz,* J. Biol. Chem., **171**:258, 1947.)

ment of his several mutant strains, Horowitz first established the following genetically controlled stages in the synthesis of methionine: $\longrightarrow$ cysteine $\longrightarrow$ homocysteine $\longrightarrow$ methionine. The conversion of cysteine to homocysteine provides an especially intriguing problem: is an extra CH_2 just "slipped into" a cysteine molecule? Chemically, this would be difficult.

Two genetically different mutant strains were found to be unable to convert cysteine to homocysteine, suggesting that perhaps the transformation that normally occurs in wild type may take place as two steps, each under the control of a different gene. When it was found that one of the strains accumulates a substance on which the other can grow, this suggestion was confirmed. The substance that accumulates was isolated and identified as cystathionine. Reference to Figure 10-4 will show you the considerable ingenuity of the biochemical mechanism by which genes solve the problem of conversion of cysteine to homocysteine. The mechanism does not involve "slipping in" a CH_2 group; it involves simply condensing a 3-carbon-sulfur chain with a 4-carbon chain, then splitting the sulfur from the three carbon atoms with which it was originally associated.

Genes and Enzymes

You have just seen how the study of experimentally produced biochemical mutant strains of Neurospora provides evidence that genes control the fundamental reactions by which nutrients are converted into the various chemical constituents of the organism. Studies of this kind have been carried out with many organisms, particularly among the bacteria, fungi, and algae, and variants similar to those produced in Neurospora have been found. Mostly these variants are characterized by requirements for particular vitamins, amino acids, or other essential growth substances. But mutants have also been found that are deficient, for example, in nitrogen fixation, in carbon dioxide fixation, in chlorophyll synthesis, or in various aspects of carbohydrate metabolism. There can be little question, then, that most of the chemical reactions making up the normal metabolism of the cell involve some sort of genic control. Now that this has been established, a more compelling question is: *how* do genes control these biochemical reactions?

The reactions that genes are known to control are frequently difficult to carry out in the laboratory. The organic chemist can duplicate many of them, but often only when he utilizes drastic conditions, for example of temperature or pressure, which are not found in cells, and are in fact generally incompatible with the maintenance of life. Yet these same reactions go on in cells under much less extreme circumstances. As is well known to students of biology, this is possible because the cellular reactions are promoted by efficient biological catalysts called *enzymes*. The general conclusion is almost inescapable that genes must control their appropriate biochemical reactions through the mediation of the enzymes that catalyze these reactions.

The One Gene-One Enzyme Hypothesis. Enzymes are generally characterized by a high degree of *specificity* of action; that is, for the most part, each enzyme is only active in promoting a single kind of biochemical transformation. Many individual genes have been shown to have this same sort of specificity for biochemical reactions. The large number of examples among many organisms in which a particular gene mutation has been found to give rise to a growth-factor requirement by blocking a single reaction in a biosynthesis is especially suggestive. Chiefly on the basis of this kind of finding, Horowitz phrased a *one gene-one enzyme hypothesis*, as follows: "A large class of genes exists in which each gene controls the synthesis of, or the activity of, but a single enzyme."

The one gene-one enzyme hypothesis should not be taken to imply that a single gene starting from scratch can produce an enzyme. Enzymes are complex substances, composed at least in part of protein. Presumably, the constituents of an enzyme molecule are built up through stepwise processes in which many genes take part. In fact, we know that the amino acids that make up protein are synthesized in this fashion. The one-to-one relationship is to be thought of as existing between an enzyme and the gene that imparts the molecular structure that determines the enzyme's specificity.

Because many instances accumulated to show that a given gene determines whether a particular biochemical reaction can be carried out, and because such reactions are known to be under the immediate control of enzymes, the inference was that the genetic control of metabolic properties must be accomplished through the specific control of enzymes by genes. An obvious test of the inference is to ask whether genes associated with specific biochemical reactions affect the particular enzymes that control those reactions. Indeed, one can go back to Garrod's time to get the beginning of an answer. You will remember that alcaptonurics lack an enzyme present in the liver of normal individuals, where it carries out the degradation of homogentisic acid. But the more recent study of nutritional mutants in microorganisms provides much more massive evidence for the determination of enzymes by genes. The genetically controlled biosyntheses of arginine and methionine in Neurospora, already familiar to you, provide examples. One of the steps in the conversion of citrulline to arginine involves the intermediate compound argininosuccinic acid. Certain mutant strains that require arginine for growth, but that cannot use citrulline, lack the enzyme argininosuccinase. Strains not carrying the mutation associated with lack of argininosuccinase have the enzyme and can convert argininosuccinic acid into arginine. The same specific relationship involving gene, enzyme, and biochemical reaction emerges when one compares a Neurospora strain having the mutant gene that blocks the conversion of cystathionine to homocysteine with strains carrying the normal allele (Fig. 10-4). If one collects mycelium of a strain carrying the normal allele, grinds the mycelium, makes appropriate extracts of the mycelial protein, and then incubates the protein in a test tube with cystathionine, homocysteine is produced. A similar protein extract of the mutant mycelium is inactive in promoting the conversion of cystathionine to homocysteine. The mutant lacks the enzyme cystathionase II; strains with the normal allele produce this enzyme.

The same sort of picture emerges from the biochemical genetic study of other organisms, from bacteria to man. Specificity of relationship between gene and enzyme appears to be a general biological phenomenon. You must not suppose, however, that the invariable effect of mutation is to remove an enzyme. As the result of different mutations within the same functional unit of heredity in Neurospora, the enzyme tryptophan synthetase, extensively studied by D. Bonner and co-workers, can show either inactivity or various alterations in activity. Some tryptophan synthetase mutants produce a specific protein that is enzymatically inactive but that by sensitive biological and chemical tests is remarkably like tryptophan synthetase. In still other mutants, no protein homologous with tryptophan synthetase can be isolated. It is worth noting here that enzyme differences can sometimes be detected between strains isolated from nature and the genetic site of the difference mapped to a particular chromosomal site. This has been done by N. H. Horowitz for the Neurospora enzyme tyrosinase.

Apparent Exceptions to the One Gene-One Enzyme Theory. Some instances have been found of single-gene mutations giving rise to dual or multiple growth-factor requirements. One of these, in Neurospora, involves a mutant strain having a nutritional requirement for the amino acids methionine and threonine. Close analysis of the situation has shown, however, that the biosynthesis of these amino acids has a common step—the compound homoserine is a precursor for both methionine and for threonine. When minimal medium is supplemented only with homoserine, the nutritional requirements of the mutant strain are satisfied.

A few experimental findings suggest that more than one gene may be involved in the control of the activity of a single enzyme. So far, when such cases have been explored in detail, they have been found not to violate the basic concept of gene-protein relationships with which you have become familiar. An example is the genetic control of the hemoglobin molecule in man. Two different genes in adult man are known to control the structure of the hemoglobin molecule. However, study of the hemoglobin from mutant individuals reveals that in fact the different genes control the structure of the different polypeptide chains of the molecule. One gene controls the structure of the alpha chain, another gene the beta chain. These findings suggest that the one gene-one enzyme hypothesis might be further refined to a *one gene-one polypeptide* hypothesis. A possible objection to the one-to-one relationship arises from the discovery that certain genes may control the amount of enzyme primarily determined by another gene, or control the time at which the enzyme is made during the life cycle of the organism. However, these genes, which regulate and modulate the action of enzyme-determining genes, do not play a role in determining the structure of the enzyme. Thus these regulating genes need not appear as exceptions to our one-to-one hypothesis. However, they do force us to recognize another class of genes—*regulatory* genes. We will discuss the activities of these genes shortly.

Another point should be made. Although most of the studies of gene action have concerned the control of enzymes, we should not think that all genes act by controlling the structure of polypeptide chains of enzymes. Certainly many genes produce proteins whose cellular function is other than enzymatic. Examples are provided by virus genes that determine the structure of the polypeptide chains that make up the coat of the virus.

Gene Control of Protein Structure

Let us now consider how a gene controls the specificity of structure of a polypeptide chain. A first major step toward understanding how a gene controls the structure of a protein has been the result of work with the hemoglobin molecule of man. Before discussing the genetic control of hemoglobin structure, however, we must describe a hereditary disease in humans called *sickle-cell anemia*.

Under the right conditions for observation, red blood cells of certain people show sickle shapes, oat shapes, and other eccentric variations from the normal disc shape. This phenomenon, called *sickling*, is found in a fairly high proportion of American Negroes. In many instances, sickling seems not to have serious consequences for an affected individual. However, some persons whose blood cells sickle are afflicted with a severe hemolytic anemia that markedly reduces their life span. This condition is called *sickle-cell anemia*. The milder condition of sickling without anemia is called *sickle-cell trait*. These differences appear to depend on a single pair of alleles, such that:

$$Hb^S\ Hb^S \longrightarrow \text{sickle-cell anemia}$$
$$Hb^S\ Hb^A \longrightarrow \text{sickle-cell trait}$$
$$Hb^A\ Hb^A \longrightarrow \text{normal}$$

Characteristic molecular differences among normal persons, those with the sickle-cell trait, and sickle-cell anemics can be shown if solutions of their hemoglobins are separately examined on the basis of migration in an electrical field. Hemoglobin, the oxygen-carrying component of the blood, is a complex molecule consisting of protein and an iron-containing compound. Under appropriate experimental conditions in an electrophoresis apparatus, the hemo-

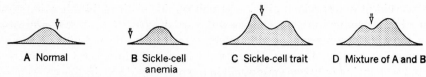

A Normal   B Sickle-cell   C Sickle-cell trait   D Mixture of A and B
anemia

Carbonmonoxyhemoglobins in phosphate buffer *p*H 6.90

Figure 10-5. *The migration of hemoglobins in an electrical field. Using the arrows as points of reference, you can see that under certain conditions the hemoglobin* (A) *of normal persons migrates toward one pole and the hemoglobin* (B) *of sickle-cell anemics migrates toward the other. The hemoglobin of persons with sickle-cell trait behaves much the same as a mixture of* A *and* B. (*After Pauling, Itano, Singer, and Wells*, Science, **110:***545, 1949.*)

globin of normals migrates toward one pole, while that of sickle-cell anemics carries an opposite electrical charge and migrates toward the other pole. The hemoglobin of persons who are heterozygous for the sickling gene separates into fractions, with one fraction moving toward each pole; it behaves much as a mechanical mixture of hemoglobins taken from the two homozygous types. These relationships are shown in Figure 10-5.

In recent years the sequence of amino acids in the polypeptide chains of hemoglobin has been worked out. The alpha chains of sickle-cell hemoglobin have the same structure as in normal hemoglobin but the beta chains differ from the normal hemoglobin in the replacement of one amino acid by another (valine in the Hb^S hemoglobin, glutamic acid in the normal) at one particular point in the chain. In a number of other genetically determined abnormal hemoglobins that have been studied, the aberrant type differs from the normal in only one amino acid at one point in either the alpha or beta chains. At least some mutations, then, exert their effect on the protein by replacing, at some point, one amino acid with another. In the instance of sickle-cell hemoglobin, such a slight change in primary structure has a profound effect upon the properties of the protein and in turn upon the individual. The tertiary structure of the protein, which gives the protein its functional specificity, is generally believed to be determined largely by its primary structure or amino acid sequence. A slight change in the primary structure can result in an important change in the folding of the protein, and consequently in its functional properties.

A number of detailed studies are being made of the changes induced in a protein by mutations of the gene that determines its structure, for example, a gene that controls the structure of the enzyme tryptophan synthetase in *E. coli.* As with the locus for the corresponding Neurospora enzyme, many mutations result in the production of no detectable protein closely related to tryptophan synthetase. Other mutations result in the production of homologous proteins that have no enzymatic activity. In most cases, these proteins are identical in structure to the wild-type enzyme except for the replacement of a particular amino acid at one specific point in the polypeptide chain. Still other mutations result in the production of an enzyme with partial activity but more thermolabile than the wild-type enzyme, or with different electrophoretic mobility, or changed in some other detectable way. These mutations, also, in general produce their effect by changing only one amino acid in the polypeptide chain. We can summarize the emerging picture as follows:

1. Mutations that abolish or alter enzyme function result in either the production of no protein or of structurally modified protein.

2. If a protein is formed by the mutant, it differs from the wild-type protein in primary structure only in one amino acid at some specific point in the polypeptide chain.

Exceptions to these conclusions are known, and certainly more will be found in the future. However, they are almost certainly correct in principle. How do we account for these conclusions? The fact is that before these findings were made they were predicted on the basis of *coding theory*. We know that the genetic information in the gene is stored in DNA, and it was suggested to you that this information is in the form of a code for the structure of protein. Both protein and DNA are linear polymers of smaller elements. We can readily imagine that the sequence of amino acids in a protein is coded by the sequence of nucleotides in the DNA. With twenty different kinds of amino acids and only four different kinds of nucleotides in DNA, the code must be such that more than one nucleotide codes for one amino acid, just as the Morse code has more than one dot or dash for most of the individual letters of the alphabet. The simplest code we could devise would be a "triplet" code, a scheme by which three nucleotides code for each amino acid. We could postulate, for example, that TTT stands for the amino acid phenylalanine, TCG for serine, TTG for leucine, and so on. Thus the sequence TTTTCGTTG in some region of a particular gene would code for the sequence phenylalanine-serine-leucine in the corresponding portion of the polypeptide chain determined by that gene.

In the chapter on mutation you saw evidence that some mutations result in the replacement of one nucleotide by another in the DNA. A mutation in the gene under discussion might result in a change of the above sequence to TTTT*T*GTTG. The protein made by the mutant gene would have at the corresponding point in its structure the amino acids phenylalanine-*leucine*-leucine. The result would be a mutant protein identical to the wild-type protein except for the replacement of one amino acid by another. As we have outlined the coding system, it permits mutation to result in a change to a triplet that is not the code word for any amino acid—a "nonsense" mutation. We might imagine this type of mutation to be of the sort that results in no gene-specific protein being made or, at most, protein fragments. We might suppose small deletions to have a similar effect.

Further evidence for the theory that nucleic acid coding determines amino acid sequence comes from work with the tobacco-mosaic virus. The sequence of

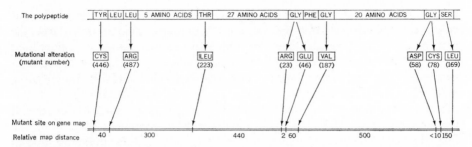

Figure 10-6. *A map of part of the gene controlling the A polypeptide of the tryptophan synthetase of* E. coli. *Only a segment of the polypeptide for which the complete sequence of amino acids has been worked out is shown. The relative recombination distances of the mutant sites are given under the genetic map. The relative order of the sites was verified by three-factor crosses. (After Yanofsky, Carlton, Guest, Helinski, and Henning*, Proc. Natl. Acad. Sci. **51:***266, 1964.)*

amino acids in the protein of the virus coat has been determined. If the RNA of the virus is treated with nitrous acid, mutations are produced. As mentioned in Chapter 8, the kinetics of mutant formation with nitrous acid suggests that the physical cause of the mutations is a change in a single nucleotide. When the coat proteins of the various mutants are analyzed, most of the differences from wild type are replacements of only one amino acid by another. The simplest and most plausible interpretation is that a change of a single nucleotide in the RNA of the virus can result in the change of one amino acid in the protein.

A prediction of this theory of gene action which has found experimental verification is that the gene and its polypeptide product are colinear. That is, the order and relative positions of mutations in the gene are directly related to the order and relative positions of corresponding amino acid substitutions in the polypeptide. This has been found to be the case for the A polypeptide of the tryptophan synthetase enzyme of *E. coli* (see Fig. 10-6) and for the head protein of phage T4.

Having in mind this general picture of how the gene works, let us consider some of the familiar properties of genes, and see how we can explain them.

Dominance. The recessiveness of mutations is understandable on the basis that in most instances the wild-type gene makes an active enzyme, but a mutant allele does not. In the simplest instance we would expect the heterozygote to make one-half the amount of enzyme formed by the wild-type homozygote. If this half amount of enzyme is sufficient to carry out the needed metabolic function, we have complete dominance; otherwise partial function (partial dominance) is expected. We also see clearly the reasons for certain cases of hierarchical dominance among multiple alleles. In the mouse, a given locus can exist in one of three allelic states that determine hair color—C = black coat color, c^h = himalayan trait (white body and black points, much like the markings of a Siamese cat), and c = albino (white coat color). C is dominant to c^h and c, and c^h is dominant to c. Biochemical studies show a pigment-promoting enzyme system that is fully active in C/C individuals. By contrast, c^h/c^h individuals manufacture a thermolabile enzyme system with the result that pigment is only manufactured in the parts of the body that are cool enough for the enzyme to have activity. No active enzyme system is formed by c/c individuals. The complete body color of C/c^h individuals is, therefore, due to the formation of active enzyme all over the body under the direction of the C gene. In c^h/c individuals, thermolabile enzyme is formed at the body extremities in spite of presence of the c gene, showing c^h to be dominant to c.

We have presented a concept of dominance that implies that the dominant allele is always the active form of the gene while a recessive is inactive or only partially active. However, exceptions to this rule can be expected to arise out of complex interactions of a kind known to exist in biological systems. As a hypothetical case, imagine a mutation that changes an enzyme in such a way that it has greater affinity than does the wild-type enzyme for their common substrate, but unlike the wild-type enzyme, the modified enzyme destroys substrate without converting it to the normal product of the reaction. Such a mutant gene would be dominant to wild type, if the modified enzyme competed successfully with wild-type enzyme for substrate, with the consequence that little or no normal reaction product were formed. Moreover, dominance is often perceived in characteristics of the organism far removed from the primary function of the gene. Under such circumstances, we can suppose that many loci participate in establishing dominance relationships. Our knowledge of gene function gives a basis for understanding dominance but does not preclude the existence of unusual or even unpredictable dominance relationships.

Suppressor Mutations. In the chapter on mutation we described suppressor mutations. These are apparent phenotypic reversions of a mutant to wild type, or near wild type, but with the suppressor mutation being at a different site from that of the (forward) mutation that accounted for the original mutant characteristic. Two classes of suppressor mutations are known: those that occur within the same gene as the original mutation, and those that are in different genes. Some of those that occur within the same gene can be interpreted as coding for a second amino acid substitution in the protein with the result that the twice-changed protein is more active than that produced by the original mutant. Such cases have been found and analyzed to the level of amino acid sequence. Other intragenic suppressors of a different nature are described later in the chapter. Suppressor mutations in genes at other loci than that of the gene in which the original mutation occurred might be thought to offer material for critical tests of the one gene-one polypeptide theory. Superficially at least, we might suppose that such a suppressor mutation would have the effect of restoring the wild-type structure to the mutant enzyme. However, in some cases, the mutant enzyme is inactive because it is more sensitive to some metabolic inhibitor of enzyme activity than is the wild-type enzyme. The suppressor acts by removing the inhibitor, permitting the mutant enzyme to function, and thus restores a normal phenotype. In still other cases, a suppressor acts by modifying a different biochemical pathway in such a way that the metabolic block induced by the original mutation is compensated for and is no longer deleterious to the organism. Nevertheless, in some instances suppressed strains produce at least some protein of the type formed by the normal allele of the suppressed mutant. The mechanism of action of this class of suppressor is not yet determined but it is thought to involve the protein-synthesizing system of the cell. Further studies of this class of suppressor mutations should tell us more about the genetic control of the mechanics of protein synthesis.

Intragenic Complementation. In Chapter 9, dealing with gene structure, we described intragenic complementation and showed that it could be used as the basis for drawing functional maps of a locus. How is intragenic complementa-

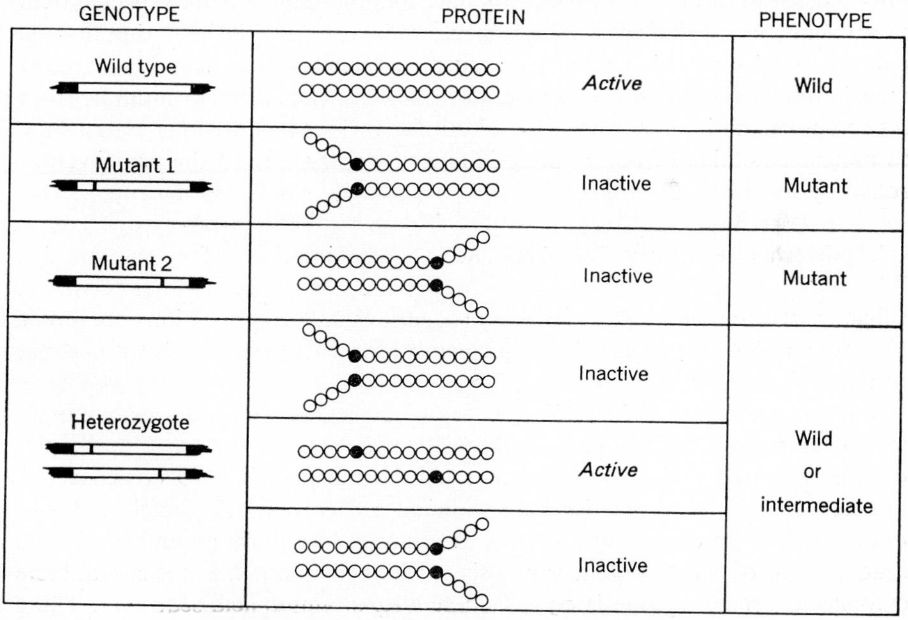

Figure 10-7. *The mechanism of intragenic complementation. The active protein is a dimer (an aggregate of two identical polypeptide chains). In the heterozygote, the two mutant types of polypeptides form random dimers. The hybrids are active.*

tion explainable at the gene-protein level? The most widely accepted theory to account for intragenic complementation is as follows. Certain genes control the structure of enzymes that are composed of two or more identical polypeptide chains. In a diploid or in a heterokaryon, the different polypeptide chains formed under the control of two different allelic genes may associate to form the *multimeric* enzyme molecules. If the two different alleles are both mutant, hybrid enzyme formed by the aggregation of the two different kinds of polypeptide chains may be active even though homogeneous multimeric protein produced by one allele or the other is not active (Fig. 10-7). In support of this theory we may point out that in some instances the enzyme produced as the result of intragenic complementation differs from either of the homologous mutant proteins *and* from the wild-type enzyme. Furthermore, under suitable conditions, mixtures of enzymatically inactive protein extracts from two complementing mutants have been found to aggregate *in vitro*, forming protein with enzymatic activity similar to but not identical with that of the wild type. By labeling the two different mutant polypeptides with radioactivity, it has been possible to show that both in fact are present in the active aggregates. Finally, the theory accounts plausibly for the existence of the noncomplementing mutants that have been found at the locus of each functional gene so far studied. Nonsense mutants are expected to form no polypeptide chains, and therefore are not expected to be "complementers." Various findings, then, are in accord with the notion that the enzyme activity resulting from intragenic complementation occurs because of association of the different polypeptides made by two different mutant genes. As complementation maps are worked out for various loci, we may expect them to reflect something of the nature of the tertiary structure of the proteins conrolled by these loci.

The Mechanics of Gene Function. The DNA coding theory for gene function is beginning to receive considerable experimental verification and has predictive value as well. As with many theories, exceptions and complications are found However, if the theory has value, these exceptions will fall into place and help us to refine the theory.

We might next ask, if the coding theory is correct, how does the gene manage to impart to the protein its structural specificity? What are the actual mechanics of translation of the information in DNA to the protein? To consider this question we will leave genetics for a moment and consider biochemical studies of the mechanism of protein synthesis. It has been known for a number of years that the site of most protein synthesis in the cell is not the nucleus but the cytoplasm. Protein synthesis occurs on small particles called *ribosomes*. These particles contain protein and RNA. It has also been known for a long time that RNA synthesis in the cell is correlated with protein synthesis. This knowledge suggested to many workers that RNA might be the intermediate between the gene and its protein product, and that perhaps the RNA acts as a "messenger" of genetic information, carrying the information from the chromosome to the ribosome, the site of protein synthesis. Many types of experimental results obtained from the living cell suggest this explanation. Evidence has also come from studies of protein synthesis *in vitro*. Proteins can be synthesized in a test tube if certain components of the living system are present, in particular, ribosomes, an energy-generating system, amino acids, and RNA. A number of elegant experiments have shown that ribosomes can be thought of as nonspecific protein-synthesizing "factories" and that the specificity of a given protein is not conferred by the ribosomal RNA but by a special RNA called *messenger* RNA. If RNA from a RNA-containing phage, f2, is added to a protein-synthe-sizing system from the bacterium *E. coli*, protein very similar to the coat protein of the phage is synthesized. This experiment shows that the specificity for protein synthesis is *not* carried by the ribosome, but that the phage RNA in some manner directs the specificity of the proteins synthesized by the bacterial ribosomes. Enzymes have been found that synthesize RNA using DNA as a template. The RNA so formed has a sequence of bases complementary to that of the DNA strand from which it is copied. Such enzymes appear likely to be those involved in the synthesis of the messenger RNA. Several lines of evidence indicate that *in vivo* messenger RNA is a copy of only one of the two DNA strands of the double helix.

Although the picture is far from complete, much is known about the way in which proteins are synthesized on the ribosomes. Still another form of RNA—transfer, or soluble RNA—is involved in protein synthesis. Under the action of specific enzymes, the various free amino acids are attached to specific species of transfer RNA. In this form the "activated" amino acids are brought to the ribosome, released from the transfer RNA and attached to the growing point of the polypeptide chain through a peptide bond. The ribosomes actually move

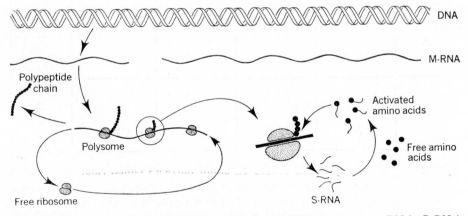

Figure 10-8. *The mechanism of protein synthesis. M-RNA is messenger RNA; S-RNA is soluble RNA. At right: detail of circled portion at left.*

along the messenger-RNA molecule, "reading" it, so to speak, as the polypeptide chain is being synthesized. More than one ribosome can be reading the same messenger. This gives rise to *polysomes*—messenger molecules with a number of ribosomes attached. When the ribosome reaches the end of the messenger, it falls off and the completed polypeptide chain is released (Fig. 10-8).

Perhaps the most exciting experiments in this field are those directed toward breaking the genetic code. Chemical methods are available for synthesizing artificial RNA polymers. Polypeptides can be synthesized *in vitro* if these artificial RNA polymers, in place of cellular RNA, are added to the ribosomes. If polyuridylic acid (a polymer of RNA containing only UUUUU. . . .) is added, a polypeptide containing only phenylalanine is formed! If the code is a triplet code, this experiment means that the code word for phenylalanine is UUU (or, in the DNA, TTT, or its complement, AAA). By using various kinds of mixed

Table 10-3. THE PROBABLE GENETIC CODE DICTIONARY

1st base ↓	← 2nd base →				3rd base ↓
	U	C	A	G	
U	PHE	SER	TYR	CYS	U C
	LEU	SER	(NONSENSE)	TRYP	A G
C	LEU	PRO	HIS	ARG	U C
	LEU	PRO	GLU-NH$_2$	ARG	A G
A	ILEU	THR	ASP-NH$_2$	SER	U C
	MET	THR	LYS	ARG	A G
G	VAL	ALA	ASP	GLY	U C
	VAL	ALA	GLU	GLY	A G

Source: This code is derived from work of M. Nirenberg and others. (See *Proc. Natl. Acad. Sc.,* **53**:1161, 1965.)

Note: A triplet code is assumed. The third base in the triplet can, in some cases, be anything and, in most cases, either one of the two purines or either one of the two pyrimidines. Note that the effects of mutations can easily be found: for transition mutations stay within one quadrant and move sideways for changes at the second base, up or down for changes at the first or third base. A key for the amino acid abbreviations is found in Figure 5-1, page 128.

RNA polymers of known composition and examining the composition of the polypeptides made in their presence, RNA code words for a number of amino acids have been deduced (Table 10-3). This work still presents a number of problems and complications, but the main features seem clear. The results give powerful support to the coding theory of gene action.

Is the code in fact a three-letter code? That a code word includes more than one nucleotide has been demonstrated by studies on the tryptophan synthetase of *E. coli*. Instances have been found of mutations, separable by recombination and thus at different sites, that cause alterations in the same amino acid (see Fig. 10-6). Indirect but elegant genetic experiments by Crick and his collaborators suggest that the code *is* a three-letter code. As was mentioned in Chapter 8, it is thought that acridine-induced and many spontaneous mutations in the *rII* genes of phage T4 are either additions or deletions of single base pairs. Such mutations often are found to revert by second-site mutation. That is, the revertant is a double mutant, the second mutation being similar to the first and both within the same gene. Detailed analysis of a variety of such mutations shows that these mutations can be classed as either $(+)$ or $(-)$. The double mutants that have a wild-type phenotype are always $(+)(-)$, never $(+)(+)$ or $(-)(-)$. Triple mutants of various types have also been constructed. The only ones that have a wild-type phenotype are $(+)(+)(+)$ or $(-)(-)(-)$ (Fig. 10-9).

Crick and his colleagues interpret this in the following way. Protein synthesis starts at one end of the messenger and proceeds along it, translating three bases

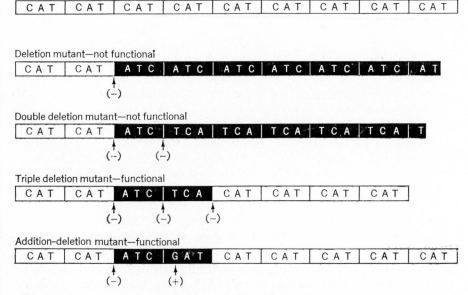

Figure 10-9. *The effects of addition and deletion mutations.*

at a time into protein. If by mutation a base is added or deleted, from that point on the "reading frame" is so shifted that the rest of the message is misread. However, in the double mutant $(+)(-)$ or $(-)(+)$, only the section between the two mutations is misread—the addition and deletion bring the message into register again. For the cases studied, the assumption is that alterations in the protein in the section between the two mutations is not severe enough to render the protein nonfunctional. If, then, the code is a three-letter code, three $+$'s or three $-$'s will also bring the message into register. It is proposed that the code is a three-letter (or else a multiple of three) code and that the translation mechanism "reads" triples by threes, starting at a fixed point.

At this point we can organize the information just presented into a summary of the mechanism of gene action.

1. A gene is a specific linear sequence of nucleotides in DNA.
2. This sequence is a code for designating the sequence of amino acids in a polypeptide chain of a particular protein. The gene and its polypeptide product are colinear. The code is probably a triplet code.
3. The gene functions through the production of messenger RNA, which is a faithful transcription copy of the DNA. (Only one DNA strand is copied.) This messenger RNA associates with the ribosomes, which carry out protein synthesis, and serves as a template for the formation of a specific polypeptide chain.
4. Many mutations are replacements of one nucleotide by another in the DNA chain. Such an altered code word may signify a different amino acid, resulting in the production of a protein with a single amino acid substitution. Or a nucleotide replacement may result in a "nonsense triplet," a combination that is not a code word for any amino acid, making it impossible for a complete polypeptide chain to be formed.

The Modulation of Gene Action. Finally, we will consider another problem, raised earlier. What controls the expression of genes? If one considers the development from a fertilized egg to the fully differentiated form of a multicellular organism, it is clear that all genes do not act all of the time. Let us cite a specific example. In the human fetus the hemoglobin is composed of four polypeptide chains, two alpha chains and two gamma chains. Near the time of birth, the production of gamma chains stops and instead beta polypeptide chains, which were not formed earlier, are produced. After birth, the hemoglobin normally contains only alpha and beta chains. It would seem that near the time of birth a gene controlling the production of gamma chains is "turned off" and the gene controlling the production of beta chains is "turned on." If this is so, some kind of regulatory mechanism for these genes must exist. Clearly, phenomena of this kind must be involved in differentiation and development.

Much of our present knowledge concerning the genetic basis of the regulation of enzyme synthesis has come from studies in bacteria. We will first describe the general features of these enzyme systems, then discuss several in some detail, and, finally, present a model—the *operon*, which F. Jacob and J. Monod have devised to explain the regulation of these systems.

Induction and Repression. Two types of control over enzyme synthesis have been found to occur in bacteria. Some enzyme systems appear to be *inducible.* That is, the bacterium does not synthesize the enzymes in question unless the substrates for the enzymes are present. Most systems of this type are *catabolic*; that is, they are involved in the degradation of some substance derived from the environment and utilized as a source of energy and of molecular fragments for synthetic processes. Enzymes that break down sugars provide good examples. Other enzyme systems are *repressible.* The enzymes of such a system normally occur in the cell, but if large amounts of the end product of the system are present in the growth medium, the synthesis of those enzymes ceases. Repressible enzyme systems are usually *anabolic;* that is, they are synthetic systems, for example, the series of enzymes involved in the synthesis of arginine. Wild-type bacteria, if grown in the absence of arginine, synthesize the enzymes described on page 291. If the bacteria are grown in the presence of large amounts of arginine these enzymes are not formed.

From the many studies on enzyme systems in bacteria, most, if not all, appear to be under controls of the kinds described. It is the *synthesis* of the enzymes that is affected by changes in concentration of metabolites.

The most intensively studied inducible system is the lactose system of *E. coli.* The presence of the sugar lactose affects the synthesis of these different enzymes: β-galactosidase, an enzyme that cleaves lactose into its two components, galactose and glucose; β-galactoside permease, a protein that specifically "pumps" lactose into the cell; and galactoside-transacetylase, an enzyme whose *in vivo* function is not yet known.

None of these enzymes is made by the cell in the absence of lactose. However, the presence of lactose results in a rapid synthesis of all three enzymes. This induction can be brought about by other compounds (inducers) similar in structure to lactose, even by compounds that cannot be degraded by β-galactosidase. Mutations that affect the structure of these enzymes have been found. The mutants behave as expected on the basis of our notions of the gene. Mutations affecting the structure of the β-galactosidase (z^- mutations) and the permease (y^- mutations) map to adjoining segments of the genome. Mutants of the z^- and y^- types complement each other in diploids, indicating that z^- mutations and y^- mutations are in different genes, which determine the structures of these two enzymes (Fig. 10-10). Mutations affecting the transacetylase enzyme have not yet been studied.

A clue as to the genetic basis of the induction phenomenon comes from studies of certain *constitutive* mutations. These mutations, whose effect is to permit the synthesis of all three enzymes, even in the absence of inducer, are clustered at a third locus, near z and y, called the *i* locus. The *i* mutations are inferred to be in a separate gene since *i* alleles complement with z and y alleles.

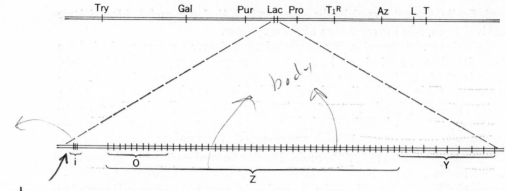

Figure 10-10. *A map of the lactose operon.*

Studies with diploids show that i^+ (inducible) is *dominant* to i^- (constitutive). It seems, then, that when the i locus is *defective* (the i^- recessive) enzymes are formed, whether inducers are present or not.

These various findings have been explained by Jacob and Monod as follows. The z and y genes control the structure of the enzymes of this system. The enzymes would always be produced by the cell were it not for the i^+ gene, which manufactures a *repressor* that prevents the z and y genes from functioning. This repressor is specific for the z and y genes, since no other enzyme system in the cell seems affected by mutations at the i locus. The inducer metabolite (in this system, lactose) acts by combining with the repressor, inactivating it and permitting the z and y genes to function. Although the hypothesis may at first seem unnecessarily complicated, a large body of evidence has accumulated to support it. Furthermore, you will see that the hypothesis fits well with the findings concerning the repressible systems. The basic suggestion is that a regulatory gene (in this particular example, the i gene) acts by repressing the activity of structural genes. The initiation of structural gene function is accomplished by interference with the repressive action of the regulatory genes.

What about the genetic regulation of repressible enzyme systems? Let us take as an example the group of enzymes concerned with the metabolic pathway of arginine synthesis (see again Fig. 10-3). In *E. coli*, in the presence of arginine no enzymes of this pathway are formed. (Other metabolites of the pathway, such as citrulline, will not interfere with enzyme synthesis.) If arginine is removed, the enzymes are immediately synthesized. As you have come to expect, each enzyme involved in the arginine pathway is determined by a particular gene, which determines the structure of the enzyme. A mutation in a given gene results in the absence of activity for the particular enzyme corresponding to the gene, but such mutations do not change the repressibility of the other enzymes in the pathway. However, mutations in a gene distinct from any of the genes determining the structure of the enzymes concerned render the whole series of enzymes "de-repressible." That is, in such regulatory mutants all the enzymes concerned with the synthesis of arginine are produced regardless of the presence or absence of arginine. In Jacob and Monod's theory this situation is explained as follows. The active regulatory gene produces a repressor of all the genes of the arginine pathway. The repressor only acts when coupled to a co-repressor, arginine. This explains the role of arginine in repression. In the de-repressed strain, the repressor is rendered inactive. One might predict that the repressible state of the gene is active, and thus dominant over the de-repressible state. Although this prediction has not been tested for the arginine system, comparable allelic states of the regulatory gene that controls the enzymes of the tryptophan pathway have been tested. The repressible state is indeed the active state.

The inducible and repressible enzyme systems, although exact opposites in activity, in fact have a common explanation. Both systems of structural genes are prevented from functioning by the regulator gene. Certain metabolites present in the specific pathways can either activate (co-repressor) or inactivate (inducer) the repressor.

The Operon. So far we have demonstrated the existence of regulatory genes, which may act on groups of genes in a negative way, that is, to prevent their activity. How do such regulatory genes work? From what is known of the mechanism of gene action described earlier, we would suppose that the repressor either specifically combines with the genes, preventing the formation of messenger RNA, or somehow combines with the messenger RNA after it is formed, and prevents it from associating with ribosomes to act as a template for the formation of protein. A second class of regulatory mutations in the β-galactosidase system gives us insight into the mechanism of repression. These are *operator* mutations. A special class of constitutive mutants, called o^c mutants, has been found. They are located at one end of the z gene distal to y. Strains that are o^c, like i^- strains, are constitutive. However, o^c mutations *only* affect z and y genes in the *cis* configuration with them. As an example let us consider a diploid, $i^+o^cz^-y^+/i^-o^+z^+y^-$. Such a strain manufactures permease (the product of the y^+ gene) constitutively, but β-galactosidase (the product of the z^+ gene) is formed only upon addition of inducer (Fig. 10-11). Thus an o^c mutation only affects the activity of those genes with which it is physically associated, while the i^+ gene acts via a cytoplasmic gene product.

Another class of o mutation has been found, designated o^0. The mutant strains make none of the three enzymes, either constitutively or after induction. An o^0 mutant allele affects only the activity of genes physically associated with it (for example the diploid $i^+o^0z^+y^+/i^-o^+z^-y^+$ makes only the permease, and only after induction, never the β-galactosidase). Some of the o^0 mutations map very close to the o^c mutations. They are not deletions of the z and y genes since they revert to o^+ and show recombination with y and z mutations.

The existence of these two types of operator mutations suggests that the region of the genome in which they occur is concerned with the site of action of the

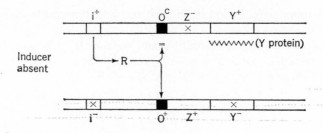

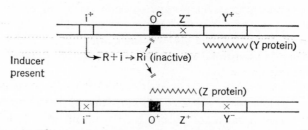

Figure 10-11. *The mechanics of operon function. Genotypic examples are described in text.*

repressor substance. The o^c mutations are interpreted as modifications of this site in such a way that the repressor can no longer interact with the operator. The operator, then, controls the responsiveness of the group of genes with which it is associated (Jacob and Monod call this association of genes the *operon*) to the repressor produced by the regulatory gene. The operator is perhaps part of the *z* gene since some *o* mutations result in altered *z* protein. One can imagine several models to explain operator function. The synthesis of messenger RNA might be initiated at the *o* end of the operon and the repressor might combine specifically with the *o* site, preventing the initiation of messenger RNA synthesis. Alternatively, the messenger RNA formed by the genes of the operon might remain physically associated (that is, there might be only one RNA messenger formed for the whole operon). Under this hypothesis the repressor combines with the *o* end of the messenger molecule preventing it from functioning in the ribosomes. At the time this book is being written the chemical nature of the repressor substances and their exact mode of action on the genes whose activities they affect is unknown.

Let us make an interpretative summary of what we have learned about the regulation of enzyme synthesis in bacteria.

1. The activities of genes that determine the structures of enzymes concerned with a particular metabolic pathway are coordinately affected by certain metabolites of the pathway.

2. Some metabolites may act to "turn off" the genes (co-repressors) or "turn on" the genes (inducers).
3. These *effector* metabolites function by interacting with the product of a regulatory gene whose role is specifically to repress the activity of the structural genes. The metabolites may activate (co-repressors) or inactivate (inducers) the repressor substance.
4. The site of action of the repressor is the operator or the product of the operator. The gene or group of genes controlled by a given operator is termed an operon.

A by-product of the interpretation is a satisfying explanation of the fact that in bacteria many systems of enzymes that are closely related in function are determined by structural genes that lie next to one another in the chromosome. These groups presumably constitute operons. Nevertheless, genes related in biochemical function are not always grouped. For example, the genes controlling the structure of the enzymes concerned with arginine synthesis are scattered throughout the genome of *E. coli*. If the operon theory holds for this system, each gene must have its own operator, each, of course, being responsive to the same repressor substance. In organisms other than bacteria, an unclustered distribution of genes controlling related biochemical functions appears to be the rule rather than the exception. Thus, the operon theory as proposed by Jacob and Monod is not applicable in its simplest form to all enzyme systems of all organisms. Whether this particular mechanism for modulating gene action is widely used in modified form or is restricted to bacteria remains to be determined. In any case, the modulation of gene action is a ubiquitous phenomenon. As we will discuss at greater length in Chapter 12, an understanding of this phenomenon is prerequisite to the solution of major problems concerning differentiation and development.

Keys to the Significance of This Chapter

Metabolism consists of interrelated chains of biochemical reactions. Where gene action has been successfully analyzed in terms of biochemistry, genes appear to function by controlling individual steps in these reaction chains. These individual steps are accomplished through the action of enzymes.

The structure and specificity of each enzyme is determined by a different gene. Each gene contains the code for the specific sequence of amino acids in a polypeptide chain. Mutations cause the alteration or loss of the polypeptide chain, resulting in loss of enzyme activity and, consequently, in a specific block in a biochemical pathway. Many mutations cause single amino acid substitutions in the protein structure.

In bacteria the synthesis of most enzymes is modulated by the presence or absence of specific metabolites. This regulation of gene action is accomplished through the action of specific regulator genes on the activity of the genes that specify the structure of enzymes. The modulation of gene action is ubiquitous— however, the underlying mechanisms of this control are as yet unknown in organisms other than bacteria.

REFERENCES

Baglioni, C., "Correlations between Genetics and Chemistry of Human Hemoglobins." In *Molecular Biology*, edited by J. H. Taylor. New York: Academic Press, 1963. (A detailed account of genetic and biochemical studies on human hemoglobin.)

Cold Spring Harbor Symposia Quant. Biol., Volume **26**, 1961. *Cellular Regulatory Mechanisms*. (Many articles on the regulation of protein synthesis, particularly in bacteria.)

———, Volume **28**, 1963. *Synthesis and Structure of Macromolecules*. (Many articles concerning the mechanism of protein synthesis, gene control of protein structure, and regulation of gene action.)

Garrod, A. E., *Inborn Errors of Metabolism*. London: Oxford University Press, 1909. (This book and its second edition, published in 1923, represent the beginnings of biochemical genetics.)

Haldane, J. B. S., "The Biochemistry of the Individual," in *Perspectives in Biochemistry*. Cambridge: The University Press, 1937. (A brief but brilliant and prophetic treatment of the scope and meaning of biochemical genetics.)

Ingram, V. M., "How Do Genes Act." *Scientific American*, January 1958. Available as Offprint 104 from W. H. Freeman and Co., San Francisco. (An account of the studies on the structural differences between hemoglobins of different heritable types.)

Jacob, F., and Monod, J., "Genetic Regulatory Mechanisms and the Synthesis of Proteins." *J. Mol. Biol.*, **3**:318–356, 1961. (The statement of the operon hypothesis.)

Wagner, R. P., and Mitchell, H. K., *Genetics and Metabolism*. New York: Wiley, 1964. (A text emphasizing the biochemical aspects of genetics.)

Wright, S., "The Physiology of the Gene." *Physiol. Revs.*, **21**:487–527, 1941. (Cogent and penetrating. The original treatment of "Theories of Dominance and Factor Interaction," pp. 514–520, is particularly important.)

QUESTIONS AND PROBLEMS

10-1. Define or specifically identify the following:

A. E. Garrod   inducible enzyme
minimal medium   operon
enzyme   one gene-one polypeptide hypothesis
messenger RNA

10-2. In many areas of genetics, studies of man have been rather unproductive as compared with studies of certain other organisms. What may account for the fact that human genetics has been relatively productive of information bearing on our understanding of relationships of genes and biochemistry?

10-3. Would you expect that alcaptonurics given large amounts of phenylpyruvic acid in their food would excrete increased amounts of homogentisic acid in their urine? Given *p*-hydroxyphenylpyruvic acid? Maleylacetoacetic acid? Explain your answers.

10-4. Penicillium does not require biotin in its culture medium as Neurospora does. Is this a reason for thinking that biotin plays no role in the cellular biochemistry of Penicillium? Amplify your answer.

10-5. Why is Neurospora not suitable for studying the biosynthesis of biotin by means of mutant strains?

10-6. Among mutants of Neurospora, two are known that require choline supplements to their media. In each strain, the growth-factor requirement is the effect of a single gene mutation. Strain 1 can grow if either monomethylaminoethanol or dimethylaminoethanol is substituted for choline. Strain 2 can grow on the dimethyl compound but not the monomethyl; however, it does accumulate this latter compound in media in which it grows. Suggest an outline scheme for the biosynthesis of choline in Neurospora.

10-7. Assume that among microorganisms in general the pattern of biosynthesis of thiamine is as follows.

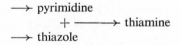

All microorganisms utilize thiamine in their metabolism. Some can make it according to the scheme outlined above; others must have it provided among their nutrients. Among the latter, certain ones are able to grow if provided with a particular fraction of the thiamine molecule. Thus Mucor is able to grow either on thiamine or on its thiazole fraction; Rhodotorula either on thiamine or on its pyrimidine fraction; Phycomyces either on thiamine or on a combined supplement of pyrimidine plus thiazole; while Glaucoma has a specific requirement for the intact thiamine molecule. Show how gene mutation might readily explain each of the different growth-factor requirements noted above.

10-8. Suppose that you have been conducting a search for biochemical mutant strains of Neurospora. After treating wild type with nitrogen mustard, you isolate a strain unable to grow on minimal medium. You can find no single substance which permits growth when added to minimal medium, but the addition of two separate compounds does satisfy the growth requirement of this strain. What biochemical genetic situations might account for your finding?

10-9. A Neurospora mutant was found initially to require a supplement of leucine to the minimal medium. However, after long laboratory culture it was found sometimes to grow on media lacking leucine. Among possible explanations of this phenomenon are: (a) back-mutation of the mutant gene to wild type; (b) mutation at another locus that "suppresses" the effect of the original mutant gene; (c) physiological "adaptation" involving no genetic change; (d) contamination of the mutant cultures with wild-type Neurospora. Outline a series of systematic experiments you might perform to determine which of these possibilities is the correct explanation.

10-10. Substances sometimes accumulate behind metabolic blocks that result from gene mutation. How might this provide insight into the fact that a single gene may have multiple phenotypic effects?

10-11. Suggest two other types of biochemical genetic situations that would provide possibilities for multiple phenotypic effects resulting from a single gene mutation.

10-12. Different mutations have been found to result in different substitution involving the same amino acid of the tryptophane synthetase enzyme of *E. coli*. A specific glycine residue in the wild-type enzyme is replaced by glutamic acid in one mutant, arginine in another. Recombination analysis indicates that the mutational sites bringing about these changes are separable by recombination. How is this observation explained by the triplet-code theory? What would you predict concerning the phenotype of the double-mutant recombinant (see Table 10-3)?

10-13. Bacteriophage T4 has a gene that controls the structure of an enzyme called lysozyme, which initiates lysis of the infected bacterial cell. This enzyme molecule consists of only one polypeptide chain. Would any mutations in this gene be expected to show intragenic complementation? Why?

10-14. In the bacterium *Salmonella typhimurium* the genes controlling the structure of seven of the enzymes of the histidine biosynthetic pathway are located adjacent to one another in the genome. The synthesis of these enzymes is coordinately repressed by excess histidine. Most mutations in this region result in the loss of only one specific enzyme. However, mutations that occur at one end of this region result in the loss of all seven enzymes, in spite of the fact that one can demonstrate that the structural genes for at least six of the seven enzymes are intact. What is the counterpart of these mutations in the β-galactosidase system?

CHAPTER 11

Extrachromosomal and Epigenetic Systems

We can take it as adequately established that a great number and variety of inherent individual differences are controlled by genes in chromosomes. These differences obey the laws of Mendelian inheritance. But they do not prove that chromosomes are the sole vehicles of inheritance, or even necessarily that they are the only ones of great significance. We need to ask: *Are there other bases of biological inheritance besides the chromosome-borne genes? If so, how important are they, and what kinds of laws do they obey?*

Our questions have been phrased in such a way that, potentially, almost any part of a cell other than the chromosomes may be implicated: membranes, including the nuclear membrane; the cell wall; various things included in the cytoplasm, such as mitochondria or plastids; and even entities as yet unknown. You might then suspect this chapter to be a kind of catchall, and indeed to the best of our knowledge it is—a catchall for a variety of phenomena presently not referrable in the ordinary way to genes in chromosomes. *A priori* the pattern of transmission for a phenotype based on any of these various potential extrachromosomal systems cannot be predicted with certainty. Our most feasible procedure is simply to examine whether instances can be found in which transmission is such that a chromosomal basis is either ruled out or at least is unlikely. Where such instances occur, we can examine whether particular extrachromosomal systems are implicated.

To rely primarily on ruling out chromosomal inheritance in order to establish extrachromosomal inheritance is a negative approach that suffers in comparison with the positive means by which Mendelian heredity was associated with chromosomes. Quite properly we can make rigorous demands before accepting a case for extrachromosomal inheritance as being conclusive. Under these circumstances, a few guidelines stated at the outset will be helpful, even though their fuller development will necessarily await the context of particular examples. What, then, are some of the kinds of observations that may suggest extrachromosomal inheritance? Our answer here is meant as neither a formal nor an exhaustive list of criteria.

1. *Differences in reciprocal cross results.* When one follows the transmission of characteristics based on chromosomal heredity, the results of reciprocal crosses are ordinarily identical. You are already aware, however, of at least one exception to this statement. Characteristics due to sex-linked genes, like those in the X-chromosome of man or Drosophila, need not show identity of transmission in reciprocal crosses.

2. *Maternal inheritance.* A characteristic form of difference in the results of reciprocal crosses is maternal inheritance, where progeny show the characteristics of their female parent. If chromosomal differences can be ruled out, maternal inheritance usually implies transmission through the cytoplasm. This is because the female gamete ordinarily provides vastly more cytoplasm to the zygote than is provided by the male gamete. It must be remembered, however, that the male gamete is not devoid of cytoplasm.

3. *Nonmappability.* If the chromosomes of an organism are well mapped, a characteristic based on chromosomal heredity should show linkages and should be able to be mapped in reference to other gene-controlled characteristics. Failure to find linkages, of course, may mean no more than an insufficient marking system for various parts of the various chromosomes.

4. *Nonsegregation.* Segregation is typical of Mendelian heredity. Failure to show segregation under appropriate circumstances may indicate extrachromosomal heredity.

5. *Non-Mendelian segregation.* When segregation occurs but in a fashion inconsistent with the segregation of chromosomes, the result may suggest that nonchromosomal determinants account for the observed phenotypic variations.

6. *Indifference to nuclear substitution.* When a heritable characteristic persists in the presence of nuclei known to have been associated with alternative characteristics, one should wonder whether nuclear genetic material has control over the characteristic.

7. *Infection-like transmission.* When a heritable phenotype is transmitted without there having been transmission of nuclei, it may seem unlikely that chromosomes control that phenotype. However, your knowledge of bacterial *transformations* will suggest to you that transmission in the absence of nuclear migration may be an insufficient criterion for heredity based outside the chromosomes.

Congenital Disease. Our consideration of extrachromosomal inheritance will be easier if we first look at one or two clear examples of that with which we are *not* primarily concerned. In certain lines of mice, almost all the females, generation after generation, die of mammary cancer. In other lines of mice, over a sequence of generations the incidence of this kind of cancer is low. When reciprocal crosses are made between members of the two lines, the outcome of a cross in the first and in later generations depends largely on the characteristic of the female parent. The situation is clarified when baby mice from lines showing respectively high and low incidence of the cancer are removed from their mothers at birth and are nursed by foster mothers of the opposite type. The results show that mother mice from a line with high incidence for mammary cancer transmit through their milk an agent that later causes mammary cancer in mice that have nursed on them. This agent, called a "milk factor," acts like an infective agent, and meets in some respects the criteria for a filterable virus. Medicine has encountered similar situations before, in the form of diseases contracted by embryos or infants through infections carried by their mothers. Syphilis is a familiar example. Such *congenital* diseases, although transmitted from generation to generation, are not truly *inherited;* they are acquired through agents external to the developing individual, and the agents are never integrated as part of the genetic system of the host organism.

Why have we concerned ourselves, even briefly, with an example in which transmission of a phenotype is not based on a genetic system of the organism that shows the phenotype? The answer is that you will later be confronted with other instances where maternal transmission is involved but where that which is transmitted appears to be an integrated part of the normal hereditary apparatus of the mother. Thus the example of mammary cancer in mice enables the making of important distinctions, which might not otherwise emerge.

Maternal Influence. In the example just discussed, the characteristic, mammary cancer, whose pattern of transmission is maternal rather than chromosomal, is the result of an agent external to the mother as an individual. Since the characteristic is not a property under direct control of the mother's own genetic apparatus, mammary cancer is not pertinent as a possible example of extrachromosomal inheritance in the sense conveyed in the introduction to this chapter. But a basic function of a mother is to provide, through the egg or across the placenta, materials of her own elaboration for use by the embryo. A few individual differences are known to depend on the extranuclear transmission of this kind of material.

Perhaps the clearest example is concerned with the color of larval skin and eyes in the meal moth, Ephestia. The basic difference here is a simple Mendelian alternative, such that

$$A = \text{pigment}, \qquad a = \text{no pigment}.$$

These are ordinary, chromosome-borne alleles. The A allele controls the production of a diffusible, hormonelike substance of known chemical composition (*kynurenin*) involved in pigment synthesis. When the a allele is homozygous, this "hormone" is not elaborated.

A female of genotype *Aa* forms eggs, half of which carry the *A* allele, and half the *a*. If the female is mated with an *aa* male, the progeny are of two genotypes, equal in frequency: *Aa* and *aa*. Now the *aa* offspring have no means of elaborating the *A* hormone, since they lack the *A* allele. They do, however, develop some pigment, as larvae, in their skin and eyes. They "fade" as they grow older, and the effect disappears in the next generation.

The *Aa* mother includes in her eggs some of the *A* hormone elaborated in her own body. This substance, present by "maternal influence" in the *a* eggs as well as the *A* ones, enables the *aa* offspring to develop some pigment. But, being unable to elaborate a continuing supply of the hormone for themselves, the *aa* individuals dilute and use up the supply passively transmitted to them from their mother, and the effect is therefore only a *transient* one.

A presumably similar but less concrete example is provided by the direction of coiling in the shells of certain snails. Snail shells may coil in either of two directions, clockwise or counterclockwise. These are commonly distinguished by the terms *dextral* and *sinistral;* if you hold a shell so that the opening through which the snail's body protrudes is facing you, this orifice may be either on your right (dextral) or on your left (sinistral). The difference is the same as that between a "right-handed" and a "left-handed" screw. Different species may be either dextral or sinistral, and within some species, races may differ in this regard. The race difference may be investigated through routine genetic techniques; the investigation has been facilitated by the fact that the species most studied (*Limnaea peregra*) is monoecious and can reproduce either by crossing or by self-fertilization (Fig. 11-1).

When reciprocal crosses are made, the F_1 progeny show the same direction of coiling as did their mothers. But F_2's produced by self-fertilization of the F_1 all coil dextrally; and when the F_2's are in turn self-fertilized, each produces a uniform progeny, three producing dextral progenies to each one that produces sinistral progeny.

This rather puzzling situation is simply explained by postulating that the direction of coiling of the embryonic shell is impressed on the egg cytoplasm by the genotype of the diploid oöcyte from which the egg came. The dextral-sinistral alternative depends on a pair of alleles in which the allele for dextral is dominant. The F_1 shells in either of the reciprocal crosses then all agree with their homozygous mothers; the F_2 shells are all dextral, since their mothers are heterozygous; while the F_3 *progenies* reflect the expected 3:1 "phenotypic" ratio of the F_2 mothers. This explanation is the simplest that can be offered for the situation, although rare exceptions to the rule in the species most carefully investigated suggest that other conditions may also sometimes operate to modify the direction of coiling.

This example probably represents the same type of "maternal influence" as does the Ephestia case discussed above. In snails, the direction of coiling is determined by the first two cleavage divisions. The pattern of early *cleavage* (the

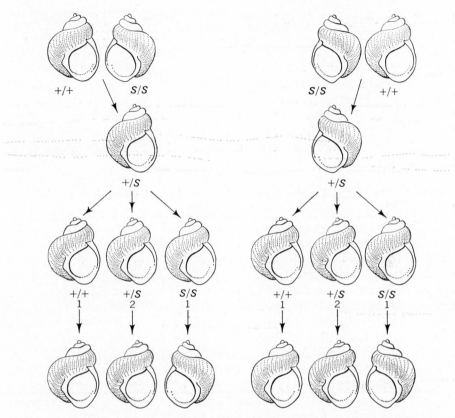

Figure 11-1. *The direction of coiling in the shells of certain snails depends on the mother's genotype, rather than on the genotype of the individual itself.* Left, *eggs produced by a homozygous dextral snail, fertilized by sperm from a homozygous sinistral individual, develop into dextral progeny. These heterozygous progeny, when self-fertilized, produce only dextral progeny in turn. But the three genotypes in this generation are reflected in the progenies produced in the next generation, again after self-fertilization.* Right, *the reciprocal cross is different, in ways that confirm the maternal influence on direction of coiling. (After Sturtevant and Beadle,* An Introduction to Genetics, *W. B. Saunders, 1940, p. 330.)*

first divisions of the fertilized egg) is apparently incorporated into the egg before the meiotic divisions occur, and the fact that the dominant allele of a pair may be lost in the polar bodies as the egg goes through meiosis, therefore, does not prevent this allele from controlling the cleavage pattern. The material basis of the egg's organization responsible for this behavior remains unknown.

In both instances, that of Ephestia and that of the snails, maternal influence occurs, and the expression of phenotype is not strictly correlated with the immediate presence of appropriate genes. Nevertheless, the ultimate control of phenotype by chromosomal genes is easily enough perceived. The reason it is perceived is that the maternal influence is indeed transient; the relevant chromosomal segregations can be identified because they are no more than a generation removed from the unorthodox phenotypic effects. The transience of the maternal influence on phenotype indicates that the material accounting for it has no powers of generating more substance like itself. In this regard, it clearly differs from genic material. *Persistence,* or the lack of it, then, may be a useful criterion for distinguishing between maternal influence and possible extrachromosomal hereditary systems.

Cytoplasmic States. If cells from a given strain of Paramecium are injected into a rabbit, an antiserum is produced which acts against other cells of the same kind. The antigen-antibody relationship is easily detected when paramecia carrying a particular antigen are placed in a drop of fluid into which homologous antibody has been introduced. In this circumstance, the paramecia first move erratically; then movement ceases; the cilia stop beating; and finally death occurs. Different strains of Paramecium may show antigenic differences among them, as demonstrated by applying the procedures just described.

Extensive investigations, primarily by T. M. Sonneborn and by G. H. Beale, have revealed fascinating relationships between these antigens and cellular heredity. (Before considering these relationships you may wish to review the life cycle of Paramecium, presented in Chap. 4.) The main facts to be considered initially are the following. (1) Within a stock of Paramecium, derived from a single homozygous animal, various lines can be established, each characterized by a different antigen, for example antigen A, B, D, H, or J. (2) In general, a single animal at a given time carries only one of the kinds of antigens to which we are referring. (3) The particular antigenic property carried by a line of paramecia is subject to environmental influence. Within stock 90, cultures raised at low temperatures show antigen S, at intermediate temperatures antigen G, at high temperatures antigen D. Antigenic type can also be influenced by other environmental factors, which the investigator can control (for example nutrition). (4) As shown in Figure 11-2, reciprocal crosses give different results. Members of an exconjugant pair, despite their genetically identical nuclei, are of different phenotype, each showing an antigenic property in agreement with the source of cytoplasm for the cell. Autogamy shows no influence over transmission of the antigenic property. And when cytoplasmic bridges form between conjugating animals the pattern of transmission is different than when such bridges are not formed.

The facts as so far presented provide no hint that chromosomal genes are acting; on the contrary, cytoplasmic heredity is strongly indicated. Nevertheless, the antigens of Paramecium are quite specifically under genic control, as can be seen by following an experimental sequence carried out by Beale. As already

Without cytoplasmic
exchange

With cytoplasmic
exchange

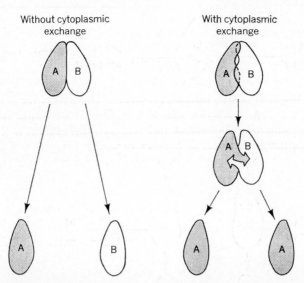

Figure 11-2. *When paramecia of antigenic types A and B are mated without cytoplasmic exchange, each of the exconjugants, despite genetically identical nuclei, shows a different antigenic property, the same property that the cell showed before conjugation. A similar mating, but with cytoplasmic exchange, yields exconjugants each of the same antigenic type.* (*After Sonneborn. In Baitsell, Science in Progress,* Seventh Series, *Yale University Press, 1951, p. 193.*)

stated, stock 90 at low, intermediate, and high temperatures produces respectively antigen S, G, or D. Stock 60 at corresponding temperature levels produces serologically related but distinguishable antigens that can be called 60S, 60G, and 60D. A cross of a 90G with a 60G animal gives hybrid exconjugants that produce a mixture of the 90G and 60G antigens. Thus corresponding but nonidentical antigens do not show the mutual exclusion characteristic of, say, the G and D antigens. If the hybrids are allowed to undergo autogamy, the resulting animals are either pure 90G or pure 60G, the numbers of the two types giving a 1 to 1 ratio. These observations say clearly that the 60G and 90G antigens are due to the action of members of an allelic pair that segregate in the typical fashion of chromosomal genes.

The situation is further elaborated by a particularly elegant experiment summarized in Figure 11-3. Beale mated 90G with 60D animals and then systematically moved the F_1 animals (vegetatively reproduced derivatives of the exconjugants) from one critical temperature to another. At each temperature, the animals showed a mixture of the 90 and 60 antigens appropriate to that temperature, in particular, G antigens at 25°C and D antigens at 29°C.

No cytoplasmic genetic system needs to be invoked to interpret the results. They can be interpreted rationally on the basis of interaction between chromo-

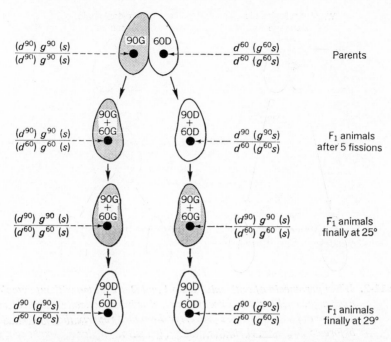

Figure 11-3. *Stocks 60 and 90 of* Paramecium aurelia *carry different alleles for loci* d, g, *and* s, *each of which controls the production of a particular antigen. At a given time, the genes at only one of these loci are able to express themselves. The state of the cytoplasm, which can be determined by temperature, determines which locus is able to express itself. In the diagram, the antigens are designated by the large numbers and letters within the animals. Genes are designated by small letters in italics, with parentheses enclosing the genes that are not expressing themselves. Different cytoplasmic states are shown by shading or absence of shading. (From Beale,* Genetics, **37**:69, 1952.)

somal genetic determinants and the cytoplasm, as follows. Stocks 60 and 90 include genes corresponding to the different antigens that they may produce. These genes occupy different loci that can be designated *s*, *g*, and *d*. Each stock has a characteristic allele at each of these loci. In homozygotes only one kind of antigen is produced at a time, but in heterozygotes two corresponding antigens are produced, these being the antigens determined by the allelic pair for the locus that is active at that time. Among loci *s*, *g*, and *d*, the one that is active at a given time depends upon the state of the cytoplasm. The cytoplasmic state is predictably modifiable by temperature, and thus temperature can be used to determine which of the loci will be active in producing antigen. In contrast to instances of maternal influence already examined, the cytoplasmic states are persistent, although reversible.

What are the different cytoplasmic states? How can we account for their persistence, if we do not invoke a self-replicating system in the cytoplasm, a sys-

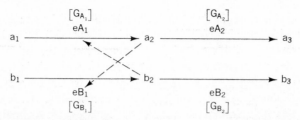

Figure 11-4. *The Delbrück model for alternative steady states. Symbols preceded by* G *designate genes; symbols preceded by* e *designate enzymes; symbols* a_1, b_1, *etc., designate metabolites in chains of reactions. Operation of either of the reaction chains excludes operation of the other by means of a specific inhibition. Cessation of operation by one chain permits the other to operate, and to tend to perpetuate itself at the expense of the first. (After Delbrück,* Unités biologiques douées de continuité génétique. *Edition du Centre Nat. Rech. Sci., Paris, 1949.)*

tem that has genetic properties? The first question cannot be answered at all at this time. In the case of the second question, a model suggested by M. Delbrück provides a basis for a possible kind of answer. We will consider the model in its simplest form. Suppose, as is shown in Figure 11-4, that a cell has two reaction chains $a_1 \longrightarrow a_2 \longrightarrow a_3$ and $b_1 \longrightarrow b_2 \longrightarrow b_3$, each step in the chains being promoted by an appropriate enzyme. Suppose further that a set of interactions between the two chains is such that when metabolite a_2 reaches a certain level of concentration it inhibits the action of enzyme B_1; conversely, when metabolite b_2 reaches a certain concentration it inhibits enzyme A_1. Under these conditions, operation of one of the reaction chains would exclude the other, and whichever chain of reactions happened to be operating at a given time would tend to perpetuate itself. If, however, the concentration of the metabolite acting as inhibitor were sufficiently decreased, for example as the result of an environmental effect, the alternative reaction chain might begin to operate. This alternative chain would then have the same potential for self-perpetuation as had the chain it supplanted.

The production of antigen in Paramecium can be readily visualized in terms of the kind of alternative steady states defined by the Delbrück model. Whether or not the model really applies to the case at hand, it provides a conceptual system under whose terms phenotypic change may occur and persist indefinitely without a corresponding change in genetic constitution.

Cytoplasmic Particles

We have emphasized repeatedly the particulate nature of the genetic material carried in chromosomes. Are there extrachromosomal particles that are in some sense the analogues of genes?

Killer Paramecia. Some paramecia, called *killers*, are the source of a particle, called P, that kills other, *sensitive* paramecia. In a brilliant series of investigations, Sonneborn and his co-workers have defined the basis of the killer characteristic and have worked out the system by which it is transmitted. To be a killer, a paramecium must have a gene K, which segregates in typical chromosomal fashion, and a complement of cytoplasmic particulate material called kappa. Sensitive animals are those that lack kappa. Animals of genotype kk lose their kappa if they have any initially, and are unable to generate it. If kappa is present in a cell with the genotype KK or Kk, it continues to be produced except under rare circumstances. However, once lost from a cell with a K gene, kappa does not reappear unless more kappa is introduced from another cell. Within a line of descent, kappa is transmitted through the cytoplasmic line. When cytoplasmic bridges are formed at conjugation, kappa can pass along these bridges.

Kappa is particulate, and in the cytoplasm of a killer animal kappa particles can be seen with the aid of a microscope. At least three kinds of particles are included within the kappa system. These are (1) P particles, the killer agents already described, which are found in the culture fluid of killer paramecia; (2) N particles, which are found in the cytoplasm of killers and which can divide transversely to produce more kappa particles; and (3) B particles, also found in the cytoplasm, which are relatively large and structurally complex. B particles arise from N particles, and B particles give rise to P particles. Thus the various components of the kappa complex may be viewed as different developmental stages of the same entity.

Kappa contains DNA, a substance whose ability to carry genetic information we already know, and kappa is mutable. At this point we seem to be dealing with a cytoplasmic genetic particle, a particle comparable to a gene except in the matter of its geography. A closer look at kappa dispels any such notion. Electron microscopy reveals that a kappa "particle" has small amounts of a fairly typical cytoplasm of its own and that it is bounded by a membrane similar in structure to that of the bacterial cell wall. The DNA in a kappa particle is diffusely distributed; RNA is present as well. Quite clearly kappa is either a parasite or a symbiote, in either case a peculiar one, and an organism in its own right. Its relationships with its host are fairly sophisticated at the genetic level, but one can scarcely maintain that kappa is part of the genetic machinery of a paramecium. Kappa is much nearer to the congenital diseases we mentioned in the earliest portion of this chapter than it is to an extrachromosomal genetic system of the kind at which our exploration has been aimed. In this context, it is worth mentioning that a particle related to kappa, called lambda, has been successfully cultured in vitro and then used to reinfect paramecia without losing its capacity to kill sensitive animals.

Hereditary parasitic infection is not confined to Paramecium. Another favored object of genetic study, Drosophila, can provide us with a similar instance. In certain strains of Drosophila the females, called SR for sex-ratio, have progeny that are almost all females—this is the result no matter what the genotype or phenotype of the male parent. When the rare sons of an SR female are mated to a normal female, the characteristic does not appear in the immediate progeny nor does it reappear in subsequent generations. The inference to be drawn is that the SR property is transmitted through the cytoplasm of the egg and not by way of the chromosomes. Such an inference is almost certainly correct, but again microscopy suggests that the transmitted material is a parasitic organism. The hemolymph of SR females, which can be used as a source of experimental infection, characteristically includes great numbers of minute spirochetes.

PLASTID INHERITANCE

At least in certain cells, the cytoplasm of plants (other than fungi and blue-green algae) includes large, complex particles called chloroplasts. By means of the green pigment chlorophyll, chloroplasts absorb radiant energy and then utilize a battery of enzymes to carry out photosynthesis. The structure of plastids, the pigments contained in them, and their enzyme systems, all can be affected by mutation. Many of these differences that arise by mutation follow straightforward chromosomal patterns of transmission, and the mutant genes are mappable by standard procedures. Plastids, then, are not free of control by the chromosomal genetic apparatus.

Mature plastids seem to arise from structures that are less differentiated, called proplastids, which are able to divide and thereby increase in number. In the sense that plastids seem only to arise from preexisting structures they are self-replicating. Moreover, certain plastid differences are not transmitted along chromosomal lines. In view of these observations, and since plastids appear to include DNA, convergent evidence indicates that plastids may have a genetic machinery of their own, a machinery that lies outside the chromosomes.

Variegated Four-o'clocks. Almost everyone is familiar with variegated plants —plants that have areas of pale green or white in their otherwise normally green leaves. Sometimes the variegation is quite symmetrical and uniform; in other cases it is irregular, and whole branches may carry normal green leaves, while others are entirely pale or white, or mixtures of pale or white areas and dark-green areas.

In the four-o'clock, *Mirabilis jalapa*, controlled pollination of flowers borne on these three kinds of branches of the same plant has given the provocative results summarized in Table 11-1.

You will notice that the type of pollen used is not important; pollen grains from pale, green, and variegated branches behave the same. The determining factor is the female's contribution. Seeds borne on pale branches produce only pale plants; those borne on green branches produce green plants, while those

Table 11-1. PROGENY OF A VARIEGATED FOUR-O'CLOCK.

Pollen from Branch of Type	Pollinated Flowers on Branch of Type	Progeny Grown from Seed
Pale	Pale	Pale
	Green	Green
	Variegated	Pale, green, and variegated
Green	Pale	Pale
	Green	Green
	Variegated	Pale, green, and variegated
Variegated	Pale	Pale
	Green	Green
	Variegated	Pale, green, and variegated

borne on variegated branches segregate in irregular ratios of pale, green, and variegated.

The results can be interpreted on the basis of two different kinds of plastid, or plastid progenitor, transmitted only through the female and distributed in a variegated plant as follows.

Type of Branch	Type of Plastid Maternally Transmitted
Pale	Pale
Green	Green
Variegated	Pale and Green

Seeds born on a pale branch would therefore include only the pale plastid type; those on a green branch only the green; and those borne on a variegated branch might include either pale or green or a mixture of the two types. Neither the constitution of the male gametophyte nor the nuclear genetic constitution of the fertilized egg would be involved in the control of this type of inherent variation. Here, then, is evidence of a cytoplasmic self-replicating particle, a "vehicle of heredity" that is extrachromosomal. The evidence would be even better if, instead of merely hypothesizing them, we could actually observe individual cells that contain a mixture of the plastid types. Such an observation would confirm the freedom of the plastid from certain kinds of nuclear control. In the four-o'clock the observations that have been made are not entirely convincing. But for a similar case in the snapdragon, the existence of cells containing mixed plastids seems adequately demonstrated (Fig. 11-5). And since recently the occurrence of DNA in plastids seems to have been demonstrated, these

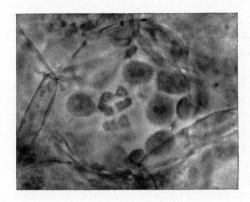

Figure 11-5. *A "mixed cell" containing two genetically different types of plastids: large, normal green plastids with normal grana and starch grains; small white plastids that have arisen by mutation. The cell was found in a plant of* Antirrhinum majus, status albomaculatus, *line Gatersleben I.* (*Courtesy of R. Hagemann:* Plasmatische Vererbung. *Jena: VEB Gustav Fischer Verlag, 1964.*)

particles include material with known potential for replication and for the storage of genetic information.

Iojap Maize. Corn plants homozygous for the recessive gene *ij* show green and white striping. When such plants are used as pollen parents in crosses with normal green, all the offspring are green. Iojap plants used as females in crosses with normal give rise to rare all-white progenies, or to all-green progenies, or to progenies that include green, white, and striped plants. In progenies of the latter kind, the phenotypic ratios are highly variable. The critical finding is that striped and white individuals appear among those F_1 progeny derived from an iojap maternal parent. The variants appear despite the F_1 chromosomal genotype $+/ij$. Put in another way, the dominant normal gene is not able to control a phenotype acquired under the action of its recessive allele.

The basis of phenotype in iojap is the plastid. In striped plants, the green tissue has normal chloroplasts; the white tissue has minute colorless plastids. The pattern of transmission of the phenotype is most readily explained on the basis of plastid transmission. Striped plants in the F_1 presumably arise from egg cells containing both normal and aberrant plastids; green plants and white plants arise from egg cells having only the one appropriate plastid type in their cytoplasm.

Even the substitution of a second normal allele at the iojap locus is not sufficient to correct the plastid abnormality. Thus the plastid alteration acquired under genotype *ij ij* is not reversible, at least not by normal genetic constitution at the iojap locus. The iojap example might seem to tell two stories, one to the effect that plastids have genetic autonomy and another to the effect that chloroplasts are subject to the action of nuclear genes. These stories are not necessarily inconsistent. One expects interaction among the components of an organism.

Plastid Inheritance in Oenothera. Many species of Oenothera have the peculiarity that in general they transmit their chromosomes as entire genomes, so that gametes formed by a plant contain either all those chromosomes transmitted by the plant's pollen parent or all those chromosomes derived from the plant's maternal parent. This unusual situation is made possible by a complex series of reciprocal translocations whose effect is to produce a large ring at the appropriate stage of meiosis. (See Fig. 7-10 for the simplest form of the Oenothera situation.) Within the ring, chromosomes of maternal and paternal origin occur alternately, and at metaphase I alternate members of the ring move to the same pole. The chromosomes that are inherited in a block because of these arrangements are known as *complexes*, and particular species of Oenothera are often described in terms of their component complexes.

The peculiarities of chromosomal genetics in Oenothera enable investigators to manipulate genotype with great precision and control. A number of investigators, among whom the German geneticist O. Renner should be mentioned as particularly outstanding, have taken advantage of these peculiarities to produce enlightening studies of plastid inheritance. We are able to give but a single example here.

The appropriate interspecific cross can give rise to a hybrid made up of a *curvans* complex from *Oenothera muricata* and a *Hookeri* complex from *Oe. Hookeri*. Depending on which species is used as female parent, the phenotype of the same chromosomal hybrid has one of two distinct aspects. Where *Oe. muricata* has been the maternal parent, the hybrids are green. The reciprocal cross gives rise to plants that are yellow and unable to survive. Apparently, *muricata* plastids are able to develop normally in the presence of a *curvans-Hookeri* nucleus but *Hookeri* plastids are not.

Even more interesting are events that occur in the green hybrids whose plastids are derived from *Oe. muricata*. In the course of time these green plants show variegation in the form of yellow sectors whose phenotype is that of the hybrids

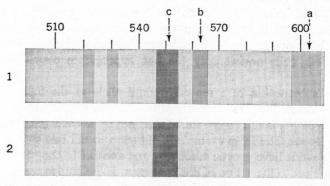

Figure 11-6. *The cytochrome spectra of big (1) and of little (2) yeast differ markedly. Note particularly the absence of cytochromes a and b in little yeast. (From Ephrussi*, Nucleo-Cytoplasmic Relations in Micro-Organisms, *Oxford University Press, 1953, p. 23.)*

that derive their plastids from *Oe. Hookeri*. Renner's interpretation was that the yellow patches are indeed due to the presence of *Hookeri* plastids. He supposed that a few plastids are carried to the hybrid by way of *Hookeri* pollen, that they multiply as the plant develops, and that somatic segregation of plastids out of cells containing the two types produces yellow sectors. Some of the yellow sectors produce flowers. On pollination of such flowers, one obtains seed, and subsequently obtains plants whose chromosomal constitution is *Hookeri-Hookeri* or *curvans-Hookeri*. The former are green plants, presumably because *Hookeri* plastids are put back into combination with a nucleus with which they are compatible. In other words, the plastids are still *Hookeri* plastids and have not been basically altered while under the influence of a genome that does not permit their normal development. The seed with *curvans-Hookeri* nuclei again give rise to plants that are yellow and die.

MITOCHONDRIA AND OTHER CYTOPLASMIC PARTICULATES

In addition to plastids, the cytoplasm includes other organelles. Some of these, particularly the mitochondria, appear to have genetic significance. Moreover, much in the way that Mendel was led to infer genes, other investigators have been led by their experimental results to infer as yet unidentified cytoplasmic particles as participants in hereditary systems.

Little Yeast. Baker's yeast, like Neurospora, is an ascomycete. Sexual reproduction involves (1) formation of a diploid zygote by the fusion of haploid cells of opposite mating type, (2) meiosis, and (3) the formation of an ascus that includes four haploid spores, not linearly arranged, which represent the four products of a meiosis. Since the four spores of an ascus may be separated, isolated, and cultured vegetatively, tetrad analysis in yeast is possible. The asexual reproduction of yeast is by budding, and under the right conditions either haploid or diploid cells may reproduce asexually for an indefinite period of time.

If you were to start with a single cell of baker's yeast and allow it to propagate asexually, you could easily demonstrate that the resultant culture is not homogeneous. When a sample of the cells of such a culture is spread out on solid medium, so that each cell grows into a colony, a fraction of the colonies, perhaps 2 percent, is found to be markedly small in comparison with the rest. Under aerobic conditions the cells found in these little colonies grow more slowly than normal cells and produce only small colonies like themselves. *Little* yeast of the kind described here never reverts to its *big* alternative.

Boris Ephrussi and his co-workers have shown that little yeast is unable to carry out respiratory metabolism. The cells lack cytochromes a and b (Fig. 11-6) and the enzyme cytochrome oxidase. Other important enzymes are found in abnormally low concentration. In a single remarkable step, a sudden heritable change, normal yeast appears to lose the capacity to carry out a whole set of highly important chemical functions. The "mutation" from big to little, which occurs with a spontaneous frequency that is high in terms of the usual standard for gene mutations, can be induced in 100 percent of cells that are allowed to grow in dilute solutions of acridine dyes. Careful experiments, in which the buds of yeast grown in acridines were dissected off the parent cells and grown separately, have shown that acridines *induce* rather than *select* the little mutations.

Little yeast, then, has many exotic attributes. It continues to be bizarre when crossed with normal yeast. When little cells are mated with bigs, only normal diploid zygotes and only normal haploid progeny are recovered. No trick of chromosomal genetics seems able to account for the disappearance of the little characteristic. If standard gene markers are introduced into the parent cultures for a big × little cross, these markers segregate normally, members of an allelic pair being recovered in a 1:1 ratio within an ascus (Fig. 11-7). Therefore, one can conclude that the nuclear genetic apparatus is not functioning aberrantly. Not only does the little characteristic fail to appear in hybrid progeny of the cross big × little but it fails to reappear when such progeny are repeatedly backcrossed to a little parent. The little trait, although perfectly heritable in vegetative reproduction, does not have the attributes of a characteristic controlled by a gene.

If you refer again to Figure 11-7, you will see Ephrussi's hypothesis to account for the curious behavior of little when crosses are made to big. He suggests that normal cells carry a cytoplasmic component that is either lacking or inactive in littles. When fusion occurs between a normal and a little cell, the active cytoplasmic component of the normal parent is automatically introduced into the common zygote cell. Assuming this cytoplasmic component replicates itself, one visualizes readily how it would be included in each of the ascospores, which are simply cut out of the cell that produces them. Under such a system, Mendelian segregation for the little trait would not be expected to occur, as indeed it does not.

The hypothesis is simplest and explains the most if the active cytoplasmic component of normal cells is visualized as occurring in particulate form. If one supposes these hypothetical particles to exist in relatively limited numbers, the problem of accounting for spontaneous and induced *little* mutations is not formidable. Given a small number of particles, their total exclusion from a yeast bud might easily happen by chance. And the enormous efficiency of acridines in inducing littles in a growing culture would be an inevitable result if acridines had the effect of inhibiting the replication of the active particles. One of the acridines, acriflavine, does have the appropriate effect in another system where actual particulate elements can be observed. It inhibits the replication of an

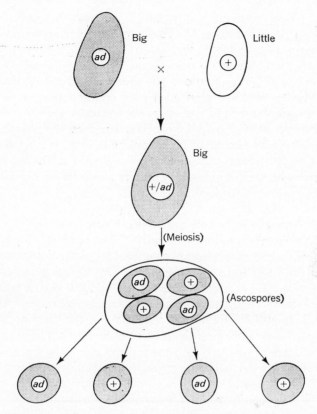

Figure 11-7. *When big and little yeast are mated, the zygote is big, and so are all the progeny ascospores. At the same time, the allelic pair +/ad segregates as expected in the ascospores. Shading of cells designates the normal cytoplasm, which determines the big phenotype. Lack of shading designates the abnormal cytoplasm of littles. We can suppose that when mating occurs, normal cytoplasm from the big parent is introduced into the fusion cell, where it replicates and becomes a component of each of the ascospores, thus accounting for their normal phenotype and for the absence of segregation for the little attribute.*

organelle, the kinetoplast, in Trypanosomes and the bacterial sex factor F to be described later.

If the difference between big and little yeast is a matter of self-replicating particles in the cytoplasm, what might these particles be? One can do no more than make an educated guess, but such a guess must point to *mitochondria,* particles that exist in almost all kinds of living cells and that are the seat of the biochemical functions with which the little mutation interferes. In fact, electron microscopy by Yotsuyanagi has shown that the mitochondria of little yeast cells are structured in abnormal ways. Moreover, evidence has accumulated to show that mitochondria include DNA, the material that codes genetic information. These observations, however, do not necessarily prove that mitochondria are the cytoplasmic particles in question. The aberrancy of mitochondria in little yeast is conceivably no more than additional manifestation of the effects of another sort of cytoplasmic particle as yet undetected.

Kinds of little yeast other than the one just described have been found. One of these, phenotypically similar to the cytoplasmic little, is a standard mutant that follows typical chromosomal patterns of segregation. Others, cytoplasmic littles, in crosses with normal impose the mutant character on a variable fraction of the zygotes.

Cytoplasmic Variants in Neurospora. *Poky* Neurospora is the analogue of little yeast. On solid medium it grows erratically but generally slowly. Like little yeast it is cytochrome-deficient and has an unusual pattern of transmission.

If you refer back to Figure 4-5, you will see that by using conidia from one parent culture to fertilize protoperithecia of the other, and vice versa, a reciprocal cross can be accomplished. By analogy with higher organisms, the conidium with its limited amount of cytoplasm is the male gamete; the protoperithecium, relatively passive and with a vastly greater bulk of cytoplasm, includes the "female" reproductive cells. In Neurospora, the zygote is the same, and in terms of Mendelian genetics the ascospore populations should be the same, for both crosses of a reciprocal pair. However, note the following result. The transmission of poky clearly follows the maternal, or cytoplasmic, line.

PROTOPERITHECIAL PARENT		CONIDIAL PARENT	PROGENY ASCOSPORES
wild type	×	poky	all wild
poky	×	wild type	all poky

A number of other cytoplasmic variants have been studied in Neurospora. One of these, characterized by markedly slow germination of the spores and called SG, fulfills many of the tentative criteria for extrachromosomal inheritance outlined at the beginning of this chapter. SG, like poky, shows reciprocal cross differences, is maternally inherited, and does not segregate at meiosis. Appropriate crosses of SG with strains carrying marker genes for all seven chromosomes of Neurospora have failed to show linkage of SG to a chromosome. SG is also indifferent to nuclear substitution, as shown in two ways. First, by extensive backcrossing (Fig. 11-8), the *N. crassa* nucleus with which SG was originally associated has been replaced by the nucleus of another species, *N. sitophila*. At the end of the backcross sequence, SG in the presence

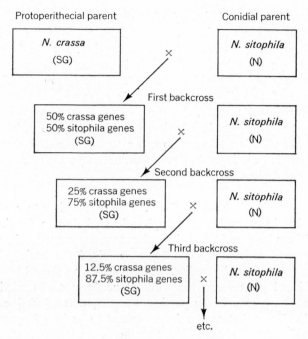

Figure 11-8. *If characteristics of the cytoplasm are transmitted only through the maternal parent, and are independent of the nucleus, then repeated backcrossing according to the pattern illustrated should ultimately give cells with the cytoplasm of the original maternal parent but with nuclei identical with those of the recurrent, male parent.* Neurospora crassa *having the maternally inherited variant character SG was involved in such a backcross sequence with normal* Neurospora sitophila, *used as recurrent, male parent. After 10 generations of backcross, when the nuclei should have accumulated all, or almost all, of the genome of* N. sitophila, *progeny cultures were still uniformly SG.* (*Based on Srb,* Cold Spring Harbor Symposia Quant. Biol., **23,** *1958.*)

of an essentially wild-type nucleus from *N. sitophila* retains all its original variant properties. A second means of showing the indifference of SG to nuclear substitution has utilized techniques permitted by the phenomenon of heterokaryosis, described in Chapter 4. If hyphae of an SG strain carrying a gene for adenine requirement fuse vegetatively with hyphae of a strain with normal cytoplasm and carrying an albino marker, isolates from the heterokaryotic mycelia include SG albino cultures and cytoplasmically normal cultures with an adenine requirement. Under the conditions of the experiment, then, nuclei from a strain with normal cytoplasm are unable to redirect SG cytoplasm into normalcy.

Complementary action of cytoplasms has been demonstrated by T. Pittenger.

Using the heterokaryon technique, he has obtained vegetative fusions between two cytoplasmic variants of independent origin. Each of the variants has an aberrant cytochrome spectrum and is slow growing, but the fusion mycelium achieves wild-type growth rate although its cytochrome spectrum remains abnormal. Presumably the rapid growth rate of the fusion mycelium is due to the coexistence of two different cytoplasmic systems (a *heteroplasmon*) that complement each other's action. The fact that the cytochromes remain aberrant in the rapidly growing mycelium indicates that mixing the variant cytoplasms has not resulted in the formation of normal cytoplasmic components. Moreover, when the heteroplasmon is used as protoperithecial parent in crosses, wild-type ascospores are not recovered.

Other Cytoplasmic Variants in Fungi. The recent literature of genetics is particularly rich in examples of cytoplasmic variation in fungi. We cannot discuss all of them here, and the various phenomena are not readily summarized, perhaps because geneticists have not yet perceived the basic principles that underlie extrachromosomal heredity. Nevertheless, you should be aware of one or two additional phenomena that do not conform to Mendelian genetics. One of these phenomena is the persistent segregation of distinct phenotypes out of fungal cultures that can be demonstrated to be *homokaryotic*, that is, to have nuclei of only one genotype. The genus Aspergillus provides several instances of persistent segregation out of homokaryons. Such segregations have been accounted for by the assumption of inherent differences within the cytoplasm. A second striking phenomenon, that resembles infection, can be well illustrated from G. Rizet's investigations of the mold *Podospora anserina*. In this mold, haploid cultures differentiated by an allelic pair s and s^S show quite different reactions when their mycelia establish contact with mycelium of a third genotype, S. When strains of genotypes s and s^S are placed in contact with each other, so that hyphal anastamoses occur, the s^S mycelium from the points of contact progressively becomes s. By the use of marker genes, Rizet has eliminated the possibility of nuclear migration from one mycelium to the other. Therefore, something in s cytoplasm, which can progress as rapidly as 7 cm/24 hrs, invades s^S mycelium and achieves the apparent alteration of an allele.

Cytoplasmic Inheritance in Chlamydomonas. The life cycle of the single-celled green alga Chlamydomonas is rather similar to that of yeast. Haploid cells are able to reproduce asexually, by mitotic nuclear divisions. Or, two haploid cells of opposite mating type may fuse to form a zygote. Following meiosis in the zygote, four haploid cells are produced, which may again reproduce asexually or under appropriate conditions may mate. Since the products of a single meiosis may be isolated, and their vegetative progenies may be observed and subjected to tests, tetrad analysis is possible.

Ruth Sager and her associates have found a number of hereditary variables

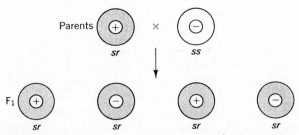

Figure 11-9. *When a streptomycin resistant* (sr) *strain of Chlamydomonas is mated with a streptomycin sensitive* (ss) *strain, all the progeny are resistant. Meantime, the mating-type alleles (+ and −) segregate 1:1 as expected with chromosomal heredity. Note the basic similarity of this result with that obtained with yeast in the experiment summarized in Figure 11-7. (From Sager and Ryan,* Cell Heredity, *John Wiley, 1961, p. 237.)*

in Chlamydomonas that fail to show segregation in a chromosomal pattern. As shown in Figure 11-9 for example, all the progeny of a cross of streptomycin sensitive × streptomycin resistant are streptomycin resistant. In contrast, the mating type alleles, *mt+* and *mt−*, give the expected Mendelian segregation of 1:1. A curiosity of the Chlamydomonas system is that in general only *mt+* cells transmit the presumed extrachromosomal factors, such as those for streptomycin resistance. It is as if all extrachromosomal factors brought to the zygote by the *mt−* parent are lost. Occasionally, however, exceptional zygotes retain extrachromosomal factors from both parental cells. Sager has made crosses involving more than one pair of extrachromosomal alternatives and has used the exceptional zygotes to test for segregation. From crosses involving two pairs of extrachromosomal determinants, for acetate requirement or non-requirement and for streptomycin sensitivity or dependence, the following important results emerge. 1) Both pairs of alternatives may be transmitted from the same exceptional zygote. (2) The pairs of alternatives segregate relatively independently. (3) Segregation of the extrachromosomal factors takes place postmeiotically. The extrachromosomal hereditary factors in Chlamydomonas have been called *nonchromosomal genes,* and indeed they seem to have substantial claim to that title.

Episomes

Whatever may be the true nature or ultimate source of control of the hereditary particles we have designated as extrachromosomal, their behavior shows various degrees of independence of the nucleus. You have seen examples in which transmission of an extrachromosomal hereditary property is independent of the transmission of chromosomes, and you have also been made aware of instances in which multiplication of an extrachromosomal property is not synchronized with multiplication of the chromosomal genetic material. Certain particles endowed with genetic continuity have been found to exist in either of two alternative states: (1) an *integrated state*, in which the particle is intimately associated with a chromosome, and multiplies in synchrony with it; and (2) an *autonomous state*, in which the particle multiplies independently of the chromosomal genetic material. The French geneticists Jacob and Wollman have defined these alternative states as the criteria for a particular kind of genetic element they call the *episome*.

The Genetic Material of Temperate Phages. Chapter 4 presented an outline of the life cycle of a temperate phage, and if you refer to that chapter you will see that the alternative states of prophage and vegetative phage fit very well the criteria for an episome. In fact it was consideration of prophage and phage in reference to similar relationships in bacterial genetics that led to the episome concept.

In a lysogenic cell, the prophage, which is the genetic material of the phage, is bound to the bacterial chromosome. This is the *integrated* state of the episome. Standard mapping experiments show that a particular kind of prophage has a particular site of attachment and behaves in all respects like regular bacterial genes. Since different prophages have their own specific attachment sites, a number of different prophages may be carried simultaneously within a given bacterial strain. Whether all prophages are actually inserted into the bacterial chromosome, or are somehow "hooked on," is not yet known. But accumulating evidence gives realism to a hypothesis, formulated by Allan Campbell, that the lambda phage of *E. coli* is inserted into the bacterial chromosome in a sequence of events that resemble synapsis followed by crossing over. Figure 11-10 summarizes a scheme of such events, in which the circular genome of lambda phage first pairs with its appropriate site in the bacterial chromosome and then by a single reciprocal recombination is inserted as a linear sequence of genetic material within the bacterial chromosome.

The *autonomous* state of the episome in this particular instance, of course, is the vegetative state of the phage—the state in which phage genetic material does not replicate synchronously with the bacterial cell. During the autonomous state of the temperate phage, infective particles are formed. They may infect and lysogenize other cells independently of the conjugation process. In these episomes, then, we have alternative states that fit quite well on the one hand chromosomal heredity and on the other extrachromosomal heredity.

The F Sex Factors in E. coli. You learned in Chapter 6 that cells of *E. coli* K12 have sexual differentiation, with F^- cells acting as recipients of genetic material at conjugation and with F^+ cells acting as donors. It was also said

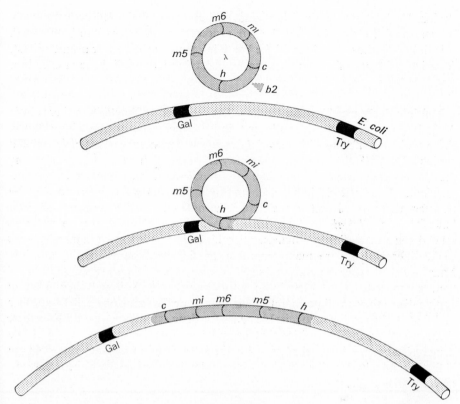

Figure 11-10. *Insertion of lambda phage into the chromosome of* Escherichia coli. *Pairing
of the phage genome with a particular site in the bacterial chromosome,
followed by a crossover-like event, results in insertion of the prophage
between the bacterial loci Gal and Try. (From Stent,* Molecular Biology
of Bacterial Viruses, *W. H. Freeman, 1963, p. 330.)*

that effective matings appear to be between F^- cells and *Hfr* cells, which arise
from F^+ by some mutation-like process. A variety of observations have led to
the conclusion that F^- cells are only able to act as recipients and not as donors
at conjugation because they *lack* a fertility factor F. One pair of pertinent
observations is that F^- cells never mutate to F^+ or *Hfr* but F^+ cells often mutate
to F^-. Moreover, F^+ cells exposed to acridine dyes, the same ones that induce
the occurrence of little yeast, lose their fertility property and become F^-.

Other observations lead us to believe that the F factor is present in both *Hfr*
cells and F^+ cells but that in the former F occurs in an integrated and in the
latter an autonomous state. On conjugation with F^- bacteria, F^+ cells transmit
their fertility property at a very high frequency but independently of the bacterial
chromosome, suggesting that the F factor is not bound to a chromosome in F^+

strains. *Hfr* bacteria, on the other hand, seldom transmit the *Hfr* or *F*⁺ property in crosses with *F*⁻ strains, but the *Hfr* property, nevertheless, can be mapped. It is linked to those portions of the bacterial genome that are transmitted last. A variety of types of *Hfr* are known, each having its own linkage association.

Multiple Resistance Transfer Factors. The resistance of bacteria to therapeutic drugs is a matter of considerable and, indeed, increasing medical importance. From patients suffering with bacterial dysentery, Japanese workers have isolated strains of *Shigella* that are simultaneously resistant to three or even four among the therapeutic drugs sulfonamide, streptomycin, tetracycline, and chloramphenicol. Quite astonishingly, such multiple resistances, as a group, can be transferred from resistant to sensitive cells, not only within the genus *Shigella* but also from *Shigella* to *Escherichia*, *Salmonella*, and to even more distantly related genera. Studies by Japanese microbiologists have shown that resistances to the different drugs are due to separable genetic determinants and that multiple transfer of resistance depends upon a *resistance transfer factor* (RTF). The RTF factors appear to exist largely in an autonomous state. Although they promote conjugation, they mediate mainly their own transfer from one cell to another and that of the readily recognizable genes for drug resistance. The RTF factors, then, are episome-like entities that promote conjugation and confer multiple drug resistance on those cells harboring them. In certain ways they resemble the F factors already described, and show complex interactions with them that are as yet poorly understood.

Colicinogens. Particular strains of *E. coli* produce colicins, proteins that kill other strains. The determinants of colicins, called *colicinogens*, behave as autonomous cytoplasmic entities that promote their own transfer to other cells. To a lesser extent, several of the different colicinogens also promote the transfer of bacterial chromosomal material from one cell to another. Again, like RTF, colicinogens resemble sex factors and have some of the properties of episomes.

Clear evidence is lacking that colicinogens or RTF factors integrate into the bacterial chromosome and thus meet all the criteria for an episome as originally defined. Perhaps such evidence will emerge. At present, it seems likely that we will come to recognize, at least in microorganisms, a large spectrum of genetic entities, some apparently confined to chromosomes, some found only in cytoplasm, others moving back and forth. These entities, in relation to the cells in which they are included, may exhibit varying aspects of parasitism, symbiosis, and integration. Their further analysis promises to be exciting and illuminating, particularly in reference to the evolution of genetic systems.

Controlling Elements

An intensive analysis of certain puzzling variations in maize led Barbara McClintock to the discovery of genetic elements, called *controlling elements*, that modify or suppress gene action and in particular instances produce mutational effects. The controlling elements are associated with chromosomal genetic material, and perhaps are part of it, but need not remain in fixed position. In fact, certain controlling elements characteristically undergo transposition from one location within the genome to another.

Perhaps the most extensively analyzed among controlling elements is an interacting pair called *Ac-Ds*. In the presence of *Ac* but not in its absence, *Ds* is associated with apparent chromosome breaks at the locus at which it resides. At that locus a dicentric and a corresponding acentric chromatid may be formed. When such an event occurs, the acentric chromatid, which is composed of the chromosomal segment extending from *Ds* to the end of the arm, is lost at mitotic anaphase. As illustrated in Figure 11-11, the appropriate use of dominant and recessive gene markers enables the detection of chromatid loss by the unmasking of recessive genes. The figure shows also the effect of increasing dosage of *Ac*. With higher doses, the abnormal events occur later in development.

Ds, in the presence of *Ac*, has other remarkable properties. It can change position within the genome, and from time to time it shows quite different linkage relations. A locus at which it resides often undergoes alteration which may be more or less stable. And *Ds* frequently modifies the action of genes at or near its location. When these are genes whose effects are known under ordinary circumstances, it can often be observed that the presence of *Ds* results in phenotypic expression similar to that determined by some mutant form of the gene. *Ds* also undergoes changes in state, that may be detected for instance by loss of ability to induce chromatid breaks or by alteration of a particular modifying effect on gene action.

Ac, besides eliciting particular responses from *Ds*, may itself control gene action at a locus where it is situated. Its presence may give rise to chromosome breaks or to mutation-like events, and it can undergo transposition.

The Dotted Gene in Maize. The dotted gene (*Dt*), first found in Black Mexican sweet corn, was long used as the classical example of a mutator gene. When gene *a*, situated in chromosome 3 of maize, is homozygous, anthocyanin pigment fails to form, and the aleurone is colorless. This situation is strikingly altered by the presence of *Dt*, a dominant that maps to chromosome 9. Plants of genotype *aaDt−* show dots of deep color in the aleurone, as if gene *a* had mutated to its dominant, active allele *A*. Since the tissue involved is triploid, one can arrange to compare several doses of the dotted gene for their effects on gene *a*.

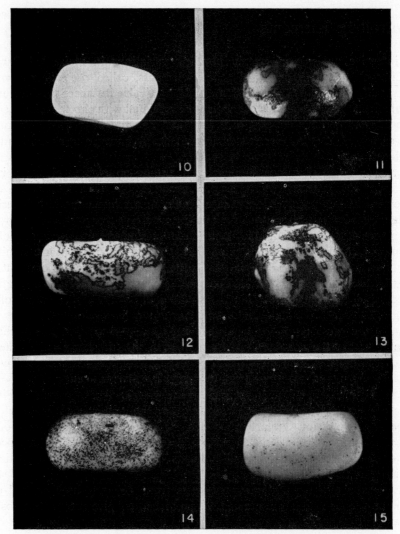

Figure 11-11. *Phenotypic effects on kernels of maize produced by breaks at* Ds. *The variegation results from chromosome breaks and a consequent loss of dominant genes that have masked their recessive alleles. In 10, no Ac is present, and the kernel is completely colorless because of inhibition of aleurone color due to the presence of dominant gene* I. *In 11 to 13 one* Ac *is present, and the breaks at* Ds *have occurred relatively early in development, so that the sectors are rather large. In 14 two* Ac *factors are present in the endosperm and in 15 three are present. With increasing dose of* Ac *breaks occur relatively later in development and the sectors are relatively smaller. (Courtesy of B. McClintock,* Cold Spring Harbor Symposia Quant. Biol., **16,** *1951, p. 23.)*

Observations by Rhoades revealed that genotype *Dt dt dt* gives an average of 7.2 dots per seed, *Dt Dt dt* 22.2, and *Dt Dt Dt* 121.9.

In the light of the *Ac-Ds* system in maize, the case of dotted has been re-examined. Allele *a*, which is stable in the absence of *Dt*, can at once be seen to respond to *Dt* in ways similar to the response of *Ds* to *Ac* when *Ds* is at some known chromosomal locus. Moreover, in different strains of corn, *Dt* has been found to occupy loci in different chromosomes, and apparently can undergo transposition. Is there, then, an element at the *a* locus that is the analogue of *Ds*? The answer is that there may be. The *a* locus is compound and subdivisible by crossing over. Some of the recombinants from within the *a* locus have properties that are *Ds*-like.

The Nature of Controlling Elements. Maize geneticists now know a number of genetic elements whose properties are similar to those of *Ac-Ds*. The nature of these elements is far from completely understood, but their existence has aroused valuable speculation. Particularly when *Ac-Ds* was first studied, it appeared to have associations with heterochromatin, long suspected to be responsible for variability in the expression of genes associated with it. To implicate heterochromatin in the origin or in the action of controlling elements is tempting but at present merely speculative. Similarly, parallels can be drawn between controlling elements and episomes, but the reality of such parallels remains open to question. More recently, McClintock has been impressed with similarities between the two-element control systems in maize and the better understood two-element control systems in bacteria (see Chap. 10). For example, one can reasonably think of *Ac* as a *regulator* and of *Ds* as an *operator*. Conceived in such terms the controlling elements are of potentially great importance in the regulation of developmental processes of the organisms in which they occur. The fact that some of the manifestations of the controlling elements are erratic does not preclude their participation in regular, normal processes of development. Until now, perhaps, controlling elements have only been detected in situations where their activities are somehow out of phase.

Paramutation

Ordinarily, when members of an allelic pair are brought together in a heterozygote, they segregate cleanly, and are recoverable in their original form. This basic principle of Mendelian genetics is violated, at least superficially, in certain instances analyzed by R. A. Brink. As a particular example, we will consider strange phenomena occurring at the *r* locus in maize. At this locus, allele r^g is a recessive that determines colorless aleurone; R^{st}, called stippled, determines fine color spots in the endosperm; R^r when homozygous gives uniformly colored endosperm, but genotype $R^r r^g r^g$ is deeply mottled.

The phenomenon that attracted Brink's attention can be illustrated by the following crosses and their results:

$r^g r^g \times R^r R^r \longrightarrow$ kernels are all darkly mottled
$r^g r^g \times R^{st} R^{st} \longrightarrow$ kernels are all stippled
$r^g r^g \times R^r R^{st} \longrightarrow$ ½ the kernels are stippled (as expected)
　　　　　　　　　　　½ the kernels are weakly pigmented (unexpected)

From the third testcross, half the progeny are expected to have darkly mottled kernels, but such kernels do not appear. Instead, those kernels expected to be $R^r r^g r^g$ seemed represented in the new group whose phenotype is weak pigmentation. The R^r allele seems to have disappeared, or rather to have been altered to a new form that can reasonably be represented as $R^{r:st}$.

The alteration of R^r is not a typical mutational event. It is associated only with specific allelic associations in heterozygotes, but when the particular conditions for the event are met it occurs invariably. The alteration is persistent, but over a sequence of generations there is a tendency toward reversion to standard phenotype. Brink has called these alterations *paramutations*. They seem to have more in common with the effects of controlling elements than with mutations interpreted as changes in the genetic code.

Cyclic Nucleo-Cytoplasmic Relationships

One of the varieties of Paramecium has mating types designated as VII and VIII. When conjugation takes place in the absence of cytoplasmic exchange, each exconjugant retains its original mating type. When conjugation with cytoplasmic exchange occurs, the exconjugants assume identical mating type. Whether this be mating type VII or VIII is temperature dependent. Except that the mating types are rather more stable, the pattern of transmission resembles that already described for antigenic types in Paramecium. However, an ingenious experiment by Sonneborn shows a nucleo-cytoplasmic relationship that the foregoing matings cannot in themselves reveal.

If VII and VIII type animals are allowed to conjugate and then are exposed to high temperatures, members of each exconjugant line include some individuals with macronuclei regenerated from fragments of the old macronucleus and others with new macronuclei. In instances where cytoplasmic exchange occurs, the determination of mating types differs between cells with new macronuclei and those with regenerated macronuclei. The paramecia with new macronuclei are all of mating type VIII. The paramecia with the regenerated type of macronucleus have the same mating type as the conjugant from which the fragments were derived. Once the mating types are determined, autogamy gives rise to no further changes.

The various foregoing observations can be summarized and interpreted as follows. (1) Mating type may be altered at conjugation as the result of cytoplasmic influence. (2) Macronuclei can exert a determining influence on mating type. (3) Although cytoplasm can affect the mating-type potential of new macronuclei, cytoplasm does not affect mature macronuclei (for example, those that arise by regeneration). In summary, cytoplasm can differentiate newly developing macronuclei in respect to their mating-type potential, but the particular determinative properties of the cytoplasm in turn depend upon nuclear action. This subtle circular interrelationship, demonstrable in Paramecium because of the presence of the macronucleus, would not be easily perceived in an organism with a less exotic genetic apparatus. In fact, many of the phenomena discussed in this chapter, which may seem to be a collection of odd facts from odd organisms, may be of common and basic occurrence but difficult to reveal except in organisms offering peculiar advantages for their detection.

The Inheritance of Preformed Structure

Occasionally, when two paramecia conjugate, they fuse and form a doublet animal, a rather amazing creature that has two complete sets of the various cortical structures, including gullet, contractile vacuole, and other organelles. As shown by Sonneborn, once it arises the doublet phenotype is remarkably persistent over a series of cell generations. Doublets that reproduce by fission give rise to more doublets. This result, of course, is not surprising, if doubleness has a genetic basis. More surprising is the transmission of the doublet character when sexual reproduction occurs. If doublets are mated with singlets, the doublet exconjugant and its asexually reproduced derivatives are all doublets. Subsequent autogamy does not change the picture. Since known marker genes are transmitted as expected through a sequence of such breeding experiments, one can conclude that the chromosomal genetic apparatus is operating normally but that its operations are irrelevant to the transmission of the doublet characteristic. Thus far we are dealing with observations similar to results obtained when killer paramecia are mated to sensitives under conditions where cytoplasmic exchange does not take place. Is, then, the doublet trait transmitted via factors in the cytoplasm?

To answer the question, the cytoplasm of singlets was marked with kappa. The killer singlets were then crossed with sensitive doublets, both under conditions where cytoplasmic transfer would occur and where it would not. As seen in Figure 11-12, cytoplasmic transfer, clearly proven by the transfer of kappa, does not alter the pattern of transmission of the doublet characteristic. Singlet exconjugants and their lineal descendants remain singlets, and the same is true for doublets. A quotation from Sonneborn summarizes the situation nicely.

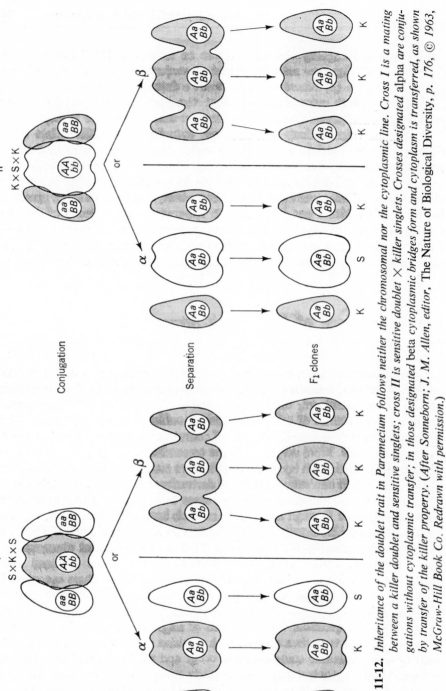

Figure 11-12. *Inheritance of the doublet trait in Paramecium follows neither the chromosomal nor the cytoplasmic line. Cross I is a mating between a killer doublet and sensitive singlets; cross II is sensitive doublet × killer singlets. Crosses designated alpha are conjugations without cytoplasmic transfer; in those designated beta cytoplasmic bridges form and cytoplasm is transferred, as shown by transfer of the killer property. (After Sonneborn; J. M. Allen, editor, The Nature of Biological Diversity, p. 176, © 1963, McGraw-Hill Book Co. Redrawn with permission.)*

"Never before in our experience had we encountered results suggesting the paradoxical conclusion that a hereditary difference was neither genotypic nor cytoplasmic in basis." The paradox is accentuated when appropriate experiments rule out a cyclic nucleo-cytoplasmic relationship of the kind presented in the preceding section.

The escape from the paradox appears to be that preformed structure, as represented in the cortical organelles, may have a kind of autonomy and genetic significance of its own, perhaps by acting as a model for orienting the molecular products of genic and cytoplasmic activity. In a sense, one is dealing with a significant inheritance of phenotype. Reinforcing some such interpretation is the finding that natural grafts of pieces of cortex can be incorporated into the host animal and then transmitted as if they possessed genetic autonomy. Vance Tartar, in the book cited among the references at the end of this chapter, has summarized extensive pioneering studies of Stentor in this context.

Keys to the Significance of This Chapter

Extrachromosomal heredity has reality, and manifests itself in almost bewildering variety. One can scarcely escape the conclusion that it plays an important role in the biology of many organisms. At the same time, many obvious questions about extrachromosomal heredity cannot presently be answered. Does extrachromosomal heredity rely on coded information? If so, does it have its own coding system, or does it rely ultimately on the coded DNA situated in chromosomes? Are the extrachromosomal systems only concerned in the expression of specific potentialities transmitted through the chromosomes or do they confer their own specific potentialities? Are the extrachromosomal systems phenotypic rather than genotypic? These questions, and others, need to be applied individually to the phenomena considered in this chapter and not to extrachromosomal heredity en masse. When real answers can be provided they may well vary from one instance to another. At the moment, for example, coded information seems a bit more likely to be found in a chloroplast than in the cortex of Paramecium. But clearly enough, both have information of some sort, and it is transmissible. In any case, codes need not be all of the same kind.

All of the systems detected on the basis of extrachromosomal transmission interact with chromosomal genes or their products. In many instances these systems seem to be concerned with the expression and integration of gene action or with determining whether a gene is active in a particular cell. They are systems, therefore, that appear to be concerned with cellular differentiation. In recognition of this, D. L. Nanney has suggested that they be called *epigenetic* systems.

REFERENCES

Beale, G. H., *The Genetics of Paramecium Aurelia.* Cambridge: The University Press, 1954. (A short book on the earlier genetic work with an organism that has provided some of the best clues to the nature of extrachromosomal heredity.)

Brink, R. A., "Paramutation at the R Locus in Maize." *Cold Spring Harbor Symposia Quant. Biol.,* **23**:379–391, 1958. (Rather detailed, but readable and authoritative.)

Brink, R. A., "Phase Change in Higher Plants and Somatic Cell Heredity." *Quart. Rev. Biol.,* **37**:1–22, 1962. (The switch from juvenile to adult-type growth in higher plants is discussed in relation to genetic and epigenetic systems.)

Campbell, A. M., "Episomes." *Advances in Genetics,* **11**:101–145, 1962. (An authoritative review.)

Correns, C., "Vererbungsversuche mit blass (gelb) grünen und buntblättrigen Sippen bei *Mirabilis Jalapa, Urtica pilulifera* und *Lunaria annua.*" *Z. Ind. Abst. Vererb.-lehre,* **1**:291–329, 1909. (The classic study of plastid inheritance.)

Ephrussi, B., *Nucleo-Cytoplasmic Relations in Micro-Organisms.* Oxford: Oxford University Press, 1953. (A small book that includes a fascinating account of little yeast and of the barrage phenomenon studied by G. Rizet in the mold Podospora.)

Hayes, W., *The Genetics of Bacteria and Their Viruses.* New York: Wiley, 1964. (Contains an excellent treatment of episomes and is generally valuable in the subject matter indicated by the title.)

Jacob, F., and Wollman, E. L., *Sexuality and the Genetics of Bacteria.* New York: Academic Press, 1961. (An excellent general work on its subject. Chap. 16 deals with episomes.)

Jinks, J. L., "Cytoplasmic Inheritance in Fungi." In *Methodology in Basic Genetics,* edited by W. J. Burdette, pp. 325–354. San Francisco: Holden-Day, 1963. (A critical review of the means for detecting and analyzing cytoplasmic heredity in fungi.)

McClintock, B., "Some Parallels between Gene Control Systems in Maize and in Bacteria." *Am. Naturalist,* **95**:265–277, 1961. (Discusses controlling elements in corn, including some not dealt with in this text. The emphasis is on comparisons with operator and regulator genes.)

Michaelis, P., "Interactions between Genes and Cytoplasm in Epilobium." *Cold Spring Harbor Symposia Quant. Biol.*, **16**:121–129, 1951. (A survey of extensive studies of nucleo-cytoplasmic relations in a higher plant.)

Nanney, D. L., "The Role of the Cytoplasm in Heredity." In *The Chemical Basis of Heredity*, edited by W. D. McElroy and B. Glass, pp. 134–166. Baltimore: Johns Hopkins Press, 1957. (Includes a comprehensible and succinct account of mating-type determination in Paramecium.)

Nanney, D. L., "Epigenetic Control Systems." *Proc. Nat. Acad. Sci.*, **44**:712–717, 1958. (A general consideration of the nature and role of extrachromosomal systems of hereditary significance.)

Rhoades, M. M., "Interaction of Genic and Non-genic Hereditary Units and the Physiology of Non-genic Function." In *Encyclopedia of Plant Physiology*, edited by W. Ruhland, **1**:2–57. Berlin: Springer-Verlag, 1955. (Outstanding review of extrachromosomal heredity, including plastid heredity. Includes cytoplasmic male sterility in plants and other topics not covered in this text.)

Sager, R., and Ramanis, Z., "The Particulate Nature of Nonchromosomal Genes in Chlamydomonas." *Proc. Nat. Acad. Sci.*, **50**:260–268, 1963. (Presents experimental evidence for nonchromosomal "genes.")

Sonneborn, T. M., "Does Preformed Cell Structure Play an Essential Role in Cell Heredity?" In *The Nature of Biological Diversity*, edited by J. M. Allen, pp. 165–221. New York: McGraw-Hill, 1963. (An enthralling consideration of certain interrelations between morphology and heredity.)

Srb, A. M., "Extrachromosomal Factors in the Genetic Differentiation of Neurospora." *Symposia Soc. Exptl. Biol.*, **17**:175–187, 1963. (Presents the experimental application of several criteria for extrachromosomal inheritance to a heritable trait in a mold.)

Tartar, V., *The Biology of Stentor*. New York: Pergamon Press, 1961. (Includes pioneering studies of the perpetuation of preformed structure, based on fusions of whole organisms and on grafting experiments.)

Wettstein, D. von, "Nuclear and Cytoplasmic Factors in Development of Chloroplast Structure and Function." *Canad. J. Bot.*, **39**:1537–1545, 1961. (Deals with plastids from a rather different point of view than that taken in this text.)

Wilkie, D., *The Cytoplasm in Heredity*. London: Methuen & Co., 1964. (A short monograph that considers cytoplasmic heredity primarily from the point of view of genetic regulatory systems.)

QUESTIONS AND PROBLEMS

11-1. Diagram reciprocal crosses of $AA \times aa$ Ephestia (p. 317), showing parental phenotypes, genotypes, and gametes; F_1 phenotypes, genotypes and gametes; F_2 genotypes and phenotypes. Distinguish phenotypes of larvae and adults in each generation.

11-2. Diagram meiosis, in both testes and ovaries, and self-fertilization in a snail heterozygous for the dextral-sinistral alternative. List genotypes and phenotypes of progeny, and indicate the phenotypes of the progeny of a subsequent generation of self-fertilization from each of these genotypes.

11-3. What was the phenotype of the heterozygous snail with which you started in Question 11-2? Explain.

11-4. It is possible to remove the ovaries of a mouse and to transplant in their place ovaries from another mouse. Such transplanted ovaries, in a high proportion of cases, become established and function normally, shedding eggs that may be fertilized in the mouse. How might this technique be used to distinguish between *transfer of materials from mother to fetus across the placenta*, and *egg transmission of cytoplasmic materials?*

11-5. Given consistent differences between two strains of mice in a particular characteristic, and a marked difference in the results of reciprocal crosses between the two strains, how might you determine whether the responsibility was owing to ordinary sex linkage, maternal influence via egg or placenta or milk, or cytoplasmic inheritance?

Hagberg, at the Institute of Genetics, Svalof, Sweden, has studied the inheritance of bitterness and sweetness in lupines. (Different kinds of lupines are used as ornamental plants and as forage crops; wild lupines decorate our countryside.)

11-6. When sweet and bitter lupines are crossed, with bitter as the pollen parent, and the plants grown from the seeds resulting from this cross are tested within a month after the seed is sown, all of the F_1 plants are sweet. F_1 plants from the reciprocal cross, grown and tested under similar conditions, are bitter. What might explain these results?

11-7. Usually, the leaves of F_1 plants more than a month old, and always the ripe seeds of these plants, whichever way the parental cross was made, are bitter. Does this help to narrow down the possibilities suggested in your answer to Question 11-6?

11-8. Hagberg observed that 40 mature F_2 plants from the cross ♀ sweet × ♂ bitter segregated as 12 sweet:28 bitter, in terms of their ripe seed. Their leaves, however, had all been bitter until they were almost a month old. Diagram the reciprocal crosses of sweet × bitter through F_2, suggesting the points in the life cycle of lupines when the alternatives sweet and bitter are affected by the genes of each generation.

11-9. By means of artificial insemination, reciprocal crosses have been effected between Shires, a very large breed of horse, and Shetland ponies. If a shire was mother in the cross, the birth weight of the young was around 50 kilograms; if a Shetland pony was mother, the birth weight of the crossbred young was 20 kilograms or less. Is cytoplasmic inheritance suggested by these observations? Are there obvious alternatives to such an explanation? In arriving at a satisfactory explanation of the reciprocal cross differences, what additional information would you like to have about the crossbred progeny horses? Suggest experiments that might clarify whether cytoplasmic inheritance is involved.

11-10. "Segregational" little yeast resembles phenotypically the cytoplasmic little yeast described in the text. When a segregational little is crossed with a normal, the diploid cells are normal. When ascospores subsequently form, each ascus has 2 normal spores and 2 spores that produce little cultures. Explain the results by the use of diagrams, designating genotypes appropriate to the observed phenotypes.

11-11. When segregational littles are crossed with cytoplasmic little yeast of the kind described in the text, the resultant diploid cells are normal. Asci formed by these diploids show 2:2 ratios of normal to mutant cells. Interpret these results. Construct a diagram that summarizes your interpretation.

In Neurospora, a chromosomal gene F suppresses the slow-growth characteristic of the poky phenotype, and makes a poky culture into a fast-poky culture, which still has, however, abnormal cytochromes. Gene F in combination with normal cytoplasm does not have a visible phenotypic effect.

11-12. A cross in which fast poky is used as protoperithecial parent and normal as "male" parent gives half poky and half fast-poky ascospores. Give a genetic interpretation of the result. What does the result indicate about the autonomy of the cytoplasmic factor in poky?

11-13. What would be the result of the reciprocal of the cross described in 11-12? Designate genotypes and phenotypes of the progeny.

11-14. Assume that you have a culture of phenotype normal both for cytochromes and growth rate, but that you are unsure whether the culture carries gene F or its wild-type allele. What cross or crosses could you make that would test whether the culture carries gene *F*? State the results of the cross if the unknown culture carries *F*. If it carries the wild-type allele of *F*.

One form of male-sterility in maize is maternally transmitted. Plants of a male-sterile line crossed with normal pollen give male-sterile plants. In addition, some lines of maize are known to carry a dominant restorer gene (*Rf*), which restores pollen fertility in male-sterile lines.

11-15. It has been found that restorer genes when introduced into male-sterile lines do not alter or affect the maintenance of the cytoplasmic factors for male-sterility. What is the kind of result that would lead to such a conclusion?

11-16. A male-sterile plant is crossed with pollen from a plant homozygous for gene *Rf*. What is the genotype of the F_1? The phenotype?

11-17. The F_1 plants in 11-16 are used as females in a testcross with pollen from a normal plant (*rf rf*). What would be the result of the testcross? Give genotypes and phenotypes, and designate the kind of cytoplasm.

11-18. The restorer gene already described can be called *Rf-1*. Another dominant restorer, *Rf-2*, has been found, and is located in a different chromosome from that of *Rf-1*. Either or both of the restorers will give pollen fertility. Using a male-sterile plant as a tester, what would be the result of a cross where the male parent was heterozygous at both restorer loci? Homozygous dominant at one restorer locus and homozygous recessive at the other? Heterozygous at one restorer locus and homozygous recessive at the other? Heterozygous at one restorer locus and homozygous dominant at the other?

11-19. L. Knudson has reported various permanent changes in chloroplasts induced in the gametophyte of the hare's-foot fern, *Polypodium aureum*, when the haploid spores giving rise to these gametophytes were X-rayed. These plants can be propagated asexually. How would you ascertain whether changes in genes or in cytoplasmic plastid primordia were responsible for the mutant-plastid characteristics?

11-20. Give an example from classical Mendelian inheritance in which reciprocal crosses give different results.

11-21. How do the general attributes of paramutation differ from those of spontaneous mutation? Of induced mutation?

11-22. Why is Paramecium more advantageous than Neurospora for studies of "nuclear differentiation?"

11-23. In *Paramecium aurelia*, killers were mated with sensitives without allowing cytoplasmic transfer. The exconjugants remained killer and sensitive respectively. From the killer exconjugants half of the autogamously produced lines were killer and half sensitive; from sensitive exconjugants autogamy resulted only in sensitives. Make an interpretative diagram of these results, designating nuclear genotypes and the presence or absence of kappa.

11-24. In another mating experiment without cytoplasmic transfer, but using different parent lines, killers were mated with sensitives. Again the exconjugants remained killer and sensitive respectively. After autogamy of killer exconjugants only killers were found, and from sensitive exconjugants autogamy again resulted only in sensitives. Make an interpretative diagram of these results, showing nuclear genotypes and presence or absence of kappa.

11-25. The same kind of parents as in 11-24 were used in another mating but in which cytoplasmic exchange took place. In this experiment both exconjugants from a mating turned out to be killer. Autogamy produced no segregations, all autogamously produced lines being killer. Again make an interpretative diagram of the results.

CHAPTER 12

The Role of Genes in Development

Genes must affect the development and physiology of the organism through their control of the specific biochemistry of the cells of which the organism is built. But new complexities arise when we turn to the role of genes in the development of multicellular organisms. These are apparent in two sets of facts, which when placed side by side seem incompatible:

The multicellular individual is typically an integrated collection of very diverse cells. These cells differ visibly, chemically, structurally, and functionally. They all originate from a single cell, the zygote.

We presume that successive mitotic divisions generally provide all the somatic cellular progeny of a fertilized egg with the same chromosomal and genic constitution. If the genes in these cells are to be considered identical, how can they be responsible for the cellular differences?

There can be little doubt that variations *among individuals* are often based on gene differences, responsible for the characteristics of individual biochemistry. But how are we to explain the variations from cell to cell that exist *within an individual?*

Development Is Epigenetic, and Genes Act in Epigenetic Systems. One of the historic advances in biological thought came with the recognition that development is not a matter of simple enlargement of a preformed germ, complete in all its parts (Fig. 12-1). Rather, development is *epigenetic*. The individual is built and continuously rebuilt anew. The details of this building and rebuilding process are determined by the materials and structures present at each stage of development and by the time and manner in which new ma-

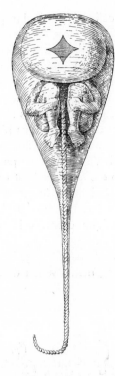

Figure 12-1. *Some early biologists thought they could see a miniature man within each sperm head.* (*After* A History of Biology, *Revised Edition*, A General Introduction to the Study of Living Things, *by Charles Singer, Henry Schuman, Inc., Publishers, p. 499; from Hartsoeker, 1694.*)

terials become available. Development, from this point of view, does not stop at some arbitrary stage in the life of the individual; it is coincident with life, and an individual that has stopped developing is dead. From the viewpoint of gene action in development, epigenesis must mean that different genes take effect at different times and places, and that their effects must depend in part on what has gone before. It all begins with the egg.

The Egg as an Organized System. Because the egg is a single cell, we may be tempted to consider it a simple, rather homogeneous blob of protoplasm except for the nucleus it contains. Nothing could be further from the truth. We know from the study of unicellular plants and animals that a high degree of internal differentiation can be achieved by single cells. Protozoa, for example, typically have visible "organelles" that perform within the single cell the specialized functions of life delegated to whole organs and organ systems in multicellular animals. There is good reason to believe that eggs are also organized into precise patterns.

Granted an inhomogeneous egg, we can observe that a division of this egg need not give identical products. The mitotic mechanism, with its virtual guarantee of *nuclear* identity in the daughter cells, offers no similar guarantee of identity in *cytoplasmic* materials. On the contrary, certain materials are regularly distributed unequally to the daughter cells at particular points in development.

The egg of a certain species of sea urchin, for example, has a subequatorial band of red pigment granules (Fig. 12-2). These granules stay in place during early development, and they can be observed and followed as indicators of the way in which the original egg cytoplasm is divided up among the daughter cells when the egg cleaves. You will note that the first cleavage plane is so oriented as to produce two similar daughter cells, and the second divisions also occur along the symmetrical axis. The next divisions, however, are at right angles to the first two, and the pigment bands record the inclusion of the part of the original egg cytoplasm marked by the pigment into the *lower* cells of the figure, while the *upper* cells are essentially free of this material. As development proceeds, these different cell types diverge further, the derivatives of each part of the original egg cytoplasm regularly making unique contributions to the developing individual.

It is possible, simply by shaking the embryo at the four-cell stage in calcium-free sea water, to separate the cells without injuring them. When this is done

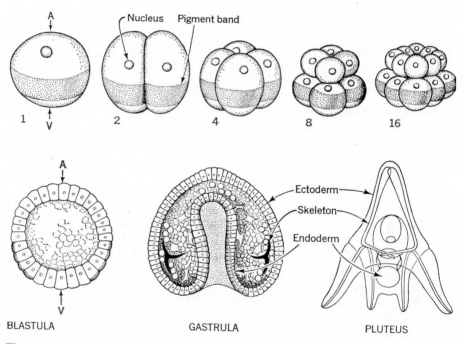

Figure 12-2. *The egg of a certain species of sea urchin has a subequatorial band of pigment. This can be traced through successive cleavage divisions, and even through the blastula and gastrula stages until the embryo becomes the type of larva called the* pluteus. (*After Barth,* Embryology, *Dryden, 1949, p. 26.*)

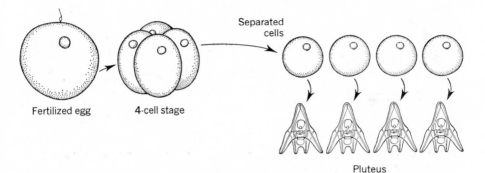

Pluteus

Figure 12-3. *Each of the four cells formed by the first two cleavage divisions in the sea urchin, when separated from the others, can form a complete embryo. (After Barth,* Embryology, *Dryden, 1949, p. 8.)*

(Fig. 12-3), each of the four isolated cells develops into a sea urchin larva that is normal except for being about a quarter of the usual size.

In contrast to this experiment, eggs can be cut in two at right angles to the symmetrical axis, and the "top" (*animal*) or "bottom" (*vegetal*) halves fertilized separately. (The position of the nucleus varies in different eggs, so it can be included in either half when the egg is cut.) The embryo that results from fertilization of the animal half (Fig. 12-4) is unable to develop the internal tissues characteristic of the normal larva; it becomes a hollow ball of cells with long cilia, with which it swims about for many days, but then it dies. The embryo that develops from the fertilized vegetal half of an egg is also incomplete and abnormal. However, it differentiates much more than, and becomes very different from, an embryo deriving from the animal half (Fig. 12-5).

The observations illustrated in Figures 12-3, 12-4, and 12-5 appear to tell a consistent story. A sea-urchin egg is organized in some material fashion, along

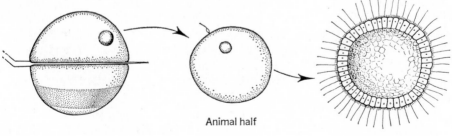

Animal half

Dauerblastula

Figure 12-4. *If a sea-urchin egg is cut in two at right angles to the symmetrical axis, the top (animal) half when fertilized develops into a hollow, ciliated ball of cells. (After Barth,* Embryology, *Dryden, 1949, p. 29.)*

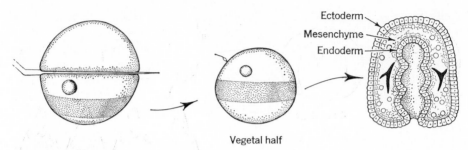

Figure 12-5. *The vegetal half of a sea-urchin egg, under the conditions of Figure 12-4, develops into a complex but incomplete embryo. (After Barth,* Embryology, *Dryden, 1949, p. 29.)*

a central axis running from the vegetal to the animal pole. This axis is at right angles to the pigment band. The first two cleavage planes run along this axis, and the four cells formed are essentially alike with regard to their visible contents, and like the egg in their organization. It need not be surprising, therefore, that any one of these four cells can give rise to a qualitatively normal embryo. The next divisions in normal development, however, are at right angles to the first two, and the resulting daughter cells are visibly different in their cytoplasmic contents. Different materials are included in the cytoplasm of the two daughter cells of these divisions, and these materials appear to play a determining role in the direction that these cells and their cellular progeny will take in differentiation.

Genes and the Organization of the Egg Cytoplasm. Our simplified discussion of the cytoplasmic organization of a sea-urchin egg is a very small sample of the work in descriptive and experimental embryology bearing on this subject. The eggs of different species differ greatly in the nature and degree of inhomogeneity so far detected in their cytoplasm. This briefly outlined example, however, is sufficient to help us understand some facts about development. It suggests that during cleavage various materials of the egg cytoplasm are separated into different presumptive parts of the developing embryo. At the same time, the immediate environments of cells in different parts of the embryo are becoming diverse. These facts offer a reasonable basis for understanding how the cells of the developing individual, assumed to be genetically identical, may nevertheless diverge in form and function.

If the organization of the egg cytoplasm has an important role in initial development, then the mechanisms through which this precise organization is achieved become important to us here. Such experimental criteria as exist suggest that the genetic constitution of the female producing the eggs plays a significant part in determining the pattern of this organization. One of the most pertinent examples is the "maternal effect" on the direction of coiling in snail shells described in Chapter 11. You will recall that a pattern appears to be imposed upon the egg by the maternal genotype, perhaps during the time when the primary oöcyte, having the same nuclear constitution as the somatic cells, is developing under the control of its own nucleus. This inherent pattern is expressed in the planes of the first two cleavage divisions, which determine whether the developing shell will coil dextrally or sinistrally. It is not affected by the genotype of the sperm that functions in fertilization of the egg, or by the genotype of the zygote.

Examples of precise genic effects on the organization of egg cytoplasm as clear as that of snail-shell coiling are rare. This is to be expected, because most mutations causing sharp changes in egg organization would have profound early effects on the developmental patterns of the embryos, and would be likely to result in embryos so aberrant as to be inviable. Perhaps the patterns are affected by numerous genes with individually small effects, and therefore are not often subject to analysis as qualitative characteristics. Perhaps also, as some recent embryological studies seem to indicate, egg inhomogeneities are much less important than other factors in triggering developmental pathways.

Inhomogeneity of Environment. As successive divisions increase the number of cells formed from the fertilized egg, the immediate environment of some of these cells will differ from that of others. For example, some will be on the surface, with relatively free access to oxygen, water, salts, and other materials reaching them through the egg membranes, and with similarly free access to a large external reservoir for the disposal of soluble wastes. Others will be covered by the surface cells and buried among their neighbors, dependent on them for whatever external materials may be passed on to them. The cells below the surface, on the other hand, will often have more generous supplies of the stored materials of the yolk. It is easy to conceive that the metabolic mechanisms established in these interior cells, under conditions of limited oxygen but virtually unlimited food supplies, are likely to be quite different from those established at the surface. Somewhat similarly, in the seed plants, physiological gradients are set up along lines radiating from the vascular supply to the developing seed.

So diverse are the environments in which different plant and animal eggs develop, and so varied are the kinds of egg organization and the patterns followed in early development, that it would be difficult to generalize about the nature of the environmental gradients that are established, or about the effects of these gradients.

Developmental Interactions of Cells and Tissues. Materials elaborated by one group of cells may diffuse to adjacent cells, or be transported by the circulation to distant tissues. Such materials play an important part in regulating the development of different tissues and organs with respect to each other, and thus in integrating the normal development of the individual as a whole.

Consider the example shown in Figure 12-6. Here a region of an amphibian embryo in the early *gastrula* stage is transplanted to abnormal positions in older embryos. You should note, first, that the region involved would normally form the eye as the embryo developed (as shown in A). If this tissue is simply removed from the embryo and permitted to develop by itself (B), it forms only a ball of undifferentiated tissue. Transplanted to various strange positions in older embryos (C, D, E), it forms a variety of complex, organized tissues. *What it forms is characteristic of the region to which it is transplanted.* Evidently, the "presumptive eye" region of the very early embryo is not fixed in terms of its

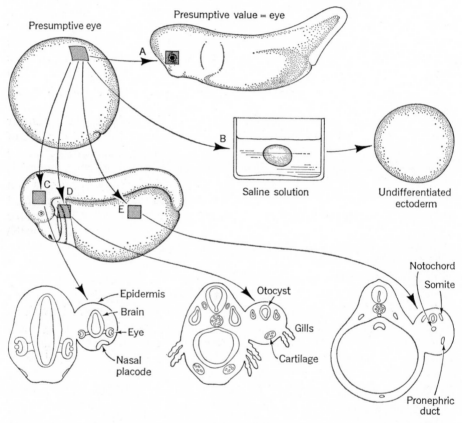

Figure 12-6. *A region of an amphibian embryo in the early gastrula stage will later normally develop into the eye (A). If this region is removed, and allowed to develop by itself, it does not differentiate (B). If it is transplanted to strange positions on older embryos (C,D,E), it can produce various organs and tissues, depending on where it is put. (After Barth,* Embryology, *Dryden, 1949, p. 74.)*

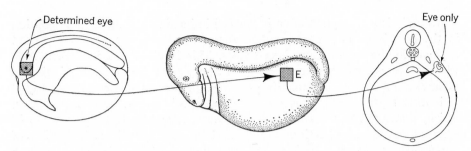

Figure 12-7. *By the time the embryo reaches the* neurula *stage* (*much later in development than the* gastrula *of Figure 12-6*)*, the eye is determined. Transplanted to position* E *on an older embryo, the eye region goes on to form an eye, regardless of its strange position.* (*After Barth,* Embryology, *Dryden, 1949, p. 77.*)

potentialities; it is capable of responding in various ways to the influences surrounding and permeating it.

This broad plasticity is lost as development proceeds; pieces of later embryos transplanted to foreign positions may develop only according to their determined patterns. For example, in Figure 12-7, the eye region of an embryo in the *neurula* stage, when transplanted to position E on an older embryo, develops only into an eye. This is in contrast to the result in E of the preceding figure, where the transplant came from the much younger gastrula stage.

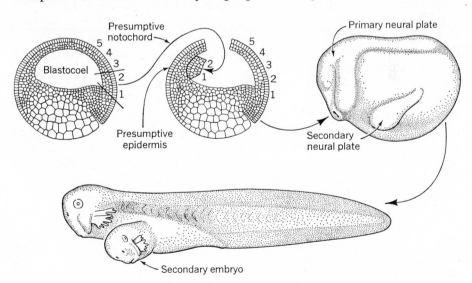

Figure 12-8. *A particular piece of one early embryo, transplanted to another, induces its host to form a secondary head in that region.* (*After Barth,* Embryology, *Dryden, 1949, p. 78.*)

Not only is a transplant often influenced in its differentiation by its position in the new host, but the transplant sometimes influences profoundly what happens in the surrounding host tissues. This is illustrated in Figure 12-8, where a piece of one very early embryo, transplanted into a strange region in another embryo of the same age, induces the formation of a "secondary embryo" in its new host. Most of this secondary embryo is composed of host tissues, which, under the influence of the transplant, are caused to differentiate in complex, organized directions not normal for their positions. This kind of interaction among embryonic tissues, whereby one group of cells evokes and appears to control the differentiation of other groups of cells, is called *embryonic induction*.

The particular kind and extent of induction accomplished by a transplant typically depend on the embryonic region from which it came, as well as on the region to which it is transplanted. The striking result just discussed, in which a secondary embryo is induced, characterizes a transplant from a special region of the early embryo called the *organizer*. This region can be subdivided, so that one can distinguish within it a "head organizer," which induces a head, and a "trunk organizer," for example. Other embryonic regions are capable of inducing other kinds of host differentiation.

The type of interaction among developing tissues we have been discussing is most important to the normal development of the individual as an integrated whole. We learn about these interactions, for the most part, by experiments that interfere with their usual course—for example, by putting tissues in embryonic environments ordinarily strange to them. This helps us to understand normal epigenetic development, in which the progressive narrowing of potentialities and the adjustment through mutual interaction proceed with regularity toward the construction of an organized individual.

Our perception of this basis for differentiation also helps us to understand other aspects of embryology and evolution. For example, human embryos, like those of other mammals, birds, and reptiles, go about the development of the adult kidney in a curious fashion. They begin by forming a fleeting vestige of an excretory organ called the *pronephros*, which probably does not function for excretion in the human embryo, but which is clearly comparable to the primitive excretory organ of the lowliest vertebrates. In the roundmouth eels and hagfishes, the pronephros is the functional excretory organ of the young, and it is partially retained by the adult. It is also found in fishes that pass through a larval period, and in the tadpoles of amphibia.

In the embryos of higher vertebrates, the pronephros is succeeded by the *mesonephros*, the functional kidney of adult fish and amphibia. In reptiles, birds, and mammals this structure is largely replaced, in turn, by the kidney (*metanephros*). Some of the parts of the mesonephros persist to become parts of the adult urogenital system.

Why is our species compelled to build, during embryonic development, two successive excretory systems that do not function in excretion, before it constructs the functional kidney? Why do we retain this record of our evolutionary history in our current development? Evidence is accumulating that the pronephros and the mesonephros successively act as essential organizers for later developmental processes in their regions. Our species has not "learned" to build a kidney or other associated structures in this region *de novo*, but only to superimpose them on an ancestral structure. We have shortened the development of the pronephros, for example, to its minimum essentials, but some kind of pronephros remains an important prerequisite to our later normal development.

We will consider this subject from a more strictly genetic standpoint later. Meanwhile, we should mention one other important category of integration already familiar to you. Our discussion up to this point has implied that the inductive and organizing interactions among groups of cells depend on the diffusion of regulating materials from *adjacent* tissues. In the *endocrine system*, however, proximity of interacting tissues is unnecessary. "Glands of internal secretion" elaborate materials that are transported to distant parts of the individual *via* the circulation, and these *hormones* evoke specific responses in particular, distant "target tissues." A similar system is important in the integration of plant growth and differentiation.

Nuclear Differentiation. In the preceding sections we have described how inhomogeneities within the egg and in the local cellular environments of the developing organism may play important roles in bringing about differentiation. From a genetic point of view, we would like to understand the mechanisms that lead a given gene to express itself in one tissue but not in another. In Chapter 11 we considered a variety of phenomena indicating that persistent transmissible cytoplasmic variation can account, to varying degrees, for differences between cells of identical nuclear genotype. Although these observations are suggestive, we do not know their significance in relation to developmental phenomena in multicellular organisms. In any case we may ask whether differentiation is entirely a cytoplasmic phenomenon, or do changes also occur in the nucleus itself? Are all nuclei capable of full gene expression in the absence of the restricting influence of a differentiated cytoplasm or a specialized intercellular environment?

The provocative nuclear transplantation experiments of T. J. King and R. Briggs illustrated in Figure 12-9 indicate that as development proceeds nuclei as well as cytoplasm become restricted in their capacities. Nuclei from differentiated cells of frog embryos are transplanted into eggs that have been "activated" to divide but whose egg nuclei have been removed or inactivated. An endoderm nucleus, for example, is found to be no longer capable of directing the normal development of an enucleated egg to which it is transplanted. Working with a different amphibian, John Gurdon observed that as cells differentiate, fewer complete blastulae and a higher proportion of abnormal blastulae are obtained from

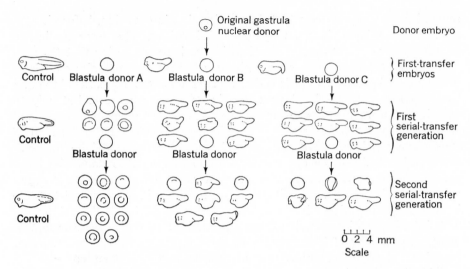

Figure 12-9. *Serial nuclear transfers in the frog,* Rana pipiens. *The first-transfer embryos developed from eggs whose own nuclei have been removed or inactivated, but which have each received a nucleus transferred from an endodermal cell of a late gastrula. These embryos achieve various types and degrees of development, suggesting that the nuclei of the different gastrula cells had differentiated, each of them providing its egg with a different development pattern. First-transfer embryos at the blastula stage provide nuclei for first serial-transfer eggs, and these in turn, when they reach the blastula stage, provide nuclei for a second serial-transfer generation. The serial-transfer embryos derived from different original donor nuclei are quite different from each other, but those derived from the same original donor nucleus are much alike. This suggests that the nuclear differences among gastrula cells are stable. (After King and Briggs,* Cold Spring Harbor Symposia on Quant. Biol., **21**:*279, 1956.)*

the transplantation of their nuclei into enucleated recipient eggs. Thus it appears that differentiation involves nuclear as well as cytoplasmic changes, and that the nuclear changes do not readily reverse when the nucleus is returned to the indifferent environment of the egg cytoplasm. There are now evidences that some of the nuclear changes encountered in nuclear transplantation experiments of this sort are gross changes in chromosome number or structure, possibly products of the experimental method rather than indications of normal developmental processes. Nevertheless, other lines of evidence also indicate that nuclear changes do occur during development. What are these changes?

Chromosomal Differentiation. As pointed out in Chapter 7, the interphase chromosomes in the salivary glands of Drosophila and other flies are polytenic, possibly consisting of 2,048 identical strands. This "amplification" of the chromonema permits its detailed observation in the optical microscope. In addition to the bands, which probably represent aggregates of replicated chromomeres, characteristic "puffs" may be seen, as illustrated in Color Plate III in Chapter 2. Unlike the chromomeres, which are constant landmarks of the chromosomes at all stages of larval development, the puffs at specific regions of the chromosomes are characteristic of particular developmental stages. They come and go as regular *temporal* features of the chromosomes. The stains shown in the color plate and other biochemical studies have revealed that these puffs are active sites of RNA and protein synthesis. In the fly Chironomus, a particular puff that normally appears prior to molting can be induced to form at earlier stages of development by the application of minute quantities of a molting hormone called *ecdysone*.

It is therefore likely that puffing is a manifestation of regional gene action. Regions of chromosomes that are not puffed are probably tightly coiled and thus inactive. The chromosomes undergo reversible changes associated with the development of the organism. These changes undoubtedly reflect functional changes.

"Looping out" of specific regions of chromosomes and associated active RNA and protein synthesis can also be seen in the so-called "lampbrush" chromosomes of newt oöcytes. It is likely that puffing is a general phenomenon. Unfortunately, in most interphase cells detailed studies of chromosome morphology and function are not possible at present.

The phenomenon of puffing suggests that differential gene action accompanying development is at least in part controlled through mechanisms that determine the specific coiling and uncoiling of regions of chromosomes. At present these mechanisms are not understood.

In Chapter 3 we mentioned another structural feature of chromosomes and chromosome parts that plays a role in differential gene activity. This is the heterochromatic-euchromatic difference. Prior to mitosis certain chromosomes or parts of chromosomes appear to condense earlier than the remainder of the chromosomal material. With isotopic labeling experiments it has been shown that such heterochromatic chromosomal regions replicate later than the normal or euchromatic material. The differences in condensation and replication are not necessarily fixed features of chromosomes or chromosome parts. For example, in most female mammals, including women, one X-chromosome is euchromatic while the other is heterochromatic. In XXX and XXXX individuals, only one X is euchromatic; all additional X-chromosomes are heterochromatic. Genetic analyses of heterozygous females in the mouse and in man indicate that the heterochromatic X-chromosome is inactive—its genes are not expressed. Each female is a phenotypic *mosaic;* in different cells either of the two X-chromosomes may be the active one. However, in any cell lineage a given X-chromosome tends to remain either euchromatic or heterochromatic. Thus this chromosomal property appears to be epigenetic in the sense of the summary to Chapter 11—it may persist for many cell generations as a stable trait.

In the mouse, genes normally located in the autosomes when translocated to the X-chromosome show *position-effect variegation*—mosaic expression associated with the new position of the gene. Evidently these genes come under the influence of the associated X region and in some cell lines become heterochromatic, but remain euchromatic and active in other cell lines. A similar phenomenon has been known in Drosophila for many years. Genes normally located in euchromatic portions of the chromosomes when transposed through translocations or inversions into heterochromatic chromosomal regions nearly always show position-effect variegation.

The two phenomena, puffing and heteropyknosis, are different features of chromosomes which dramatically illustrate that chromosomes are not just inert gene strings but are elaborate cell organelles. Like the rest of the cell, chromosomes undergo profound changes accompanying differentiation. These changes probably play critical roles in the differential expression of blocks of genes through the course of growth and development. You will find much material for thought at this point, by referring back to the sections on episomes, controlling elements, paramutation, nuclear-cytoplasmic relationships, preformed structure, and epigenetic systems in Chapter 11.

The Operon Model. Although we have suggested that puffing and heteropyknosis play important roles in the control of gene action during development, we do not yet know if these phenomena are causes or consequences of differential gene action. At present there is no clear understanding of the causal mechanisms involved in these aspects of chromosomal differentiation. However, a mechanism of gene regulation described in Chapter 10 is fairly well understood and could possibly play a very important role in development. This mechanism is the operon and its regulator gene.

With the essential elements of the operon—the operator, the regulator gene, and the co-repressors or inducers—one can construct various types of hypothetical "gene circuits," which display properties at least mimicking many of the phenomena that must occur during development. One such "circuit" is shown in Figure 12-10.

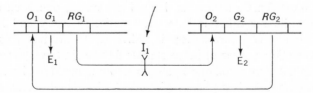

Figure 12-10. *An operon circuit that would switch the production of one enzyme off and another on.*

In this model, with operon 1 active, operon 2 is repressed from activity by the product of the regulator gene RG_1, and the cell is making only enzyme E_1. Transient contact with an inducer I_1, introduced from outside the system, temporarily blocks the repressor and operon 2 becomes active. Enzyme E_2 and repressor from regulatory gene RG_2, which blocks expression of operon 1, are now made. As a consequence the cell now makes only enzyme E_2; operon 2 is now permanently "locked on" and operon 1 "locked off." This simple circuit has the property of showing an apparently irreversible change in gene expression caused by contact with an exogenous inducer.

At present it is not known whether such mechanisms do in fact exist in organisms other than bacteria or whether they are used in any systems other than those that control biosynthetic pathways of metabolism. We can see, however, that the presence or absence of the co-repressors or inducers plays a critical role in triggering metabolic systems. We have already emphasized that development is largely determined by inhomogeneities in the environment and perhaps in the cytoplasm. These may finally be reflected in differential concentrations of various co-repressors and inducers. As in the model (Fig. 12-10), some such circuits have the property that certain gene groups are permanently turned off, accounting for the irreversibility of some developmental processes.

The operon model suggests a way in which much of the orderly sequence of development, the "turning on and off" of genes in space and time, could in fact be encoded in the genome, much like a computer program. It is possible that the operon type of mechanism is restricted to bacteria, whose genetic material does not go through a condensation phase comparable to that of chromosomes of higher organisms and thus does not have available the regulatory mechanisms associated with puffing and heteropyknosis. On the other hand there may exist many different hierarchical mechanisms for programing development at the gene level. This is clearly an area of research from which much is to be learned in the future.

Morphogenesis—The Differentiation of Complex Structures

Having considered processes by which cells with the same genetic complement can come to be different from each other, we are now in a position to examine some of the ways in which initial genic effects in particular tissues may have widespread consequences for other components of the developmental pattern. We shall survey this subject by selecting four categories of such effects.

Genic Effects on Systems of Embryonic Induction. There are two main ways in which genes may affect the integration of the developmental pattern accomplished through organizers or similar inductive mechanisms. These depend on the fact that there are two partners in any such system—the tissue that is doing the inducing and the tissue that is being induced. Thus, on the one hand, gene-controlled deviations from the normal biochemical pattern in a developing organizer may have marked effects on the ability of this tissue to organize normally the regions around itself. On the other hand, genes may affect the *competence* of tissues to respond normally to the organizing stimuli. We will describe an example only of the first of these alternatives here. Genetic effects of the second kind are most clearly illustrated through the closely similar changes in the competence of tissues to respond normally to hormones, and will be considered later.

For an example of an apparent organizer effect, we may consider one aspect of the action of a semidominant lethal in the mouse. Mice homozygous for "Danforth's short-tail" (*Sd*) have no tails at all; neither do they have a rectum or an anus, or urethra, or genital papilla; they usually lack kidneys, too, and several vertebrae of the lower part of the spinal column; and they may show a variety of additional abnormalities. When they are born alive, they die on their birthday.

The allele gets its name from Danforth's discovery of it in heterozygous mice, where its most regular and conspicuous effect is shortening or absence of the tail. We will confine our immediate attention to less conspicuous, but physiologcally more important, aspects of the heterozygotes, their kidney and ureter anomalies. Figure 12-11 shows sketches of fourteen arbitrary classes of such

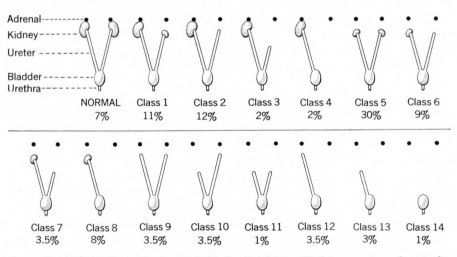

Figure 12-11. *Mice heterozygous for "Danforth's short-tail" show a variety of anomalies of kidney and ureter structure. (After S. Glueksohn-Schoenheimer, in* Genetics of the Mouse *by H. Grüneberg, Cambridge University Press, 1943, p. 202.)*

anomalies, together with the frequencies of these classes among 109 individuals of genotype $Sd/+$. You will note that 7 percent of these heterozygotes, in the population studied, appeared normal with respect to their kidneys and ureters, while the remainder showed varying degrees of reduction in size or even complete absence of one or both kidneys or ureters.

If you examine this figure carefully you will observe that several possible classes of abnormality are conspicuous by their absence. You do not find a kidney hanging in the figure by itself, as the adrenals may; wherever there is a kidney, even a small one, a complete ureter extends to it. Whenever a ureter fails to reach the kidney region, the corresponding kidney fails to develop.

In brief, developmental studies on *Sd* heterozygotes strongly suggest that the ureter, which arises as a bud from the mesonephros, acts as a kind of organizer in inducing the formation of the metanephric part of the mammalian kidney. An early effect of the *Sd* allele in heterozygotes is interference with the normal elongation and branching of this mesonephric bud. The frequent reduction or absence of the kidney results from the failure of this organizing tissue to make adequate contact with the tissue it normally induces to form the capsules and secretory tubules of the kidney.

The other effects of the *Sd* allele, particularly in homozygotes, suggest that it may interfere much earlier in development with a chain of inductive relationships necessary to the eventual normal development of the whole posterior part of the embryo. Grüneberg has emphasized the importance of such observations in understanding similar human anomalies of medical significance.

Genic Effects on Endocrine Systems. Dwarf mice homozygous for a simple recessive appear to grow normally for a short while after birth, but their growth rate soon decreases, and by about the twelfth day they have stopped growing. They begin to lose weight, and some of them die. A little later, the survivors seem to reach some kind of equilibrium and begin to grow again slowly. They eventually reach about a quarter of the normal adult weight. Their bodily proportions differ somewhat from normal—for example, their tails and ears are shorter. They are slow and inactive; their basal metabolic rate is about 60 percent of normal for mice their size. They can live longer without food than can normal mice. Both sexes are sterile. Their thyroid, thymus, ovaries or testes, and adrenal glands are infantile. Their pituitary glands are unusually small and lack a particular normal cell type, the eosinophilic cells.

Pituitary glands from normal mice can be implanted beneath the skin of dwarfs. When this is done, the dwarfs begin to grow more rapidly; they may soon reach normal size. Their thyroids, adrenals, and other infantile organs become normal, and males often become fertile.

Clearly, the essential defect of a dwarf mouse is in the "master gland" of his endocrine system, his pituitary. His many other defective tissues, glands, and organs prove themselves competent to respond to the secretions of a normal pituitary. The critical block produced by this recessive gene is evidently in the development of a particular cell type of the pituitary gland. Because this kind of cell is responsible for a key group of hormones, the consequences of this block are extensive and profound.

Numerous essentially similar examples of gene action on endocrine glands and on the production of diffusible or circulating substances might be cited. In human beings, for instance, certain extreme, inherited deviations from the normal range of stature seem to be endocrine in origin. A common type of sugar diabetes results from a primary failure in the hormonal function of the pancreas. *Cretins*, typically stunted both physically and mentally, result from thyroid malfunction.

Another important aspect of genic effects on endocrine systems remains to be discussed. Genetic differences may not only modify the kind or amount of hormones produced by the endocrine glands; they may also affect the sensitivity and the nature of the response of target tissues to particular hormones.

As an example, we may consider a common difference between male and female fowls. Cocks often show their sex in the structure of many of their feathers—the long, narrow, pointed feathers of hackles and saddle; the pointed feathers of cape, back, and wing bow; the long curving pointed "sickle feathers" of the tail; and so on. Hens usually show the contrasting feather characteristics; their feathers are short, broad, blunt and straight.

A great deal of work has been done in studying the control of these sexual feather characteristics. It is evident that the endocrine system is involved; for example, the administration of female sex hormones, or of thyroid or pituitary extracts, can evoke the development of hen feathering in a male. Patches of potentially cock-feathered skin transplanted to a hen ordinarily develop hen feathering—clear evidence that here it is the internal, endocrine environment that distinguishes the sexes. Similarly, patches of hen skin transplanted to a cock ordinarily become cock-feathered, and the removal of the functional ovary of a hen results in modification of her feathering in the direction characteristic of the cock.

There are some breeds of fowl in which there is no normal sexual difference in feather character of the sort we have been discussing. The males as well as the females are hen-feathered. In other breeds, there are some strains with cock-feathered males and others with hen-feathered males. This difference in the roosters is inherited, and can most simply be explained on the basis of a single pair of alleles, such that males of genotype *HH* or *Hh* are hen-feathered, while *hh* males are cock-feathered. Females of any genotype are normally hen-feathered.

How does the *H* allele act to control hen-feathering in cocks? T. H. Morgan, who pioneered in this work, as in Drosophila genetics, believed that the hen-feathered cocks were "endocrine hermaphrodites"; that their testes secreted female hormones responsible for their hen-feathering. This concept was based on Morgan's observation that hen-feathered Sebright bantam males, when castrated, became cock-feathered; it seemed clear, therefore, that their testes had been responsible for their hen-feathering.

Later work has suggested a different conclusion. It was found that a caponized Sebright cock that had become cock-feathered, as in Morgan's experiment, again became hen-feathered after implantation of a Leghorn testis. Leghorn males are cock-feathered. Evidently there was no significant difference between the testis of a normally cock-feathered male and that of a hen-feathered male with respect to its action on the Sebright feather development. Similarly, a castrate Leghorn male continued to be cock-feathered even after implantation of a Sebright testis.

Skin grafts studied by Danforth completed the story. He found that transplants of skin from a hen-feathered breed (*H_*) to cock-feathered hosts (*hh*) continue to develop hen feathers. This is in contrast to the results previously cited, where transplants between cock-feathered males (*hh*) and hen-feathered females (*hh*) acquire the sexual feather characteristics of their hosts.

We can summarize the situation briefly, simplifying it somewhat. Breeds that are homozygous *hh* show sexual differences in feathering that depend on the endocrine differences between males and females. Here, the genetic differences in feathering between the sexes involve mainly the effects of the sex chromosomes on the kinds and relative amounts of hormones developed, and the sexual differences in feather structure result from these initial genetic effects on endocrines. Differences between *hh* and *HH* or *Hh* roosters, on the other hand, in which those having the *H* allele are hen-feathered while *hh* males are cock-feathered, result from a genetic effect on the way the feather-producing tissues respond to hormones. The initial genetic effect here is in the skin itself, as a target organ, and not in the endocrine glands.

Genic Effects on Migrating Cells. Our discussion has dealt with genetic effects on substances that are elaborated in one part of the developing individual and subsequently affect other parts. Another rather common event in development is the differentiation, in particular regions, of whole cells that later migrate to other regions to become part of, and to affect the character of, other tissues. Examples are the *melanophores*, which may migrate from a region near the embryonic nerve cord to developing skin, hair, or feathers all over the embryo, and which produce the pigmentation of these tissues. Similarly, in many forms, the cells that are to become the ancestors of the germ cells migrate into the prospective genital regions from tissues far outside this region; and the red blood cells of adult animals evidently descend from circulating embryonic cells that later "settle down" in the various blood-forming tissues of the individual. All three of these characteristics—hair pigmentation, fertility (germ-cell number), and red blood-cell anemia—are affected by the alleles of one particularly interesting gene, *W*, in the mouse, studied in detail by Elizabeth Russell. Genes that affect wandering cells or the patterns of their migration have effects far from the original time and place at which the initial genic effects occur.

Genic Effects on the Regulation of Growth and Metabolism. We will consider two additional ways in which a gene may affect the regulation of growth and metabolism in the developing individual. One is to modify a metabolic process of direct importance to the whole organism. The second is to affect a metabolic process to some degree characteristic of a particular region at a particular time, and thus to change selectively the growth of this region relative to other parts of the individual.

As an instance in the first category, consider Figure 12-12, which contrasts a normal Drosophila larva with one showing the "meander" lethal characteristic discovered in 1943 by E. Hadorn. This third-chromosome lethal gets its name from the wandering, curving course taken by the tracheae in the affected larvae.

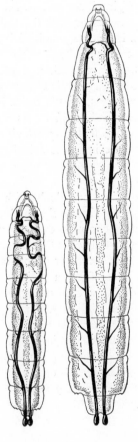

Figure 12-12. *A Drosophila larva homozygous for the "meander" lethal (left) is small in contrast with a normal larva its age, and its tracheae meander. (After Schmid, Z. Indukt. Abst. Vererb., **83**: 224, 1949.)*

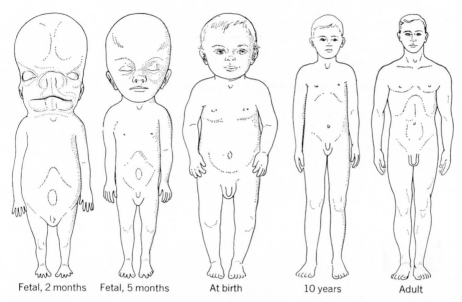

| Fetal, 2 months | Fetal, 5 months | At birth | 10 years | Adult |

Figure 12-13. *Different parts of the human body grow at different rates.*

Larvae homozygous for the lethal cannot become pupae; they start to grow normally, but stop by the third day, when they have reached about half the normal size of larvae at pupation time, and they die as dwarf, misshapen larvae.

W. Schmid has subjected the development of the meander lethal to detailed study. The effects of this gene appear to be quantitatively comparable to the effects of starvation on normal larvae. Furthermore, transplantation studies have indicated that a block in the assimilation of protein is responsible for the lethal characteristic. This would appear to be a general block indeed, but it is evident that its effects are not the same on all organs of the larva. A very few tissues, like the tracheal trunks, continue to grow after the third day; others achieve only a fraction of their normal size. This fraction is characteristic for each organ; for example the salivary glands are about one-third the size, while the testes and ovaries are about two-thirds as large, as those of normal larvae when they are four days old. The kinds of internal compensation that make it possible for certain organs to continue growing for some time after a general block has stopped the growth of others are not well understood at present.

A somewhat similar genetic effect involves a modification of a general *regulatory mechanism*, like the control of body temperature. The frizzle fowl introduced in Chapter 1 is an example.

Most of us have observed, in a general way, that the various parts of the human body grow at different rates. The sketches of Figure 12-13 illustrate this familiar fact. Growth of the head is extremely rapid in early development.

while the trunk and appendages are later in starting, and relatively slow. If these relative rates persisted, adults would be monsters indeed by our present standards, with gigantic heads and dwarf limbs and trunks. As it is, however, the rapid relative growth rate of the head slows down toward the end of fetal life, and increasing proportions of the total growth are channeled into trunk and appendages.

Many of the details of how a person looks are governed by relative rates of growth. Consider noses as conspicuous examples. Most infants seem to be more or less "pug-nosed." By adolescence, however, the nose has usually become a rather large and dominant feature. During the interval between infancy and adolescence, the nose typically enjoys a high growth rate, relative to the growth rates of other facial components.

In line with the two general categories of genetic effects on the regulation of growth phrased at the beginning of this section, we can discern two kinds of effects on nose growth. The first involves the overall growth of the individual. If his general growth slows down or stops unusually early, the proportions characteristic of earlier developmental stages may prevail. Thus, midgets usually have noses that are small, relative to the rest of their features, when compared with ordinary adult proportions in this regard. In the other direction, unusually prolonged growth, extending the period during which a relatively large fraction of total facial growth is diverted to the nose, typically results in disproportionately large noses. Thus "gigantism" in man is frequently associated with "coarse features," in which a relatively big nose plays an important part.

It is therefore apparent that genes affecting the final size that the individual reaches also secondarily affect the relative proportions of his parts. The initial genic action may be an effect on total growth, and the differences in proportions may derive from the fact that various relative growth rates prevail for different parts of the individual.

On the other hand, particular genes sometimes single out for their individual effects particular components of the network of relative growth rates. This would explain why some small people have relatively big noses, while some large people have relatively small ones.

Numerous studies, much more quantitative than our discussion of noses, have indicated that both types of genetic regulation of growth and metabolism described in this section are important to the patterns of development and evolution.

Sex Determination and Sex Differentiation. The processes leading to formation of sex differences illustrate and extend many of the principles developed in this chapter. These processes and their bases are diverse in different forms of life. You will recall that in mammals sex appears to be *determined* at fertilization by whether the zygote has a Y-chromosome; XY = male, XX = female. XO zygotes develop into females, more or less functional depending on whether one is considering the mouse or man, while XXY zygotes develop into phenotypic males. The Y-chromosome therefore acts as a trigger to direct development along one of two very different pathways—the pathways of sex differentiation; in the absence of the Y-chromosome the organism can only travel down the other pathway.

The initial pathways of sexual development are alike in the two sexes. In higher vertebrates, large *primordial germ cells* originate in extra-embryonic tissues and migrate to the region where the gonads will form. Genes can affect the numbers of these cells and their migration, and can produce sterility in either sex. Once in the epithelium of the presumptive gonad, the primordial germ cells participate in the first definitive event of sex differentiation: cords of cells, the medullary sex cords, proliferate from the epithelium. In the male the primordial germ cells are carried down with these cords, to become the germ cells of the testis. In the female, the primordial germ cells remain behind in the epithelium, and a second proliferation forms the *cortex* of the ovary with its germ cells. Evidently, some characteristic of the primordial germ cells interacting with the proliferating cord cells is affected, in mammals, by whether the cells have or do not have a Y-chromosome.

The later processes of sex differentiation are similar in principle to examples from other systems already discussed in this chapter. The developing gonad has organizing effects on other tissues in its neighborhood, and both its qualities as an organizer and the competence of other tissues to respond are affected by the genomes of the cells involved. Hormones come into conspicuous play; the medulla secretes a hormone that suppresses cortical development, while the cortex produces a hormone that may act to suppress medullary development. Specialized cells differentiate in the gonad to produce sex hormones, and these in turn interact with endocrine cells elsewhere in the body—in the pituitary, the adrenal, the uterus, even in the placenta and the fetus when a female is pregnant.

Even tissues genetically determined as female can differentiate in the direction of maleness if acted upon by male hormones. This situation arises, however, under a system of sex differentiation that is not universal for the animal kingdom. In insects, the key factors in sex differentiation are apparently intracellular rather than hormonal. Occasionally you can find Drosophila, for example, that show a sharp mosaic of sex characters. These individuals, typically male in certain portions of the body and typically female in others, are called *gynandromorphs*. The mosaic patterns often involve the sex combs, the color patterns of the abdomen, and body size. Some individuals have both male and female gonads and genitalia. These flies, when looked on as whole organisms, can scarcely be designated either as male or female. Their own state of confusion is shown vividly by the anomalies frequently displayed in their courtship patterns.

The origin of gynandromorphs in Drosophila has been successfully explained in relation to the XY mechanism for sex determination. In Drosophila as in mammals, XX individuals are females and XY individuals are males. But in contrast to mammals, XO Drosophila are males, although sterile. You will recall that once in a great while a chromosome gets lost at the time of nuclear division. If the chromosome lost is an X-chromosome in a somatic cell of a developing Drosophila determined at fertilization as female (XX), the daughter cell missing the chromosome will be male in chromosomal constitution (XO), and the other daughter cell will remain female (XX). Assuming the divisions to be regular thereafter, nuclear progeny of the former will reproduce a male chromosome complement and descendants of the latter will continue XX. The *when* and *where* of the chromosome loss determine the size and position of the male sector in the mosaic.

Beautiful genetic demonstrations that chromosome loss is involved in gynandromorphism may occur when the original fertilized egg cell is heterozygous for one or more sex-linked characters. If the X-chromosome eliminated is one carrying the dominant alleles of heterozygous gene pairs, the recessive characteristics may be expressed in the male areas from which this chromosome is missing. A diagrammatic representation of the origin of gynandromorphism and of certain of its possible phenotypic consequences is shown in Figure 12-14.

The Balance Concept of Sex. A number of lines of evidence indicate that in many groups of organisms, even ones in which males and females are ordinarily separate, all individuals have genes for both sexes. One such indication comes from instances where genetically determined males are influenced by female hormones, or vice versa. In extreme instances, almost complete sex reversal may take place. A notable chicken, reported by Crew, began life as a female, climaxing her existence as a hen by laying fertile eggs. Subsequently, "she" took up another career, evidenced first by the appearance of a male comb and the sounding of cock calls, and functioned successfully as a father. The explanation of this transformation is that in the originally normal hen the ovary was destroyed by disease. Loss of the ovary removed the inhibitory effect this organ seems to exert on the rudimentary testis found in all female birds. The testis developed, and presumably secreted male hormones, which then influenced sex differentiation in the direction of maleness.

Another indication that single organisms generally have inherent potentialities for both sexes is found in the existence of *intersexes*, individuals more or less intermediate between male and female. The Greeks had words for six sexes in people. No objective evidence bears out the number of sexes they recognized, but the implication of varying degrees of masculinity and femininity is quite in conformity with modern views. The consensus among biologists today is that there is a *balance* of male and female tendencies in the hereditary complement of an individual, and that mechanisms like the XY ordinarily serve to trip the balance in one direction or another. Sometimes, however, other factors enter onto the scales and alter or reverse the effect of the normal mechanism for sex determination.

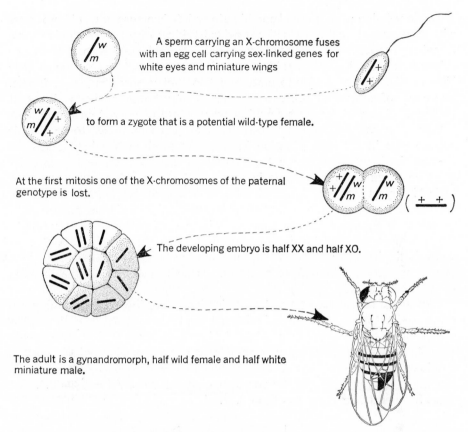

A sperm carrying an X-chromosome fuses with an egg cell carrying sex-linked genes for white eyes and miniature wings

to form a zygote that is a potential wild-type female.

At the first mitosis one of the X-chromosomes of the paternal genotype is lost.

The developing embryo is half XX and half XO.

The adult is a gynandromorph, half wild female and half white miniature male.

Figure 12-14. *The origin of gynandromorphism in Drosophila, and some of its possible phenotypic consequences. If the X-chromosome lost had been the one carrying the mutant alleles, the male portion of the adult would have been phenotypically wild type. If the chromosome elimination had occurred some time after the first mitotic division of embryonic development, a correspondingly smaller portion of the individual would have been male.*

Intersexes and Supersexes in Drosophila. Impressive support for the balance concept of sex comes from work by C. B. Bridges with Drosophila. When triploid females are crossed to normal males, some of the progeny are aneuploids, as a result of irregularities at meiosis in the mothers. Not all of the aneuploid types that might arise from such a mating are recovered, since many of the combinations are inviable. Of those that do survive, flies with two X-chromosomes and three of each kind of autosome turn out to be *intersexes*. They are variable in appearance, in general having complex mixtures of male and female attributes for the internal sex organs and external genitalia, without showing clear regional mosaicism.

The correlation of an unusual combination of X-chromosomes and autosomes with an unusual brand of sexuality provided Bridges with the foundation for a meaningful concept of the basis of sexuality in Drosophila. This concept may be briefly summarized as follows: Two sets of autosomes and one X-chromosome give a male. Two sets of autosomes and two X-chromosomes give a female. Three sets of autosomes and two X-chromosomes, however, give an intersex! Differences between the sexes would appear, then, to be based on proportions between X-chromosomes and autosomes. Y-chromosomes seem to be of little consequence in sex determination in Drosophila, since XO individuals are phenotypically male. Moreover, the unusual presence of Y's in female chromosome complements, for example in attached-X stocks, does not alter sex. A logical interpretation of sex determination in Drosophila is that X-chromosomes carry genes that are predominantly female-determining, while the predominant tendency among genes in the autosomes is toward maleness. Under normal circumstances the additional X-chromosome found in females is a decisive factor in tipping the balance toward feminine development. In individuals with one X-chromosome and two sets of autosomes the balance hangs in the direction of maleness. (See Fig. 12-15.)

For ease of summarization and analysis, the proportion between X-chromosomes and autosomes in a given individual can be formulated as follows. If X stands for an X-chromosome, and if A signifies a set of autosomes, the formula $2X\ 2A$ designates the chromosome complement of a normal diploid female. As a ratio of X-chromosomes to autosomes this can be expressed as $X/A = \frac{2}{2} = 1$. For males the derivation of the corresponding value is $X/A = \frac{1}{2} = 0.5$. The value for an intersex is $X/A = \frac{2}{3} = 0.67$.

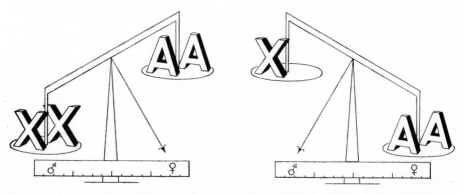

Figure 12-15. *In Drosophila, sexuality is a matter of balance between opposing factors toward maleness and femaleness. Under normal circumstances, the mechanism for sex determination is such that the additional X-chromosome in females tips the balance decisively toward feminine development. Where there is a single X-chromosome and two sets of autosomes, the balance is tipped toward maleness.*

The values 1.0, 0.5, and 0.67 doubtless have no deep intrinsic meaning with regard to sex determination. But by indicating proportionality between X-chromosomes and autosomes, they do serve as sex indices, thus pointing to femaleness, maleness, and the area in between. It is of particular interest to see whether still different X-chromosome/autosome relationships, and the sex types they determine, fall into an orderly scheme consistent with the sex index numbers already presented.

Triploid ($3X/3A$) and tetraploid ($4X/4A$) flies have been studied. The duplication of chromosome sets has some phenotypic consequences, but sexually these polyploids are females. The value for both $4X/4A$ and $3X/3A$ is 1. Since each of these types appears to be female, the result conforms with what might be expected on the basis of the index number of a diploid female.

The same cross, triploid ♀ × normal ♂, that gives $2X/3A$ progeny also produces $3X/2A$ and $1X/3A$ offspring. You have already been introduced to $3X/2A$ Drosophila, since they are one of the types produced by crosses involving attached-X females (Chap. 2). The $3X/2A$ individuals are designated as *superfemales*, the $1X/3A$ as *supermales*. These flies, unfortunately, scarcely live up to their glamorous names, and are weak, sterile, and poorly viable. In fact, their names are not intended to connote any particular kind of prowess, but were chosen to indicate that the sex indices in each case, $3X/2A = 1.5$ and $1X/3A = 0.33$, lie outside the range of values for normal female and normal male.

A few other aneuploid types of Drosophila have been reported. Among these, $2X/4A$ individuals turn out to be males, and $3X/4A$ flies are intersexes, results that fit nicely into the series established by the sex indices already described. Table 12-1 summarizes the relations of sex to chromosomes in Drosophila.

Table 12-1. THE RELATIONSHIP OF CHROMOSOME COMPLEMENT TO SEX EXPRESSION IN *Drosophila melanogaster*.

Chromosome Complement, Formulated as X-Chromosomes and Sets of Autosomes		*Ratio X/A*	*Type*
3X	2A	1.5	Superfemale
4X	3A	1.33	Superfemale
4X	4A	1.0	Tetraploid female
3X	3A	1.0	Triploid female
2X	2A	1.0	Diploid female
3X	4A	0.75	Intersex
2X	3A	0.67	Intersex
1X	2A	0.5	Male
2X	4A	0.5	Male
1X	3A	0.33	Supermale

The "Transformer" Gene in Drosophila. A recessive gene, *tra*, which is situated between 44.0 and 45.3 in the third chromosome of Drosophila, has the effect when homozygous of transforming diploid females into sterile males. Phenotypically the XX *tra/tra* flies are like normal males in that they have well-developed sex combs, normal male external genitalia, genital ducts and sperm pump, a male colored abdomen, and other of the masculine traits of Drosophila. The testes, however, are much reduced in size.

The *tra* gene, of course, may affect sex ratios quite drastically. For example, as seen in Figure 12-16, a cross between a heterozygous female and a homozygous male gives three males for every female in the progeny.

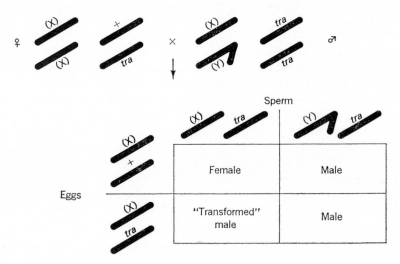

Figure 12-16. *A cross in Drosophila between a male homozygous for the* transformer *gene and a female heterozygous for* tra. *Note the sex ratio in the progeny.*

Out of an experimental cross involving *tra*, one individual was obtained that was superfemale in X-chromosome/autosome ratio and homozygous for the *tra* gene. This fly was a kind of intersex with strong tendencies toward maleness. If the single specimen is characteristic, superfemales would seem more resistant than ordinary females to modification to a male status by the *tra* gene. At least in this unusual situation, therefore, superfemales may be ultrafeminine indeed.

From our present point of view, the most significant thing about the *tra* gene is the demonstration that the normal chromosomal mechanism for sex determination can be nullified by a single gene substitution. The transformer effect is understandable when we remember that sexuality is a matter of balance, and that normal devices for sex determination are only means for weighting the scales in one direction or the other. The *tra* gene may be thought of as an extra counterweight introduced into the system for sex determination.

Balance and Sex Determination in Lychnis. In only one instance has sex determination in higher plants been studied with anything like the thoroughness applied to the corresponding problem in Drosophila. This is in the species of Lychnis, often called Melandrium. Lychnis belongs to the pink family. Its genetics and cytology, particularly with reference to sex determination, have been investigated by M. Westergaard, H. E. Warmke, and others.

The normal basis for sex determination in Lychnis is the distribution of members of an XY chromosome pair. As is true in all but a few of such instances known in higher plants (the strawberry is one exception), presence of the unequal pair in diploids determines *staminate* (pollen-bearing) plants; XX individuals are *pistillate* (egg-bearing) plants. Beyond analysis of the basic situation just described, sex expression in artificially induced polyploids and in their polyploid and aneuploid progeny has also been studied. As in mammals, the Y-chromosome of Lychnis is not the mere nonentity it appears to be in Drosophila. The Y in Lychnis is physically larger and more prominent than the X, and it definitely carries factors for maleness. Analysis of the effects of fragments of the Y-chromosome on sex expression has enabled investigators to designate some degree of localization of functions to different portions of the Y. This is seen in Figure 12-17.

Sex determination in Lychnis appears to depend upon a balance between the Y-chromosome and the X-chromosomes plus the autosomes. Upsets in the normal mechanism for balance in sexuality, resulting either from unusual

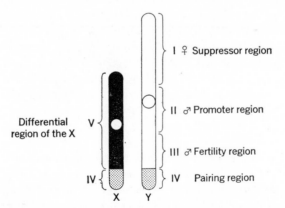

Figure 12-17. *Diagram of the sex chromosomes in Melandrium. Regions I, II, and III of the Y-chromosome do not have homologous segments in the X, and hence they make up the* differential *portion of the Y. Regions IV are homologous in the X and Y, and are pairing regions at meiosis. V is the differential portion of the X-chromosome. When I is lost from a Y-chromosome, a bisexual plant is produced. When II is lost, a female plant is produced. If III is absent, male-sterile plants with abortive anthers appear. (After Westergaard,* Hereditas, **34:***269, 1948.)*

chromosome combinations or certain outside influences, may give rise to bisexual individuals. In other respects also, sexuality in Lychnis is consistent with the general concept of sexuality being built up from several converging fields of research. There is considerable evidence that female plants carry potentialities for maleness. A most suggestive observation has been that pistillate plants infected by a particular smut fungus develop anthers.

Sex Determination in Asparagus. Garden asparagus is normally dioecious, with staminate and pistillate plants appearing in approximately equal numbers. But a tendency toward the monoecious state can be seen in the occurrence of rudimentary stamens in pistillate flowers and of rudimentary, usually nonfunctional, pistils in staminate flowers. Once in a great while, the pistils found in staminate flowers produce viable seeds, which almost undoubtedly arise from self-pollination. C. Rick and G. Hanna planted seeds originating in this way and obtained 155 staminate plants and 43 pistillate plants. The results, which closely approximate a 3:1 ratio, suggest segregation of a single allelic pair. To test this interpretation, staminate plants from the progeny indicated above were crossed with normal pistillate plants. One-third of the staminate parents produced all staminate offspring. The other two-thirds produced staminate and pistillate offspring in about equal numbers. This confirms the hypothesis of segregation at a single locus, the gene for staminate being dominant over its allele for pistillate. (As a review of Mendelian genetics, utilize gene symbols to diagram the crosses made by Rick and Hanna; verify in detail that the interpretation fits the results obtained.)

You have seen that chromosomal mechanisms for sex determination operate on the principle of the testcross, with homozygotes mating with heterozygotes to give equal numbers of two kinds of progeny. The widespread occurrence of this essentially simple sort of system testifies to its effectiveness. Precisely the same principle is utilized in asparagus, in which the segregating germinal elements that determine sex are not whole chromosomes but members of a single allelic pair.

Transformation from the Monoecious to the Dioecious State. In contrast to asparagus, corn is monoecious. The tassel is staminate; the ear is pistillate. Among the many mutant genes known in corn are several that can be used to convert corn plants into individuals that are no longer monecious but are either pistillate or staminate. For example, gene *ba* (*barren stalk*) when homozygous makes plants staminate by eliminating the ears. Gene *ts* (*tassel seed*) when homozygous converts the tassel into a pistillate inflorescence that produces no pollen. A plant *ba ba ts ts*, then, is effectively pistillate. It has no ears, but produces egg cells in the tassel. A plant *ba ba Ts Ts* is staminate. Figure 12-18 shows how one might manipulate genotypes to establish in corn a system of sex determination, complete with 1:1 ratio, analogous to that in asparagus. Such

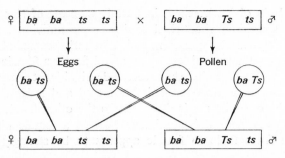

Figure 12-18. *Through the manipulation of mutant genes, corn can be transformed from a monoecious into a dioecious plant. With parents of the genotype shown, equal numbers of pistillate and staminate plants would occur in the progeny.*

an experiment is perhaps not far in principle from evolutionary processes that have actually taken place.

Sex Determination in Habrobracon. Not all genetic mechanisms for sex determination work on the testcross principle. An interesting variation is found in Habrobracon, a parasitic wasp whose genetics was studied intensively by P. W. Whiting and A. R. Whiting. Habrobracon belongs to the insect order Hymenoptera, which includes bees, wasps, ants, and saw flies. In general, in Habrobracon, the males are haploid and females diploid. More specifically, females come from fertilized egg cells, and males arise from unfertilized eggs by parthenogenesis. However, under some circumstances, diploid males do occur. It now appears that femaleness is determined by heterozygosity for members of a multiple allelic series designated as *xa*, *xb*, *xc*, *xd*, and so on. (These are not thought to be alleles in the sense of occupying a single gene locus but rather are regarded as homologous chromosome segments perhaps containing several genes.) Maleness is determined by absence of heterozygosity for members of this multiple allelic series. Thus males may be haploid, having genotypes as *xa*, or *xb*, or *xc*, or may be homozygous diploid, as *xa xa*, or *xc xc*, or *xd xd*. Heterozygous combinations like *xa xb*, or *xa xc*, or *xb xd* determine femaleness.

A possible general interpretation of the situation in Habrobracon is that members of the *x* series act in *complementary* fashion to produce femaleness. Support for this view comes from an exceptional kind of gynandromorph, called *gynandroid*, which now and then appears in Habrobracon. Gynandroids arise from binucleate eggs that have not been fertilized, and in which the two nuclei have different *x* alleles. These mosaic wasps are haploid, and therefore male. However, in the region of the genitalia, at the frontier between tissues of different genotype for *x*, small female reproductive appendages develop. A possible interpretation is that gene products peculiar to one of the *x* tissue regions diffuse for a distance into the other region and interact there with locally produced substances to give rise to female differentiation.

The Major Role of Environment in Certain Systems of Sex Determination. In the genus Crepidula (boat-shell snails), each individual normally goes through a developmental sequence in which an early asexual stage is followed by a male phase, then a transitional phase, and finally a female phase. Studies of the duration of the male phase have revealed a significant phenomenon. When individuals in the male phase are suitably mated and sedentary, their transformation to the opposite sex is deferred. On the other hand, wandering unmated "males" relatively quickly change over to a female phase. Thus the transformation from male to female appears to be strongly influenced by environment.

A related phenomenon occurs in the marine worm Bonellia. The free-swimming larval forms of this organism are sexually undifferentiated. Those individuals that settle down by themselves become females; some of the larvae attach to the bodies of adult females and differentiate as males. Here then, as in Crepidula, the direction of sex differentiation is not determined at fertilization by some genetic component that has segregated at a preceding meiosis. Instead, alternatives in sex differentiation are conditioned by environmental factors related either to association or to lack of association with other members of the species. Some plants, as well as animals, regularly rely on environmental influences to direct the course of sex differentiation. In the parasitic fungus Olpidium, for instance, environmental effects determine whether particular cells are to behave sexually or asexually in the life cycle. These examples do not imply an absence of a genetic basis for sexuality; they rather illustrate that sex, like other characteristics of developing organisms, is subject to threshold effects.

Compared to a mechanism like the XY chromosomes, the Bonellia system of sex determination may seem to be haphazard. However, as W. C. Allee has pointed out, it has the advantage of guarding against waste of the reproductive potential of isolated males. Isolated individuals in Bonellia become females. As females, they are able to determine that newcomers into their spheres differentiate as members of the opposite sex.

Complementary Genes for Sexual Fertility in Glomerella. Glomerella belongs to the same general group of fungi as Neurospora, and is well known as a plant pathogen. Almost all strains of Glomerella will mate within themselves or with any other strain. However, among many strains, differences in fertility can be detected, for example by the experimental device illustrated in Figure 12-19. As you look at the picture you will see dotted lines, which appear to divide the Petri dish into four sections like pieces of pie. The dividing lines are really rows of perithecia (fruiting bodies containing ascospores) formed at the lines of juncture of different strains of the mold being grown on agar. Mating has taken place along these lines, and the numbers of perithecia are a measure of fertility between the strains involved.

The figure illustrates what appears to be a general phenomenon in Glomerella:

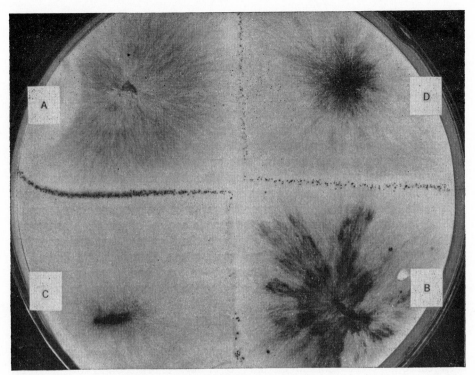

Figure 12-19. *Four different strains of Glomerella, with mating indicated by perithecial formation at the lines of juncture on the agar plate. Comparative fertility of the different crosses is indicated by the numbers of perithecia formed. For example, A × C is a more fertile combination than A × D. (Courtesy of C. L. Markert,* Am. Naturalist, **83:***228, 1949.)*

no single strain shows greater fertility than all others, in all possible crosses. For instance, strain D is more fertile than C in crosses with B; but in crosses involving A, many more perithecia are formed with C than with D. This kind of result implies that the fertility of a particular cross depends upon complementary action of different genes in the parents. That these genes may perhaps produce their effects by way of diffusible chemical substances has been demonstrated by C. L. Markert in the following way. Strains A and D normally show rather weak formation of perithecia when they are crossed. If a Petri dish is prepared in such a way that there are two layers of agar medium separated by a cellophane membrane, and strain C is inoculated on the bottom layer and A and D are inoculated together on the top, A and D, in contrast to their normal indifference to each other, cross vigorously. Strain C itself cannot penetrate the membrane. The results indicate, therefore, that C produces a substance that can penetrate the membrane and enhance fertility between A and D.

Keys to the Significance of This Chapter

Fertilization of an egg is followed by a succession of mitotic divisions through which the material of the egg cytoplasm is partitioned among many small cells. We assume these cells to be of identical genotypes. But cytoplasmic materials may be distributed unequally to the cleavage products, and dissimilar immediate environments impinge upon different parts of the embryo. The genetically identical nuclei of the individual therefore become involved in increasingly different reaction systems in various embryonic regions.

Once an initial basis for divergence has been established, regular genic effects on these differentiating reaction systems increase and regulate the divergence. The initial genic effects in this epigenetic sequence are presumably matters of cellular biochemistry. They involve the genetic control of specific biochemical reactions that can occur in some of the reaction systems that are becoming established, but not in others. Many genic effects, therefore, are limited to particular times and places in development—to times and places in which the conditions necessary for the determinative action of these particular genes are encountered.

Not only are the cytoplasm and the cellular milieu involved in and affected by these processes; the cell nuclei also develop stable changes in their capacities for gene expression. "Puffing" of particular regions of chromosomes and heterochromatic-euchromatic changes of whole chromosomes or parts of chromosomes reveal temporal patterns of chromosomal activity. The operon model suggests how the orderly program of development may be built into the genome itself.

Many genes have effects at some distance from the time and place of their initial action. Thus, genes may affect the integrating systems of embryonic induction, and of endocrines; and they may act in various other ways to influence broadly the processes of development. The processes of sex determination and sex differentiation illustrate many of the principles of the genetic control and the complex interactions of morphogenesis.

While the problems of differentiation still constitute in large part an unsolved riddle that can be dealt with only speculatively in terms of our present very limited knowledge, approaches now seem to be at hand toward the solution of important aspects of this riddle. Present indications point to the genes as the essential catalysts of epigenetic development, whose regular transmission from generation to generation is of prime significance in the control of developmental patterns.

REFERENCES

Allee, W. C., *et al.*, *Principles of Animal Ecology*. Philadelphia: W. B. Saunders, 1949. (The section on "Animal Aggregations and Sex," pp. 408–410, deals with relations of environment and sexuality, including the cases of Bonellia and Crepidula.)

Allen, C. E., "The Genotypic Basis of Sex-expression in Angiosperms." *Botan. Rev.*, **6**:227–300, 1940. (An extensive summary of the field, including a large and useful bibliography.)

Barth, L. G., *Embryology*. New York: Dryden Press, 2nd ed., 1953. (A clear and interesting elementary textbook on which part of our discussion of experimental embryology was based.)

Beermann, W., and Clever, U., "Chromosome Puffs." *Scientific American*, April, 1964. Available as Offprint 180 from W. H. Freeman and Co., San Francisco. (An excellent, readable summary of the present status of this subject, and the source of Color Plate III in Chap. 2 of this book.)

Bridges, C. B., "Sex in Relation to Chromosomes and Genes." *Am. Naturalist*, **59**:127–137, 1925. (A classical paper on sex determination in Drosophila.)

Ephrussi, B., "Chemistry of 'Eye Color Hormones' of Drosophila." *Quart. Rev. Biol.*, **17**:327–338, 1942. (A review, with references.)

Gluecksohn-Waelsch, Salome, "Physiological Genetics of the Mouse." *Advances in Genetics*, **4**:1–51, 1951. (An excellent review, discussing genic effects on the development and function of practically all organ systems.)

Goldschmidt, R., "Lymantria." *Bibliographia Genetica*, **11**:1–186, 1934. (An important monograph on the gypsy moth. Pages 10–105 are concerned with the analysis of intersexuality in Lymantria.)

Grüneberg, H., *Animal Genetics and Medicine*. London: Hamish Hamilton, 1947. (See especially Chap. 3, "Some Principles of Developmental Genetics . . . ," and Chap. 21, "The Urogenital System.")

Gurdon, J., "Nuclear Transplantation in Amphibia." *Quart. Rev. Biol.*, **38**:54–78, 1963. (A clear and thoughtful review by a substantial contributor in this field.)

Hadorn, E., *Developmental Genetics and Lethal Factors*. New York: Wiley, 1961. (Translation of the definitive work in this important field.)

Haldane, J. B. S., *New Paths in Genetics*. New York: Harper, 1942. (See especially Chap. 3, "Genetics and Development.")

Hämmerling, J., "Ein- und zweikernige Transplante zwischen *Acetabularia mediterranea* und *A. crenulata.*" *Z. Abstgs. Vererbungsl.*, **81**:114–180, 1943. (Stimulating transplantation studies on a unicellular alga, the source of Probs. 12-2 to 12-4.)

Huxley, J., *Evolution, The Modern Synthesis.* New York: Harper, 1942. (See especially pp. 525–555, "Consequential Evolution: The Consequences of Differential Development," for consideration of relative growth and related phenomena.)

Monod, J., and Jacob, F., "General Conclusions: Teleonomic Mechanisms in Cellular Metabolism, Growth, and Differentiation." *Cold Spring Harbor Symposia Quant. Biol.*, **26**:389–401, 1961. (The implications of operons and other mechanisms in developmental problems.)

Rick, C. M., and Hanna, G. C., "Determination of Sex in *Asparagus officinalis L.*" *Am. J. Botany*, **30**:711–714, 1943. (Reports the genetic mechanism for sex determination in asparagus.)

Schmid, W., "Analyse der Letalen Wirkung des Faktors *lme* (Letal-Meander) von *Drosophila Melanogaster.*" *Z. Abstgs. Vererbungsl.*, **83**:220–253, 1949. (Analysis of the meander lethal.)

Stern, C., "Two or Three Bristles." *Am. Sci.*, **42**:213–247, 1954. (Text of a general lecture—clear, perceptive, and charming.)

Sturtevant, A. H., "The Vermilion Gene and Gynandromorphism." *Proc. Soc. Exp. Biol. Med.*, **17**:70–71, 1920. (Classic observation of genic effect on hormone-like materials, the source of Probs. 12-8 to 12-10.)

Warmke, H. E., "Sex Determination and Sex Balance in Melandrium." *Am. J. Botany*, **33**:648–660, 1946. (Concise. Graphically summarized with figures and tables.)

Weiss, P., "Perspectives in the Field of Morphogenesis." *Quart. Rev. Biol.*, **25**:177–198, 1950. (A review, with many references, of problems of development and differentiation.)

Whiting, P. W., "The Evolution of Male Haploidy." *Quart. Rev. Biol.*, **20**:231–260, 1945. (A review. Presentation at a high level.)

Wright, S., "Gene and Organism." *Am. Nat.* **87**:5–18, 1953. (A broad and deep consideration of a provocative subject.)

Young, W. D., editor, *Sex and Internal Secretions.* Baltimore: Williams & Wilkins, 3rd ed., 1961. (A survey, with different subjects treated by specialists. See particularly Chap. 1, "Cytologic and Genetic Basis of Sex," by J. W. Gowen, and Chap. 2, "Role of Hormones in the Differentiation of Sex," by R. K. Burns.)

QUESTIONS AND PROBLEMS

12-1. What do the following terms signify?

cleavage	internal environment
competence	migrating cells
differentiation	organizer
embryonic induction	relative growth rates
epigenetic	target tissue
hormone	

Figure 12-20 illustrates an experiment performed by Hämmerling on two species of the unicellular alga Acetabularia. *A. mediterranea* has an intact, umbrellalike "hat,"

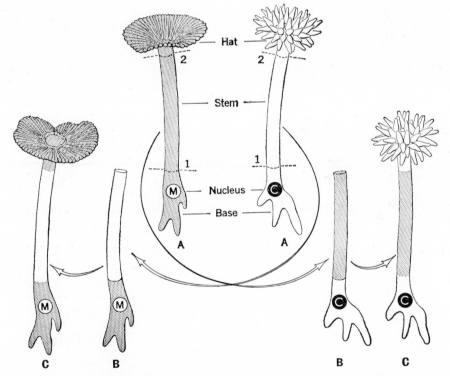

Figure 12-20. *Grafting experiments in Acetabularia have given important information about the control of patterns of regeneration. Left, grafting* A. crenulata *stem to the nucleated base of* A. mediterranea *results in the regeneration of a hat like that of* A. mediterranea. *Right, the reciprocal experiment. See Problems 12-2 to 12-4 (Diagrammatic; based on Hämmerling, Z. Abstg. Vererb.* **81:***114–180, 1943.)*

while the hat of *A. crenulata* is deeply indented. The nucleus is imbedded in the base of each of these single cells, and the hat is borne on a long cytoplasmic stem.

12-2. When the stem of *A. mediterranea* is cut off just at the base, and after removal of the hat the cut stem is grafted to a base of *A. crenulata* (Fig. 12-20B), a new hat is regenerated. Similarly, the reciprocal transplant of *A. crenulata* stem to *A. mediterranea* base regenerates a new hat. In each case, the hat regenerated has an eventual character consistent with the base from which it grows and different from the character of the intervening cytoplasmic stem (Fig. 12-20C). What do these results suggest relative to the cytoplasmic versus the nuclear control of regeneration in Acetabularia?

12-3. In Acetabularia, if the nucleus of a cell is removed, and the hat alone is then cut off, a new hat will be regenerated at once. But if the adjacent part of the stem is removed when the old hat is cut off, the enucleate cell is unable to regenerate a new hat. Assuming a "hat-forming substance" responsible for regeneration, what can you say of the origin, qualitative control, and distribution of this substance in the intact cell?

12-4. If a stem with intact hat is transplanted from one species of Acetabularia to a base of the other, the hat retains its original character. What does this fact add to your discussion in answering Question 12-3?

A technique for the study of genic effects on hormone-like materials in Drosophila has been developed and applied to good advantage by Ephrussi and Beadle. Figure 12-21 diagrams this technique and some of the results of its application. A piece of the larval tissue that would later give rise to the eye of the adult fly is transplanted to a genetically different larva. The developmental interactions between host and transplant are then observed, particularly in terms of the color developed by the transplanted eye disc as the host larva matures.

12-5. The reciprocal transplants shown in Figure 12-21A and B are typical of a large majority of Drosophila transplantation experiments. The larval disc from a wild-type fly develops the color characteristic of its own genotype, even when its differentiation and color development occur in a white-eye host. Similarly, a white-eye larval disc develops according to its own genotype, and is not influenced in any noticeable fashion by a wild-type host environment. Would you conclude that the white-eye gene determines a circulating (hormone-like) material, or something in the developing eye tissue itself?

12-6. The experiment diagramed in Figure 12-21C and D represents the kind of exception from which a good deal of information has been derived. Here again, when wild-type discs are transplanted to either vermilion (*v*) or cinnabar (*cn*) host larvae, the transplants develop autonomously into wild-type eyes. But when vermilion or cinnabar discs are put into wild-type larvae, these discs do not develop according to their own constitution; instead, *they become wild type*.

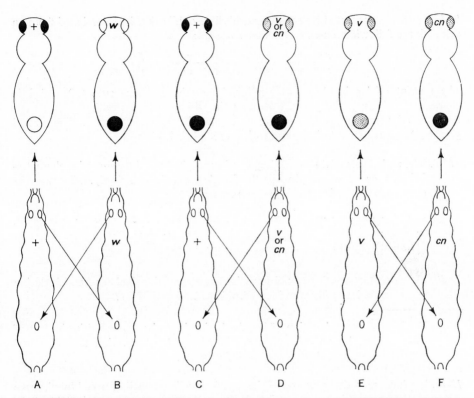

Figure 12-21. *Transplants of larval tissue have elucidated some aspects of gene action in Drosophila. See Problems 12-5 to 12-7. (After Ephrussi, Quart. Rev. Biol., 17:329, 1942.)*

Assuming that the body of the wild-type host is capable of providing the developing eye tissue with circulating or diffusing materials that compensate for the genetic blocks in vermilion and cinnabar flies, derive a detailed explanation of these experimental results.

12-7. Parts E and F of Figure 12-21 show the results of reciprocal transplants between vermilion and cinnabar larvae. Cinnabar discs developing in vermilion hosts maintain their cinnabar phenotype, but vermilion discs develop as wild type in cinnabar hosts. It is now known that these genes are concerned with sequential steps in a biochemical synthesis of hormone-like materials directly involved in eye pigment production:

$$\text{tryptophan} \xrightarrow{v^+} \underset{\text{("}v^+\text{ substance")}}{\text{formylkynurenin}} \xrightarrow{cn^+} \underset{\text{("}cn^+\text{ substance")}}{\text{hydroxykynurenin}} \longrightarrow \longrightarrow \text{pigment}$$

Using these facts, give a detailed explanation of the transplantation results.

In 1920, Sturtevant found a gynandromorph Drosophila that had developed from an egg with the following sex-chromosomal genotype:

$$\frac{sc \quad w^+ \quad ec \quad rb^+ \quad ct \quad v \quad g \quad f}{sc^+ \quad w^e \quad ec^+ \quad rb \quad ct^+ \quad v^+ \quad g^+ \quad f}$$

The mutant alleles involved, all sex-linked recessives, were: scute (*sc*), a bristle character; eosin (*w^e*), an eye-color allele at the white-eye locus; echinus (*ec*), an eye characteristic; ruby (*rb*), an eye color; cut (*ct*), a wing-shape effect; vermilion (*v*) and garnet (*g*), eye-color mutations; and forked (*f*), a bristle character.

12-8. The male parts of Sturtevant's gynandromorph showed the recessive characteristics *sc*, *ec*, *ct*, *g*, and *f*, while the female parts showed *f* only. What was the chromosomal basis for this gynandromorph?

12-9. How might the failure of the male parts of Sturtevant's gynandromorph to display the characteristic *vermilion* be explained? (There is no question that vermilion would have been recognized if it had been present.)

12-10. How do Sturtevant's early observations compare with the later transplantation experiments involving vermilion (Problems 12-6 and 12-7)?

12-11. In view of the explanation of hen-feathering in cocks developed in the text of this chapter, how would you explain Morgan's observation that castrated Sebright bantam males become cock-feathered?

12-12. We have seen that any one of the first four cells normally formed by cleavage of the sea-urchin egg can give rise to a qualitatively normal individual. In snails, on the other hand, the two cells formed by the first cleavage, when isolated, develop differently. Assuming that initial segregation of cytoplasmic components is basic to differentiation, compare the probable relation of cytoplasmic organization to cleavage in snails and sea urchins.

12-13. Identical twins in man are derived from a single fertilized egg. Is the human egg probably more like the snail egg or the sea-urchin egg (Question 12-12) with regard to its developmental organization?

12-14. The Dionne quintuplets are believed to have originated from a single egg. If cleavage results in the partitioning of cytoplasmic materials essential to the formation of a complete and integrated individual, what is the earliest cleavage division in which this critical partitioning may occur in man?

In man, a rather rare disorder involves metabolic components normally regulated by the parathyroid glands. The disorder is somewhat variable, but the basic deviations from normal metabolism of calcium and phosphorus are typically expressed in aberrant bone growth, deposition of calcium in abnormal places, particularly in certain parts of the brain and under the skin, a round head and small stature, and, frequently, in jerky involuntary movements and mental deficiency.

12-15. The disorder just described is grossly similar to that caused by failure of the parathyroids to produce a normal supply of their hormones (*hypoparathyroidism*). However, patients with the disorder under discussion prove to have very active parathyroids, and they do not respond to the administration of additional parathyroid hormones. The disorder is therefore called *pseudohypoparathyroidism*. Albright *et al.*, describing the condition in 1942, referred to it as a "Sebright bantam syndrome." Can you discern the rationale of this reference in terms of our discussion of hen-feathering in Sebright bantams?

12-16. Recent studies of a large family including several individuals affected with pseudo-hypoparathyroidism suggest that a single dominant allele, varying considerably in its final expression, is responsible for the characteristic segregating in this family. Discuss briefly the possible action of this allele.

12-17. Figure 12-14 illustrates the origin of gynandromorphism in Drosophila—the loss of an X-chromosome to give sectors of XX and XO tissues. Would this event lead to sectors of male and female cells in mice or men? Why? What chromosomal events in a mammal would lead to sectors of different sex?

12-18. Considering functional sex reversal in the fowl (p. 61), discuss whether the path taken in gonad development, and the differentiation of cells in the germ line to form sperm or eggs, are primarily functions of the constitution of the cells themselves or of the environment provided these cells during the course of their development.

12-19. Compare a system of alternative cytoplasmic steady states (Fig. 11-4) with the operon-switch model of Figure 12-10.

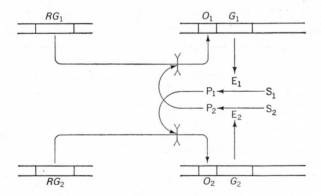

Figure 12-22. *An operon circuit involving reciprocal repression.*

12-20. In the operon circuit diagram in Figure 12-22, the product of enzyme action in each operon system acts as a repressor of the other operon. In such a model, what would be the consequences of variations in supply of the substrates S_1 and S_2 upon which the respective enzymes act?

12-21. Relate the facts that a tortoiseshell female cat has a mixture of yellow and black hairs and is heterozygous for a pair of sex-linked alleles for yellow and black respectively, to the "single active X" hypothesis for mammalian females.

12-22. Distinguish between sex determination and sex differentiation.

12-23. What are the evolutionary advantages of sexual reproduction over asexual reproduction?

12-24. Give some evidence that the Y-chromosome in Drosophila carries few, if any, genes involved in sex determination.

12-25. If through some aberrant circumstance chromosome doubling occurred in certain somatic cells of a triploid Drosophila female, would you expect these islands of hexaploid cells to differentiate as "male" or as "female" tissue?

12-26. 2X 3A Drosophila are intersexes. The addition of duplicating fragments of X-chromosomes into 2X 3A nuclei shifts the balance of sexuality. Would this shift be in the direction of maleness or femaleness?

12-27. Do you think it likely that gynandromorph human beings might occur that showed sharp mosaicism for primary and secondary male and female characteristics? Explain your answer.

12-28. What ratio of females and males would be expected in the progeny of a Drosophila cross between parents both heterozygous for the *tra* (*transformer*) gene?

12-29. Assume a chicken has undergone transformation from female to male. What would be the sex ratio of offspring from a mating between such a "male" and a female with part of her single X translocated to an autosome?

12-30. In corn, plants homozygous for gene *sk* (*silkless*) have abortive pistils, no silks, and thus are female-sterile. In addition, a gene Ts_3 is known that has much the same effect as the tassel-seed gene described in this chapter, except that Ts_3 is dominant to its normal allele. Utilizing genes *sk* and Ts_3, show how genotypes could be manipulated to produce dioecious lines of corn with a heterozygous female system of sex determination.

12-31. There is some evidence that staminate plants of asparagus outyield pistillate plants. Suggest a method by which seed could be obtained that would give all staminate plants.

12-32. Assume that in Habrobracon 25 percent of the eggs a female lays are unfertilized, and that the rest are fertilized. What would be the sex ratio in a progeny from the mating *xa xb*♀ × *xc*♂? From the mating *xa xb*♀ × *xb*♂?

12-33. Making the same assumption about percentage of unfertilized eggs as in the last question, what would be the sex ratio in a progeny from the mating *xa xb*♀ × *xa xa*♂? From the mating *xa xb*♀ × *xc xc*♂? (Assume fertility in the diploid males, although in fact they are usually highly sterile.)

12-34. In Habrobracon, gene *vl* (*veinless*) is an autosomal recessive. When making experimental crosses, how could you utilize *vl* and its wild-type allele (+) so that haploid males could be readily separated from diploid males in the cross progenies?

12-35. Gene *fu* (*fused*) in Habrobracon affects the antennae, legs, and wings. It is in the same chromosome as the locus for the multiple-allelic series concerned in sex determination, with the two loci showing about 10 percent crossing over. From the cross $\dfrac{xa \quad +}{xb \quad fu} \times \dfrac{xa \quad fu}{}$, what ratio of nonfused to fused would you expect among the female offspring? Among the diploid male offspring? Among the haploid male offspring?

12-36. Answer Question 12-35 on the basis that the male parent in the cross is <u>*xb   fu*</u>.

12-37. Answer Question 12-35 on the basis that the male parent in the cross is <u>*xc   fu*</u>.

12-38. From the point of view of survival of the species, what advantages can you see to systems of sex determination, like the XY, where the production of relatively equal numbers of females and males is genetically assured?

12-39. From the point of view of survival of the species, what advantages are there to the system of sex determination found in Crepidula?

12-40. Having in mind the material on sex determination in this chapter, can you suggest any reason why polyploidy should become established much more frequently among the species of plants than of animals?

CHAPTER 13

Genes in Populations

Up to this point, we have been mainly concerned with experimental analyses of the nature, function, and transmission of the genetic materials. These analyses have contributed to our perception of the most basic functions of life; the utility of this knowledge is not restricted to the laboratory. But if we turn now to the living world around us we find that some of the conventions of laboratory experiments do not commonly prevail in nature. Of many differences between laboratory experiments and natural populations, two are, from a genetic point of view, probably basic to almost all others. The first is that *the relative frequency of alleles at a locus under study in the laboratory is usually fixed at some convenient ratio by the design of the experiment.* Typically, two homozygous parents are crossed to begin an experiment; the alleles are therefore introduced in equal frequency. *In natural populations, on the other hand, the relative frequencies of alleles may vary greatly;* one allele at a locus may be homozygous in almost all the individuals of a population, and other alleles may be relatively rare. We need to know what effects these variations in the relative frequencies of alleles may have on the genotypes and phenotypes in the population.

The other essential difference between controlled and natural populations concerns the frequencies with which different genotypes and phenotypes mate and leave progeny. *In laboratory populations, a system of mating is usually sharply defined;* the experimenter typically first mates one homozygous class with another, then systematically self-fertilizes or intercrosses the F_1 heterozygotes or mates them back to homozygotes. In controlled plant and animal populations outside the laboratory, the breeder commonly entertains a wider variety of choices, but if he is eager to progress systematically toward the genetic improvement of his material, he adopts some regular system of mating and combines this with some form of selection. Thus, he may methodically outcross or inbreed, or adopt some other definite program based on genetic relationships among the animals or plants he selects as his breeding stock. Or he may select similar or dissimilar animals or plants and mate "likes" or "unlikes" on the basis of phenotype only.

In natural populations, much more is generally left to chance. Important also in the controlled populations of the plant or animal breeder is an appreciation of the systems of mating different from the usual simple Mendelian laboratory analysis and of the effects of mating systems on what happens to the genetic qualities of the population.

Gene Frequencies for M and N in Bedouins. One important class of simply inherited characteristics of animals and man includes antigenic characteristics of the red blood cells, identified by means of antibody test reagents. When human red blood cells are injected into rabbits, for example, the rabbits respond by producing very reactive antibodies against this foreign material. *Immune serum*, or *antiserum*, produced in this way will usually agglutinate human blood cells even when the serum is very greatly diluted. Such a serum is of little direct use in demonstrating individual differences in man. It reacts about equally well with the blood cells of all human beings, because the rabbit has elaborated antibodies against many materials shared by humans but foreign to rabbits. But by a technique known as *antibody absorption*, the serum can often be fractionated so that it will distinguish individual differences in human blood. This technique is illustrated diagrammatically in Figure 13-1.

By mixing the serum with the blood cells of a person and then centrifuging down the cells, one can remove all antibodies reacting with the "absorbing" cells. This may leave behind in the supernatant antibodies that will react with the cells of other persons. In the experiment shown in Figure 13-1, besides the general human antigens designated as "H" on the cell surface, two antigens (labeled M and N) were present on the cells used in injecting the rabbit. The rabbit elaborated three classes of antibodies *anti*-M, *anti*-N, and *anti*-H. Mixing this antiserum with cells containing M (and of course H) but not N, and centrifuging down these cells, leaves antibodies specific for N in the supernatant. Similarly, a reagent specific for M could be prepared by absorbing the antiserum with N

Table 13-1. REACTIONS OF BLOOD CELLS.

| Type | *Reagents* | | Genotype |
	Anti-M	*Anti*-N	
M	+	0	*MM*
MN	+	+	*MN*
N	0	+	*NN*

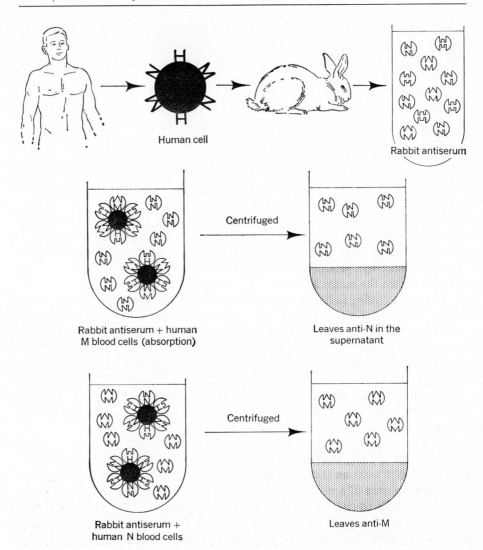

Figure 13-1. *A complex antiserum, produced by injecting human blood into a rabbit, can be fractionated by absorbing out part of the antibodies with blood cells from other persons. This often leaves antibody reagents that will react with blood from some but not all human beings.*

human cells. These two reagents could then be used to identify four kinds of human blood—with M or N, with both or with neither.

Regarding two human cellular antigens actually called M and N, and identified in this manner, an interesting fact emerges from many tests. No human blood actually lacks both of these antigens (Table 13-1).

The characteristics M and N depend on alleles; strictly speaking, we should use some common basic symbol for the alleles and distinguish them by superscripts. Instead, we shall adopt the more convenient convention of referring to the alleles in terms of the antigens they control.

A population of 208 Bedouins of the Syrian desert, when tested for M and N, was distributed as follows:

M	MN	N	Total
119	76	13	208

(Data from Boyd, *Tabulae Biologicae*, **17**:234, 1939.)

A glance at these data may suggest that the Bedouins do not conform to Mendelian laws. We are now so accustomed to thinking in terms of 1:2:1 ratios for a simple allelic difference that a ratio like 119:76:13 seems unreasonable. But if you stop to consider this point for a minute, you will see that a 1:2:1 ratio is based on an equal frequency of the two alleles in the population being described; thus, $1MM:2MN:1NN$ includes four M and four N alleles in the total of eight representatives of this allelic pair in this ideal population. All that is really clear from a glance at the Bedouin distribution is that M was considerably more frequent than N among the Bedouins tested.

We can relabel the classes as genotypes

MM	*MN*	*NN*	Total
119	76	13	208

and observe that there were $(119 \times 2) + 76 = 314M$, and $(13 \times 2) + 76 = 102N$ alleles in the total of $(208 \times 2) = 416$ representatives of this allelic pair in this population. Calculating on a decimal basis, the *frequency* of M is $\frac{314}{416} = 0.76$, and of N is $\frac{102}{416} = 0.24$. In other words, the M allele was about three times as frequent as the N allele among the Bedouins tested. A little more than ¾ of the chromosomes of the pair bearing the M-N locus carried the M allele, and a little less than ¼ the N allele.

This calculation of the relative frequency of alleles in a population illustrates what is meant by *gene frequency*, a basic concept for understanding heredity in populations.

The Binomial Distribution of Genotype Frequencies. One of the primary advantages of computing gene frequencies on a decimal or fractional basis, as was done for M and N in the Bedouin population above, is that the frequencies so computed can be treated and manipulated as *probabilities*. For example, the probability that any given chromosome of the proper pair will carry the allele M, in the above population, is 0.76. The similar probability for N is 0.24. The probabilities of the different *genotypes* in the population, if these chromosomes are combined in pairs by chance, are therefore:

$$MM = 0.76 \times 0.76 = 0.58$$
$$\left.\begin{array}{l} MN = 0.76 \times 0.24 = 0.18 \\ NM = 0.24 \times 0.76 = 0.18 \end{array}\right\} = 0.36$$
$$NN = 0.24 \times 0.24 = 0.06$$

On the basis of random combinations we should therefore expect a distribution of genotypes, among the 208 Bedouins, of about:

$$0.58 \times 208 = 121 \; MM$$
$$0.36 \times 208 = \;\; 75 \; MN$$
$$0.06 \times 208 = \;\; 12 \; NN$$

This is very close to the actual distribution of 119:76:13.

These computations demonstrate a precision in the distribution of genotypes for M and N in Bedouins that was not evident in the unanalyzed frequencies. We can state the case in more general terms:

1. If p is the frequency of a particular allele A, and q is the frequency of its alternative a, then the chance distribution of genotypes in a population will be $p^2AA + 2pqAa + q^2aa$.

By comparing this general statement with the specific example of M and N in Bedouins, you can easily see how the distribution is derived. If allele A occurs with frequency p, then when chromosomes carrying the A locus are combined in pairs at random, the probability that both members of a pair will carry A is $p \times p$, or p^2. Similarly, the probability of the aa combination is q^2, and of Aa is $pq + pq$, or $2pq$.

2. The distribution $p^2 + 2pq + q^2$ can be written in another way, $(p + q)^2$. This is an expression of the familiar *binomial*.

3. In terms of the above symbolism, $p + q = 1$. For example, the M and N frequencies in Bedouins were $p(M) = 0.76$; $q(N) = 0.24$, and $0.76 + 0.24 = 1.00$. We can eliminate the symbol p, when we wish to do so, by observing that

Table 13-2. M AND N BLOOD TYPES IN SAMPLES OF THREE POPULATIONS.

Population	Number Tested	Phenotypes Observed M	MN	N	Gene Frequencies $p(M)$	$q(N)$	Phenotypes Expected $p^2(M)$	$2pq(MN)$	$q^2(N)$
American Indian (Pueblo)	140	83	46	11	0.76	0.24	81	51	8
Brooklyn, U.S.A.	1849	541	903	405	0.54	0.46	536	925	388
Australian Aborigines	102	3	44	55	0.25	0.75	6	38	58

Source: From Boyd, *Tabulae Biologicae,* **17**:230, 235, 1939.

$p = 1 - q$. Rewritten on this basis, the distribution of genotypes becomes $(1 - q)^2 AA + 2q(1 - q)Aa + (q^2)aa$.

Binomial Distribution of Genotypes in Other Populations. You may regard it as accidental that the 208 Bedouins tested for M and N showed so nice a "binomial distribution" of genotypes. In Table 13-2 several other populations are classified for M and N types.

You should check the *expected* column yourself, to be sure that you know how to calculate gene frequencies and expected genotypic frequencies from population data. Try some chi-square tests to see how closely the observed and expected agree.[†] You will find that in all three of these populations the observed distribution of phenotypes agrees satisfactorily with that expected from probability calculations based on the gene frequencies.

Another point is evident here. Different populations show markedly different frequencies of these alleles. The Pueblo Indians, like the Bedouins, have relatively more M alleles than N alleles, but this situation is almost exactly reversed in the Australian Aborigines; there N is three times as frequent as M. In fact, the comparison of gene frequencies in human populations is one way of describing these populations genetically.

The Binomial Distribution of Genotypes as an Equilibrium Distribution. The precision with which a variety of populations fits the chance distribution of genotypes for M and N suggests that this kind of distribution provides a general basis for the genetic structure of populations. If an allele A has a frequency p in a population, and allele a has a frequency q, and if $p + q = 1$, then chance combinations of gametes through random mating in this population will give in the next generation approximately the distribution shown in Table 13-3.

The binomial distribution of genotypes is therefore approximated in a single generation of random mating, and it is maintained in successive generations. A population having this kind of genetic stability of structure is described as being in *equilibrium*. The phenomenon is often referred to as the *Hardy-Weinberg Law*, after its co-discoverers. (See the references at the end of this chapter.)

The stability asserted above is of course not a matter of absolute and immov-

---

[†] In doing chi-square tests on these distributions, you should note that, although there are three classes, there is only one degree of freedom. This is because for any given total number of people typed for M and N, only one class can be filled in at random; once this class is filled the other two are automatically committed. For example, in a sample of 100 people having 50M alleles, any number up to 25 of these people may be of blood type M (MM). To fill this class at random, suppose that 14 are MM. This means that the remaining 22M alleles are in heterozygotes, and the sample must be 14MM:22MN:64NN.

Incidentally, if you check this *hypothetical* population against the binomial expectation for the same gene frequencies, you will find that the two differ greatly. This should serve to convince you that the precision of the distribution in the *natural* populations we have been discussing reflects a significant aspect of population structure, and is not simply a result of the algebraic manipulations in which we have indulged.

Table 13-3. RESULTS OF RANDOM COMBINATIONS OF SPERM AND EGGS.

	Sperm	
	$A(p)$	$a(q)$
Eggs $\quad A(p)$	$AA(p^2)$	$Aa(pq)$
$a(q)$	$Aa(pq)$	$aa(q^2)$

$$= p^2AA + 2pqAa + q^2aa$$

able fixation. Various factors act to move or to modify this particular random equilibrium and to change the relative frequencies of genes and genotypes from generation to generation. Recurrent mutation from A to a, for example, serves to increase the value of q at the expense of p; and this may be opposed by the pressure of reverse mutation, from a to A. If we were to consider only the values of p and q for many generations, we would find that under the counterbalancing pressures of mutation and reverse mutation an *equilibrium* for the value of p is achieved, and that the point at which this value stabilizes is determined by the relative rates of mutation in the two directions for the particular locus under study. Similarly, *selection*, in which one genotype or phenotype leaves (or is permitted to leave) fewer progeny than others, acts to modify the relative frequency of alleles from generation to generation. If individuals of blood type M were sterile or partly sterile, for instance, or were barred from marrying in Brooklyn, the relative frequency of N would increase in later generations, until this population reached and even surpassed the level of the Australian Aborigines in this regard. Similarly, migration, for instance of Bedouins into Brooklyn, would change the character of the Brooklyn population. And, as we will shortly see, beneath all this the sampling nature of Mendelian inheritance itself subjects the relative frequencies of alleles to chance variations.

These are all parts of the dynamics of population behavior basic to evolutionary change and to the improvement or deterioration of both natural and controlled populations. We will discuss them in more concrete fashion later in this chapter.

Considering genotype and phenotype frequencies, still other variables enter to modify the chance binomial distributions. We have based the binomial equilibrium on random mating, in which each genotype mated with each other genotype with a frequency strictly proportional to the relative numbers of these genotypes in the population, that is, on a purely chance basis. But sometimes in nature, and often in controlled populations, there may be systematic deviations from this pattern of chance matings, so that one genotype, for example, may be more likely to mate with a similar genotype than with a different one. The effect of different systems of mating, departing from the random scheme, is often to establish stable equilibrium distributions of genotypes and phenotypes different from the equilibrium achieved under random mating.

Our choice of the *M–N* alternative for our introductory illustration was a considered one, based among other things on the following attributes of this particular characteristic:

1. Most people grow up, marry, have children, and die without ever knowing their *M–N* type. Even when the type is known, it does not influence decisions to marry or to have children. The frequencies of the different kinds of matings, with regard to *M* and *N*, are determined by the frequencies of the genotypes themselves within any freely intermarrying population.

2. Since the heterozygote *MN* is phenotypically distinguishable from either homozygote, the frequency of each allele can be directly computed from the phenotype frequencies in the population. In the more common situation in which one allele is dominant, some aspects of the computations must be less direct. We will turn to this complication now.

Dominance and Gene Frequency. A lecturer recently explained an historical change in the skull shape of an Indian population as being in large part the result of the genetic dominance of alleles for the increasingly frequent broad, short skull. In the opinion of this speaker, the genetic dominance of an allele was supposed to be reflected in the numerical predominance in a population of the phenotypic characteristic this allele controlled.

The fallacy of this opinion will be evident to you. The frequency of a phenotype in a population depends on the frequency of the allele controlling it, and this in turn has no necessary relation to the dominance or recessiveness of the allele. For example, the allele for *fragile bones with blue sclera* is dominant to its normal alternative in man, but the aberrant phenotype is only rarely encountered in human populations. This is because this dominant allele is very infrequent, while its normal alternative is common. Many similar human examples might be cited.

As an example of genotype and phenotype frequencies for a dominant-recessive alternative in a population, we can consider the simply inherited difference in human ability to taste the compound called PTC (phenylthiocarbamide or phenylthiourea). Some people find this compound very bitter; they are called *tasters* for PTC. Others (*nontasters*) find it tasteless or virtually so. The difference depends on a simple Mendelian alternative, in which the allele for taster (*T*) is dominant to that for nontaster (*t*).

If the members of your class are tested, you will probably find that about 70 percent are tasters, and 30 percent nontasters. We can therefore write:

Phenotypes:	Tasters	Nontasters
Genotypes:	$TT + Tt$	tt
Frequencies:	0.7	0.3

Now, if we can assume that the population is at equilibrium under random mating for these alleles, the genotypes should be distributed as follows.

$$\text{If } p = \text{frequency of } T$$
$$q = \text{frequency of } t$$
$$p + q = 1$$

then the genotypes occur in the following ratio:

$$p^2TT : 2pqTt : q^2tt$$

We know that q^2 (the frequency of nontasters) = 0.3.
Therefore,

$$q = \sqrt{0.3}, \text{ and } q = 0.55; \quad p = 0.45$$

The probable distribution of genotypes in the population can now be computed:

$$p^2TT : \quad 2pqTt \quad : \quad q^2tt$$
$$(0.45)^2 : 2(0.45)(0.55) : (0.55)^2$$
$$0.2 : \quad 0.5 \quad : \quad 0.3$$

The Utility of Gene-Frequency Analysis. The preceding section involves an important extension of our consideration of gene frequencies. Up to this point, we had been using rather obvious characteristics of a population to derive some information about gene frequencies. But we are now able to extrapolate from our knowledge of gene frequencies and their consequences to some characteristics of other populations that are not obvious, because the recessive allele is obscured by its dominant alternative in heterozygotes in the population. The calculation of the frequencies of homozygotes and heterozygotes among tasters is a case in point.

Suppose, for example, we should ask what kinds of children (with respect to the taster-nontaster alternative), and in what proportions, marriages between tasters would be likely to produce. Without the gene-frequency analysis, we could have observed that some of the taster parents would be likely to be heterozygous, and that whenever two such heterozygotes married, a quarter of their children on the average would be expected to be nontasters. But we could not often tell which of the taster parents actually were heterozygous, and except in rather special cases we could not make a statistical prediction for the children of tasters at all.

Knowing, however, that homozygotes (*TT*) and heterozygotes (*Tt*) should occur in the population in a ratio of 2:5, it is easy to predict the consequences of marriages between tasters with regard to this characteristic:

Probability of father's being heterozygous = $\frac{5}{7}$
Probability of mother's being heterozygous = $\frac{5}{7}$
Probability of both parents' being heterozygous = $\frac{5}{7} \times \frac{5}{7} = \frac{25}{49}$

Approximately half of the matings between tasters, therefore, will be between heterozygotes.

The other matings between tasters, *TT* × *TT* and *TT* × *Tt*, will have no non-taster children.

The matings between heterozygous tasters will be expected to have children in the ratio: $\frac{3}{4}$ tasters: $\frac{1}{4}$ nontasters.

The probability of a nontaster child from a taster $\times$ taster marriage is therefore:

Probability that a given mating is between two heterozygotes = $\frac{25}{49}$
Probability of nontaster child from this mating = $\frac{1}{4}$
Probability of nontaster child from taster $\times$ taster mating = $\frac{25}{49} \times \frac{1}{4} = \frac{25}{196}$

Taster $\times$ taster matings in our population should therefore produce children in the ratio of approximately $\frac{7}{8}$ taster: $\frac{1}{8}$ nontaster. One study reports an observed ratio of 464 taster:65 nontaster children from such matings.[†] Is this in agreement with expectation?

Confirm by the same sort of calculation that marriages of taster $\times$ nontaster should produce children in the ratio of about $\frac{9}{14}$ taster: $\frac{5}{14}$ nontaster.

We can work the above problem in several somewhat different ways. Two of these are shown below.

1. Probability that taster parent is heterozygous = $\frac{5}{7}$
   Probability of nontaster children from heterozygous taster $\times$ nontaster = $\frac{1}{2}$
   Probability of nontaster child from taster $\times$ nontaster mating = $\frac{5}{7} \times \frac{1}{2} = \frac{5}{14}$
   Probability of taster child from above mating type = $1 - \frac{5}{14} = \frac{9}{14}$.

2. Taster $\times$ Nontaster Matings

Mating	Frequency	Children	
		Taster	Nontaster
$TT \times tt$	$\frac{2}{7}$	$\frac{2}{7}$	
$Tt \times tt$	$\frac{5}{7}$	$\frac{5}{14}$	$\frac{5}{14}$
Total		$\frac{9}{14}$	$\frac{5}{14}$

One study reports an observed ratio of 242 taster:139 nontaster children from such matings.[†] Is this in agreement with expectation?

An interesting relationship is illustrated in the above examples: the frequency of recessive (nontaster) children in matings of dominant phenotype (taster) $\times$ recessive (nontaster) parents is $\frac{5}{14}$, and in matings dominant $\times$ dominant is $\frac{25}{196}$, or $(\frac{5}{14})^2$. Snyder recognized this relationship as a general rule; the frequency of recessive children in matings dominant $\times$ recessive is $q/1 + q$, and in matings dominant $\times$ dominant is $(q/1 + q)^2$. The relationship applies for traits determined by multiple recessives as well as for single recessive traits.

[†] These data are from *The Principles of Heredity*, by L. H. Snyder, D. C. Heath & Co., 1951, p. 499.

The taster characteristic in itself is of relatively little practical concern. For example, one can be a nontaster for PTC and still enjoy roast beef with mashed potatoes and brown gravy perfectly well. You will be able to see, however, how the kind of predictions made possible through gene-frequency analyses, introduced here by means of the taster-nontaster example, can sometimes be of great value in connection with simply inherited characteristics of greater human and medical concern.

Furthermore, the gene-frequency analysis provides a tool for genetic research in populations beyond laboratory control. The frequencies of phenotypes and the results of different types of matings in human populations, for example, should be consistent with gene-frequency considerations. If a given trait does not conform to expectation in this regard, and if there is no good reason to suspect that nonrandom mating or selection, or some other specified factor is responsible for the deviations, then the investigator is likely to conclude that the basis of inheritance of the trait is more complex than had at first been suspected. In fact, the only human genes that may be considered as adequately studied are those that have been subjected to detailed gene-frequency analyses and have met all the tests that human-population geneticists have devised on this basis. There are several such tests.

This field is rather complex mathematically and statistically, and it need not be explored further in this general text. Those who are interested in the research aspects of human heredity will find, in the references cited at the end of this chapter, techniques available for the distinction of different types of simply inherited characteristics, for the calculation of linkage intensities, for the measurement of penetrance, and for many related and similar problems.

Special Aspects of Gene Frequency. Before turning to other subjects, we will consider three rather special aspects of gene frequency. These involve *sex linkage, multiple alleles, and cousin marriages.*

Sex Linkage. Since one sex is haploid for genes in the sex chromosome, this sex cannot show the binomial distribution for chance combinations of chromosomes in pairs.

$$p = \text{frequency of } A$$
$$q = \text{frequency of } a$$
$$p + q = 1$$

then the equilibrium distribution of genotypes will be:

$$\text{In XY sex:} \quad pA + qa$$
$$\text{In XX sex:} \quad p^2AA + 2pqAa + q^2aa$$

You will note that the gene frequency of a sex-linked characteristic can be obtained directly from its frequency in the sex having a single X-chromosome. This should check in specific ways with the frequency of the characteristic in the XX sex. For example, the frequency of a sex-linked recessive characteristic among men should be the same as the square root of its frequency among women, since the frequency of the recessive phenotype is q in men, q^2 in women.

One consequence of this relationship is that sex-linked recessive characteristics are more common in men than in women. Among men, the recessive phenotype occurs with the same frequency as the responsible allele does, while among women the frequency of the phenotype is the square of the gene frequency. Thus, if one man in ten were color blind due to the common sex-linked gene for this characteristic, only one woman in one hundred would show the same type of color blindness. The actual figures in our population are not far from these values, although the occurrence of types of color blindness other than the common kind complicates the analysis somewhat.

Multiple Alleles. A convenient example of gene-frequency applications to a multiple-allelic series is provided by the normal human blood groups, doubtless already familiar to you. The genotypes and corresponding cell and serum types of human beings are as shown in Table 13-4. You will note that both I^A and I^B

Table 13-4. THE HUMAN BLOOD GROUPS.

Genotypes	Cellular Antigens	Serum Antibodies	Blood Group
$I^A I^A$ or $I^A i$	A	β	A
$I^B I^B$ or $I^B i$	B	α	B
$I^A I^B$	A and B	none	AB
ii	none	α and β	0

appear as dominant to their allele i since the heterozygotes $I^A i$ and $I^B i$ cannot easily be distinguished from the homozygotes $I^A I^A$ or $I^B I^B$, respectively. I^A and I^B, however, mutually lack dominance with respect to each other; the heterozygote $I^A I^B$ is easily identifiable as blood group AB. But the heterozygote $I^A I^B$ is not really an intermediate between the two homozygotes $I^A I^A$ and $I^B I^B$; it shows characteristics of both homozygotes. Current convention refers to this kind of situation as *codominance*, to distinguish it from other types of interaction between alleles.

Genetic consequences of this situation should be clear to you. An individual of blood group O, for example, might come from a mating of two individuals of blood group A ($I^A i \times I^A i$), or from two individuals of blood groups A and B respectively ($I^A i \times I^B i$), or from certain other possible kinds of matings, but neither of his parents could be AB ($I^A I^B$). This kind of analysis is the basis of application of blood grouping to legal medicine—the solution of cases of possible baby mixups and a source of evidence in cases of disputed paternity.

In a transfusion, B or AB cells introduced into an individual of group A or O would react with the β antibodies present in the serum of such individuals.

Transfusion reactions may be severe or even fatal. It is therefore vital to match donor and recipient before a transfusion is made.

A thorough discussion of the human blood groups would include several extensions of the information so far presented. For example, within the four major blood groups minor subdivisions have been recognized. The subtypes of A are called A_1, A_2, and so on. Insofar as the genetic determination of these types has been analyzed, it appears that a different allele controls each specific subtype. A complete list of the alleles at this locus would, therefore, begin I^{A_1}, I^{A_2}, and so on. Serologists now sometimes speak of a pleiades of alleles at this locus.

For gene-frequency analyses, we can:

let p = frequency of I^A
q = frequency of I^B
r = frequency of i
$p + q + r = 1$

Then the equilibrium distribution of genotypes under random mating is $(p + q + r)^2$. These are tabulated by phenotypes in Table 13-5.

Table 13-5. EQUILIBRIUM FREQUENCIES OF
HUMAN BLOOD GROUPS.

Phenotype	Genotype	Frequency
A	$I^A I^A$	p^2
	$I^A i$	$2pr$
B	$I^B I^B$	q^2
	$I^B i$	$2qr$
AB	$I^A I^B$	$2pq$
O	ii	r^2

The value of r is immediately obvious as the square root of the frequency of the O class. The values of p and q can be determined somewhat less directly. (In this determination, when we use the symbols $\overline{A}$, $\overline{B}$, $\overline{AB}$, and $\overline{O}$, we mean *the frequency of A, B, AB, and O phenotypes.*)

$$\overline{A} + \overline{O} = p^2 + 2pr + r^2$$
$$= (p + r)^2$$
$$\sqrt{\overline{A} + \overline{O}} = p + r$$

But, since $p + q + r = 1$, $\quad p + r = 1 - q$

therefore $1 - q = \sqrt{\overline{A} + \overline{O}}$

$$q = 1 - \sqrt{\overline{A} + \overline{O}}$$

Similarly, $p = 1 - \sqrt{\overline{B} + \overline{O}}$

As an example of a population apparently in equilibrium for the human blood groups, we can turn again to Brooklyn (Table 13-6).

Table 13-6. GENE FREQUENCIES FOR HUMAN BLOOD GROUPS IN BROOKLYN.

Individuals Tested	Individuals in Blood Groups				Gene Frequencies		
	O	A	B	AB	$p(I^A)$	$q(I^B)$	$r(i)$
1849	808 (43.7%)	699 (37.8%)	259 (14.0%)	83 (4.5%)	0.24	0.10	0.66

Source: From Boyd, *Tabulae Biologicae*, **17**: *166*, 1939.

The calculation of p, the frequency of I^A, is as follows:

$$p = 1 - \sqrt{\bar{B} + \bar{O}}$$
$$= 1 - \sqrt{0.14 + 0.437}$$
$$p = 0.24$$

You can check the values of q and r for yourself. You will note that the allele I^B is relatively infrequent in Brooklyn. It is even rarer in American Indians, with a frequency of only 0.007 in Pueblo Indians. But it is more common in Orientals; in Canton, for example, it is more frequent than is allele I^A. The human blood groups provide a good anthropological tool in the genetic description of human populations.

In such a city as Brooklyn, racial, religious, economic, and other barriers split the total population up into subgroups among which mating is not at random. Insofar as there are differences in the frequencies of the blood-group alleles among these partial *isolates* in the population, we might expect deviations from the random-mating equilibrium in the total population. Such deviations, however, are not statistically evident in the data cited in Table 13-6.

Cousin Marriages. Another interesting gene-frequency consideration in human populations concerns matings between relatives. In the Western world, about one marriage in 200 is between cousins; this is the closest form of inbreeding practiced with significant frequency in our population.

If two people (call them John and Mary) are essentially unrelated, their children will have two parents, four different grandparents, and eight different great-grandparents. But if John and Mary are first cousins, their children have only six different great-grandparents; one pair is duplicated in their pedigree (Fig. 13-2).

Now, suppose that one of the common great-grandparents, say great-grandfather 2 (GGF$_2$), was heterozygous for a rare recessive defect. John's mother (GM$_1$) had one chance in two of inheriting this particular allele from him, and

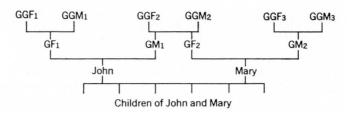

Figure 13-2. *John and Mary are first cousins.*

the probability that John himself has it is $\frac{1}{4}$ ($\frac{1}{2}$ for the probability that his mother has it $\times$ $\frac{1}{2}$ for the probability that if she has it, John inherited it from her $= \frac{1}{4}$, the probability that both events occurred). Similarly, the probability is $\frac{1}{4}$ that Mary is heterozygous for the same recessive defect, tracing back to GGF_2. The probability that *both* John and Mary are heterozygous is $\frac{1}{4} \times \frac{1}{4} = \frac{1}{16}$, and if they are both carriers, their children have an average expectation of one chance in four of showing the recessive defect in the homozygous condition. The probability of a child homozygous for the defect is therefore $\frac{1}{16} \times \frac{1}{4} = \frac{1}{64}$.

On the other hand, if John and Mary were unrelated, it would be very unlikely that they would both be carriers of the same rare recessive defect.

We can cite an example of the effect of this situation. Only about one in 20,000 Western Europeans is an albino. The allele for albinism is therefore very rare. About 15 percent of the albinos in this population come from first-cousin marriages; in other words, the incidence of first-cousin marriages among the parents of albinos is very much higher than it is in the population generally.

Note carefully that we did not say that 15 percent of cousin marriages produce albinos! On the contrary, albinos are, of course, very rare even among the children of first cousins. Only when we look at the figures the other way around and notice that the frequency of cousin marriages is relatively high among the parents of the rare albinos do we detect the effect of this mild form of inbreeding in human society. In fact, the rarer the allele we are considering, the more likely it is that the allele depends in large part on matings between relatives to reach homozygosity and therefore phenotypic expression in the population. A good method of estimating the frequency of a rare recessive is to determine the incidence of cousin marriages among the parents of children homozygous for the defect.

Cousin marriages represent a very slight form of inbreeding. They have little effect, compared with brother-sister matings, for example, in increasing general homozygosity. They are not, of course, necessarily "harmful," and some of the most prominent men and women in history have had parents who were cousins. But there are many different rare recessive defects in human germ plasm, carried for the most part hidden in the heterozygous state. In view of the increased likelihood that any one of these defects will crop up in the progeny of cousins, an individual does doubtless run a somewhat greater risk of aberrant children if he marries a first cousin than he does if he marries an unrelated person. On the other hand, of course, desirable recessive traits are also more likely to come to expression in the children of related individuals.

There is in most societies a general repugnance to matings between relatives, and this often extends so far as to discourage marriages between people as distantly related as cousins. It has been suggested that part of this repugnance is biological and is based on unsystematic observations of an increased frequency of abnormal children among the progeny of related people. Probably, however, other social and economic considerations are at least equally responsible for the taboos against marriages between relatives.

Changes in Gene Frequencies

Now that we have defined the concept of gene frequency, and have seen how alleles distribute themselves into genotypes of different frequencies in natural populations, we can turn to the elementary processes of evolution, the *changes* in gene frequency that may occur from one generation to another. When we speak of changes in a population, from a genetic viewpoint, we are usually speaking primarily of changes in gene frequencies. These changes are affected by mutation, selection, random fluctuations, meiotic drive, and migration.

MUTATION—THE ORIGIN OF HEREDITARY VARIATIONS

If changes in the relative frequencies of alleles are to contribute to the evolution of populations, alleles must obviously be there to start with. If segregation and recombination are to result in individual differences, there must be alternatives present to segregate and recombine. From these and other points of view, the phenomenon most fundamental to evolutionary change is the occurrence of the sudden, random, and generally unpredictable changes in germ plasm called *mutations*.

We have already discussed in sufficient detail the characteristics of the process of gene mutation (Chap. 8), and we need now only to capitalize on our familiarity with this process by applying it to evolution.

The Origin of New Alleles. Gene mutations are important in population changes for two rather different reasons. First, they provide the working materials for other factors affecting evolution. If selection is to operate, for example, there must be alternatives to select. Second (and this is our immediate concern here), the process of mutation itself is a force affecting gene frequencies. Whenever gene A mutates to an allele a, the frequency of A is a bit reduced in the population, and the frequency of *a* is correspondingly increased. Over a long period of time, if *A* keeps changing to *a*, and if no other forces modify the effect of this process, *A* will disappear from the population, and *a* will replace it.

It is as though we had a sack full of red marbles in the classroom and every year we removed a red marble and replaced it with a white one. It would take a long time, but eventually another class and another professor might find that there were only white marbles in the sack. Under the slow but inexorable pressure of recurrent mutation, a population may change drastically with regard to the characteristics controlled by the mutating gene.

We know that mutation is not only *recurrent;* it is often also *reversible*. Allele *a* is in turn changing back to *A*, with measurable frequency. Our model should therefore be modified; instead of drawing out a red marble and replacing it with a white one every time, we draw out any marble and replace it with one of the other color. At first, if we start with all red marbles, we are far more likely to draw red marbles out and to replace them with white. But as white marbles become more and more common in the sack, we are more and more likely to draw a white one out nearly as often as a red. Eventually we arrive at a point at which there are equal numbers of red and white marbles in the sack. From this point on we make no further progress in changing our population of marbles; for each time that we withdraw a red marble and replace it with a white one, there will be a compensating time when we do the opposite.

Returning from marbles to genes, we note that a population in this state is at *equilibrium* with respect to the relative frequencies of the two alleles under the pressures of reversible, recurrent mutation. Genes, however, are different from the marble models in that the mutation rates of alleles in the opposing directions are not often equal. In other words, mutation is usually much less likely to replace *a* with *A* than it is to make the opposite substitution. We can show this relationship symbolically:

$$A \underset{\underset{v}{\longleftarrow}}{\overset{\overset{u}{\longrightarrow}}{}} a$$

A mutates to *a* at rate *u*; *a* back-mutates to *A* at rate *v*, and *u* and *v* are not necessarily equal.

Under these conditions, if *q* is the frequency of *a* in any generation, and $(1 - q)$ is the frequency of *A*, the *change* in the frequency of *A* due to mutation, for the next generation, will be:

the *addition* of more *A* alleles to the extent vq
the *subtraction* of *A* alleles to the extent $u(1 - q)$

The gene frequencies will be at equilibrium under mutation pressures when the additions just balance the subtractions. In other words:

$$vq = u(1 - q)$$
$$vq + uq = u$$
$$q(v + u) = u$$
$$q = \frac{u}{u + v}$$

An example will help to make this more concrete. If A mutates to a twice as frequently as a back-mutates to A,

$$u = 2v$$

then the value of q (the frequency of a) at equilibrium under mutation pressures will be: $q = \dfrac{u}{u + v}$; $q = \dfrac{2v}{3v}$; $q = \frac{2}{3}$.

In other words, the population will become stable with respect to the frequencies of the mutating alleles when the frequency of A is 0.33 and that of a is 0.67.

We can summarize this algebraic consideration of the effects of mutation pressures on the changing qualities of populations.

1. Recurrent mutation tends to spread an allele through a population and to change the relative frequencies of alleles from generation to generation.
2. This pressure toward genetic change is modified by the reversible nature of mutation. Under reversible, recurrent mutation, a population approaches an equilibrium with respect to the relative frequencies of alleles—a stable point beyond which mutation alone does not change the population. The point at which this stability is reached depends on the relative magnitudes of the component mutation rates.
3. This equilibrium must not be confused with the equilibrium discussed earlier in connection with random mating. That was a relative stability in the frequencies of *genotypes* and *phenotypes* on the basis of assigned gene frequencies. We are dealing with a more basic kind of stability here—an equilibrium of the gene frequencies themselves, under opposing mutation pressures.

The fact that a gene has numerous mutational possibilities complicates the picture somewhat. In our discussion, we have let A represent one allele at a locus—usually the wild-type allele—and a all other possible alleles. The rates u and v are therefore each made up of various component rates.

Chromosomal Mutations. In addition to changes in single genes, the general term *mutation* includes a wide variety of changes in chromosomal structure and number. These also have profound significance to evolution, and many interesting and pertinent studies have been based on them. We considered some of the evolutionary aspects of variations in chromosomal number in Chapter 7.

One kind of chromosomal mutation is probably important in providing material for new genes. You will recall that small "repeats" of genic material occur at some points in some chromosomes, as in Figure 7-6, page 198. This process of duplicating small sections of germinal material may be a first step in the origin of new loci. At first, the duplicated material is likely to have the same function as it did when single; it may represent only one gene, duplicated. But through mutation at the duplicated regions, the repeats may diverge so that in time a duplicated portion fulfills a new function, different from that of its "twin" on the chromosome. We would then clearly be dealing with two genes where one had been before.

The formation of adjacent "repeats" is not the only way in which new functions can be assumed by genes. J. A. Weir has pointed out that a part of the evolutionary progress of a species may be dependent on the biosynthetic capacities of other species. Compare yourself with Neurospora, for example. No insult is intended when we say that it is questionable which of you, the reader or the mold, comes off better in a comparison. Neurospora is a great deal more self-reliant, biochemically, than you are. The wild type grows nicely on inorganic salts, a carbon source, and biotin, while you require many vitamins, amino acids, and other compounds that Neurospora makes for itself. But you can think and/or play football, for example—activities at which Neurospora is notoriously incompetent. The nutritional jobs that the wild-type alleles do for Neurospora, are done for you by the genes of the plants and animals you eat. Somewhere back in your evolutionary past, these genes have been spared to do other jobs. No doubt a part of your proud superiority over the mold results from the genteel nutritional parasitism on other living things in which your ancestors indulged, and which you now must practice in order to live.

SELECTION—THE IMPACT OF ENVIRONMENT ON GENE FREQUENCIES

The adaptations of organisms to their environment are many and marvelous. When you consider, for instance, the variety of special characteristics of desert plants that make it possible for them to survive and multiply in their very dry environment, you can scarcely help wondering how these nice adjustments have come about. Some naturalists and philosophers have speculated that inherent variations in plants and animals may occur in anticipation of particular ends, and may in fact be directed by these ends. Others have suggested that the environment may mold or modify germ plasm in an adaptive and causative way. They have spoken of the acquisition of adaptive characters under the direct impress of the environment, and of the subsequent inheritance of these adaptive, acquired characteristics.

We are now reasonably sure that both the teleologists and the advocates of the inheritance of adaptive, acquired phenotypes are wrong. This conviction is based on two complementary considerations, a negative one and a positive. On the negative side, no one has been able to detect a mechanism whereby a peculiarity in the ordinary environment can direct a responsive, adaptive change in the germ plasm of the plants and animals we know best. Environment does indeed affect phenotypes; it affects the way genes express themselves, but it does not place an adaptive impress on the genes themselves.

The positive aspect of our conviction that environmental modifications of phenotypes are irrelevant to evolutionary changes in germ plasm is based on the fact that adaptation seems to be adequately explainable in terms of what we do now know to be true about heredity. It is explainable on the basis of the natural selection of spontaneous, inherent variations—the random mutations we have just been discussing. We need now to consider selection in relation to adaptive evolution. Artificial selection, through which man directs the evolution of plants and animals under domestication, will be discussed later.

An Example of the Effect of Selection on Gene Frequencies. To illustrate how selection affects the genetic character of a population, we can set up an extreme example. Suppose that we begin with a population at equilibrium under random mating, in which just half the individuals in the population show a recessive trait. The distributions of phenotypes, genotypes, and gene frequencies are as shown in Table 13-7.

Table 13-7. A POPULATION IN WHICH HALF OF THE INDIVIDUALS SHOW THE RECESSIVE PHENOTYPE.

Phenotypes		*A–*	*aa*
Phenotypic Frequencies		0.5	0.5
Genotypes	*AA*	*Aa*	*aa*
Approximate Genotypic Frequencies	0.09	0.42	0.5

Note: Approximate Gene Frequencies: p (frequency of A) = 0.3; q (frequency of a) = 0.7.

Now suppose that individuals of genotype *aa* are unable, in a particular environment, to mate or to leave any progeny. (This is what we mean by an *extreme* example; it can be described as complete selection against a recessive phenotype.)

The effective breeding population is then reduced to two genotypes, *AA* and *Aa*. They occur in the ratio 0.09*AA*:0.42*Aa*; or, in terms of the breeding population they represent, $\dfrac{0.09}{0.09 + 0.42} = 0.18AA$; $\dfrac{0.42}{0.09 + 0.42} = 0.82Aa$.

If these genotypes (constituting a single phenotype) mate at random in becoming the parents of the next generation, and if all of the types of matings can be assumed on the average to be equally prolific, the pattern of the next generation will be determined as shown in Table 13-8.

Table 13-8. RESULTS OF RANDOM MATINGS AFTER ELIMINATION OF RECESSIVE PHENOTYPE.

Type of Mating	Frequency	Progeny (Frequencies)		
		AA	*Aa*	*aa*
AA × *AA*	$(0.18)^2 = 0.03$	0.03		
AA × *Aa*	$2(0.18)(0.82) = 0.30$	0.15	0.15	
Aa × *Aa*	$(0.82)^2 = 0.67$	0.17	0.34	0.17
Total	1.00	0.35	0.49	0.17

You can verify that this new generation again approximates a binomial distribution of genotypes, as though it were again in equilibrium under random mating. But a fundamental change has occurred in this population. The frequency of allele *a* has *decreased* from its value of 0.7 in the previous generation to a little more than 0.4 in this generation; and there has been a corresponding *increase* in the frequency of *A*. In terms of phenotypes, the change is even more impressive; the recessive phenotype has declined from a frequency of 0.5 to about 0.17 in a single generation of complete selection against this trait.

Other Kinds and Degrees of Selection. Selection against a dominant allele can be even more effective. In the extreme case, preventing all the carriers of a dominant allele from mating or leaving progeny can completely eliminate this allele from the population in a single generation. It can return to the population only through mutation or migration.

On the other hand, except in the common case of lethals, natural selection is not often as intense as the examples we have been considering. Instead of one genotype (or phenotype) being prevented from leaving *any* progeny in the next generation, much more commonly one phenotype will be only a little less effective in reproduction than others. And the situation may be complicated by the fact that heterozygotes may not be quite the same as homozygous dominants— that is, dominance may not be quite complete.

As the intensity of selection varies, the effectiveness of selection in changing the population varies too. Intense selection can result in very rapid change. Even mild selection, in time, may spread a desirable gene through a population, or virtually eliminate an undesirable one. But under very mild selection, the time required for these ends may be more than the time available, even on a geological scale. In other words, as an allele approaches neutrality in terms of its selective value to the organism, selection becomes less and less important compared with other pressures affecting the frequency of the allele in the population.

Dependency of Selection on Gene Frequencies. In populations at equilibrium under random mating, the proportion of individuals carrying a recessive allele

Table 13-9. DISTRIBUTION OF A RECESSIVE ALLELE BETWEEN
HOMOZYGOTES AND HETEROZYGOTES.

	Genotype Distribution			Ratio $\dfrac{Aa}{aa}$	Value of $\dfrac{Aa}{aa}$
Value of q	AA	Aa	aa		
0.9	0.01	0.18	0.81	$\dfrac{0.18}{0.81}$	0.22
0.5	0.25	0.50	0.25	$\dfrac{0.50}{0.25}$	2.0
0.1	0.81	0.18	0.01	$\dfrac{0.18}{0.01}$	18.0
0.01	0.9801	0.0198	0.0001	$\dfrac{0.0198}{0.0001}$	198.0

hidden in the heterozygous condition, compared with those homozygous for the recessive allele, *increases* as the frequency of the allele *decreases* (Table 13-9).

When *a* is frequent relative to *A*, for example $q = 0.9$, there are about five times as many homozygous recessives as there are heterozygotes in the population. But when *A* and *a* are equally frequent, there are only half as many homozygous recessives as there are heterozygotes. And when *a* is so rare as to have a frequency of 0.01, about 198 individuals in the population will carry it hidden in the heterozygous condition for every individual that shows the recessive trait.

You can see how this would influence the effectiveness of selection against a recessive trait. When the trait is common, the responsible allele is so distributed among phenotypes that it is very vulnerable to selection. As we have seen, a single generation of complete selection against a recessive trait having a phenotypic frequency of 0.5 can reduce its frequency to 0.17. But in the next generation, a larger proportion of the representatives of the undesired allele in the population will be carried by heterozygotes, where it is screened from selection

Table 13-10. EFFECTIVENESS OF SELECTION AGAINST A RECESSIVE
PHENOTYPE AT DIFFERENT LEVELS OF INITIAL FREQUENCY.

Generations of Complete Selection Against a Recessive Trait	Frequencies of the Trait in Populations Having Different Initial Frequencies			
0	0.005	0.050	0.500	0.990
1	0.004	0.033	0.172	0.249
2	0.004	0.024	0.086	0.111
3	0.003	0.018	0.051	0.062
4	0.003	0.014	0.034	0.040
5	0.003	0.011	0.024	0.028

Source: From Snyder, *Milbank Mem. Fund Quart.*, **26**:*328*, 1948.

by its dominant alternative. Instead of eliminating the undesired allele entirely, as one might expect from the remarkable progress in the first generation, another generation of complete selection will only decrease the incidence of the recessive trait from 0.17 to about 0.09. In the next generation the incidence will be about 0.05, then 0.03, then 0.02. And from this point on, progress in eliminating the trait, even with complete selection against the recessive phenotype, is very slow. Table 13-10 summarizes the effects of five generations of complete selection against a recessive trait at several different levels of initial frequency.

This suggests that, with very rare alleles, extremely intense selection over long periods of time is required to change a population significantly.

You may recall from our discussion of the taster trait that the expected frequency of recessive progeny in dominant × dominant matings is $\left(\dfrac{q}{1+q}\right)^2$. With complete selection against the recessive phenotype, only dominant × dominant matings occur; the frequency of the recessive type therefore decreases in one generation from q^2 to $\left(\dfrac{q}{1+q}\right)^2$. It may readily be shown that in two generations this frequency is $\left(\dfrac{q}{1+2q}\right)^2$, and in n generations the frequency of the recessive type is decreased to $\left(\dfrac{q}{1+nq}\right)^2$. This expression provides insight into the rate of progress under this kind of selection, and its relationship to allele frequency.

Equilibrium Between Mutation and Selection. What happens when mutation of gene A to its recessive allele a, occurring at a given rate u, is opposed by a selective disadvantage of a? This is doubtless a very common situation. Here we introduce the concept of *relative fitness;* the ability of a recessive type to survive and reproduce is reduced by a fraction s as compared with the dominant phenotype.

GENOTYPE	FREQUENCY	RELATIVE FITNESS
AA	p^2	1
Aa	$2pq$	1
aa	q^2	$1 - s$

The allele frequencies are at equilibrium when new mutations to a are balanced by the eliminations of a by selection. The fraction of mutant genes eliminated is sq^2, and at equilibrium:

$$u = sq^2$$

$$q^2 = \frac{u}{s}$$

For example, if the mutation rate $A \longrightarrow a$ is $1/100{,}000$, and s is 0.1, at equilibrium q^2 is $1/10{,}000$ and q is $1/100$. If you try substituting other values for q

and *s* in the above equations you will get some idea of how high a mutation rate it would take to maintain a recessive phenotype in a population with moderate frequency in the face of strong selection against it, and, conversely, how small a selection pressure it takes to keep a phenotype relatively infrequent even when the mutation rate is rather high—if the frequency depends only on a balance between selection and mutation.

Maternal-Fetal Incompatibility. An important kind of selection is illustrated by the Rh blood groups in man. The Rh characteristic of major clinical significance depends on a simple Mendelian alternative, in which individuals of genotype *RR* and *Rr* have the Rh-substance on their red blood cells, while *rr* people lack this specificity and are Rh-negative. When an Rh-negative woman is married to an Rh-positive man, their child may inherit an *R* allele from its father and therefore make the Rh substance. This material sometimes gets across the placenta and evokes an antibody response in the mother. The antibodies may in turn get back to the fetus and cause hemolytic anemia of the newborn child, or *erythroblastosis fetalis*. (Fig. 13-3).

Rh incompatibility apparently has serious consequences in about one human birth in 200 or 300 (in North America). Through the development of a transfusion technique for replacing the reactive blood of an affected baby with nonreactive blood, and through other recently developed techniques, medicine can now save the potentially erythroblastotic children that are born alive. There are several distinguishable forms of the *R* and the *r* alleles, making up a complex multiple-allelic series.

Each zygote eliminated by maternal-fetal incompatibility of the Rh sort is a heterozygote; it must have inherited an *r* allele from the mother, who is negative, and an *R* allele from its father as a basis for the incompatibility reaction. The effect of this kind of selection on a population depends on the relative frequencies of the alleles. If they are equally frequent in the population, equal numbers

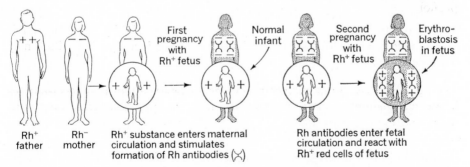

Figure 13-3. *Second or later Rh-positive babies of Rh-negative mothers are sometimes severely anemic, because antibodies that react with the baby's blood cells may be developed by the mother and passed across the placenta to the fetus.*

of the two alleles are eliminated by the selection against heterozygotes, and the population remains on the knife-edge of an equilibrium. But if one allele is more common than the other, the elimination of equal numbers of the two alleles represents the loss of a greater *proportion* of the less frequent allele, and the next generation will have a more extreme difference in the allelic frequencies. For arithmetic clarity, let us consider an extreme example. Suppose there are 600 "+" alleles of a particular gene with an incompatibility effect in a population, and 400 "−" alleles ($p = .6$; $q = .4$). If 100 heterozygotes were to be eliminated from this population, this would leave 500 "+" alleles and 300 "−" alleles. The relative frequency of the two alleles would then have changed to .625 and .375. Similarly, if the original allele frequencies were .4 for the "+" allele and .6 for the "−", the change in the next generation would be to .375 and .625. Under such a selective pattern, the population would be expected to move rather rapidly toward the fixation of one allele and the loss of the other, not because of any direct physiological advantage of the allele fixed, but only through the operation of the incompatibility mechanism.

A similar kind of maternal-fetal incompatibility has been thought to apply to the normal human blood groups. It has seemed most evident when type O mothers have A or B embryos, but not to be limited to this combination. In a type O mother, anti-A and anti-B antibodies in her serum may sometimes interfere with the development of an A or B zygote in her body. Such zygotes may be eliminated very early; because they are such early resorptions or abortions, they are not detected as embryonic deaths. The magnitude of this incompatibility load on our population is debatable at present.

This kind of selective system also illustrates important interactions among different genes. For example, type O Rh-negative women are rarely sensitized by or react against the Rh antigens of type A or B fetuses that are Rh-positive. Probably the normal anti-A and anti-B antibody of the O mother disposes so efficiently of fetal cells expressing the A or B antigen that they do not effectively reach the mother's antibody-forming tissues and are therefore unable to stimulate the formation of anti-Rh antibody. Thus, A-B-O incompatibility in the direction fetus ⟶ mother suppresses an Rh incompatibility reaction, but it is just in these maternal-fetal combinations that A-B-O incompatibility itself may become important.

Genetic Polymorphism. Under the conditions of selection so far discussed, we would expect a favored allele to approach fixation in a population and then be maintained at a high frequency but kept short of complete fixation by recurrent mutation to other forms of the gene. At this point, we might well ask why populations do in fact retain so much genetic heterogeneity. Why, for example, are the A-B-O blood group alleles kept in our population at intermediate levels of allele frequency? Why have we not virtually all become type O or A or B? Why is the allele for sickle-cell hemoglobin, so patently disadvantageous in the homozygote, maintained at a relatively high level of frequency in many populations? Many other examples will occur to you.

There is a complex of answers. The Rh-A-B-O interactions mentioned above illustrate one kind of explanation. If we were all alike for A-B-O blood types we would become more vulnerable to Rh and other types of incompatibility. Selection does not operate on isolated genes; it operates in general on phenotypes, the products of sets of genes acting in coordinated patterns in diverse environments.

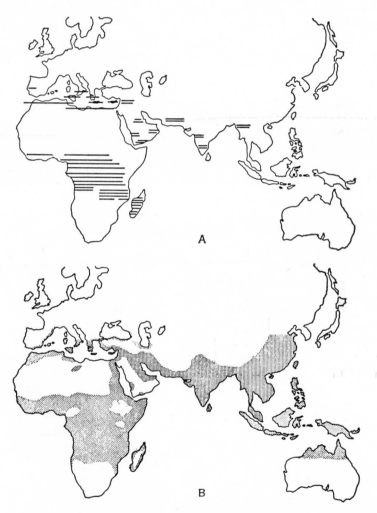

Figure 13-4. *Distribution of sickle-cell hemoglobin (bars in Map A) and falciparum malaria (shaded area, Map B).* (*From Motulsky,* Human Biology, **32**:43, 45, 1960. *By permission of the Wayne University Press,* © 1960.)

But intergenic interactions do not provide all, and perhaps not even most, of the basis for the maintenance of genetic heterogeneity in populations. Even at a particular locus, and restricting attention to only a single pair of alleles, the three possible genotypes may have "fitness" values such that the heterozygote leaves more progeny than either homozygous type, and both alleles are therefore maintained. The simplest example is probably sickle-cell hemoglobin in man. There is good reason to believe that the presence of the abnormal hemoglobin provides a condition unfavorable for the growth or maintenance of malarial parasites in the red cell. This might be a relatively indirect effect; for example, the reduced life span of these red cells may put them out of phase with the malarial organism's growth cycle in the host. In any case, in an environment that includes endemic malaria, particularly of the fatal falciparum type, individuals with only normal hemoglobin in their red cells are at a serious disadvantage. At the same time, individuals homozygous for sickle-cell hemoglobin are at another kind of disadvantage—they have sickle-cell anemia. The heterozygotes are at much less disadvantage in either respect. Compatibly with this interpretation, the "gene geography" for relatively high incidences of sickle-cell anemia coincides with areas of the world in which falciparum malaria is endemic (Fig. 13-4). In the United States, the subpopulation groups in which sickle-cell hemoglobin is still common are those that derive from ancestors who came from malarial regions.

When both homozygous classes are at some disadvantage relative to the heterozygotes, selection can be described by assigning two relative fitness values, as in the following example of sickle-cell anemia. (The allele symbols stand for hemoglobin A, the normal adult hemoglobin, and hemoglobin S, sickle-cell hemoglobin.)

PHENOTYPE	GENOTYPE	RELATIVE FREQUENCY	RELATIVE FITNESS
normal	$Hb^A \, Hb^A$	p^2	$1 - t$
sickle-cell trait	$Hb^A \, Hb^S$	$2pq$	1
sickle-cell anemia	$Hb^S \, Hb^S$	q^2	$1 - s$

The proportion of the total Hb^A alleles eliminated by the reduced fitness of normal people in a malarial environment is $\dfrac{tp^2}{p}$, and the proportion of the total Hb^S alleles eliminated by sickle-cell anemia in each generation is $\dfrac{sq^2}{q}$. At equilibrium, these proportions balance:

$$\frac{tp^2}{p} = \frac{sq^2}{q}$$

$$p = \frac{s}{s + t}; \qquad q = \frac{t}{s + t}$$

Assuming that Hb^s Hb^s is effectively lethal ($s = 1$), and that at equilibrium in a region of endemic malaria $q = 0.1$, what do you estimate to be the relative fitness ($1 - t$) of normal individuals, compared with the heterozygote?

Answer: $\dfrac{t}{1 + t} = 0.1$; $t = \frac{1}{9}$; $1 - t = \frac{8}{9}$

Malaria seems to have been an important selective factor with regard to a number of other genes as well—alleles that promote red-cell destruction. The global distributions of the sex-linked recessive allele for glucose-6-phosphate dehydrogenase deficiency (Fig. 13-5), which causes red-cell hemolysis in re-

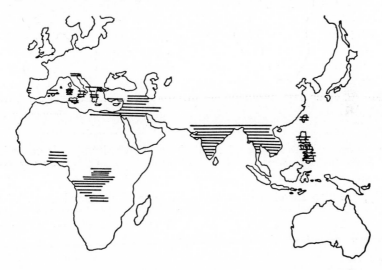

Figure 13-5. *Distribution of glucose-6-phosphate dehydrogenase deficiency. (From Motulsky,* Human Biology, **32**:50, *1960. By permission of the Wayne State University Press, © 1960.)*

sponse to particular drugs or articles of diet (especially the broad bean—which causes the condition specifically known as favism), and of thalassemia, another type of hemolytic anemia (Fig. 13-6), also coincide quite closely with an epidemiological map of malaria. It even seems that Rh incompatibility, still another type of hemolytic effect, fits into this scheme in certain areas.

But such simple situations as a direct relationship between a particular pair of alleles and a particular environmental agent are probably less common than are more complex interactions. For example, the A-B-O blood groups, besides their incompatibility effects, seem to be related in subtle ways to the incidence of a variety of diseases. (Fig. 13-7). The basis for these relationships between A-B-O groups and various diseases is not known. Some of the diseases typically occur too late in life to have much effect on the reproductive potential of the individual. Others, such as duodenal ulcer, may have some effect, because they

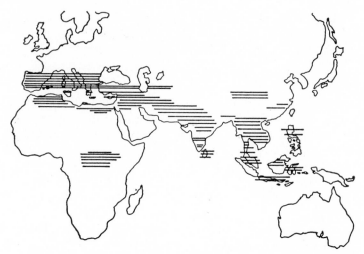

Figure 13-6. *Distribution of thalassemia. (From Motulsky,* Human Biology, **32***:47, 1960. By permission of the Wayne State University Press,* © *1960.)*

sometimes occur in young people, but the selective situation is complicated. For example, it is known that the alleles at the *secretor* locus, which determine whether the blood group substances will be found in the watery secretions of the body, are also correlated with the incidence of duodenal ulcer. Diet is a

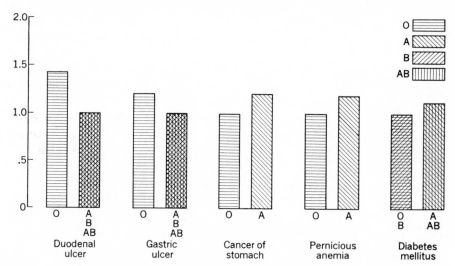

Figure 13-7. *The relative incidence of duodenal ulcer in people of blood group O, as compared with people of other blood groups, is about 1.4:1. Other disorders seem also to differ in frequency in different blood groups, as shown. (After Stern,* Principles of Human Genetics, *W. H. Freeman, 1960, p. 625.)*

significant variable in determining whether ulcers appear, and the mode of life of the individual—the amount of stress in his existence—is also critical. Many interactions remain to be examined and established, before even the relatively straightforward polymorphism of the A-B-O groups can be explained. Undoubtedly many other polymorphisms are still to be recognized.

Situations in which selection acts against the various homozygous classes, often at different times and places and in different ways, confer a greater net fitness on heterozygous types, and lead to "balanced polymorphism" in natural and experimental populations as well as in man. Th. Dobzhansky, B. Wallace, and a group of geneticists at Sao Paulo, Brazil have contributed particularly interesting and significant studies of these phenomena. They have pointed out that the territory occupied by a species is often a mosaic of many *microenvironments*, with different genetic types better fitted to particular niches. Heterozygous genotypes are often found to have the highest fitness averaged over all environments, even though they may not be most fit in any single niche.

Genetic Load and Genetic Death. Genetic heterogeneity in a population produces many individuals that have less than maximal fitness. The reduction in fitness has been called *genetic load;* it may be considered as the price a species pays for a vital advantage, the ability to respond to a changing environment and to occupy various environmental niches. Several components of this price are recognized. The *mutation load* represents the extent to which the population is impaired by recurrent mutation; most mutations are disadvantageous, and are eliminated in time by selection because they impair fitness. But the ability to mutate is obviously worth the price—the price to be paid for the possibility of evolving. The *segregation load* results from the maintenance, because of a net advantage to heterozygous types, of alleles that are unfavorable in homozygous condition, as discussed in the preceding section. The cost is in lost homozygotes, constantly being produced by mating among heterozygotes; the gain is sheer survival. The *recombination load* comes from the maintenance of favorable gene combinations, which when broken up through genetic recombination become less advantageous. The segregation and recombination loads occur only in sexually reproducing species and are therefore the price paid for the evolutionary advantages of sexual reproduction. The *incompatibility* load was discussed earlier with reference to red-cell antigens in man; when polymorphism maintained by selection involves antigens, its price may be the risk that offspring, inheriting from their fathers an antigen differing from their mothers', will be eliminated by their mothers' immune responses. At least two other forms of "loads" have been given names, but will not be discussed here.

All these loads together represent the cost to the population of the ability to adapt through Mendelian mechanisms—to have at hand a store of alleles, combinations of genes, and physiological and morphological variations—in short, the cost of being able to evolve. There are, of course, limits on the load a population can carry; when too many progeny fail to survive to reproduce in turn, the population faces extinction. How large the load can be is in part dependent on the size and structure of the population itself; as we will see in the next section, large and freely interbreeding species can retain much genetic heterogeneity, while a small closed population, such as a colony of mice being systematically inbred, can maintain very little. In our own species, it has been estimated that the typical person carries from three to five "lethal equivalents" and about four "detrimental equivalents." That is, the typical load adds up to about the same ultimate reproductive effect as if each of us were carriers of three to five lethal heterozygous genes and four detrimentals causing such effects as childhood abnormalities. Geneticists find most worrisome the contemplation of factors in our environment tending to increase this load, such as increasing radioactivity or certain chemical mutagens suggested for use in pest control. Evaluations in this area are difficult, because they involve weighing many kinds of social, political and economic values along with biological ones. Even the purely biological values are not easy to estimate.

RANDOM FLUCTUATIONS AND THE EFFECTS OF POPULATION SIZE

In the preceding section, we considered mainly the mechanisms by which a population responds to its environment in adaptive ways through changing gene frequencies under the impact of selection. Other factors are at work as well in the changing genetic structure of populations, some of them indifferent to the fitness of the genotype. One of these is the phenomenon of random *drift* in gene frequencies, most important in populations of limited size.

Sampling Errors in Mendelian Inheritance. Throughout our discussion of gene frequencies in evolution we have assumed that we were dealing with populations of infinite size, in which the sampling of the genes of a parent population always achieved the ideal, expected values of Mendelian inheritance. But we are now familiar enough with the laws of chance to know that this is an unrealistic assumption for populations of finite size.

Suppose that in a breeding population of 1,000 individuals, 500 males and 500 females, mating in pairs and having two offspring each, a new mutation m has appeared in one female. The ideal expectation is shown in Table 13-11.

We would expect, therefore, that the next generation under these arbitrary conditions would again have 999 homozygotes to 1 carrier of m. But in fact we know that the two offspring of the original heterozygous female would not always show the 1:1 ratio. On the contrary, *both* of these offspring might happen to get m from their mother, or, with equal probability, perhaps *neither* of them would get m. The chance is $\frac{1}{4}$ that the allele will double in frequency the next generation, $\frac{1}{2}$ that it will remain the same, and $\frac{1}{4}$ that it will be lost, simply as

Table 13-11. MENDELIAN EXPECTATION FOR THE TRANSMISSION
OF A NEW MUTATION.

Mating Types ♀ × ♂	Number of Matings	Progeny	
		MM	Mm
MM × MM	499	998	0
Mm × MM	1	1	1
Total	500	999	1

a result of the sampling nature of Mendelian inheritance. The probability of gain or loss in allele frequency is enhanced by the fact that not all parents have the same number of progeny. Which types may have more progeny and which fewer than the average in the population are also subject to chance variation; these sampling errors have been shown to be as important as the fluctuation of allele samples from matings of heterozygotes as a source of *drift variance* in allele frequencies.

If allele *m* remains in the population, it is subjected to the same kind of sampling errors in the next generation; it might in successive generations "drift" toward higher and higher frequencies. But if it is lost in the sampling process in any generation, it cannot be replaced simply by a compensating variation in sampling in another generation; it can only be replaced by the rare recurrence of the same mutation, again to be subject to the risk of going over the edge of zero frequency through the vagaries of chance.

This risk of loss from the population is so great for alleles at very low frequency that in a finite population only a little more than one in ten mutations conferring no selective advantage will still be present after 15 generations. Of course, if an allele is beneficial, it has a somewhat better chance of surviving, but R. A. Fisher calculated that with a 1 percent selective advantage only about 271 mutations out of 10,000 will be retained after 127 generations. Many mutations are lost within a few generations, regardless of their selective value. Those that do persist may spread by chance fluctuations in frequency through a population, even though they may be selectively neutral, or sometimes even in the face of adverse selection.

These fluctuations are not limited to very rare alleles, although the fluctuations are most critical in connection with very infrequent alleles because the process is not in itself reversible when an allele is lost (or fixed through the loss of its alternative). Even when $p = q = 0.5$, chance may achieve $p = 0.55$, $q = 0.45$ in a generation, and similar fluctuation may occur in the same or the opposite direction in another generation. The probable magnitudes of such chance deviations are related to the size of the breeding population, as we shall see in the following section.

An Effect of Population Size. The significance of chance fluctuation is related to the number of representatives of an allele being sampled, as well as to the frequency of the allele. If an allele has a frequency of 0.01 in a population of 5,000 individuals, there are actually 100 representatives of this allele in the population. Almost all of them are in heterozygotes. Some of these representatives will be lost in polar bodies as heterozygous females form eggs; some will be lost in sperm that never fertilize eggs; some will be lost in individuals that die before they can reproduce. But the alternative of the allele in question is subject to the same chance losses in heterozygotes; and with almost 100 individuals carrying the infrequent allele there is a good chance that positive and negative chance variations will come near cancelling each other out.

An allele with a frequency of 0.01 in a population of 50 individuals is in a much more precarious position. There is only one representative of this allele in this population, and it exists in a single individual. If this individual fails to reproduce or leaves only a few offspring, or if the offspring do not survive, for reasons quite independent of their possession of this allele, the allele may well be lost. There is little opportunity for positive and negative chance variations to compensate for each other, and the drift in gene frequency may be extreme.

Similarly, at other levels of gene frequency in a population the absolute number of alleles being sampled affects the probability of wide deviation. If $p = q = 0.5$, a population with the absolute numbers $50A:50a$ alleles might very well give a chance sample of $41A:59a$ in the next generation. This would represent an insignificant deviation from the expected 1:1 ratio; that is, it is reasonably probable as a result of chance alone. The gene frequency would change, then, to $p = 0.41, q = 0.59$; and there would be no reason to expect that it must drift back the next generation. In fact, random drift in gene frequency is on the average cumulative from generation to generation; the amount in one generation may not be much, but in several generations frequencies may drift widely.

In a population with the absolute numbers $500A:500a$, a similar deviation to $p = 0.41, q = 0.59$ would mean a sample of alleles $410A:590a$. But this is very unlikely, as you can confirm with a chi-square test. In fact, a deviation to $p = 0.47, q = 0.53$ in the large population is less probable than a deviation to $p = 0.41, q = 0.59$ in the population of 100 alleles.

The sizes of breeding units in nature are often small; an individual is likely to mate with another born nearby, and the total population is often broken up into a large number of relatively small subpopulations, interbreeding to some extent but still largely self-sufficient and separate. Furthermore, the winter populations of many species are very small, and chance may play a great part in determining which of the large number of individuals of a summer population will successfully "overwinter" to become the parents of the next summer's population. Many ecological observations demonstrate great summer-to-summer changes as a result of drift during the "population bottleneck" represented by the small winter populations.

Some species show cyclic types of fluctuation in population size over a period of years. Species of limited mobility (snails, for example) are likely to be broken up into small, practically isolated breeding units. The homing phenomenon, the restricted breeding range, the phenomenon of "territory" in animal sociology, and many other factors decrease the size of effective breeding populations and therefore increase the significance of nonadaptive drift in the genetic history of these populations.

The Inbreeding Effect in Small Populations. The fixation or loss of alleles in an inbred population is really a consequence of this same kind of drift. Inbreeding in itself, of course, does not change gene frequencies for the population as a whole; it only results in a relative increase in homozygous genotypes as compared with heterozygotes. But the practical limit on the size of an inbred population automatically results in the frequent selection of homozygous individuals to become the parents of the next generation; and the result of the sampling error in a small population is the rapid elimination of alleles under intense inbreeding. Systematic inbreeding in controlled populations is, of course, to be distinguished from random mating in small populations; in the former homozygosity is increased because of nonrandom mating, while in the latter the zygotes are present in Hardy-Weinberg proportions.

The most extreme example of inbreeding is the maintenance of a line by the self-fertilization of an individual in each generation. The size of the breeding population is 1. A given heterozygous gene pair under this condition has a 50-50 chance of being fixed as either homozygote in each generation; and this is going on for all the heterozygous loci in the population at once. About half of the heterozygosity is therefore automatically lost in each generation; the process of fixation and loss of alleles goes on so rapidly that selection cannot hope to keep up with it. About all that a corn breeder can do under this system, for example, is to produce a large number of inbred lines, and to choose, from among the survivors of seven or eight generations of self-fertilization, those that happen through chance to have arrived at homozygosity for alleles that will be useful in further breeding operations.

The process of fixation and loss is slower under less intense inbreeding; even continued brother-sister mating offers considerable opportunity for conscious selection of the particular genotypes that should be used to continue the inbreeding program. As the size of the breeding population increases, the likelihood of chance mating between close relatives decreases, and the inbreeding effect in large populations becomes almost negligible. Sewall Wright has calculated that in a population of N breeding individuals about 1 in $4N$ alleles are lost and about 1 in $4N$ are fixed in each generation. Under certain breeding conditions the rate of fixation and loss may be smaller.

The "Isolate Effect"—Another Effect of Population Size. In human populations, rare alleles often occur in "pockets" of relatively high frequency; they are not uniformly distributed over a state or a country. In many areas of the world, but in America to a lesser extent than was true earlier, a small rural population, centered in a village, may show a remarkably high incidence of polydactyly, or webfeet, or fragile bones and blue sclera, for example—characteristics that are rare in the general population. The high frequency of particular disorders and the total absence of others reflects the fact that the population traces back to a relatively small number of individuals, among whom particular alleles were present or absent.

Dahlberg has called attention to an important consequence of the breakdown of such partial *isolates* in human populations. If there are N individuals in Community A, a given locus is represented 2N times. Assume that there are c representatives of an undesirable recessive allele in this population. The frequency of this allele will be $\frac{c}{2N}$, and the incidence of the recessive phenotype at equilibrium will be $\left(\frac{c}{2N}\right)^2 = \frac{c^2}{4N^2}$.

Now suppose that the boys and girls of Community A and of Community B, a nearby town of equal size, come to represent a freely intermarrying population. If the undesirable recessive allele is absent from Community B (as is likely if this allele is rare in the general population), and if the effective population size becomes 2N individuals, the frequency of this allele becomes $\frac{c}{4N}$, and the incidence of the recessive phenotype, in a single generation of random marriage, becomes $\frac{c^2}{16N^2}$. The recessive phenotype is now only one-quarter as frequent in this new, merged population as it was in the smaller population with which we began. If the allele is rare, many generations of complete selection against it would have been required to effect a similar result in the original population. However, isolate breakdown leads to a decrease in recessive phenotype frequencies only if the gene frequencies differ in the merging populations, as in the example we have discussed. The primary basis for the effect is the reduction in allele frequency in the new, larger population unit.

MEIOTIC DRIVE

Another mechanism affecting the genetic composition of populations has only recently been clearly recognized. It is based on systematic deviations from the Mendelian principle that a heterozygote *Aa* forms, through meiosis, two types of gametes *A* and *a*, in equal numbers. The section on preferential segregation in Chapter 7 is relevant here; you will recall that corn plants heterozygous for an abnormal chromosome 10 tend to include the knobbed chromosome preferentially in the functional megaspore. Somewhat similar situations are known in Drosophila. In females heterozygous for a pair of structurally different chromosomes, one of the two chromosomes may be systematically retained in the egg while the other is extruded into the nonfunctional polar bodies. The physically shorter chromosome of a pair with unequal lengths is more frequently included in the functional egg nucleus. Females heterozygous for chromosomal deficiencies or translocations, or carrying attached-X or other compound-X chromosomes, often also display preferential segregation. Even in the male, *sex ratio* effects have been known for many years to result from abnormal spermatogenesis leading to degeneration of all sperm carrying the Y-chromosome in particular cases, or the X-chromosome in others.

The term *meiotic drive* has been coined to emphasize the evolutionary effects of inequalities in chromosomal transmission. Regardless of their individual selective advantages or disadvantages, the genes on a chromosome favored by preferential segregation make a disproportionate contribution to the next generation. Meiotic drive, therefore, can result in increases in frequency of alleles opposed by even strong selection, or decreases in frequency of alleles that would otherwise be favored. Of course, the mechanisms that lead to meiotic drive are themselves subject to selection.

MIGRATION

Our consideration of the mechanisms of evolution up to this point can be fitted into a framework something like this: Species containing numerous individuals may be subdivided into breeding units of various sizes. Within each of these units, the processes of adaptive evolution through natural selection in the particular environment of the unit, and of nonadaptive change through mutation pressures and chance drift in gene frequencies, progress to some degree independently of the changes going on in other units. Varieties inherently different from each other therefore develop within the species, and these may represent the usual first stage in the origin of new species.

But if the subunits of a population continue to "pool" their genes through migrants that act as genetic canals among units, the varieties must remain to some degree similar. Unless a unit is reproductively isolated from an adjacent one, the two populations evolve *together*, at least to some extent; they cannot diverge to the point of consistent and uniform differences represented by distinct species. The role of migration (or perhaps better, from some points of view, the role of *isolation*) in the origin of species is therefore an important one. It is also important through its modification of the effective sizes of breeding populations, discussed in a preceding section, and through its effects on potential recombinations of genes, to be discussed below.

Migration and Recombinations of Genes. Some time before the nature of Mendelian heredity was understood, the importance of isolation in evolution had been emphasized. It was generally believed that the interbreeding of inherently distinct groups of individuals resulted in a dissolution of their differences, and that there was a sort of blending effect in inheritance, so that an intermediate, invariable hybrid type replaced the diverse parent types. The origin of species—of clear-cut and discontinuous groups of plants and animals—was therefore believed to require the reproductive isolation of groups of individuals from each other, so that these groups might diverge and remain distinct. This concept was regarded by some biologists as the key to organic evolution.

You will recognize that the concept contained both fact and fallacy. We now know that inheritance is not a blending process, but is a matter of segregation and recombination of discrete particles. Hybridization does not result in the formation of invariable blends; on the contrary, it can provide for the emergence in later generations of novel forms through the recombinations of alleles that differed in the parent types. On this level, migration among inherently different subunits of a population leads to an increase, not a decline, in variation in the total population. It gives selection a wider variety of phenotypes upon which to operate and therefore facilitates adaptive evolution. And it provides an excellent balancing mechanism against the tendency of small subpopulations to become homozygous through the chance fixation or loss of alleles.

In all these respects, the pre-Mendelian emphasis on isolation as a factor in organic evolution was at variance with what we now know about the machinery of heredity. But in another regard isolation was and still is properly emphasized: The divergence, in time, of subunits of a parent population, until two or more of these subunits reach a point of essential discontinuity and are recognizable as distinct species, depends on the reproductive isolation of these subunits. Insofar as evolution is indeed a matter of the "origin of species," some form of isolation is basic to this process.

Isolating Mechanisms. There are in nature a great many mechanisms effective in preventing the pooling of the genes of diverging groups and therefore in permitting the development of genetic discontinuity among these populations. These isolating mechanisms act at two levels: to prevent the mating of individuals from two populations, or the fertilization of eggs of one population by sperm from another; and to prevent the reproduction of hybrids even when such hybrids are produced. Included in the first category are various geographical, ecological, psychological, and physiological barriers to cross-fertilization. The second category includes the formation of inviable, aberrant, or sterile hybrids. We discussed certain of the bases for hybrid sterility, with particular reference to chromosomal behavior in meiosis, in Chapter 7.

Keys to the Significance of This Chapter

The extension of Mendelian genetics to the patterns of inheritance in populations is important to many branches of biology, particularly those dealing with human, agricultural, and natural populations. The primary differences between these populations and the usual laboratory populations of the geneticist involve variations in the relative frequencies of alleles and in the patterns of mating.

The relative frequencies of alleles can be computed as gene frequencies and treated as probabilities. Under random mating, a population approximates in a single generation an equilibrium distribution of genotypes and phenotypes for an autosomal allelic pair. This binomial distribution of genotypes provides a basis for analysis of the genetic structure of populations, and suggests useful tools for research and applications in the field of human heredity. Gene-frequency analyses can be extended to such special aspects of genetics as sex linkage, multiple alleles, and cousin marriages.

Population changes in gene frequency under migration, random fluctuation, meiotic drive, selection, and mutation pressures are fundamental to evolutionary change. Large populations are typically broken up into smaller subunits, varying in size and in degree of isolation. Since migration provides some degree of genetic continuity throughout the large population, new mutations and new combinations of genes, when they confer advantage on their possessors, can be capitalized upon by the population generally, and there is adequate occasion for adaptive evolution under selection. But at the same time the random fixation and loss of genes continues in the subunits at rates inversely proportional to their sizes. Nonadaptive changes, especially in characteristics near neutrality in selective value, can therefore accumulate in the subunits, and the subunits can also diverge as they become adapted to particular, different niches in the general environment.

When complete isolation occurs, cutting some of the subunits off from the parent population, the "leveling" process of migration stops. Divergence continues in the isolated subunits, and new species come into being.

REFERENCES

Crow, J. F., *Genetics Notes*. Minneapolis: Burgess Publishing Co., 5th ed., 1963. (Notes for genetics lectures, including good brief surveys of population genetics and selection, Chaps. 19 and 22.)

————, "Selection." In *Methodology in Human Genetics*, edited by W. J. Burdette. San Francisco: Holden-Day 1962. (Competent and interesting treatment of the analysis of selection in human genetics.)

Dahlberg, G., "Genetics of Human Populations." *Advances in Genetics*, **2**:69–98, 1948. (Consideration of factors affecting the genetic structure of human populations, with particular emphasis on isolates.)

Darlington, C. D., *The Evolution of Genetic Systems*. Cambridge: Cambridge University Press, 1939. (Cytological aspects of evolution.)

Darwin, C., *On the Origin of Species by Means of Natural Selection, or the Preservation of Favoured Races in the Struggle for Life*. New York: Appleton, 1860. (The classic, still well worth studying.)

Dobzhansky, Th., *Genetics and the Origin of Species*. New York: Columbia University Press, 3rd ed., 1951. (Thorough, well written, and interestingly presented survey.)

Falconer, D. S., *Introduction to Quantitative Genetics*. New York: Ronald Press, 1960. (Includes excellent coverage of population genetics and selection.)

Fisher, R. A., *The Genetical Theory of Natural Selection*. Oxford: Clarendon Press, 1930. (A pioneer work in the quantitative treatment of the effects of selection.)

Haldane, J. B. S., *The Causes of Evolution*. New York: Harper, 1931. (See especially the appendix, "Outline of the Mathematical Theory on Natural Selection.")

Hardy, G. H., "Mendelian Proportions in a Mixed Population." *Science*, **28**:49–50, 1908. (With Weinberg's [*op. cit.*], the classic statement of the binomial distribution of genotypes.)

Horowitz, N. H., "On the Evolution of Biochemical Syntheses." *Proc. Nat. Acad. Sci.*, **31**:153–157, 1945. (Consideration of how complex biosynthetic sequences under gene control may have been established.)

Lerner, I. M., *The Genetic Basis of Selection*. New York: Wiley, 1958. (An excellent textbook at a high level.)

Li, C. C., *Population Genetics*. Chicago: University of Chicago Press, 1955. (A clear and interesting elementary textbook.)

Mourant, A. E., *The Distribution of the Human Blood Groups*. Springfield, Illinois: C. C. Thomas, 1954. (Data and discussions of frequencies of the blood groups in various human populations.)

Muller, H. J., "The Guidance of Human Evolution." *Perspectives in Biology and Medicine*, **3**:1–43, 1959. (A thoughtful and provocative essay.)

Patterson, J. T., and Stone, W. S., *Evolution in the Genus Drosophila*. New York: Macmillan, 1952. (A superlatively thorough and interesting account of much that we have been unable even to mention, this book will richly repay careful study.)

Race, R. R., and Sanger, R., *Blood Groups in Man*. Oxford: Blackwell, 4th ed., 1962. (An up-to-date and readable coverage of the field. The basis of some of the examples in this chapter.)

Shapiro, H. L., *The Heritage of the Bounty*. New York: Simon & Schuster, 1936. (An anthropologist's account of Pitcairn through six generations. See Problems 13-12 to 13-17.)

Simpson, G. G., *The Major Features of Evolution*. New York: Columbia University Press, 1953. (Draws from genetics and other disciplines in a broad and clear analysis of organic evolution.)

Stern, C., *Principles of Human Genetics*. San Francisco: W. H. Freeman and Co., 2nd ed., 1960. (Especially Chap. 10, "The Hardy-Weinberg Law," and Chaps. 28 and 29 on selection for material similar to that covered in this chapter.)

Wallace, B., and Srb, A., *Adaptation*. Englewood Cliffs: Prentice-Hall, 1961. (An interesting elementary paperback, much of its content relevant to this chapter.)

Weinberg, W., "Über den Nachweis der Vererbung beim Menschen." *Jahreshefte Verein f. vaterl. Naturk. in Württemberg*, **64**:368–382, 1908. (See also Hardy, *op. cit*. An important part of this paper is translated in C. Stern's, "The Hardy-Weinberg Law." *Science*, **97**:137–138, 1943.)

Weir, J. A., "Sparing Genes for Further Evolution." *Iowa Acad. Sci.*, **53**:313–319, 1946. (Presentation and discussion of the point made on p. 412.)

Wright, S., "On the Roles of Directed and Random Changes in Gene Frequency in the Genetics of Populations." *Evolution*, **2**:279–294, 1948. (Essentially a reply to criticism, with references.)

————, "Evolution in Mendelian Populations." *Genetics*, **16**:97–159, 1931. (A classic of the population-genetics approach to evolutionary mechanisms.)

QUESTIONS AND PROBLEMS

13-1. What do the following terms signify?

binomial	adaptation	isolating mechanism
gene frequency	evolution	natural selection
random mating	inbreeding effect	random drift in gene frequencies
meiotic drive	isolate effect	relative fitness

Thalassemia major is a severe anemia, usually fatal in childhood and rather frequent in Mediterranean populations. Thalassemia minor is a very mild anemia, often difficult o detect at all.

✓ 13-2. Among people of southern Italian or Sicilian ancestry now living in Rochester, New York, thalassemia major occurs in about one birth in 2400, and thalassemia minor in about one birth in 25. Extrapolating these frequencies to a population of 10,000, the distribution is approximately as shown in the following tabulation:

Thalassemia major	Thalassemia minor	Normal
Th Th	*Th +*	*+ +*
4	400	9596

Verify that the frequencies of the *Th* allele and its normal alternative in this population are about 0.02 and 0.98 respectively.

√ 13-3. Does the population approximate the binomial distribution of genotypes expected from this gene frequency? $,0004$ $,0392$ $,9554$

$4es$

Consider the group of alleles controlling positive reactions to the primary Rh reagent as a single alternative to the group of alleles controlling negative reactions to this reagent. Thus,

$$R- \text{ = Rh-positive}$$
$$rr \text{ = Rh-negative,}$$

for the Rh grouping of greatest clinical significance. Assume a population in which 16 per cent of the individuals are Rh-negative. This is a convenient figure in the arithmetic to follow, and is not far from the frequency in our own population.

Erythroblastosis resulting from Rh incompatibility has in the past represented a form of selection, since failure of a proportion of children of a particular genotype to survive affects the frequency of the alleles responsible for their inviability.

$,16 = f(rr)$ $f(r) = \sqrt{.16} = .4$ $f(R) = .6$

√ 13-4. Assuming the population to be at equilibrium under random mating for this alternative, what is the calculated frequency of the *r* allele? Of *R*?

$.16(rr)$ $,36(RR)$ $,48(Rr)$

√ 13-5. What is the calculated distribution of genotypes in this population?

√ 13-6. A man in the above population is Rh-positive and is married to an Rh-negative woman. What is the probability that the man is heterozygous (*Rr*)? If he is heterozygous, what is the probability that their first-born child will be Rh-negative? What is the probability that the first-born child of an Rh-positive man and an Rh-negative woman will be Rh-negative?

$Prob$ of $Heter = .57$ of

"   " $1^{st} born Rh^- = .5$

$.28 + .43 = .71$

13-7. In our population, about one birth in 250 is erythroblastotic. Affected infants are heterozygous, since they are Rh-positive infants born to Rh-negative mothers. Assume for the purposes of this problem that these infants are born alive but do not survive.

a. How many *R* alleles would be found in a population of 1,000 newborn babies the next generation?

b. How many *r* alleles?

c. How many heterozygotes would be eliminated? 4

d. How many *R* alleles would be eliminated? 4

e. How many *r* alleles would be eliminated? 4

f. Confirm that the effect of selection in this circumstance would be to reduce slightly the frequency of *r* in each generation and to increase correspondingly the frequency of *R*.

g. What would you expect to be the effect of long-continued selection of this sort?

h. Why do you suppose that the Chinese, a population of long, relatively unmixed history, are practically homozygous *RR*?

13-8. Repeat the calculations of Problem 13-7, but beginning with a population in which the initial frequencies are reversed (i.e., *p*, frequency of *R*, = 0.4; *q* = 0.6). In which direction will the allelic frequencies move under selection in this population?

13-9. Present opinion is that the gene frequencies for Rh alleles in the North American white population are the result of relatively recent intermixtures of populations in which one or the other allele had been much more predominant. Why?

13-10. Color blindness occurs in about 8 percent of men in the North American white population, but in only about 4 percent of Negro men. Assuming sex-linked recessive inheritance, what frequencies of color blindness would you predict for women in each of these populations?

13-11. Actually, about 1 percent of North American white women, and about 0.8 percent of Negro women, are color-blind. How might you account for the excess of affected females found in both populations, compared with your prediction in Question 13-10?

In 1790, several British mutineers from H. M. S. *Bounty*, together with a few men and women native to Tahiti, established a settlement on Pitcairn Island. This new island population remained largely isolated for several generations.

The preceding statements represent historical fact; the following are in part deviations from the true history of Pitcairn, as far as it is known, for the purposes of these problems. Assume that:

1. Families were founded by six men from the *Bounty*, of whom three were blue-eyed, two brown-eyed but heterozygous for blue, and one homozygous brown-eyed; and that there were two Tahitian men and eight women, all homozygous for brown.

2. The blue eyes-brown eyes alternative depends on a single autosomal allelic pair with brown (*B*) dominant, and no modifiers of eye color capable of obscuring this alternative were segregating in this population.

3. Intermarriage among the residents of Pitcairn Island was at random with respect to eye color, with approximately equal numbers of descendants surviving from each type of marriage.

13-12. What were the gene frequencies for the eye-color alleles among the six white men, two Tahitian men, and eight Tahitian women who, we have assumed, established families on Pitcairn Island? Was this population at equilibrium in this regard at the time of its establishment?

13-13. What genotypic and phenotypic proportions would you expect to prevail in this population after it had reached equilibrium?

13-14. Fill in the blank spaces in the following table, illustrating the predicted results of the first generation of marriages on Pitcairn Island.

TYPE OF MARRIAGE		NUMBER	NUMBER OF PROGENY†		
♀	♂	OF MARRIAGES	*BB*	*Bb*	*bb*
BB × *BB*		3	6	0	0
BB × *Bb*		2	2	2	0
BB × *bb*					
	Total	8	8	—	—

† Assuming two children per marriage.

Is the population at equilibrium after one generation? Is this inconsistent with the statement that equilibrium is achieved in one generation of random mating? Why?

13-15. Fill in the blank spaces in the following table, representing marriages among the immediate progeny of the original settlers of Pitcairn Island. (We assume that the eight *BB* individuals in the preceding table were equally distributed between the sexes, and make similar assumptions for the other classes; and that the ideal chance distribution of mating types was actually realized even in this small population. Note, however, that we need not postulate brother-sister marriages here.)

TYPE OF MARRIAGE ♀ ♂	NUMBER OF MARRIAGES	NUMBER OF PROGENY† BB	Bb	bb
BB × BB	2	4	0	0
BB × Bb	2	2	2	0
Bb × BB	—	—	—	—
Bb × Bb	—	—	—	—
Other types	0	0	0	0
Total	8	9	—	—

† Assuming two children per marriage.

Is the population now at equilibrium?

13-16. Fill in the following table for still another generation of random marriages on Pitcairn Island, utilizing fractions and probabilities rather than absolute numbers of children. Is equilibrium maintained under the assumptions we have made? What advantages has the fraction-and-probability method?

TYPE OF MARRIAGE†	FREQUENCY	TYPES OF PROGENY BB	Bb	bb
BB × BB	$(\%_{16})^2$	$81/_{256}$	0	0
BB × Bb	$2(\%_{16})(\%_{16})$	$54/_{256}$	$54/_{256}$	0
BB × bb	$2(\%_{16})(\%_{16})$	0	$18/_{256}$	0
Bb × Bb	———	$9/_{256}$	—	—
Bb × bb	———	—	—	—
bb × bb	———	—	—	—
Total	———	—	$96/_{256}$	—

† Note that the sexes are not distinguished in listing the mating types.

13-17. Dr. H. L. Shapiro, in 1935, six generations after the original settlement, found that about 5.6 percent of the residents of Pitcairn Island had "light" eyes. Assume that the eye colors of the original settlers, and the inheritance of eye color in this population, were approximately as we have described them. Assume also that the "light" eyes as classified by Shapiro can be taken for our present purposes to represent the "blue" eyes of our series of problems. List several likely deviations from our further assumptions and predictions that might help to explain the eye color of the present-day population.

At the California Institute of Technology 171 students of General Biology found their blood groups to be distributed as follows:

Blood Group	O	A	B	AB	Total
Number	77	60	24	10	171
Percentage	45%	35%	14%	6%	100%

13-18. What blood-group frequencies would be expected among 171 students, if the distribution characteristic of Brooklyn (Table 13-6) represented ideal expectation? Do Cal Tech sophomores differ significantly from the expectation so defined? (Compute p in a χ^2 test using the nearest whole numbers in the *expected* class).

13-19. What are the frequencies of the I^A, I^B, and i alleles among Cal Tech sophomores? Does this population appear to conform to a random, equilibrium distribution, as far as you can judge from the distribution of phenotypes?

13-20. Assuming that Cal Tech men marry at random women from a population like their own, what types of children, and in what proportions, may be expected by Cal Tech men of blood group A, married to group O women? To group A women?

13-21. In our discussion of first-cousin marriages in this chapter, we computed the probability that both John and Mary (Fig. 13-2) would be heterozygous for the same rare recessive carried by GGF_2. Compute similarly the probability that both John and Mary will be heterozygous for a rare recessive carried by their common grandmother, GGM_2. Now, what is the probability that a child of John and Mary will be homozygous for both the GGF_2 recessive and the GGM_2 recessive? For neither?

13-22. Suppose that GGF_2 was heterozygous for two independent, rare recessives not present in any other member of his generation in the pedigree. What is the probability that a child of John and Mary would be homozygous for one or the other or both of these?

About one normal allele in 30,000 mutates to the sex-linked recessive allele for hemophilia in each human generation. It is difficult to observe reverse mutation of a human recessive essentially lethal like the allele for hemophilia. Assume for purposes of the following problems that one h allele in 300,000 mutates to the normal alternative in each generation. In terms of the formulation on page 410, this can be described as follows:

$$+ \xrightleftharpoons[v]{u} h, \text{ where } u = 10v$$

13-23. What gene frequencies would prevail at equilibrium under mutation pressures alone in these circumstances?

13-24. Check your answer to Problem 13-23 by verifying the following:
   a. At the calculated equilibrium frequency, assuming a population with a total of ten million alleles at this locus, about 900,000 of these would be the + allele, and about 9,100,000 would be the *h* allele.
   b. In the above circumstance, about thirty of the + alleles would mutate to *h* in each generation, and this would be balanced by the mutation of about thirty *h* alleles to +.

13-25. The conclusions in the above two questions are obviously at variance with the fact that *h* remains rare relative to its normal allele. Why does *h* not become increasingly frequent?

13-26. Suppose that a quick, easy, and effective method of stopping or preventing bleeding in hemophilia should be developed. (Some methods are in fact at hand, but at present they are not effective over long periods of time.) Under the conditions postulated in Problems 13-23 to 13-25, what might you expect to happen to the frequency of hemophilia in successive generations?

13-27. Carry the complete selection against a recessive trait, illustrated in Table 13-8, another generation, confirming that the frequency of the recessive phenotype will be reduced from 0.17 to about 0.09.

13-28. Would you expect a sex-linked recessive to be more or less vulnerable to selection than an autosomal recessive? Why? *More if lethal*

13-29. Although breeders of purebred black-and-white cattle have selected completely against the autosomal recessive red phenotype for many generations, there is no tangible evidence of current progress toward the elimination of this rare allele from the black-and-white breeds. Why?

13-30. Suppose that a method is developed to identify cattle carrying the recessive red allele in the heterozygous condition. What change in the effectiveness of the selection described in Question 13-29 would follow the application of this method to the selection of breeding stock in black-and-white breeds?

13-31. Drosophila trapped in nature and brought in to be bred in the laboratory are often found to be heterozygous for recessive lethals. Why have these lethals not been eliminated by natural selection?

13-32. What characteristic of the usual laboratory mating pattern renders the lethals of Question 13-31 more vulnerable to selection in the laboratory than in nature? Why?

13-33. Animals and plants under domestication typically display a great deal more genetic variation than do comparable populations in the wild. How may the evolutionary forces and mechanisms we have discussed be involved in this contrast between domestic and natural populations?

13-34. What common kind of selection leads to an equilibrium at some intermediate level of gene frequency, rather than to the loss from the population of an allele being selected against?

13-35. There is a suggestion that beef cattle heterozygous for a particular recessive dwarfism have been selected, in preference to homozygous normals, to be used for breeding stock. Such heterozygotes may on the average be more efficient producers of the better cuts of meat than are homozygous normal cattle. Homozygous dwarfs are of course very undesirable. In qualitative terms, what would you expect to be the equilibrium situation for this pair of alleles?

Assume a generally rare kind of recessive mental defect in man, the recessive allele being present with a frequency of 0.2 in Community A but absent from Community B of equal size.

√ *13-36.* If Community A is at equilibrium, what proportion of the people in it will show the recessive trait?   .04

√ *13-37.* If Communities A and B come to represent a freely intermarrying population, what proportion of the people in this new, single population will show the recessive trait at equilibrium? What was the incidence of the recessive trait in the two communities taken together, before their amalgamation?   .02

√

13-38. Estimate, by comparison with Table 13-10, about how many generations of complete selection against the recessive trait would have been necessary to reduce its frequency to the extent achieved by the breakdown of the isolate under the assumptions made in Problems 13-36 and 13-37.   5

13-39. In contrast to their importance in plant evolution, changes in numbers of whole chromosomes or sets of chromosomes are generally believed to have been relatively unimportant to the evolution of higher animals. Discuss the probable relevance of the following statements to this opinion:

a. Animals frequently depend on a genic balance mechanism for sex determination.
b. Animals do not often self-fertilize, so that a spontaneous tetraploid, for example, would usually have to mate with a diploid.
c. The higher animals do not often reproduce asexually.

13-40. We have observed that mutation may be induced by a variety of external agents. Do you regard such induced mutations as examples of the inheritance of acquired characteristics? Why?

13-41. Show algebraically that for a single gene difference in a population mating at random, the proportion of the recessive type among the progeny of recessive ×

dominant phenotype matings is $\dfrac{pq^3}{p^2q^2 + 2pq^3} = \left(\dfrac{q}{1+q}\right)$, and that among the progeny of dominant $\times$ dominant phenotype matings the proportion with the recessive phenotype is $\dfrac{p^2q^2}{p^4 + 4p^3q + 4p^2q^2} = \left(\dfrac{q}{1+q}\right)^2$.

13-42. Show algebraically that after n generations of complete selection against the recessive type, the frequency of the recessive phenotype will be $\left(\dfrac{q}{1+nq}\right)^2$, where q is the original frequency of the recessive allele.

The fear is sometimes expressed that medical advances that allow genetic defectives to reproduce will lead to the genetic deterioration of mankind. In this context:

13-43. Suppose that a medical advance permits the eradication of malaria as a significant disease of man. What is likely to happen to the incidence of sickle-cell anemia? Would this medical advance lead to genetic deterioration or improvement?

13-44. Suppose that a medical advance makes it possible for some sickle-cell anemics to reproduce—say s changes from 1 to 0.5 in the example on p. 421. Ignoring muta-tion, what will be the equilibrium distribution of allele frequencies expected
    a. if t remains $\frac{1}{9}$, as calculated in the example?
    b. if t changes to 0, as assumed in problem 13-43?
What can you say about the genetic effects of these medical advances under the above assumptions? How may considerations of equilibrium between mutation and selection modify your answer, over periods of a few or many human generations? Is the common apprehension that medically saved defectives will reproduce geometrically and soon come to predominate in the population valid, in your opinion?

CHAPTER 14

Quantitative Inheritance

Mendel's genius as an experimenter led him to choose, as parents for his crosses, individuals that differed by sharply contrasting alternative characteristics, and to keep careful numerical counts of the progenies with reference to the readily definable characteristics on whose inheritance he focused his attention. Much of the progress since Mendel's time in elaborating the nature and functions of the germ plasm has depended on his precedent in this regard—on categorizing and counting attributes that can be described as *qualities* of individuals. Even understanding the genetics of populations, as treated in the preceding chapter, has been based in large part on statistical analyses of qualitative characteristics.

Many of the most important attributes of plants, animals, and man, however, cannot be adequately described as "sharply contrasting alternative characteristics"; they are differences of degree along continuous scales of measurement, and are best expressed in such terms as inches, pounds, bushels, or I.Q.'s. These variations are *quantitative* in nature, and characteristics showing them are called *quantitative characters*. They cannot be identified with one or two major genes, but rather depend upon the action and interaction of several genes. Typically, they are subject to considerable phenotypic modification by environment.

Examples of quantitative characters include stature in man, egg-laying in chickens, yield of grain in cereal crops, and milk production by dairy cattle. Such characteristics are of great practical importance, and an understanding of their inheritance is a primary objective wherever useful application of genetic principles is a goal.

The Statistical Approach

The description and analysis of quantitative characters require special tools provided by the branch of mathematics called statistics. We must turn briefly to this field before proceeding with the genetic analysis of quantitative characters.

The Arithmetic Mean. You are already familiar with, and often utilize, one of the central concepts of statistics. In describing phenomena subject to considerable variation, you commonly refer to things that are true "on the average," or to *mean* or *average* values. You have used such terms as "mean annual rainfall," and no doubt you have calculated your "grade-point average." You know that to compute your average grade in genetics to date, for example, you need only add up the individual grades and divide the sum by the total number of grades added. The resulting quotient is the arithmetic mean, or average. If the individual grades do not "count" equally, a weighted average is used. If your midterm examination grade is to be given three times the weight of a single short quiz, you simply add in three times the midterm grade, and count this as three grades in determining the weighted average.

You may also have had occasion to use a shortcut approximation to the arithmetic mean, when some large number of values was to be averaged. This involves first grouping the data into arbitrary classes and counting how many observations fall within each class. The mean is obtained by taking the sum of the products of these frequencies by their respective midclass values and then dividing by the total number of observations. The treatment of the values in Tables 14-1 and 14-2 illustrates this method of calculation.

Table 14-1. INDIVIDUAL WEIGHT GAINS OF ONE HUNDRED SWINE
OVER A TWENTY-DAY PERIOD.

3, 7, 11, 12, 13, 14, 15, 16, 17, 17, 18, 18, 18, 19, 19, 19, 20, 20, 21, 21, 21, 22, 22, 23, 23, 24, 24, 24, 25, 25, 25, 26, 26, 26, 26, 27, 27, 27, 28, 28, 28, 29, 29, 29, 29, 30, 30, 30, 30, 30, 30, 30, 30, 30, 30, 31, 31, 31, 31, 32, 32, 33, 33, 33, 33, 33, 34, 34, 34, 35, 35, 35, 36, 36, 36, 37, 37, 38, 38, 39, 39, 39, 40, 40, 41, 41, 41, 42, 42, 42, 43, 43, 44, 45, 46, 47, 48, 49, 53, 57

Source: The values, slightly modified from experimental data, are from Snedecor's *Statistical Methods*, Iowa State College Press.

In Table 14-1, the individual weight gains of 100 swine over a 20-day period are arranged in order of magnitude from the weight gain of the pig that gained only 3 pounds to that of the one that showed a gain of 57 pounds. In Table 14-2, these values are collected in a *frequency distribution*, representing eleven classes with a "class interval" of 5 pounds. The class midpoints (midclass values) are placed at successive multiples of 5 pounds. You should check this table carefully,

Table 14-2. FREQUENCY DISTRIBUTION OF THE WEIGHT GAINS IN TABLE *14-1.*

Midclass value (pounds), v	5	10	15	20	25	30	35	40	45	50	55
Frequency, f	2	2	6	13	15	23	16	13	6	2	2
(fv)	10	20	90	260	375	690	560	520	270	100	110

$$\text{Mean} = \frac{\Sigma(fv)}{N} = \frac{3005}{100} = 30.05 \text{ pounds}$$

and also the calculation of the mean gain in weight, to be sure you understand what has been done with the original measurements of Table 14-1.

Other Measures of Central Values. The arithmetic mean is by far the most useful single statistic for describing the "central tendency" of a normal population. For certain purposes, however, two other central values, the *mode* and the *median*, are utilized. The mode is simply the *most frequent value* in the population; thus the modal class in the population of weight gains we have been considering is 30 pounds. The median is the *middle item* in an array; in any sample there are as many values above the median as below it. For the weight gains we are considering, 30 pounds, again, is the median value. It is not an attribute of all populations that the mean, mode, and median coincide.

The *geometric mean* is a special average value useful in certain cases. You will have noticed that the arithmetic mean of two different numbers is midway between them on an additive scale. Thus the arithmetic mean of two and eight is five; it is the middle term in an arithmetic series (2, 5, 8) in which the difference between successive terms is three. The geometric mean is a similar midpoint, but on a multiplying (geometric) scale. The geometric mean of two and eight is four; it is the term between two and eight in a geometric series (2, 4, 8) in which there is a constant of multiplication rather than of addition. The geo-

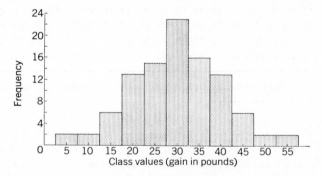

Figure 14-1. *A histogram of the weight gain values summarized in Table 14-2 shows the familiar bell-shape of a normal distribution. (After Snedecor,* Statistical Methods, *4th ed., Iowa State College Press, 1946, p. 56.)*

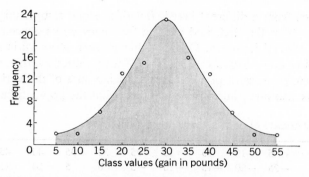

Figure 14-2. *If the frequency distribution in Table 14-2 is plotted as a smooth curve, one obtains a close approximation of a normal curve. (After Snedecor,* Statistical Methods, *4th ed., Iowa State College Press, 1946, p. 56.)*

metric mean of two numbers is the square root of their product. We call attention to the geometric mean because it describes some of the data of quantitative inheritance better than does the arithmetic mean.

Normal Distributions. The values in Tables 14-1 and 14-2 are good representatives of a kind of distribution called a *normal* distribution. If you consider the frequency distribution in Table 14-2, you can scarcely fail to be impressed with its symmetry. The central class, in which the mean, the mode, and the median of the population coincide, is the high point of a balanced distribution of frequencies that falls off regularly and symmetrically in both directions. This can most easily be pictured graphically, as in Figures 14-1 and 14-2. In these figures, the class values are arrayed along the baseline of a graph, and the frequencies are placed on the vertical scale. In both the histogram of Figure 14-1 and the smooth curve of Figure 14-2 the familiar bell-shape of a normal distribution is evident.

The distribution is familiar because it is experienced commonly. The heights of a sample of boys of a given age are normally distributed around a mean value in much this same way. The intelligence quotients of a large and randomly selected sample of people also show a normal distribution around the average. A great number of similar examples, from both the living and the inanimate worlds, could be cited. Because of the common occurrence and regularity of normal distributions, their precise description in mathematical terms is of great value in characterizing and comparing quantitative population data. We cannot approach this problem with mathematical rigor here, but we can and should consider those conclusions—the formulas—derived from rigorous mathematical treatment that are most important to statistical analysis.

The Standard Deviation. The average gain of 30 pounds that characterized the sample of swine we considered earlier might be quite closely approached in a sample taken from a different breed. But this second sample might be much more uniform than the first, and show a frequency curve concentrated much more closely around its mean. Or, it might be much more variable, with the curve spreading more broadly (Fig. 14-3). Mean values by themselves, then, give incomplete pictures of populations; some measure of the variation within populations is also essential to understanding what the populations are like.

Table 14-3. CALCULATION OF THE AVERAGE DEVIATION.

Midclass value	5	10	15	20	25	30	35	40	45	50	55
Deviation, d	-25	-20	-15	-10	-5	0	5	10	15	20	25
Frequency, f	2	2	6	13	15	23	16	13	6	2	2
(fd)	-50	-40	-90	-130	-75	0	80	130	90	40	50

$$\text{Average deviation} = \frac{\Sigma(fd)}{N} = \frac{775}{100} = 7.75$$

Note: We ignored the "$+$" and "$-$" signs of the deviations in order to compute this *average deviation;* if this were not done, the positive and negative deviations would compensate for each other, and the average deviation, for this or any other distribution, would approximate 0.

How can such variation be expressed? One reasonable method is represented in Table 14-3. We can compute the *average deviation from the mean* simply by multiplying all the class differences from the mean by their respective frequencies, and dividing by the total number of classes.

(Confirm the following average deviations for the values upon which curves A and B in Figure 14-3 are based:

MIDCLASS VALUE	FREQUENCIES	
	DISTRIBUTION A	DISTRIBUTION B
5		3
10		6
15		9
20	2	11
25	21	13
30	54	16
35	21	13
40	2	11
45		9
50		6
55		3
Average Deviation	2.5	10.1

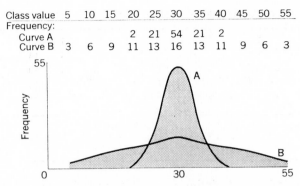

Class value	5	10	15	20	25	30	35	40	45	50	55
Frequency:											
Curve A				2	21	54	21	2			
Curve B	3	6	9	11	13	16	13	11	9	6	3

Figure 14-3. *Frequency curves for hypothetical populations more uniform (curve A) and more variable (curve B) than the population illustrated in Figure 14-2, but with the same mean values.*

It should be clear that the magnitude of the average deviation does in fact reflect the variability within each sample.)

We have introduced the average deviation as a measure of variability, or "spread," because you can easily see how this statistic is related to the population characteristic it is supposed to measure. In practice, the average deviation is seldom calculated; another kind of average, which you will now find relatively easy to compute and understand, is used instead. This is the *standard deviation* (σ). It is essentially the *square root of the average squared deviation*. Expressed in symbols this would be

$$\sigma = \sqrt{\frac{\Sigma(fd^2)}{N}}$$

By reasoning more complex than we are prepared to detail, mathematicians can show that $(N - 1)$ is a more reasonable divisor to use in computing the standard deviation for a sample than is N. The formula actually used to measure variation, therefore, is

$$\sigma = \sqrt{\frac{\Sigma(fd^2)}{N - 1}}$$

Table 14-4 shows the computation of σ for the weight-gain values of Table 14-1.

(Verify that for distribution A (p. 446), $\sigma = 3.8$, and for distribution B, $\sigma = 12.5$.)

You should realize that the values in these illustrations permit maximal ease of calculation. Usually, the mean will not coincide with the midpoint of the central class, but will be some other value. It could be, for example, 28.5 or 31.1, for other samples of weight gains from the same population we have been considering. In such cases, the deviations are not nice even numbers to be squared, and calculations become something of a chore. Again, there exist

Table 14-4. CALCULATION OF THE STANDARD DEVIATION.

Midclass value	5	10	15	20	25	30	35	40	45	50	55
Frequency, f	2	2	6	13	15	23	16	13	6	2	2
Deviation, d	-25	-20	-15	-10	-5	0	5	10	15	20	25
d^2	625	400	225	100	25	0	25	100	225	400	625
(fd^2)	1250	800	1350	1300	375	0	400	1300	1350	800	1250

$$\sigma = \sqrt{\frac{\Sigma(fd^2)}{N-1}} = \sqrt{\frac{10175}{99}} = 10.1$$

shortcut methods (and calculating machines) to make the computations some-what easier, but we will leave such devices to your further experience with statistical methods.

The Standard Error. We have now realized a primary aim of practical statistics—the reduction of a mass of quantitative data to a few precise values that picture faithfully the original data and render them susceptible to com-parison and analysis. Instead of the 100 separate figures for weight gains shown in Table 14-1, we now have, representing this sample:

$$N = 100; \qquad \bar{x} \text{ (the mean)} = 30.05; \qquad \sigma = 10.1$$

We have a better picture of the sample than we could have comprehended in the original array, and are in a much better position to compare the sample with others.

However, our statistical descriptions up to this point are largely matters of convenience. We have arrived at a shorthand way of describing a *sample,* by computing its mean and its standard deviation, but have not yet considered the main purpose for which the sample is taken. This is to get as reliable an idea as possible of the population that the sample represents. In almost any quanti-tative experiment, a certain limited number of observations is made as a sample of a general situation. These observations provide true values for the sample investigated. But the larger problem is: "What does the sample tell about the general situation which exists under the conditions investigated?" The values we have computed are *statistics;* their significance is that they are estimates, or more or less reliable samples, of true and stable values in much larger parent populations. It is because the statistics provide information about the larger population from which the sample was drawn that the experiment has meaning outside the sample it comprises.

How is the mean of a sample related to the mean of the population from which it was taken? If different samples are drawn from the same population, the means of these samples will not all be the same, but will vary around the population mean. Most of them may be near the population mean; a few will show considerable deviation from it. If we were to plot such sample mean values as a frequency distribution, we would find that they are normally distributed around the true mean. This kind of distribution is in turn subject to description in the terms now familiar to us: there is an average value for the means that coincides with the population mean, and there is a standard deviation of the means that is related to the amount of variation in the population and to the adequacy (size) of the samples.

Mathematical analysis establishes that the standard deviation of this distribution of means (called the *standard error* of the mean) can be estimated from the standard deviation of any sample as follows:

$$S.E. = \frac{\sigma}{\sqrt{N}}$$

where N is the size of the sample and σ is its standard deviation. The standard error gives an indication of how much the means of other similar samples drawn from the same population might be expected to vary. In any normal distribution (and this is true of the normal distribution of means to which the standard error applies), about two-thirds of the individual observations have values within one standard deviation above or below the mean of the distribution. About 95 percent fall within two standard deviations, and over 99 percent within three, above or below the mean (Fig. 14-4).

It is customary to describe any sample in terms of its *mean value plus or minus the standard error*. Thus in the sample of weight gains, we may summarize by noting that $N = 100$, $\bar{x} = 30.05$, $\sigma = 10.1$, and $S.E. = 1.0$, while 30.05 ± 1.0 gives an indication of the sample's reliability as a representation of 20-day gains in swine of the breed used and under the conditions of the experiment.

There are, of course, many other aspects of statistics that are important in biology and genetics. Tests for the significance of differences between samples, or between observation and hypothesis, are as useful in quantitative inheritance

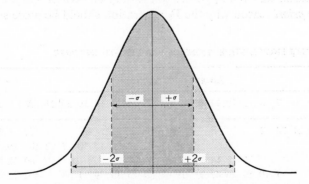

Figure 14-4. *A normal distribution, showing proportions of the distribution that are included between $\pm 1\sigma$, $\pm 2\sigma$, or $\pm$ more than 2σ, with reference to the mean.*

as they were in dealing with the ratios of qualitative characteristics. If you go on with advanced study in genetics or some other biological science, you will encounter more of these statistical techniques and will gain increasing appreciation of their utility. At present, equipped with a few simple statistical tools, we can proceed to a consideration of quantitative inheritance.

The Inheritance of Quantitative Characters

Significant aspects of quantitative inheritance emerge when crosses between short-eared and long-eared parental lines of corn are carried to the F_2. An extensive study of this kind has been reported by R. A. Emerson and E. M. East. Their data are found in Table 14-5. Your first glance at the data will confirm that ear length is a quantitative character, in that measurements of individual ears, instead of falling readily into discrete classes, show continuous variation.

The pertinent difference between the parental types, Tom Thumb popcorn (parent 60) and Black Mexican sweet corn (parent 54), can be seen most readily in a comparison of their respective means for ear length, which are 6.63 and 16.80 centimeters. Within each parental line, however, there is some variability in ear length. And this is also true of the F_1 and F_2 populations. Viewed as histograms, shown in Figure 14-5, all these different populations fall into rough approximations of normal distributions. If we assume each parental line to be homozygous for most genes affecting ear length, it would seem that a good deal of the variability within lines must be due to environment. Furthermore, members of the F_1 populations should be genetically alike. Here also, then, we must suppose that most of the variation around the mean is derived from outside influences. And, naturally, if ear length is subject to alteration by environment, there is no reason why the F_2 generation should be exempt. On the other hand, there is no *a priori* reason why the F_2 generation should be more susceptible to

Table 14-5. FREQUENCY DISTRIBUTIONS OF EAR LENGTH IN MAIZE.

	Length of Ear, cm																						
	5	6	7	8	9	10	11	12	13	14	15	16	17	18	19	20	21	N	$\bar{x}$	σ	S.E.
Parent 60:	4	21	24	8														57	6.632	.816	.108
Parent 54:							3	11	12	15	26	15	10	7	2			101	16.802	1.887	.188
$F_1(60 \times 54)$:				1	12	12	14	17	9	4								69	12.116	1.519	.183
F_2:			1	10	19	26	47	73	68	68	39	25	15	9	1			401	12.888	2.252	.112

Source: After Emerson and East, *Nebraska Research Bull. 2*, 1913.

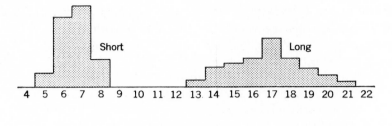

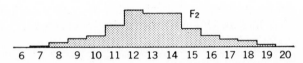

Figure 14-5. *Distributions of ear length of corn (in centimeters) in the parental lines and* *F_1 and F_2 generations studied by Emerson and East. The vertical axes* *represent percentages of the different populations. (From Sturtevant and* *Beadle,* An Introduction to Genetics. *W. B. Saunders Co., 1940, p. 265.)*

environmental influence than the parental generations or the F_1. The increase in variability in F_2, therefore, might be taken as preliminary evidence for genetic segregation and recombination.

With the complicating factor of environmental influence in mind we can make several observations about these data, which are typical of findings in many similar studies.

1. The mean of the F_1 is approximately intermediate between the means of the long- and short-eared parents.
2. The mean of the F_2 is similar to the mean of the F_1.
3. The F_2 is appreciably more variable than the F_1, as shown by the σ values in Table 14-5 and also by the "spread" of the histograms in Figure 14-5.
4. The extreme measurements in the F_2 overlap well into the distributions of parental values.

Now, do these basic observations make sense in terms of genetics as we presently understand it? Obviously, no simple genetic situation already familiar to you entirely explains the observations. (You should ask yourself why, for instance, it is an inadequate hypothesis to suppose that the ear length difference of the parental types depends on different members of a single allelic pair that shows incomplete dominance.)

The Multiple-Gene Hypothesis. The general type of result just summarized can be reasonably well explained in terms of the multiple-gene hypothesis, originally called the "multiple-factor hypothesis." This hypothesis, which represents one of the significant advances in genetic thought, we owe particularly to East and to a Swedish geneticist, Nilsson-Ehle. Since it was first proposed around 1910 the multiple-gene hypothesis has been refined and amplified by a large number of workers in several fields. Its basic tenets, however, remain as cornerstones in our understanding of quantitative hereditary phenomena.

In its simplest form, the hypothesis proposes that many aspects of quantitative inheritance may be accounted for on the basis of the action and the segregation of a number of allelic pairs having duplicate and cumulative effects without complete dominance. This situation may be visualized in terms of a specific example.

A Hypothetical Model for Quantitative Inheritance. For the sake of demonstration and argument, assume the following kind of model experiment involving plant height, which, like ear length, is a quantitative character.

1. A cross is made between two true-breeding parent plants, one relatively tall and the other relatively short.
2. Under the conditions of the experiment, environment is assumed to be so uniform that it is not responsible for variation in plant height.
3. Genetic differences for height of the parents are assumed to reside at three independent loci, with the tall parent being homozygous for allelic forms designated by capital letters, $XX\ YY\ ZZ$, and the short plant being homozygous $xx\ yy\ zz$.
4. The alleles designated by small letters are inert with reference to plant height.
5. Each allelic form designated by a capital letter acts in such fashion as to contribute 3 inches of height in any plant in which it is present.
6. Each of the parents, apart from loci represented by X, Y, and Z, has the same genotype for plant height, and this is expressed by 60 inches of growth.

Under the assumptions we have just made the short parent ($xx\ yy\ zz$) is 60 inches tall. The tall parent is 78 inches. An F_1 hybrid ($Xx\ Yy\ Zz$) would be 69 inches high, which is exactly intermediate between the parents. And, as shown in Table 14-6, the mean height of F_2 plants would equal the height typical of the F_1; however, the F_2 population would show the effects of segregation by considerable variability in plant height.

Several of the results predicted for our hypothetical experiment conform with the actual findings of Emerson and East with reference to the inheritance of ear length in corn. These include the intermediacy of the F_1 as compared to the parents, the greater variability in F_2 than in F_1, and the fact that the F_2 and F_1 mean values are approximately equal.

However, we must recognize at the same time that several of the assumptions on which our model experiment is based are more or less unreal. Certainly, in practice, the influence of environment on phenotypic expression could not be eliminated. Complete lack of dominance at each locus seems unlikely. Our assumption that the several genes work to identical effect by adding equal amounts to plant height is likewise a vast oversimplification. And, finally, the assumption that only three allelic pairs act to produce height differences may be far removed from the complexity of most real instances of quantitative inheritance. Actually, of course, our model might be made more meaningful if we assumed a larger number of genes involved, each having a smaller individual influence on plant height. If this assumption were made, the F_2 population of the model would be modified in the direction of more nearly *continuous* variation, and would thus be more typical of quantitative inheritance. As the model stands, F_2 individuals fall into discrete classes, represented by 60, 63, 66, . . ., 78 inches.

On the whole, then, although our model seems to explain a good deal, its

Table 14-6. A HYPOTHETICAL MODEL EXPERIMENT FOR ILLUSTRATING A SIMPLE MULTIPLE-GENE INTERPRETATION OF SIZE INHERITANCE.

Individual contributions of genes are assumed to be

$X = Y = Z = 3$ inches $\qquad\qquad$ $x = y = z = 0$ inches

The residual genotype gives a plant 60 inches tall

Tall Parent (78 inches) $\qquad\times\qquad$ Short Parent (60 inches)

$XX\ YY\ ZZ$ $\qquad\qquad\qquad\qquad$ $xx\ yy\ zz$

F_1: $Xx\ Yy\ Zz$ (69 inches)

F_2	GENOTYPE	HEIGHT	F_2	GENOTYPE	HEIGHT
1	$XX\ YY\ ZZ$	78	2	$Xx\ yy\ ZZ$	69
2	$XX\ YY\ Zz$	75	4	$Xx\ yy\ Zz$	66
2	$XX\ Yy\ ZZ$	75	1	$xx\ YY\ ZZ$	72
2	$Xx\ YY\ ZZ$	75	2	$xx\ YY\ Zz$	69
4	$XX\ Yy\ Zz$	72	2	$xx\ Yy\ ZZ$	69
4	$Xx\ YY\ Zz$	72	4	$xx\ Yy\ Zz$	66
4	$Xx\ Yy\ ZZ$	72	1	$XX\ yy\ zz$	66
8	$Xx\ Yy\ Zz$	69	2	$Xx\ yy\ zz$	63
1	$XX\ YY\ zz$	72	1	$xx\ yy\ ZZ$	66
2	$XX\ Yy\ zz$	69	2	$xx\ yy\ Zz$	63
2	$Xx\ YY\ zz$	69	1	$xx\ YY\ zz$	66
4	$Xx\ Yy\ zz$	66	2	$xx\ Yy\ zz$	63
1	$XX\ yy\ ZZ$	72	1	$xx\ yy\ zz$	60
2	$XX\ yy\ Zz$	69			

Number of active alleles	0	1	2	3	4	5	6
Height value (inches)	60	63	66	69	72	75	78
Frequency	1	6	15	20	15	6	1

limitations must be understood. Models are conscious oversimplifications, devised to increase one's understanding of situations too complicated to be analyzed directly. Our hypothetical experiment is not meant to duplicate reality, but to provide insight into reality.

A Naturally Occurring Model for Quantitative Inheritance. Artificial models for quantitative inheritance of the kind we have examined are not based on ideas suddenly snatched from thin air. A good deal of their strength depends upon the fact that they are strongly suggested by the analysis of certain real genetic situations. A classical example of the latter is provided by the genetics of kernel color in wheat, studied extensively by Nilsson-Ehle.

Starting with true-breeding white kernel and red kernel parents, Nilsson-Ehle found that in different strains of wheat different F_2 ratios could prevail. In some strains, segregations of 3 reds to 1 white were observed; these are clearly instances of a single gene difference. In other strains, however, among the F_2 individuals the numbers of reds greatly exceeded those expected on the basis of a 3:1 relationship. Certain F_2 populations showed close approximations to a ratio of 15 red to 1 white; others seemed to be best described by a ratio of 63:1. The 15:1 can be explained as a modification of a 9:3:3:1 ratio, the assumption being that only the double-recessive individuals among the segregants are white. Likewise, the 63:1 ratio can be ascribed to the segregation of three independent pairs of alleles in the F_2, where only the triple recessives are white. Nilsson-Ehle was able to confirm these explanations through the analysis of the results of crosses designed to test the genotypes of F_1 and F_2 individuals.

These experiments, and others like them, are important because they show us clearly that multiple genes in the same organism may affect the same character. But Nilsson-Ehle's determination of the genetics of kernel color in wheat goes even farther in promoting our understanding of quantitative inheritance. The F_2 ratio of 15 red to 1 white, for example, becomes highly instructive when you realize that not all the red kernels are of the same intensity of color, and that this variation can be related directly to genotype. Table 14-7 describes how color intensity in wheat kernels depends upon the number of genes present that are active in promoting color. Thus kernel color, which on superficial analysis would seem to be a perfectly discontinuous character, on close study assumes many of the aspects of quantitative inheritance. Of course, the color variation is not quite continuous, since the different reds can be classified, but it is this very fact that permits direct analysis of the situation in terms of particular genes.

In populations where three gene pairs for kernel color are segregating, the different color classes are much more difficult to distinguish. As a consequence, a much closer approximation to continuous variation exists. You can well imagine that if even more such gene pairs were segregating in a population, and especially if there were an appreciable environmental effect on kernel color, class differences would become imperceptible, and a typical state of continuous variation would prevail.

Table 14-7. THE RELATIONSHIP OF GENOTYPE TO COLOR INTENSITY OF WHEAT
KERNELS SEEN IN CROSSES CARRIED TO AN F_2.

Parents: $R_1R_1R_2R_2$ $\times$ $r_1r_1r_2r_2$
(Dark Red)   (White)
F_1: $R_1r_1R_2r_2$ (Medium Red)

F_2	GENOTYPE	PHENOTYPE	
1	$R_1R_1R_2R_2$	Dark red	
2	$R_1R_1R_2r_2$	Medium-dark red	
2	$R_1r_1R_2R_2$	Medium-dark red	
4	$R_1r_1R_2r_2$	Medium red	15 Red
1	$R_1R_1r_2r_2$	Medium red	to
2	$R_1r_1r_2r_2$	Light red	1 White
1	$r_1r_1R_2R_2$	Medium red	
2	$r_1r_1R_2r_2$	Light red	
1	$r_1r_1r_2r_2$	White	

Summary of Phenotypes

Dark red	Medium dark red	Medium red	Light red	White
1	4	6	4	1

The Nature of the Genes Affecting Quantitative Characters. We have already said enough about the multiple-gene explanation of quantitative inheritance to indicate that it is at least a good working hypothesis. As such, it can be used as the basis for further exploration of the phenomena it begins to explain. A fundamental idea in the hypothesis is that back of quantitative characters are genes. Presumably these are genes in chromosomes, much like genes we have studied before except that their individual contributions to phenotypic differences are smaller. Indeed these individual contributions are usually completely obscured by the effects of the genotype as a whole and by the influences of environment. K. Mather has called such genes *polygenes*. He thus distinguishes them from *major* genes, which are readily identifiable because of the pronounced effects of their individual functions.

The fact that we make terminological distinctions between "multiple" genes and "major" genes does not mean that there is no overlap of the two categories or no area between them. We have discussed how the genes for kernel color in wheat provide one kind of intermediate instance.

Duplicating Effects Produced by Major Genes and Multiple-Gene Systems. The phenotypic effect usually ascribed to the operation of a particular multiple-gene complex may be more or less duplicated by the consequences of the action of a single major gene. The normal, continuous range of stature in man, for instance, seems almost certainly to depend upon multiple-genic effects. The stature of a very short person may usually be accounted for by the combined influences of a good many different genes. However, a single gene for dwarf stature may give rise to the same end result, at least insofar as height in inches is concerned.

The Possible Role of Major Genes in Multiple-Gene Systems. There seems to be no good reason why certain genes should not serve a dual capacity by simultaneously affecting both quantitative and qualitative characters. Some evidence does exist that this kind of situation occurs. In white clover, for example, two independent dominant genes interact to cause mottling and lesions of the leaf blades, a "qualitative" difference from the normal smooth green. In addition, however, the dosage of the dominant genes that interact to give mottling has a pronounced effect on leaf number, which is generally considered a quantitative characteristic of the plant.

Linkage Shown by Multiple-Gene Complexes. Finding the linkage relations of members of a multiple-gene complex is no simple matter. The primary difficulty, of course, is that members of such systems seldom produce individually identifiable phenotypic effects. There are, however, several more or less well-established instances of linkage between major genes and multiple genes.

In 1923, Karl Sax found that color in beans, which is under the immediate control of a single gene, shows linkage with bean weight, a quantitative character. Similarly, Lindstrom has shown that size of tomatoes, as determined in weight, is linked with skin color. In Drosophila, major genes in all four chromosomes appear to be linked to different genes for egg size, which is quantitative in expression. And in the mouse, the gene for brown coat color is apparently in the same chromosome as genes for a number of size characters, including adult weight and length of bone in the hind limb.

An enlightening study of polygenic inheritance has been reported by J. Crow. A strain of Drosophila resistant to DDT was developed by growing a large population of the flies in a cage painted on the inside with the insecticide, in amounts that increased irregularly over a prolonged interval. Male progeny from crosses between susceptible (control) and resistant strains were backcrossed to the parental strains; because there is no crossing over in male Drosophila, the segregation of whole chromosomes in these backcrosses could be followed by means of easily visible mutants marking the X-chromosomes and the two large pairs of autosomes. Each of the major chromosomes was found to make a significant contribution to resistance, as indicated by a regular correlation between the number of chromosomes from the resistant strain and the percentage of survival of flies exposed to DDT (Fig. 14-6). Close analysis using crossover markers succeeded in identifying particular chromosomal regions in which dominant resistance factors were located, but the evidence indicated that there are very probably several genes on each chromosome concerned with this polygenic trait.

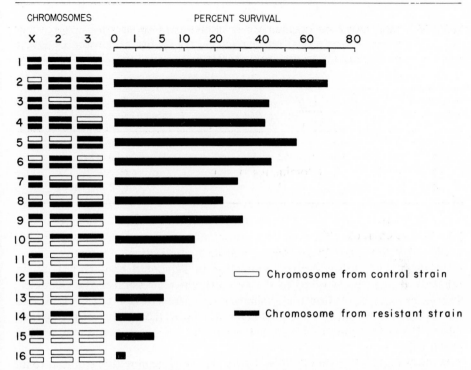

Figure 14-6. *Resistance to DDT in Drosophila is correlated in a particular backcross with the number of chromosomes from the resistant parental strain. (From Crow, Annual Rev. of Entomology, 2:228, 1957.)*

Estimating the Number of Members in a Multiple-Gene System. Knowledge of the number of genes involved in the expression of quantitative characters would facilitate the development of new and better methods for investigating phenomena of quantitative inheritance. Such knowledge could also be put directly to use by plant and animal breeders, who must often deal with the practical problem of predicting probabilities for recovering, from a segregating population, individuals showing the extreme forms of a quantitative character. For example, how often in a population of dairy cattle showing segregation for butterfat production can one expect appearance of the best kind of producer? What is the probability that in the progeny of hybrid tobacco plants the breeder will be able to find at least one individual as tall as the tallest of the original parents? Obviously the solution of such problems depends in large part on the number of genes involved.

A bit of simple reasoning provides a way for estimating the number of genes operating in a given multiple-gene system, when the system approaches the simplicity of the models we discussed earlier. Go back to Table 14-7, which summarizes one of Nilsson-Ehle's naturally occurring models of quantitative

Table 14-8. PROBABILITY OF OCCURRENCE OF INDIVIDUALS MANIFESTING AN EXTREME EXPRESSION OF A QUANTITATIVE CHARACTER, UNDER CONDITIONS IN WHICH INDEPENDENTLY SEGREGATING ALLELIC PAIRS HAVE DUPLICATE, CUMULATIVE EFFECTS.

Number of Allelic Pairs	Number of Segregating Alleles	Fraction of Population Showing an Extreme Expression of the Character
1	2	$(\frac{1}{2})^2 = \frac{1}{4}$
2	4	$(\frac{1}{2})^4 = \frac{1}{16}$
3	6	$(\frac{1}{2})^6 = \frac{1}{64}$
4	8	$(\frac{1}{2})^8 = \frac{1}{256}$

inheritance. You see that under conditions of two independently segregating allelic pairs that show no dominance and have duplicating, cumulative effects, one-sixteenth of the segregating population in F_2 should be dark red, or, in other words, can be expected to show an extreme expression of the character. The generalized prediction under similar circumstances is that $(\frac{1}{2})^n$ describes the expectation for an extreme phenotype when n = the number of segregating alleles. Table 14-8 shows the essential relationships for one, two, three, or four allelic pairs.

With this table, it is simple to apply the kind of reasoning that can lead to an estimate of the effective number of gene pairs in a multiple-gene system. In brief, if an F_2 population shows approximately $\frac{1}{64}$ individuals of each of the extreme phenotypes, it can be supposed that three segregating allelic pairs are involved. If approximately $\frac{1}{256}$ of such a population were the frequency of one of the extreme types, we might conclude that four pairs of alleles affecting the character had segregated. Unfortunately, the method we have described is of little use when five or more allelic pairs are involved. From a variety of lines of evidence, geneticists have been led to believe that many quantitative characters represent the composite influence of genes at more than 10—perhaps occasionally more than 200—loci.

Other, more complicated ways of estimating gene number for quantitative characters have been formulated and are available in the genetic literature. But no methods so far devised are able to take sufficiently into account the complications of overlapping effects of the environment, linkage among members of a multiple-gene system, differences in dominance relationships among such members, and individual variations in expression of the effects of gene action. Most attempts to determine number of multiple genes, then, must be characterized as brave efforts rather than as successfully completed missions. However, the picture is not entirely discouraging. Probably for some characters, reasonable approximations of the numbers of multiple genes in operation have already been attained. And if geneticists are never able to do better than give orders of magnitude for these numbers, still the information will be helpful in the solution of some of our most important genetic problems.

Gene Action in Multiple-Gene Systems. We have already indicated that a weakness in our hypothetical model for quantitative inheritance is the assumption of simple additive effects of the action of all the operative genes involved. The assumption at first sight is attractive, largely, perhaps, because it is simple. It also appears to have some force, because it is suggested by the experiments of Nilsson-Ehle with kernel color in wheat, which have long served as models and as standard points of departure for the investigation and interpretation of quantitative inheritance. Neither of these reasons is sufficient for generalizing the assumption, however, in face of the fact that many of the data of quantitative inheritance cannot be easily reconciled with a theory of additive action. Moreover, duplicate genes of the kind involved in kernel color of wheat may be somewhat atypical, and perhaps relatively rare. You will remember that among wheats hexaploid and tetraploid species as well as diploids are found. It seems reasonable to suppose that the presence of two and three allelic pairs for red kernel color in wheat may be accounted for by reduplications of genomes. If this is the case, relationships among these duplicate genes are probably not typical of multiple-gene associations in general.

The data of a good many studies of quantitative inheritance indicate that gene substitutions may have geometric rather than arithmetic effects. In other words, the genes seem to contribute their effects not so much by adding or subtracting constant amounts but rather by multiplying or dividing the effect of the residual genotype by some constant amount. The average value of a quantitative character in F_1 in such a system more nearly approximates the

Table 14-9. MEAN FRUIT WEIGHT IN GRAMS OF DIFFERENT TOMATO CROSSES.

Larger Parent	*Smaller Parent*	*Large* P	*Small* P	F_1	*Geometric Mean*	*Arithmetic Mean*
PARENTS DIFFERING GREATLY IN SIZE						
Large Pear × Red Currant		54.1	1.1	7.4	7.4	27.6
Putnam's Forked × Red Currant		57.0	1.1	7.1	7.7	29.0
Tangerine × Red Currant		173.6	1.1	8.3	13.2	87.3
Devon Surprise × Burbank Pres.		58.0	5.1	23.0	17.2	32.5
Honor Bright × Yellow Pear		150.0	12.4	47.5	43.3	81.2
PARENTS OF MORE NEARLY THE SAME SIZE						
Peach × Yellow Pear		42.6	12.4	23.1	23.0	27.5
Dwarf Aristocrat × Peach		112.4	42.6	67.1	69.5	77.5
Albino × Honor Bright		312.0	150.0	160.0	217.0	231.0

Source: After MacArthur and Butler, *Genetics,* **23:**254, 1938.

geometric than the arithmetic mean between the parents. The geometric mean is closer to the mean of the smaller parent than it is to that of the larger parent, while the arithmetic mean is midway between the two parental means. Of course, an F_1 mean closer to that of the smaller parent might only indicate that factors for smallness were more often dominant than factors for largeness. An analysis of F_2 segregations is necessary to distinguish this situation from geometric genome effects.

In Table 14-9, which summarizes some genetic studies of size in tomato fruits, you will find that wherever the parents differ greatly in size the geometric mean of the parent values provides much the better fit to the F_1 average values. For example, in the first combination summarized in the table, let *LP* symbolize one Large Pear genome, and *RC* stand for a Red Currant genome. Then, compare the mean values, $RC/RC = 1.1$ grams, $LP/RC = 7.4$ grams, and $LP/LP = 54.1$ grams. These values provide a geometric series in which the effect of substituting each *LP* genome is to multiply by approximately 7. If you do this multiplication, you will obtain a series $1.1:7.7:53.9$, which closely resembles the values found experimentally.

Another kind of approach to the problem of multiple-genic action has been made by H. H. Smith. His method is to study the effects on quantitative characters of adding single extra chromosomes into hybrids between different species of Nicotiana and also into the parental species themselves. The character studied most extensively is length of the corolla of the flower, which is admirably adapted to quantitative genetic analysis because of the ease of taking accurate measurements and because corolla length is relatively insensitive to environmental influence.

Different blocks of multiple genes, as identified by the particular chromosome added to a diploid complement, have different effects on the expression of quantitative characters. However, for the genes in a particular chromosome the effects are characteristic and usually conform to the concept of geometric action. For example, chromosome 1 of *Nicotiana langsdorffii* when added to the F_1 interspecific hybrid *N. langsdorffii* × *N. sanderae* has the effect of reducing corolla length to 0.8 of that found in normal hybrids (that is, those not trisomic). Almost the identical result is obtained when this same chromosome 1 is added either to *N. sanderae*, which has relatively large flowers, or to *N. langsdorffii*, which has relatively small flowers. In both instances, the extra chromosome acts to multiply the normal mean by the constant 0.8; it cannot be acting in arithmetic fashion to subtract a constant quantity.

The Status of the Multiple-Gene Hypothesis. Your own reaction as a student must be that the overall picture of multiple-genic action is exceedingly complicated. Professional geneticists would certainly concur in this judgment. At the present time, the physiological significance of results like those we have just discussed is entirely unknown. There are other serious gaps in our information that are only bridged temporarily by the expedient of drawing analogies from the behavior and action of major genes. What one can say with fair confidence is that different groups of multiple genes present different problems. Multiple genes can act either in plus or minus directions with reference to quantitative inheritance. In some instances they appear to act arithmetically, in others geometrically.

Inbreeding

Deleterious Effects Associated with Inbreeding Generally speaking, people tend to associate inbreeding with unfavorable effects. There are many expressions of this attitude. Some belong in the nebulous realm of old wives' wisdom, but others are spoken with more authority. Most of us are aware, for example, of church or governmental regulations whose purpose is to prevent marriages between relatives of a certain closeness of kinship. All of the states in this country forbid marriages of nearer relatives than first cousins, while some forbid first-cousin marriages as well.

Reinforcing the attitude that inbreeding, on the whole, is something to be avoided are the observations and judgments of the early plant and animal breeders. And in the field of pure biological science, we can find such statements as the one by Charles Darwin, in 1862, that nature "abhors perpetual self-fertilization."

What has led to the idea that inbreeding is to be avoided? For one thing, inbreeding seems to suffer in contrast with its antithesis, outbreeding, which in present popular stereotype is usually represented by such mighty examples as hybrid corn and the mule. Darwin, thinking in terms of adaptation and evolution, was particularly impressed by the numerous devices among organisms that tend to encourage or to ensure some degree of outbreeding. We have mentioned before that in many plants, and especially among higher animals, the different sexes are always found in separate individuals. This situation precludes self-fertilization, thus serving as effective insurance against inbreeding in its most intense form. Still more impressive is the fact that among hermaphroditic organisms, a number of means for promoting cross-fertilization are found. In certain hermaphrodites, self-fertilization is impossible because gametes of the different sexes mature at quite different times; in others, *hereditary self-incompatibility* favors outbreeding. Figure 14-7 shows a type of genetic mechanism for self-incompatibility that is widely distributed among plants. Even where self-fertilization can take place, ways of forwarding cross-fertilization are strikingly frequent. In orchids, for example, complex structural adaptations of the floral parts foster insect pollination from one flower to another.

Besides these general indications of a superiority of outbreeding over in-

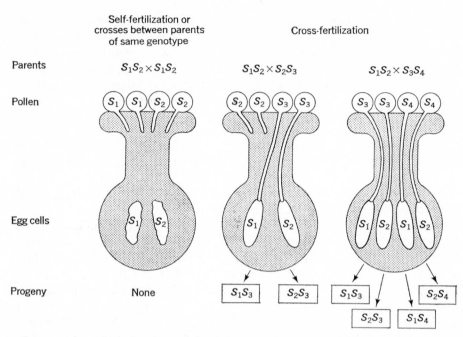

Figure 14-7. *In a number of groups of plants, including clovers, cherries, tobacco, and evening primroses, multiple allelic series determine compatibility in sexual reproduction. Pollen tubes carrying a given incompatibility allele fail to grow properly in stylar tissue carrying the same allele for incompatibility. Cross-compatibility depends upon genotype, and, as shown in the figure, some, all, or none of the pollen of a given plant may function effectively in the stylar tissue of another. Incompatibility alleles are usually designated S_1, S_2, S_3 and so on. At least forty different alleles for incompatibility are known in red clover. (After Crane and Lawrence, J. Pomol. Hort. Sci., 7:290, 1929.)*

breeding, there are numerous specific instances where inbreeding appears to give rise fairly directly to unfortunate biological consequences. For example, we can consider briefly what happens when corn plants are self-fertilized, and then their progeny are self-fertilized, and their progeny's progeny, and so on for a number of generations. Typically, after a few generations the material separates into distinct lines that become more uniform following each self-pollination. Plants with unexpected deleterious characters are likely to appear, such as white seedlings, virescents, yellow seedlings, and dwarfs. Many of the lines die out. Those that survive show a general decline in size and vigor that can be described in the kind of measurements that we have previously utilized for quantitative characters. This is illustrated in Figure 14-8, where you can see graphed the yield of three lines of corn over a period of thirty generations of self-fertilization. Notice the leveling of the yield after about twenty generations. In this same experiment, uniformity of height for each line was attained somewhat earlier. By the twentieth generation, each line had become constant for all visible characters, except for variation that could be ascribed to environment.

Somewhat similar results have been noted when certain other plants are inbred. Animal breeders, too, have observed that "weaknesses" may appear following intensive inbreeding. We can appreciate, then, why inbreeding has been thought to be biologically undesirable. And we should pursue the matter by asking two questions. Is inbreeding as such directly accountable for the biological evils often associated with it? If not, what is the relationship between inbreeding and its apparently deleterious effects? In part, these questions have already been answered in Chapter 13, in a somewhat arbitrary way and in relation to single allelic pairs. Here we will approach the matter more concretely and in relation to phenomena of quantitative inheritance.

You might begin to answer the first question for yourself if you were to take a survey of the typical life histories of various organisms. You would find, perhaps to your surprise, that in many successful groups of plants, self-fertilization is the habitual means of reproduction. You would probably conclude that if oats, peas, beans, and tomatoes, for example, flourish under generation after generation of intensive inbreeding, the practice of inbreeding as such can scarcely be judged harmful.

Your conclusion might be reinforced in various ways. Examination of human family histories would reveal that by no means does inbreeding always lead to disaster. Much the same lesson can be taken from the experiences of breeders of animals. If you own a fine purebred dog, you need not be astonished to find a good deal of common ancestry in its pedigree. From planned experiments, too, there is abundant evidence that inbreeding does not always produce harmful effects. A clear demonstration of this fact is that vigorous lines of albino rats have been maintained after more than a hundred generations of brother-sister mating.

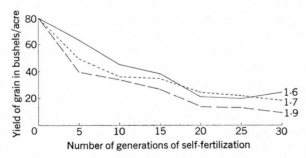

Figure 14-8. *Effect of thirty generations of continued inbreeding on the yield of three different lines of corn. (After Jones,* Genetics, **24**:463, 1939. *Courtesy of Connecticut Agricultural Experiment Station.)*

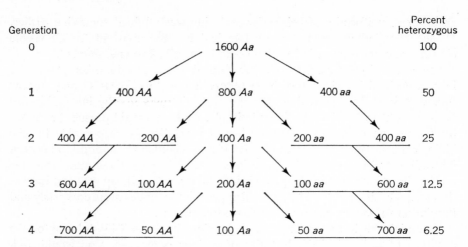

Figure 14-9. *Reduction in heterozygosity over four generations of self-fertilization.*

Inbreeding and Homozygosis. In summation, we can say that deleterious effects seem to follow more frequently on inbreeding than could be expected from mere chance; nevertheless, inbreeding is not in itself necessarily harmful. Further clarification of the situation calls for a closer and more precise analysis. We must turn from a description of end effects to a study of genetic mechanisms.

One outstanding genetic effect of inbreeding accounts for many of the other effects associated with it. Inbreeding results in *homozygosis*, or, if you will, the homozygous state at numerous genetic loci. Let us first reconsider this principle in its simplest form, by seeing what happens as the result of self-fertilization in organisms heterozygous for a single pair of alleles, *A-a*. You know that from *Aa* selfed we expect a progeny of ¼*AA*, ½*Aa*, and ¼*aa*. For that half of the progeny which is *Aa*, reproduction through self-fertilization will again give rise to ½ heterozygous offspring and ½ homozygous, with equal numbers of *AA* and *aa* individuals being expected. For the half of the progeny that is either *AA* or *aa*, however, self-fertilization can produce only offspring that are genotypically identical with their parents. Over a series of generations, then, assuming heterozygous parents to begin with, we might expect that the proportion of heterozygotes would be reduced by half in each succeeding generation. Correspondingly, there should be an increased frequency of homozygotes.

Perhaps you can see this principle more readily after examining Figure 14-9, which shows the results of self-fertilization over a period of four generations. Notice that for our simple model we have made the assumption that each genotype reproduces equally well, a situation not always found in actuality. And if you wonder why we chose 1,600 individuals to start with, it should be explained that the number is an arbitrary one chosen to permit the expected progeny to come out in simple whole numbers. The results speak for themselves.

Beginning with 1,600 individuals, all *Aa*, four generations of self-fertilization will produce a population with 15 homozygous individuals for every heterozygote. A continuation of self-fertilization over succeeding generations would further reduce the frequency of heterozygotes.

For simplicity's sake we have begun our discussion of inbreeding and homozygosis in relation to single allelic pairs. Except where organisms are already highly inbred, however, most individuals are doubtless heterozygous for many allelic pairs. The effects of inbreeding operate on all the genetic loci, so that quantitative characters as well as characters determined by major genes are subject to its influence.

What we have said about self-fertilization and homozygosis applies directly but to somewhat lesser degrees to other forms of inbreeding. Brother-sister matings continued over a number of generations also result in increasing homozygosis, but somewhat more slowly. Less drastic forms of inbreeding are correspondingly less efficient in producing homozygosis. These differences are brought out in Figure 14-10.

Bringing Recessive Characters to Light Through Homozygosis. We now have at hand an essential clue to why inbreeding is associated with deleterious consequences. In a noninbred population, deleterious recessive genes may often be concealed by their normal dominant alleles. Inbreeding, however, favors segregation into homozygotes, so that if deleterious recessive alleles are carried in a population they quickly come to light.

Natural populations in which inbreeding is not the rule do indeed carry deleterious recessive factors in fairly high frequencies. Th. Dobzhansky and co-workers have investigated wild populations of *Drosophila pseudoobscura* and found that the majority of individuals carry deleterious recessive mutants. That much the same situation holds in corn was indicated in our earlier discussion

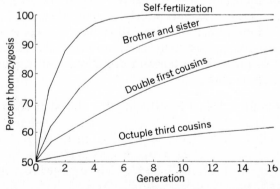

Figure 14-10. *The percentage of homozygosis in successive generations under various forms of inbreeding, as indicated by the closeness of relationship of the parents. (After Wright,* Genetics, *6:172, 1921.)*

of inbreeding in this organism. And a fact you will remember from Chapter 13, that a large fraction of human individuals showing rare, recessive defects come from cousin marriages, helps to round out a general picture.

The revelation of deleterious recessive characters through inbreeding is so striking a phenomenon that it may easily assume undue significance, even when its general genetic basis is understood. To be sure that you are not being led astray, consider carefully the following points:

1. Not only *deleterious* recessives come to light through inbreeding but other recessives as well.
2. Inbreeding does not favor an increase in the number of recessive alleles; it merely gets them placed where they can be detected phenotypically. One way to assure yourselves in this matter is to turn back to Figure 14-9, where you will see that in the original parental generation 1,600A and 1,600a alleles were present. In the last generation shown in the diagram, the proportions, and in our case the actual numbers, of alleles are exactly the same. Only the distribution of alleles among genotypes has changed.
3. With reference to homozygosis, the effects of inbreeding are the same for dominants as for recessives. Inbreeding may have less spectacular consequences for dominants, however, because they have never been phenotypically submerged and thus cannot be brought suddenly to light.

Inbreeding and the Fixation of Genetic Characters. We have said that continued inbreeding results in homozygosis. Another way of putting the same idea, with a slight but important shift in emphasis, would be to say that inbreeding results in the fixation of genetic characters. To place the argument in more specific form, assume a group of organisms heterozygous for two gene pairs (*AaBb*). Inbreeding might result in the formation of four homozygous lines—*AAbb*, *aaBB*, *aabb*, and *AABB*. With reference to the characteristics determined by these genotypes, the lines would be true-breeding within themselves, barring the possibility of mutation. If a greater number of heterozygous loci were involved in the first place, the same principle would still hold. After sufficiently long and intense inbreeding, the population would become separated into genetically distinct groups, each uniform within itself. This effect of inbreeding has implications of prime importance in evolution, and in plant and animal breeding as directed by man.

Inbred Lines and Selection. It is fortunate that almost as soon as widespread investigations in Mendelian heredity began, brilliant experimentation provided a solid basis for an understanding of selection. In the early 1900's, a Danish plant scientist, W. Johannsen, began experiments to test some of the ideas about selection which were then current. Like Mendel, he turned to a common garden plant for his experimental material; also like Mendel, he had the happy faculty of being able to follow a trail of experimental facts to logical and significant conclusions.

Table 14-10. THE EFFECTS OF SIX YEARS OF SELECTION IN
PURE LINE NO. 19 OF THE PRINCESS BEAN.

Harvest Year	Average Weight of Selected Parent Seeds		Average Weight of Progeny Seed	
	Lighter Seeds	Heavier Seeds	Lighter Seeds	Heavier Seeds
1902	30 cg	40 cg	36 cg	35 cg
1903	25	42	40	41
1904	31	43	31	33
1905	27	39	38	39
1906	30	46	38	40
1907	24	47	37	37

Note: In each generation, the lightest and the heaviest seeds were selected for propagation.
Source: Data from Johannsen, *Elemente der Exacten Erblichkeitslehre.* Jena: Gustav Fischer, 1926.

In a notable sequence of experiments carried out with the Princess variety of garden bean, Johannsen investigated the possibilities of selecting for weight of seed. Starting with a mixture of seed obtained from many different plants, he found initially that progenies derived from heavier seed in general had greater average seed weight than progenies obtained from lighter seed. This result indicated that selection had been effective.

To explore the problem further, Johannsen refined his experimental approach. He chose 19 seeds, each derived from a different mother plant, and grew them into 19 progeny plants. These produced their own lots of seed, which were kept separate. Within each seed lot, weights varied from one individual to another, falling into arrays roughly approximating normal distributions. Johannsen propagated his 19 separate lines by selecting from each seed lot the heaviest and the lightest seeds. This procedure was followed for several generations, with the kind of result shown in Table 14-10. In short, *there was a remarkable tenacity of weight averages within each line, generation after generation, no matter whether the line was reproduced by its heaviest or by its lightest seed.* After six generations of selection in Line 19 the smallest parent seed produced a progeny with an average weight of 37 centigrams, while the largest parent seed also produced a progeny averaging 37.

Johannsen's results are perfectly understandable if we remember that beans are normally self-fertilizing and if we recall that inbreeding leads to homozygosis. The parent beans with which Johannsen started his second series of experiments were, therefore, each homozygous. Progeny of any individual bean could not show genetic segregation; they constituted what Johannsen named a *pure line.* Differences among members of a single progeny could be due only to environment. *Selection within pure lines, then, is ineffective because it is biologically meaningless.* Each generation is built on exactly the same genetic base and has precisely the same capacities for variation in response to environmental influence as had its parents.

The reason that Johannsen had some success in selecting for size in his first series of experiments is also apparent. The population out of which he made his original selections was not a pure line but a mixture of pure lines, and so presented genotypic as well as phenotypic variation. By separating the genotypes with relatively greater inherent capacities for expression of size, it was possible to raise the mean size of a group of progeny to a level higher than the average for the initial population. Johannsen's experiments are schematically summarized in Figure 14-11.

The great contributions of these experiments were: first, to distinguish effectively between heritable and nonheritable variation; second, to emphasize that inbreeding does indeed lead to genetic homogeneity; third, to demonstrate that selection does not create variation; and fourth, to reaffirm, on the other hand, that selection within a group that is genetically diverse may change the character of subsequent populations. In addition, Johannsen's studies demonstrate in a striking way the sensitivity of quantitative characters to environmental influence, and point to the hopelessness of trying to deal with quantitative

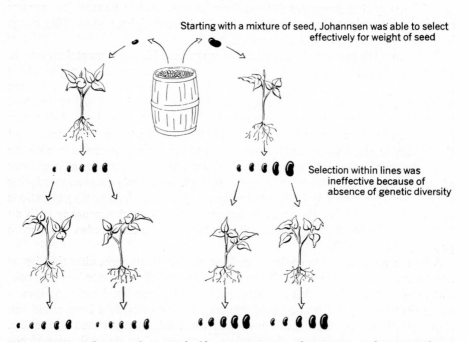

Figure 14-11. *Johannsen distinguished between genetic and environmental sources of variation, and showed that only genetic variation is subject to effective selection.*

variation in living things unless heredity and environment are adequately distinguished and placed under some sort of control.

Hybridization of Inbred Lines

Hybrid Vigor in Corn. Reference to Figure 14-8 will recall to you that inbreeding corn over a period of generations leads to successive reductions in vigor. Sometimes, inbred lines die out entirely. If they do not, a time comes when continued self-fertilization is accompanied by no increase in deleterious effects. Presumably this stabilization occurs when homozygosity is reached. You will not suppose, of course, that such a characteristic as yield will ever be entirely constant, even in a homozygous line. Yield, like most quantitative characters, is sensitive to fluctuations of environment, and planting seasons differ sufficiently from one year to the next to provide a basis for appreciable variation in phenotypic expression.

If two different inbred lines of corn are crossed, the hybrid progeny display *heterosis*. They are almost always strikingly more vigorous than their parents. Usually such hybrids are vigorous by any standards. But to clear up a fairly common misconception, it should be said that hybrid corn plants are not unique in vigor or even markedly superior to the best plants of open-pollinated origin. In fact, certain open-pollinated plants are superior to many hybrids. The basic reason for the success of hybrid corn in our agricultural economy is that members of an F_1 hybrid group show a uniformity of high-level performance not found in open-pollinated varieties. After the F_1 generation, however, characters like height and yield are not maintained at so high and uniform a level. In a study by N. P. Neal the yield of ten different corn hybrids was compared in the F_1 and F_2 generations. The first-generation hybrids gave an average yield in bushels per acre of 62.8; the average yield for the same hybrids in F_2 was only 44.2. These results are typical. They should remind you of facts that emerged earlier, in our discussion of quantitative inheritance uncomplicated by heterosis. You will remember that for quantitative characters in general, the F_2 is much more variable than the F_1, and encompasses a whole spectrum of variation. Increased variability after the F_1 results from genetic segregation and recombination.

Explanations of Hybrid Vigor. Heterosis is a phenomenon that is at once intriguing and practically important. It is manifested in different groups of organisms, being by no means confined to corn, or even to plants. No doubt what is called hybrid vigor in various groups of organisms is not everywhere the same phenomenon. But these various manifestations doubtless have much in common, and a satisfactory explanation of one will aid considerably in understanding others.

All attempts to explain hybrid vigor stem from one basic fact. This is that the vigor is found associated with the heterozygous state. Most genetical theories designed to explain heterosis fall into one of two categories.

Explanations Based on Interaction of Alleles. A number of geneticists have proposed, in one way or another, that heterozygosity *per se* is essential for heterosis. Reduced to simple terms, theories of this kind say that if there are the alleles a_1 and a_2 for a single locus, the heterozygous combination a_1a_2 is superior to either of the possible homozygotes, a_1a_1 or a_2a_2. The implication is usually that alleles a_1 and a_2 do separate things, and the sum of their different products, or some interaction product between them, is superior for vigor to the single products produced by either allele in the homozygous state. Recall intragenic complementation; the plant breeders' term for the concept is *overdominance*.

Explanations Based on the Interaction of Different Dominant Genes. Many geneticists have felt that heterosis does not require overdominance, but that it can be rather simply explained in terms of ordinary dominance of genes relatively favorable for vigor and the corresponding recessiveness of genes unfavorable for vigor. The reasoning behind this second kind of explanation can be most readily seen if we return to the effects of inbreeding.

We have already discussed how, in groups of organisms that are normally not inbred, deleterious recessive mutant genes may accumulate because they are masked by dominant normal alleles. Deleterious dominant mutations tend to be eliminated from populations rather rapidly because they are immediately and continuingly subjected to adverse natural selection. Among normally inbred groups, even deleterious recessives do not accumulate to any great degree. Inbreeding leads to homozygosity, and thus in these groups unfavorable recessives are subject to much the same sort of pruning through natural selection as occurs for deleterious dominants everywhere. This continual process of pruning away undesirable recessive genes in naturally self-fertilized organisms accounts for the fact that such plants as oats can maintain a level of vigor apparently as high as is found among naturally cross-fertilized plants.

If one makes a cross between unrelated inbred lines of corn, it is likely, indeed almost inevitable, that at particular loci the parent lines will differ depending on whether dominant or recessive alleles have become homozygous through the inbreeding process. If we were to designate five loci in a hypothetical cross between inbreds, a representative situation might be as follows:

Inbred I   *aaBBCCddEE* × Inbred II   *AAbbCCDDee*

F₁   *AaBbCCDdEe*

If the different recessive alleles are even mildly unfavorable to vigor, the hybrid, having the relatively favorable dominant alleles at more different loci than is true for either inbred, should be more vigorous than either parental line. A look at the hypothetical cross will also help to make it clear how it is that uniformity is a characteristic of hybrid corn. Since the parental inbred lines typically are homozygous, their progeny must be *genetically uniform*. You realize, of course, that this is not the same as saying the progeny are *homozygous*.

The above general explanation of heterosis is different in a fundamental way from the first kind of theory we discussed. Since in the second case it is assumed that heterozygosity is largely incidental to the phenomenon of hybrid vigor, it should be possible to obtain lines of corn that breed true for the vigor found in particular hybrids. For instance, the hypothetical F_1 individuals *AaBbCCDdEe*, if intercrossed or self-fertilized, should give some progeny of the genotype *AABBCCDDEE*. Continued failure, in actual experiments, to find true-breeding lines as vigorous as F_1 hybrids might be taken as evidence in favor of theories based on allelic interaction. For if the heterozygous condition as such accounts for hybrid vigor, then this kind of vigor could not be expected to be stabilized in homozygous lines.

In corn, fixing the vigor of F_1 hybrids into true-breeding lines has not so far been possible. It has been pointed out, however, that the recovery of lines homozygous for all the favorable alleles for a multiple-gene character may pose a Herculean task for the plant breeder. If expression of vigor is influenced by genes at, say, thirty different loci in a hybrid, the probability of recovering one particular homozygous combination among the progeny of this hybrid is small. Singleton has calculated that it would require a land area more than 2,000 times the total land area of the earth to grow enough corn plants to have an even chance of obtaining one such homozygous combination, even if there were independent recombinations among the thirty loci. Moreover, since quantitative characters are susceptible to environmental influence, the detection and preservation of a particular genotype might be immensely difficult. Finally, linkages among favorable dominant genes and unfavorable recessives would further reduce the probability of recovery of multiple dominants. Since many loci are likely to be involved in expressions of vigor, such linkages are to be expected. At present we cannot tell whether fixing heterotic vigor into homozygous lines of corn is impossible or whether it is only an infinitely arduous task.

Our present judgment of the different explanations for hybrid vigor must be in the nature of a compromise. There are rather good indications that either allelic interaction or a complex of dominant linked genes may feasibly account for heterosis as we see it, for example, in corn. In fact, there appear to be no obvious reasons why in given instances both systems should not operate simultaneously in producing heterotic effects. However, true understanding of the causes of heterosis must wait for a better understanding of gene action, of complementation, of dominance, and of other fundamental genetic phenomena, particularly as they relate to quantitative inheritance.

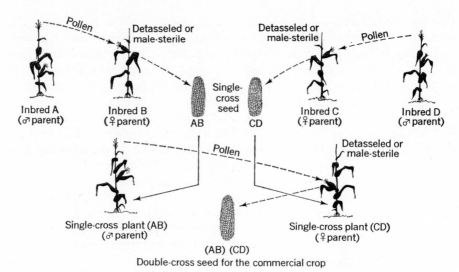

Figure 14-12. *Hybrid seed for field corn is usually produced through double crosses.*

The Practical Utilization of Heterosis. It should be emphasized that not all organisms show heterosis, at least in any obvious way. In corn, where heterosis may be strongly manifested, not every hybrid is favorable from the agricultural standpoint. Hybrids that do well under some conditions or in some localities may be poor in others. Again, as in other areas of plant breeding, the breeder who wishes to utilize heterosis must work with specific aims and specific problems in mind.

The performance of a hybrid depends upon its genetic constitution, and therefore stems from the hereditary makeup of its parental inbreds. The fundamental task of the corn breeder is to develop suitable inbred lines. For this work, the effectiveness of selection for particular characteristics, the accurate measurement of these characters, and the degree of certainty with which their heritability can be predicted are all highly important. Corn breeders are continuing to evolve special methods for getting at these various problems. In the final analysis, however, the real test of the worth of an inbred is its *combining ability*, that is, its performance in combination with other inbreds in producing hybrid corn. No shortcuts have yet been found that eliminate the necessity for careful selection and stringent testing of inbred lines for their ability to produce good hybrids.

The Production of Hybrid Corn Seed. Important practical considerations for corn breeders have arisen in connection with hybrid corn seed production, now a large and flourishing industry. If seed is used after the first generation following a cross, segregation takes place, and most of the advantages of hybrid corn are lost. The farmer needs, therefore, to obtain a new supply of hybrid corn seed each year.

When hybrid seed is produced simply by crossing two inbred lines, the maternal parent plant, if typical of inbreds, is small and produces small, low-yielding ears. The price of such seed necessarily comes high. To meet this difficulty, hybrid seed for the farmer is generally produced through double crosses, as illustrated in Figure 14-12. This device permits seed for distribution to the farmer to be produced on the large and uniform ears of single-cross plants.

Other Instances of the Practical Utilization of F_1 Hybrids. First-generation hybrids are beginning to be used in commercial crops other than field corn. Sweet corn hybrids are planted extensively. In sweet corn, single rather than

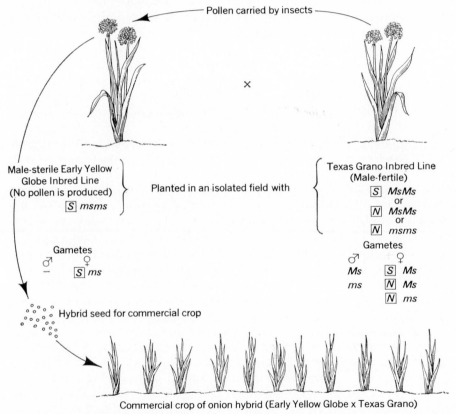

Figure 14-13. *Controlled hybridization in onions may be achieved by utilizing a genetically determined male sterility in the inbred line to be used as the female parent in the cross. This male sterility in onions results from interaction between a cytoplasmic factor* S *and a recessive nuclear gene* ms. *All plants with* N *(normal) cytoplasm produce viable pollen, and all plants with gene* Ms, *either homozygous or heterozygous, are also male fertile. Male-sterile plants are* S *ms ms. (After Munger, based on work of Jones and Clarke,* Proc. Am. Soc. Hort. Sci., **43:***193, 1943.)*

double crosses are the more commonly used, largely because the sweet corn breeder must put unusual emphasis on uniformity. This crop requires harvesting at a particular stage of maturity, either for canning or for commercial sale, and it is desirable to harvest all the ears in a field at one time. In terms of the high value of the product, the cost of single-cross seed in sweet corn is not excessive, particularly since parental inbreds of reasonable vigor have been developed.

Breeding methods following the general outlines of modern corn improvement programs are successful with onions. However, onions are not well suited for the mass production of hybrid seed. The reason arises from the fact that both male and female organs occur within the same flowers. The emasculation process necessary to control hybridization requires such painstaking work that the expense prohibits commercial production.

In recent years, breeders have made ingenious use of inherited male sterility in onions to produce inbred lines that do not require emasculation. The male-sterile character has been introduced into various standard varieties like Crystal Wax, Yellow Bermuda, Yellow Globe Danvers, and Sweet Spanish. Inbred lines have been established, and certain hybrids show exceptional promise. California Hybrid Red No. 1 is used rather widely. It is larger than either of its parents and combines the relatively early maturity of one with the delayed bolting habit of the other. Figure 14-13 illustrates a method utilizing male sterility in the production of hybrid onion seed on a commercial scale. A similar method, using cytoplasmic male sterility and "restorer genes" to correct this sterility at the appropriate point, is now commonly used to make detasseling unnecessary in hybrid corn seed production (cf. Problems 11-15 to 11-18, p. 350).

Other plants in which first-generation hybrids show promise from the breeder's standpoint include cucumbers, squash, and pine trees. Yankee Hybrid summer squash has an advantage over competing nonhybrid varieties because of its relatively early productiveness. Particular interspecific combinations of pines seem to manifest heterosis in various ways, including greater vigor of growth and superior seed germination. Pine hybrids are being tested extensively in the national forests of California and elsewhere.

Systems of Mating in Animal Breeding

In many respects, plant and animal breeding can be said to direct the evolution of domestic populations. We described in Chapter 13 some of the ways in which natural selection, mutation, migration, and drift effect changes in evolving populations or counteract each other to stabilize them. The animal breeder substitutes, to a greater or lesser extent, his own criteria of selection for the natural selection of wild populations. He can control, to a degree at least, *which* animals shall leave progeny in his herd or flock and, within limits, *how many* progeny they shall leave.

The breeder cannot at present control mutation, although he may be able, by selection and control of the mating pattern, to take advantage of such desirable mutations as do appear, or to weed out, more quickly than nature commonly does, undesirable ones. He can control migration; the purchase of new sires or breeding females admits new alleles and new gene combinations into the herd. On a broader scale, the establishment of closed "pure-breed" registers is the creation of an artificial isolating mechanism that prevents the breed from pooling its genes with outside populations. By controlling migration in these ways, and by controlling the mating pattern, breeders can regulate the effective breeding sizes of their populations, and can thus to some degree control the rapidity and significance of drift in these groups.

Relationship. Control of selection in his herd or flock and control of its mating pattern are the two primary tools with which an animal breeder may expect to make progress. These two tools, although they are in fact quite different from each other, are both so intimate a part of any breeding program, and so necessary to each other, that we need to look rather closely to see that they are really separate items for consideration. We have said that *selection involves the choice of which animals shall leave progeny in the herd and, within available limits, how many progeny these selected animals shall leave. The choice of a mating pattern is the determination of which of these selected animals shall mate with which. This can practically be summed up, from a genetic point of view, by observing that it involves control of the extent to which the parents of a forthcoming generation shall be genetically related to each other.*

Genetic relationship is obviously a variable quantity for which we need some kind of quantitative measure. All animals are related to some degree, because they have evolved from common ancestors. All the members of a breed are related; commonly almost all of the individuals trace back, more or less directly, to a few foundation animals. Exceptionally popular or outstanding individuals have had substantial genetic influence on each breed, and living animals are related through their common descent from such individuals. Most of the members of a herd are often even more closely related, tracing back, for example, to the limited number of herd sires and female families existing in the herd during the past few generations. Second cousins are related to a degree; first cousins even more closely related but less closely than parents and offspring or full sibs (sisters and brothers).

Figure 14-14 shows two degrees of full-sib relationship. In Figure 14-14A, individuals A and B are full sibs, having common parents C and D. Similarly, in Figure 14-14B, individuals A' and B' have common parents C' and D'. But A and B are really less closely related than are A' and B', because the parents of the latter are in turn related (half-sibs), while no relationship is shown for

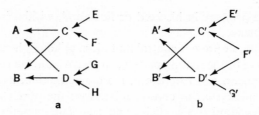

Figure 14-14. *A and B are full sibs, having the same parents, C and D. Similarly, A' and B' are full sibs. But A' and B' are more closely related than are A and B.*

C and D. A and B have gene samples from four different grandparents, while A' and B' depend on only three.

Relationship on this scale can be measured by the *coefficient of relationship*, which depends on the knowledge that each individual receives a sample half of the genes of each of his parents. The simplest way of calculating the coefficient of relationship for two individuals is from arrow diagrams like those in Figure 14-14. One need only count the number of "sample-halving" processes that separate the related individuals from a common ancestor in the diagram, and use this sum as the exponent of one-half to arrive at the contribution of that ancestor to the relationship of the individuals in question. The total relationship is the sum of the separate contributions of all the common ancestors.

In Figure 14-14a, a path from A to B involving two "sample-halving" processes can be traced through C:

$$A \leftarrow C$$
$$B \swarrow$$

and another through D:

$$A \nwarrow$$
$$B \leftarrow D$$

The contributions of these paths are then:

$$A \leftarrow C \longrightarrow B = (\tfrac{1}{2})^2 = 0.25$$
$$A \leftarrow D \longrightarrow B = (\tfrac{1}{2})^2 = 0.25$$

and the total relationship of A and B is $0.25 + 0.25 = 0.5$.

In Figure 14-14b, these same paths occur, but in addition there are two through F':

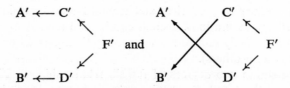

Each of these involves four "sample-halvings," so that the total relationship of A' and B' is $0.5 + (\tfrac{1}{2})^4 + (\tfrac{1}{2})^4 = 0.625$.

The coefficient of relationship really measures the extent to which animals may have genes in common because of their descent from the same ancestral individuals. This figure is relative to the amount of diversity in the population concerned, and to the average relationship in this population. We have already pointed out that all the members of a breed are to some extent related and have many genes in common. The coefficient of relationship measures the extent of genetic similarity arising from specific relationships in excess of this common store of similarity.

Inbreeding. A mating system that involves the breeding together of individuals *more closely related than the average for the population concerned* comprises some form and degree of inbreeding. The degree of inbreeding in any particular mating is therefore a function of the relationship between the animals mated. For our present purposes, we can define the inbreeding coefficient for any individual as approximately one-half of the relationship between its parents. Thus, if A and B in Figure 14-14 were to be mated, their progeny would be inbred to the extent measured by an inbreeding coefficient of 0.25. If A' and B' were mated, their offspring would have an inbreeding coefficient of about 0.31.

We have observed earlier that the primary genetic effect of inbreeding is to reduce heterozygosity in the population. The inbreeding coefficient measures the extent to which heterozygosity may be expected to be reduced in any individual as a consequence of relationship between his parents. As is true of the relationship coefficient, the absolute genetic value of this measure depends on the extent of diversity in the population concerned. If a random-bred individual in a given breed is on the average heterozygous at 400 loci, then an individual with an inbreeding coefficient of 0.25 is estimated to be heterozygous at about 300 loci. But if a random-bred individual in another breed is on the average heterozygous for 4,000 loci, then an individual with this same inbreeding coefficient is probably heterozygous for about 3,000 allelic pairs. On the whole, we recognize this situation when we observe that one breed is "more uniform" than the other. But we do not have at hand accurate measures of the number of heterozygous loci in any domestic population. Such measures as have been applied have sometimes suggested that the statistical estimation of heterozygosity after inbreeding may be in error, perhaps because unknown mechanisms may act to maintain heterozygosity in excess of that predicted from the inbreeding coefficients. Much remains to be learned in establishing that such mechanisms really exist and, if they do, in ascertaining their nature.

Types of Mating Patterns. Reference has been made to the utility of inbreeding in bringing to light, and rendering vulnerable to selection, genetic qualities that might otherwise remain largely hidden in heterozygotes. We have also emphasized the uniformity within inbred lines and the uniformity and vigor of certain plant hybrids between inbred lines. The possibility of equivalent progress in animal improvement through the use of similar techniques has prompted a great deal of research into the effects and potential utility of close inbreeding and crossing in animals.

The investment in an individual animal is of course much greater than that in a plant, and one cannot speak lightly of the establishment of numerous inbred lines of animals, most of them defective or weak, in order to select a few lines that might survive the random fixation of genes that comes with close inbreeding. On the other hand, inbreeding approximating the intensity of self-fertilization is in any case impossible in domestic animals; and the rapidity with which random fixation or loss of genes occurs becomes much less as the intensity of inbreeding decreases. Even under a system that calls for successive generations of full-sib or parent-offspring matings, the most intense inbreeding possible in farm animals, the rate of drift may be slow enough so that intense selection can keep up with it. (Recall our discussion of this subject in Chap. 13, p. 427.)

This means that in animal breeding one need not depend, as one does in corn, on selecting from among a large number of inbred lines the few in which chance has fixed a desirable combination of genes. Selection may be effectively applied as the inbreeding continues, and the resultant lines may be less a product of chance than of design. Under these conditions, large-scale inbreeding programs in swine are now under way, and there is indication that the uniformity and vigor of hybrid progeny will prove to be of practical value. Hybrid poultry are well established in commercial poultry breeding enterprises.

Individual breeders, however, can seldom afford to take the chances involved in the rapid establishment of truly inbred lines. They must experiment with milder inbreeding if they are to take advantage of the desirable genetic effects of this mating system without risking its undesirable consequences. A form of inbreeding of longstanding value in animal improvement in this regard is known as *linebreeding*. This is an effort to intensify the contribution of particular admired individuals to the herd or flock while keeping the relationship among contemporary individuals at a minimum. There is much opportunity for the exercise of ingenuity and originality in achieving this end.

The use of "family systems" of mating is usually another method of attaining the same kind of objective. But "families" are in general so poorly defined that it would be difficult to discuss this technique briefly. Similar objections apply to the discussion of "blood lines" and their use in mating systems.

Some mating systems attempt the opposite of inbreeding, and breed together individuals less closely related than the average for the population concerned. As one might expect, insofar as such systems increase heterozygosity they result in improved individuals with uncertain breeding value. An extreme of such "outbreeding" is involved in *crossbreeding*, in which the parents are of different breeds. Gains from such systems are usually debatable, and such gains as may occur must be weighed against the costs of replacing the breeding stock from pure breeds in each generation.

Other mating patterns may ignore genetic relationships and be based only upon phenotypic similarity or dissimilarity (*assortative* or *disassortative mating*). Except in simple and infrequent circumstances, when the trait depends on only one or two pairs of genes and the degree of assortative mating is very high, such systems appear to have little effect on homozygosity or heterozygosity. Assortative mating approaches its full possible effect quite rapidly, so that after a few generations this system loses much of its potentiality for further progress. Its effects disappear very rapidly if the system is abandoned. Disassortative mating is sometimes useful when the desired type is a heterozygote (for example, maintaining blue Andalusian fowl by mating blacks with whites).

Bases for Selection

In modern plant-breeding practice, principles of selection are utilized in various and sometimes complex ways. We cannot hope to review them in detail here. However, you should be aware that techniques of selection can be adjusted to the demands of different objectives and to the problems peculiar to particular kinds of plants. In the following paragraphs, two general techniques of selection, *mass selection* and *pedigree selection*, will be introduced. We will then turn to techniques of selection in animal breeding.

Mass Selection. One of the most common techniques of mass selection is simply to rogue out undesirable types and save seed from the rest. This practice is effective as a means of purging genetic deviants that, either through mutation or some other process, have become contaminants in a variety. In contrast to searching out undesirable plants and preventing their reproduction, breeders may take the more positive approach and look for especially desirable plants, seeds, ears, or the like, and then confine reproduction to this choice material.

Under the right circumstances, mass selection can produce appreciable results with fair rapidity. For example, in the important forage plant big bluestem it was possible, by five generations of selecting desirable plants as maternal parents in open-pollinated lines, to increase the leaf area of plants in their first season of growth by more than twelve times. In general, however, mass selection has the disadvantage of producing changes relatively slowly, because the breeder utilizing this method has too few experimental controls on his material. In merely selecting among naturally cross-pollinated plants, there is no control over the pollen source, which may be on the average genetically inferior to the selected maternal parents. Lack of strict genetic control accentuates the difficulty—always a great one for the breeder—of untangling differences due to environment from those due to heredity.

Pedigree Selection. Many of the disadvantages of mass selection are minimized in *pedigree selection*. Under this method, individual plants and their progenies are kept separate for study and for breeding purposes. Selection is based upon the comparative performances of these pedigreed lines. Such an approach enables the breeder to control and to define the genetic properties of his material. As a consequence, estimating environmental influences becomes less difficult.

Pedigree selection has been practiced with great success among habitually self-pollinated plants. Here the effect is isolation of different pure lines. When this is accomplished, the breeder's task is to find out by repeated testing whether any of the isolates are better for his purpose than varieties already in existence. This is by no means as easy as it sounds, because the differences with which the breeder works may be relatively small, and the problem of demonstrating them satisfactorily under a wide range of field conditions may be a long-term project.

For example, in oat breeding new progenies are usually started from the seed of a single panicle. This original isolation of lines is followed by preliminary tests of one to three or more years. During this time, yield is measured, and observations are made on stiffness of straw, disease resistance, and other characteristics. Promising lines are then subjected to larger-scale tests in experimental plots for several more years. Finally, lines that emerge successfully from these earlier trials are studied in different field plots over a period of time. If, after all this examination and cross-checking, a line proves to be superior to varieties already in use, the breeder may decide to release it as a new variety. Before this can be done, however, seed must be multiplied for planting by the farmer. The total process may take from eight to fifteen years or more. This same general procedure has been followed in the breeding of most other small grains, including wheat and barley. Pedigree selection as a breeding method is more expensive than mass selection. It has paid huge dividends many times, however, and doubtless will remain a standard breeding technique under appropriate circumstances.

When pedigree selection is applied to crops that are normally cross-pollinated, artificially controlled self-pollination is carried on within the pedigreed lines whenever feasible. The effect of continued inbreeding is to produce homozygosity and, therefore, uniformity within selected lines. Moreover, as homozygosis occurs, undesirable characters like albinistic, dwarf, or twisted seedlings appear, and may be selected against.

With some normally cross-pollinated crops, pedigree selection may result in the development of superior, true-breeding varieties. Recall, however, that continued inbreeding of many habitually cross-fertilized plants is attended by a general loss in vigor that cannot be attributed to the effects of any small number of genes. Where this is true, mass selection may present advantages over pedigree selection. Through mass selection it is often possible to achieve varietal uniformity for certain key characteristics and still retain sufficient heterozygosity for the maintenance of vigor.

Pedigree selection within self-pollinated lines is being practiced more and more, even with normally cross-fertilized crops that show a marked decline in vigor when inbred. Here the purpose of inbreeding is not to produce varieties directly usable by crop growers but to establish uniform lines of predictable behavior, from which particular lines may be selected to serve as the parents of superior hybrids.

Techniques of Selection in Animal Breeding. Probably the most important kind of question to confront the breeder of any class of livestock is, "How should I decide which animals to save for breeding stock and which to cull from my breeding population? Or, if I buy sires or breeding females from outside my own herd, flock, or kennel, how can I best assure that the animals I buy will raise the level of performance of my population in the future?"

Two rather different sources of information about an individual are useful in evaluating his potential breeding worth. One is his own performance or productivity. The other depends on his genetic relatives—ancestral, collateral, and descendant.

Individual Performance. Estimating the breeding value of an animal from his own productivity has long been recognized as a keystone of animal improvement, epitomized in the familiar rule of thumb "Like begets like." Modern animal breeding does not fail to emphasize the value of this guide to selection, but it qualifies the old rule a bit, and places it on a more quantitative basis.

In the discussion to follow, we will use the term *phenotype* not to denote simply the way an animal looks, but to include any characteristic that is discernible from the consideration of the individual by himself. Thus we will include a production record achieved by an individual as part of his phenotype, and will assume that this phenotypic characteristic, like others, is based upon a system of genetic potentialities interacting with the environment.

If the characteristics in which we are interested depended on uncomplicated, additive gene effects, then the phenotype of an animal would be a statistically reliable index of his transmitting ability. (Recall our discussion of this kind of quantitative inheritance earlier in this chapter.) But the hard fact is that few of the economic characters of animals are really so uncomplicated, and selection on the basis of phenotype alone is likely to make mistakes from a genetic point of view. These mistakes arise from at least three sources: *environmental effects, dominance,* and *epistasis.*

Environmental Effects. The conditions under which animals are grown, fed, and evaluated affect profoundly the records of productivity they achieve. Environment varies from stall to stall, from season to season, from farm to farm, and from region to region. Even the prenatal environment of the developing animal varies from one dam to another, and from one pregnancy to another in the same dam. Among the most critical problems of the animal breeder are those that deal with the evaluation of these environmental variables and their effects, and with the distinction between inherent merit and the fortunes of environment.

Probably the clearest result of our increasing awareness of these environmental sources of error is the number of "corrections" that are nowadays applied in the standardization of records of performance. In dairy cattle, for instance, statistics are available that make it possible to estimate the mature equivalent of the performance of a young cow; to correct various lactation periods to a constant (usually 305-day) basis; to convert different numbers of daily milkings to a twice-a-day schedule; to correct for the lesser reliability of single or short-term records compared with several records on the same individual; and so on. These corrections are all of unquestionable value in the attempt to eliminate clearly nongenetic sources of error in the selection of individual cows.

But many of the more subtle effects of environment are not subject to such straightforward statistical control. Even when all the corrections are made, it is probable that two individuals of similar inherent merit will achieve different records of performance, or that two individuals of different inherent levels of ability may end up with similar records. When the breeder's basis for selection depends upon individual performance alone, therefore, he is likely sometimes to mistake environmental advantage for genetic superiority; and to the extent that this happens, his selection will be unavailing in the genetic improvement of his stock.

Dominance. A quite different source of error in estimating the transmitting ability of an animal from phenotypic data alone involves dominance deviations from simple additive inheritance. You will recall that under the additive scheme the substitution of a given "active allele" A, for a, was assumed to make a constant addition to the phenotype, whether this substitution were to change genotype aa to Aa, or Aa to AA. In other words, the additive scheme is based on a complete lack of dominance at each locus under consideration, so that the heterozygote is exactly intermediate between the two respective homozygotes.

Now if A is dominant to a, the additive scheme in its ideal form cannot apply. Replacing A for a in the genotype aa may have a certain measurable effect, but the same replacement in genotype Aa will have less effect—no effect at all if dominance is complete. The heterozygote may not be intermediate, but may approach the level of the homozygous dominant. Or the heterozygote may even exceed the level of either homozygote.

Some of the effects of dominance on the precision of phenotypic selection will be discussed in connection with the other sources of error in the section on the selection differential and regression, on the following page.

Epistasis. Just as dominance at any particular locus results in a deviation from the simple additive scheme, so also interactions among genes at different loci result in epistatic deviations. To cite an easily recognized effect of this sort, we can recall epistasis in color inheritance. If an animal has the dominant allele of albinism, which permits the development of color, then gene substitutions at other loci may affect noticeably the quality or distribution or quantity of pigment. But if the animal is an albino, substitutions at other loci affecting color produce no phenotypic effect; no pigment is present to enable other aspects of color diversity to come to expression. There is good reason to believe that in many other important characteristics similar gene interactions occur. In the example of albinism, the effects of epistasis are evident; we would not think of selecting for color characteristics in an albino line. But epistasis in economic characteristics may be much more subtle.

The Selection Differential and Regression. If, from a reasonably large original population, a number of individuals that excel in a particular attribute are chosen to be parents of the next generation, these selected individuals have a higher average value for the characteristic concerned than has the population from which they come (Fig. 14-15). The difference between the mean of those selected and the population mean is called the *selection differential*. It is a measure of the intensity of selection.

You can easily see how environmental effects, dominance, and epistasis reduce the precision of selection based only on phenotype. If the characteristic under selection were determined entirely by additive gene effects, the mean of the selected parents would be a good measure of their average genotype. Different individuals would vary, of course, in the number of "active alleles" they possessed, and in the ways in which the "active alleles" at different loci were combined. Segregation and recombination of these alleles would result in genetic variation among the progeny. But *the mean of the next generation would be the same as the mean of the selected parents,* and in successive generations rapid progress could be made in the improvement of the population (Fig. 14-16). The ultimate level reached would depend on the amount of genetic diversity present in the original population or introduced into it by bringing in "new blood" from outside.

When environmental effects and dominance and epistatic deviations are

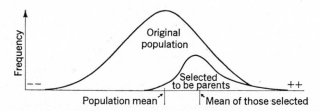

Figure 14-15. *On a scale running from* (− −) (*most undesirable*) *to* (+ +) (*most desirable*) *for a trait under selection, the mean of selected parents is higher than the average of the population from which the parents came.* (*After Lush,* Animal Breeding Plans, *Iowa State College Press, 1945, p. 146.*)

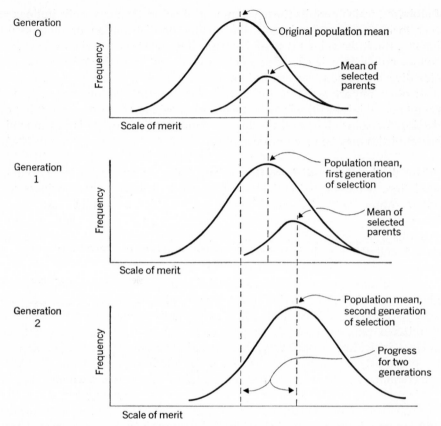

Figure 14-16. *If the trait under selection were determined entirely by additive gene effects, the mean of a generation would be the same as the average of its selected parents, and rapid progress would be made in improving the population.*

present, this kind of selection becomes less effective. Among the individuals selected to be parents, a fraction will be animals with only ordinary or even inferior genotypes that have been fortunate in their environment and thus have achieved levels of performance high enough to mislead the breeder. Similarly, some animals superior in genotype will, through misfortune in their environment, be left out of the breeding program, and their desirable alleles will be lost. Even among the genotypically superior and selected individuals, undesirable recessives will be present to segregate out in later generations. This will result in progeny that average less than the value predicted by simple additive inheritance (without dominance or recessiveness). And if gene interaction (epistasis) is involved in phenotypic superiority, Mendelian recombination will effect the breaking-up of many favorable, selected gene combinations, again with the result that the progeny may average less than the mean of their selected parents.

For these reasons, the history of a herd or flock under phenotypic selection generally looks something like Figure 14-17. Instead of showing the average value of their selected parents, the progeny regress from this value toward the original population mean, and in fact average only a little better than the population from which their parents came.

This phenomenon, in which the progeny of selected parents slip back toward the average of the population from which the parents were chosen, has long been known. It was early encountered, for example, in studies of the heights of children of unusually tall or short men and women, in which it was observed that the children were, on the average, less extreme than their parents. In this connection the phenomenon was called *regression*. Today, the term regression has a broader statistical connotation, but it is still applicable in its original sense to the problems we are discussing.

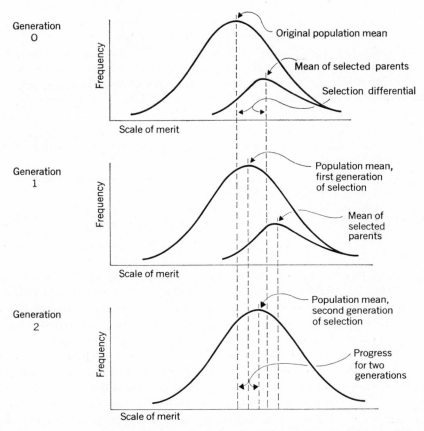

Figure 14-17. *The progeny of selected parents do not usually average as well as the parents did, but regress toward the mean of the population from which the parents came.*

Type and Performance. In many classes of livestock, a primary basis for individual selection is conformity to the ideal type of the breed. Usually this ideal is rather strictly defined, and to a more or less conscious extent it represents an esthetic standard—the standard of beauty for the breed. Breeders who compete for prizes in the show circuits enter their animals in what are essentially beauty contests, in which the animals that in the opinion of the judges conform most closely to the breed type win ribbons and prizes. Several of the breed organizations undertake *type classification* of registered animals, and the degree of conformity to type becomes a matter of permanent record for each animal.

There are at least three reasons why conformity to type may be regarded as a justifiable consideration in selecting breeding animals. First, to a great many breeders the esthetic value of owning and working with beautiful animals is one of the primary satisfactions of husbandry. Ruggedly materialistic dairymen, for example, may regard the size of the milk check as the only important aspect of dairying, but to many breeders of dairy cattle the occupation has other values beyond economic evaluation. Second, in the animal-breeding world as it is at present constituted, conformity to type has a very real economic aspect. Few breeders who depend for a significant part of their margin of profit on the sale of breeding stock can afford to ignore the looks of their animals, lest they achieve productive animals that few buyers want because they are ugly. The type classification of an animal or his record in the show ring is often reflected in his cash value, and in the value placed on his relatives and progeny. Third, it is often asserted that there is a relationship between type and productivity, so that selection on the basis of type is in fact selecting for higher or more persistent production.

The first two reasons above are matters of things-as-they-are, and each breeder must evaluate them for himself in terms of his own objectives. The third is a point that we may discuss briefly. Probably in all classes of livestock, some of the grounds for type classification are related to the productive merit of the animal. This is probably particularly true of animals grown for meat or for wool or fur, because the potential productivity of the animal is displayed for all to see and to judge. In dairy cattle or in poultry, on the other hand, it is not equally clear that the butterfat or egg production of an individual can be discerned in its appearance. Some of the characteristics observed in judging individuals, while they may not show up in immediate production records, are nevertheless important to the animal's practical, continued performance. No matter how productive, a cow whose udder has broken away from her body to drag on the ground, suspended by loose sheets of skin, is probably not a good cow under most farming conditions. Other similar examples might be cited.

The main difficulty in using type as a means of selecting for productivity, however, is that many of the aspects of type have no clear relationship to productivity, and a few standards of type appear even to be negatively correlated with production. *The effectiveness of selection in any herd or flock of limited size varies in inverse fashion with the number of criteria for culling or saving individuals.* If a dairyman wishes to set up as his single goal the increased butterfat percentage content of his milk, by rigid selection and ignoring all else he should be able to progress toward this goal rather rapidly. If he aims toward increased total butterfat yield, he has two criteria for selection: total yield of milk and butterfat percentage content. These two are negatively correlated; he will be tempted to save some high producers of low-test milk and some low producers of high-test milk. If he computes a single index for selection, the total number of pounds of butterfat per lactation, he will progress somewhat more rapidly than if he selects on the basis of the two components independently.

Now if in addition to this index the breeder must consider the balance and the attachment of the udder, he may be forced to discard some highly productive animals that do not meet this additional standard, and to save some well-uddered heifers that would not otherwise be used in the breeding program. He thereby reduces the effectiveness of his selection for butterfat yield alone. If the animals must also have straight backs and legs, broad muzzles, and the right amount of white spotting in the right places, this breeder must find himself compromising more and more with his primary objective. It is not really relevant to justify selection on the basis of muzzle, back, legs, and udder, on the grounds that these criteria may be correlated with productivity. Such correlations, if they exist, are generally low, and certainly much less direct and reliable as measures of butterfat productivity than the figure for butterfat yield itself.

This must not be understood to mean that it is unwise to include type as a criterion for selection. As we pointed out earlier, under many circumstances such characteristics as straight backs or the right amount and distribution of white spotting are entirely legitimate ends in themselves, even though they may be independent of productivity. Each breeder must decide for himself what his goals will be, and then select toward them in as objective a manner as possible, with the firm recognition that as he complicates his criteria for selection he slows down his progress toward any single standard of success. He must be sure that the criteria he uses are really important to him, and he must try to combine these criteria according to their relative importance and the degree to which they are inherited in some kind of overall index for selection.

Pedigree, Relatives, and Progeny. We have observed that the phenotype of an animal, including all of the aspects of his individual performance, is only an uncertain guide to his breeding value, and that selection on the basis of performance alone is likely to be accompanied by mistakes from a genetic point of view. Some of these mistakes can be avoided if other estimates of the genetic merit of the animal, in addition to those suggested by careful evaluation of his phenotype, can be introduced into the selection index. The obvious way to get such additional information is by studying the ancestors, collateral relatives, and descendants of the individual.

Selection on the basis of pedigree and relatives has for a very long time been a valuable adjunct to animal breeding. Its essential rationale is that we can often get at least some idea of the quality of germ plasm from which an individual comes by considering the achievements of his ancestors and relatives. Without discussing the technique in detail, we can make three simple observations.

1. It is indeed true that valuable information regarding the potential breeding worth of an animal can be gained from proper consideration of an informative pedigree.

2. The standards for reliable information about the ancestors and relatives of an animal must be as objective and as pertinent as those for evaluating the performance of the individual himself. An informative pedigree should include numerous unselected records on closely related individuals; it should provide as complete and reliable as possible an indication of the genetic background of the individual.

3. As a student of genetics, you are now well aware of the sampling nature of inheritance, and of the regular shuffling and reshuffling of genes between sexual generations. You know that "like" does not automatically "beget like," and that the progeny of highly select parents may be expected to regress toward the mean of the population from which the parents came. In predicting the probable accomplishments of the progeny of an individual from those of his parents, two opportunities for segregation and recombination must be taken into account: that which gave rise to the individual and that which will produce his progeny. In any heterozygous population the opportunities for genetic variation in this situation are very great. For this reason, selection on the basis of pedigree alone is risky, while selection that combines critical pedigree information and a consideration of relatives, with a competent evaluation of the individual himself, is much less likely to go awry. It should go without saying that the significance of pedigree information decreases very rapidly as the ancestors or relatives concerned become more remote in their relationship to the individual under consideration. Uniform excellence in the immediate ancestors and relatives of an individual is much more trustworthy than is the appearance of outstanding individuals several generations back in his pedigree.

Progeny tests also provide important information about the breeding merit of an animal. They amount to estimating the productive level of transmission of an individual by evaluating a sample of his actual progeny. You can see how such tests would help to correct errors in selection based on phenotype alone; for example, if subtle environmental effects had given a genetically mediocre individual a misleading appearance of excellence, even a small sample of his progeny would be likely to betray his true, inherent mediocrity.

Observations on relatives and progeny are especially useful in evaluating characteristics that are limited in their expression to one sex. One cannot detect directly, for instance, the genetic potentialities for milk production in a bull or for egg production in a cock. One is therefore limited in selecting males for such attributes to estimates based on pedigree, relatives, and progeny plus whatever correlation there may be between the male's physical conformation and the genetic potentialities for productivity he may transmit. In this situation, four or five milking daughters of a bull, from dams of known productivity, could tell as much about the breeding merit of the individual as could the individual's own record of butterfat production. The record of the bull's dam is worth about as much, in terms of information on the bull, as records on four or five half-sisters.

Today, when artificial insemination is making it possible for particular bulls to produce very large numbers of offspring, the wisdom of using sires proved good by progeny tests and by careful observations on relatives has become evident to almost all breeders. A mistake in selecting a sire for an insemination ring can be very costly; the use of information on relatives and progeny can render such mistakes less likely.

The simplest measure of the transmitting merit of a sire from progeny tests is the "mid-parent index." If the progeny can be assumed to have inherited a sample-half of their dams' inherent productive capacities, and similarly a sample-half of their sire's, then

$$\frac{\text{Dams}}{2} + \frac{\text{Sire}}{2} = \text{Daughters.}$$

The words refer to averages for the categories indicated. Solving for *Sire*, which is the unknown we want to estimate,

Sire = (twice the average of the daughters) minus (the average of the dams).

Obviously, this calculation is based on the assumption that the inherent productive capacity of daughters and dams is competently measured by their performance. We have already discussed the limitations of such an hypothesis in individual cases; here, however, since we are dealing with averages of a number of values, some of the chance errors may be expected to be compensated by others in the opposite direction. Such statistical correction factors as are available can of course be used to standardize the dam and daughter records and to make them comparable. Nevertheless, a mid-parent index is at best an approximation, more or less precise depending on the number of dam-daughter pairs it includes and the nature of the tests it is based on. Like the individual's own record, some regression must typically be allowed for in evaluating selected individuals. For these reasons, more sophisticated procedures than calculating the "mid-parent index" are commonly used by professional animal breeders.

In particular instances the loss of time in obtaining progeny test information on superior males may more than offset the additional information to be obtained from such tests. Poultry provide a good example. The evaluation of cockerels on the basis of their sisters' production up to January 1 of their pullet year is probably sufficiently accurate to justify cockerel selection for extensive use in their first breeding season, and it is probably unwise to wait for progeny to come into production before cockerels are selected. Of course, mature cocks with progeny in production can be further evaluated on the basis of these progeny.

Heritability and the Selection Index. We have suggested that the most effective guide to selection would be the computation of an *index* that combines

Table 14-11. REPRESENTATIVE HERITABILITY VALUES FOR VARIOUS CHARACTERISTICS OF FARM ANIMALS.

Animal	Characteristic	Heritability
Swine	Litter number at birth	0.2
	Litter number at weaning	0.2
	Weaning weight of litter	0.4
	Individual weaning weight	0.1
	Individual 180-day weight	0.3
	Conformation score	0.2
Sheep	Birth weight	0.3
	Weaning weight	0.3
	Yearling weight	0.4
	Yearling clean fleece yield	0.4
	Face covering	0.6
	Neck folds (Columbia, Corriedale, and Targhee)	0.1
Beef cattle	Birth weight	0.5
	Weaning weight	0.3
	Weight at 15 months	0.9
	Rate of gain in feed lot	0.8
	Slaughter grade	0.5
	Carcass grade	0.3
Fowl	Egg number	0.3
	Viability	0.1
	Production index	0.1
	Sexual maturity	0.3
Dairy cattle	Single annual butterfat record	0.2
	Single annual milk record	0.2
	Single official type classification	0.3
	Single annual butterfat percentage	0.5

Source: Derived from summaries by Warwick, Henderson, and Lerner.
Note: Determinations of heritability in various studies differ considerably; these are selected values, corrected to one decimal place.

in a single figure as many as possible of the complex bases for deciding whether to cull an animal or to use it in the breeding program. This objective estimate of the breeding worth of an individual would include the information from all the sources of evidence about him, each source weighted according to its estimated value. Pedigree, relatives, performance, and progeny, when they are available, should all be considered. Such an index should weigh all the characteristics to be selected, each according to its relative importance to the breeder's conscious objectives. These objectives should be as clearly defined and as simple as possible, because increasing the number of criteria for selection markedly decreases the rate of progress toward any single goal.

Another consideration should also be taken into account in weighing the characteristics to be selected This is the degree to which each of them is *heritable*—that is, the extent to which variation in successive generations is predictable in terms of genetic control and is therefore subject to selection in straightforward fashion. For most practical purposes in selection, this straightforward genetic control refers to variation that can be treated as the result of additive gene effects. Estimation of the heritability of characteristics important in domestic animals is a major subject for research in this field. It involves precise experimental design and rather powerful statistical methods, and we cannot elaborate the techniques of estimating heritability here. But we should say a few words about the existence of measures of heritability and about their importance.

It is generally found that only a fraction of the total variation in any generation, with respect to any economic characteristic, is identifiable as genetic in its origin. Table 14-11 lists representative heritability values for a variety of characteristics. We can describe the practical significance of these values in two rather different ways.

First, the figure for heritability is a measure of the importance of heredity, relative to environment and unpredictable interactions, in determining the characteristic designated. For example, about 90 percent of the variation in weight at fifteen months in beef cattle can be regarded as additive, genetic in its control (in a particular population, under the conditions of a particular

Table 14-12. USE OF THE HERITABILITY VALUE FOR AVERAGE INDIVIDUAL 180-DAY WEIGHT IN SWINE TO ESTIMATE PERFORMANCE OF PROGENY OF SELECTED PARENTS.

Average of Herd	Selected Individuals	Selection Differential	Heritability	Expected Performance of Progeny
180 lb	195 lb	15 lb	0.3	184.5 lb

Source: After Warwick, in *Breeding and Improvement of Farm Animals,* by Rice and Andrews, McGraw-Hill Book Co., 1951, p. 663.

experiment). But only 10 percent of the variation in individual weaning weight in swine (again, in a particular experiment) can be so regarded.

Second, the figure for heritability is an estimate of the amount of regression to be expected among the progeny of selected parents. As a concrete example, consider Table 14-12. You will see that the use of the heritability value for 180-day weight in swine provides a measure of expected genetic gain in terms of a given intensity of selection. Highly heritable traits are easily subject to effective selection; slightly heritable ones require intense selection to provide significant genetic progress.

Knowledge of the heritability of selected traits is therefore of obvious value in deciding what emphasis should be put on these traits in selection. Estimates of heritability are, however, subject to some variation in themselves, because they are affected by the amount of environmental fluctuation in the herd or flock in which they are determined, and by the system of mating and the degree of genetic heterozygosity within that herd or flock.

Keys to the Significance of This Chapter

Variation in some of the most important characteristics of organisms, such as size and shape, is not expressed in discrete steps. Individuals therefore do not fall into distinct classes with respect to such characteristics, and can only be reasonably described by quantitative measurements.

Effective summarization and manipulation of the data of quantitative inheritance require the use of appropriate statistical techniques. Beyond this, it is desirable to employ special mathematical methods for establishing the reliability of the techniques used in studying quantitative inheritance.

Typical instances of quantitative inheritance apparently depend upon the composite activity of multiple genes, which are individually unidentifiable because their separate effects are relatively insignificant to the phenotype. These genes appear not to differ in principle from genes with major effects. In fact, a given gene may exert a major effect on one character and at the same time act as a member of a multiple-gene complex affecting another trait of the organism. Different multiple-gene complexes appear to vary considerably in the number of individual components. As few as three loci may be involved in the ordinary range of expression of some quantitative characters, but hundreds of loci may be involved in the expression of others.

The nature of gene action in multiple-gene systems remains rather obscure at the present time. But it is clear that multiple genes do not operate entirely through simple additive effects. In fact, many cases of quantitative inheritance seem to show evidence for geometric rather than arithmetic action.

The two most important tools available to the breeder are selection and control of the mating pattern.

The mating system of a group has an important influence on its genotypic composition. Because inbreeding promotes *homozygosis*, it sometimes is accompanied by unfavorable consequences for members of the group in which it occurs. This happens especially in populations which normally are not inbred, and which carry deleterious recessive genes whose effects are masked by more favorable dominant alleles. That inbreeding as such is not deleterious is proved by the fact that for several large and successful groups of plants inbreeding is the habitual means of reproduction. In these groups, deleterious recessives cannot accumulate under the masking influence of dominant alleles.

Phenotypic variation occurs even within lines that have been long inbred. Unless mutations have occurred, this variation is usually environmental in source and cannot be utilized as the basis for effective selection. Only genetic variations are subject to selection.

Heterozygosity is often accompanied by vigor.

By constructing an index for selection, which weighs the minimum number of characteristics that can represent the goals of his breeding program in terms of the relative economic importance and the heritability of these characteristics, a breeder may render his selection of breeding stock more objective and effective. Pedigree and relatives, individual performance, and progeny tests can provide important guides for the selection of individuals.

REFERENCES

Allard, R. W., *Principles of Plant Breeding*. New York: Wiley, 1960. (A complete, modern text.)

Brewbaker, J. L., *Agricultural Genetics*. Englewood Cliffs: Prentice-Hall, 1964. (A clear, comprehensive paperback.)

Burton, G. W., "Quantitative Inheritance in Pearl Millet." *Agr. J.*, **43**:409–417, 1951. (Presents a wealth of data on quantitative inheritance. Utilized for Questions and Problems for this chapter.)

Charles, D. R., and Smith, H. H., "Distinguishing between Two Types of Gene Action in Quantitative Inheritance." *Genetics*, **24**:34–48, 1939. (An amplification of the characteristics of arithmetic as compared with geometric gene action in multiple-gene systems.)

Crow, J. F., *Genetics Notes*. Minneapolis: Burgess Publishing Co., 5th ed., 1963. (Chaps. 20–22, on inbreeding, statistical analysis of quantitative characters, and selection are especially valuable at this point.)

——, "Genetics of Insect Resistance to Chemicals." *Ann. Rev. Entomology*, **2**:227–246, 1957. (An interesting review with many references, and the source of information on the polygenic basis of DDT resistance in Drosophila mentioned in this chapter.)

Davenport, C. B., "Heredity of Skin Color in Negro-White Crosses." *Carnegie Inst. Wash. Pub.* 188, 1913. (The details of a study of the only quantitative character in man "for which a reasonably well-founded specific hypothesis of multifactor inheritance has been proposed" [see reference to Stern, below].)

East, E. M., "Heterosis." *Genetics*, **21**:375–397, 1936. (A review article. Examines certain of the conflicting ideas on heterosis, and gives an idea of the development of thought in this field.)

——, and Jones, D. F., *Inbreeding and Outbreeding*. Philadelphia: Lippincott, 1919. (A classic in genetics, having much more than historic interest, although written some decades ago.)

Emerson, S., "A Physiological Basis for Some Suppressor Mutations and Possibly for One Gene Heterosis." *Proc. Nat. Acad. Sci.*, **34**:72–74, 1948. (The ingenious experimental analysis of a situation that has important implications for our understanding of heterosis.)

Falconer, D. S., *Introduction to Quantitative Genetics*. New York: Ronald Press Co., 1960. (More than a simple introduction, this book presents clearly the sophisticated modern techniques and their bases.)

Johannsen, W., *Ueber Erblichkeit in Populationen und in reinen Linien*. Jena: Gustav Fischer, 1903. (Often ranked with Mendel's paper as one of the cornerstones of modern genetics. Reprinted in *Classic Papers in Genetics*, edited by J. A. Peters. Englewood Cliffs: Prentice-Hall, 1959.)

Lerner, I. M., *The Genetic Basis of Selection*. New York: Wiley, 1958. (An advanced book, clear and comprehensive.)

Li, C. C., *Population Genetics*. Chicago: University of Chicago Press, 1955. (Chaps. 12 and 13 deal with inbreeding and relationship.)

Lush, J. L., "Genetics and Animal Breeding." In *Genetics in the 20th Century*, edited by L. C. Dunn, pp. 493–525. New York: Macmillan, 1951. (Primarily an evaluation of the important interactions between formal genetics and applied animal breeding, with interesting factual data.)

Mangelsdorf, P. C., "Hybrid Corn: Its Genetic Basis and Its Significance in Human Affairs." In *Genetics in the 20th Century*, edited by L. C. Dunn, pp. 555–571. New York: Macmillan, 1951. (A general treatment, well described by the title.)

Mather, K., "Polygenic Inheritance and Natural Selection." *Biol. Rev.*, **18**:32–64, 1943. (A meaningful treatment of many of the fundamental concepts of quantitative inheritance.)

Müntzing, A., "On the Causes of Inbreeding Degeneration." *Arch. Julius Klaus-Stiftung Vererbungsforschung, Sozialanthropol. u. Rassenhyg.* Supplementary volume to vol. **20**:153–163, 1945. (A brief and readable essay, in English, touching on major aspects of inbreeding and heterosis.)

Pearl, R., and Surface, F. M., "Selection Index Numbers and Their Use in Breeding." *Am. Naturalist*, **43**:385–400, 1909. (A classic paper presenting early recognition of the importance of constructing an index for selection, and suggesting indexes for poultry and sweet corn.)

Stern, C., *Principles of Human Genetics*. San Francisco: W. H. Freeman and Co., 2nd ed., 1960. (Pages 351–358 are a summary and discussion of the material in the reference to Davenport, above.)

Wright, S., "The Effects of Inbreeding and Crossbreeding on Guinea Pigs." *U.S. Dept. Agr. Bull.* No. 1090, Professional Paper, 1922. (Extensive data illustrating the decline in vigor and the fixation of characteristics resulting from inbreeding.)

———, "Systems of Mating." *Genetics*, **6**:111–178, 1921. (A series of five papers, written while Wright was with the U.S.D.A. These papers provide the foundation of our current appreciation of the processes of both evolution and animal improvement.)

———, "The Analysis of Variance and the Correlations Between Relatives with Respect to Deviations from an Optimum." *J. Genetics*, **30**:243–256, 1935. (A technical paper, highly significant, with particular reference to the probably common situation in which the favored class is an intermediate type.)

QUESTIONS AND PROBLEMS

14-1. What do the following terms signify?

additive gene effects	discontinuous variation
arithmetic mean	double cross
coefficient of inbreeding	frequency distribution
coefficient of relationship	genetic variability
continuous variation	geometric mean

heritability	outbreeding
heterosis	overdominance
homozygosis	pedigree selection
hybrid vigor	progeny test
inbreeding	pure line
inbred line of corn	qualitative character
male sterility	quantitative character
mass selection	regression
median	selection differential
midclass value	selection index
mode	single cross
multiple-gene hypothesis	standard deviation
natural selection	standard error
normal distribution	system of mating
open-pollinated variety	

The results of certain crosses between Negroes and whites are given to a fair approximation by a model that assumes that skin color differences may depend upon allelic substitutions at two independent loci. Whites can be designated by the genotype *aabb*; Negroes by the genotype *AABB*. The active alleles for pigmentation, *A* or *B*, are supposed to have independent cumulative effects upon the intensity of coloration. Thus an individual with any two active alleles (*e.g.*, *AAbb* or *AaBb*) is darker than an individual with only one active allele but lighter than an individual carrying three active alleles. Under this scheme, individuals may fall into one of five classes: white, light brown, medium brown, dark brown, or black. Use this interpretation, which incidentally does not hold satisfactorily for all color differences emerging from white-Negro crosses, in answering the next four questions.

14-2. How does the preceding interpretation of skin-color inheritance compare with Nilsson-Ehle's analysis of the genetic basis of kernel color in wheat?

14-3. Is it proper to refer to skin color in man as a quantitative character? Elaborate.

14-4. Can the first-generation progeny of white-black matings produce black offspring? White offspring?

14-5. Does every marriage between two medium-brown individuals offer the possibility of progeny showing either lighter or darker skin colors?

14-6. In experiments conducted by Wexelsen, in Norway, the mean internode length in spikes of the barley variety Asplund was found to be 2.12 mm, and of the variety Abed Binder, 3.17 mm. The mean of the F_1 derived from a cross between these varieties was approximately 2.7. The F_2 gave a continuous range of variation from one parental extreme to the other. Analysis of the F_3 population indicated that in F_2, 8 out of the total of 125 individuals were of the Asplund type, giving a mean of 2.19 mm. Eight other individuals were similar to the parent Abed Binder, giving a mean internode length of 3.24. Which does internode length in spikes of barley seem to be—a qualitative or a quantitative character? Why?

14-7. From the information given in the previous question, how many gene pairs involved in the determination of internode length appear to have been segregating in F_2?

Burton has made extensive studies of quantitative inheritance in pearl millet, an important pasture crop in the southeastern United States. Following is a frequency distribution for number of leaves per stem in a pearl millet cross. Utilize these data in answering the next five questions.

	9	10	11	12	13	14	15	16	17	18	19	20	21	22	23	N
				NUMBER OF LEAVES PER STEM											
P 16			1	17	55	66	34	5								178
P 782				3	6	6	16	17	23	50	29	20	12	4	1	187
F_1			5	5	28	51	58	27	5							179
F_2	1	8	37	144	303	454	452	322	153	31	8					1913

14-8. For each of the populations shown in the preceding frequency distribution calculate $\bar{x}$, σ, and S. E.

14-9. Compare the different σ values you obtain. What might explain the differences?

14-10. Calculate the arithmetic mean between the two parental means, and also the geometric mean. Which of these values is closer to the observed F_1 mean value?

14-11. From the fact that one of the parental extremes (P 782) was not recovered in an F_2 population of almost 2,000 individuals, what general conclusion may you draw about the number of genes determining leaf number?

14-12. Why must your answer to the preceding question remain tentative?

Burton has also studied the inheritance of head length in pearl millet crosses. Utilize the following frequency distributions for answering the next four questions.

	2	3	4	5	6	7	8	9	10	11	12	13	14	15	16	N
				MIDCLASS VALUES IN INCHES FOR HEAD LENGTH											
P 16					3	38	72	12	1							126
P 18	1		7	40	50	15	2									115
F_1									1	88	66	1				156
F_2			5	5	11	20	54	120	358	576	524	245	90	15	1	2024

14-13. For each of the populations shown in the preceding frequency table calculate $\bar{x}$, σ, and S. E.

14-14. Is the relatively large σ value for F_2 expected? Justify your answer.

14-15. Compare the $\bar{x}$ values of the F_1 and F_2 populations with the parental means. Is the picture you obtain typical of cases of quantitative inheritance you have studied so far? If not, in what respects does it differ?

14-16. Is the type of gene action assumed in the hypothetical model for quantitative inheritance discussed in this chapter adequate to explain the results detailed above? Discuss.

Utilize the following information in answering the next six questions. Assume a situation where genes *A*, *B*, *C*, and *D* have duplicate, cumulative effects and are independently inherited. Each of these genes contributes 3 cm height to the organism when present. In addition, a gene *L*, always present in the homozygous state, contributes a constant 40 cm of height. Neglecting variation due to environment, an organism *AABBCCDDLL*, then, would be 64 cm high, and one *aabbccddLL* would be 40 cm.

A cross is made *AAbbCCDDLL* × *aaBBccDDLL* and carried into F_2.

14-17. How would F_1 individuals compare in size to each of the parents?

14-18. Compare the mean of the F_1 with the mean of the F_2.

14-19. What proportion of the F_2 population would show the same height as the *AAbbCCDDLL* parent?

14-20. What proportion of the F_2 population would show the same height as the *aaBBccDDLL* parent?

14-21. What proportion of the F_2 population would breed true for the height shown by the *aaBBccDDLL* parent?

14-22. What proportion of the F_2 population would breed true for the height characteristic of F_1 individuals?

14-23. Would a law banning marriages between individuals and their step-parents be well founded on genetic principles?

14-24. In some lines of corn, monoploids occur with a frequency of about one per thousand sporophytes. Occasionally these monoploids produce pollen and egg cells containing complete genomes. It is possible through self-pollination of a monoploid plant to obtain a diploid line derived from a monoploid, and to maintain the line through further selfing. Would selection within such a diploid line, for size, for example, be likely to be effective? Explain your answer. Of what use might such diploids be?

14-25. What is a circumstance under which inbreeding does not have deleterious consequences?

14-26. Demonstrate that inbreeding *per se* does not favor an increase in the frequency of recessive alleles in a population, but affects only the distribution of the alleles among genotypes.

14-27. Which would be more likely to carry a fair number of deleterious recessive genes: (a) a natural population representing a line of organisms that has been reproducing vegetatively over a number of generations, or (b) a natural population representing a line in which self-fertilization has been habitual? Justify your answer.

14-28. Discuss the implications of the fact that it has not so far been possible to isolate lines of corn that breed true for the high degree of vigor and uniformity found in hybrid corn. Include a consideration of possible reasons for this failure.

14-29. Corn breeders have carried out extensive programs of *convergent improvement*. Under this practice, an F_1 hybrid is backcrossed to each parental inbred over a succession of generations. At the same time, selection is carried on within each backcross line for favorable characters of the nonrecurrent parent. Which would be more homozygous, an F_1 hybrid between the original inbreds or an F_1 hybrid between "recovered inbreds" resulting from the backcross program?

14-30. Results of convergent improvement programs, referred to in the previous question, indicate that hybrids between "recovered inbreds" are at least as vigorous as hybrids between "original inbreds." Does this result suggest that heterozygosity as such determines heterosis? Discuss.

14-31. What are some of the specific advantages for the plant breeder of working with a plant that reproduces asexually as well as sexually?

14-32. In what respects does natural selection operate in favor of the objectives of plant-improvement programs? Make your answer as specific as possible.

14-33. Does natural selection always work in the same direction a plant breeder might wish? Amplify your answer with examples.

14-34. Name two specific circumstances under which the breeder may have to deal with very large populations of plants in order to achieve an objective.

14-35. Discuss how particular instances of plant improvement may have rather far reaching consequences for society, apart from merely giving a greater abundance or finer quality of products.

14-36. Until very recently, the artificial induction of mutations has played almost no role in plant-improvement programs. Why do you suppose plant breeders have not generally resorted to this well-known technique of experimental genetics? In other words, do you see any difficulties about utilizing mutagens in plant-breeding practice?

14-37. In what sorts of instances, and for what kinds of objectives, do you think the artificial induction of mutations might be of practical use in plant-improvement programs?

14-38. What are some of the difficulties the breeder may have in working with polyploids as compared to diploids?

14-39. Polyploids are often larger and more vigorous than their diploid analogues. Why not, as part of every breeding program, simply double the chromosome number of all available varieties?

14-40. Under what circumstances may the breeder find techniques of test-tube embryo culture useful?

14-41. What is the breeder's purpose in inbreeding normally cross-pollinated crop plants that show reduction in vigor when they are inbred?

14-42. Under what circumstances might a breeder wish to make *wide* crosses as a step toward plant improvement?

14-43. In oat breeding, for example, which would be more effective for dealing with characters that are sensitive to environmental influence: (a) mass selection, or (b) pedigree selection? Why?

14-44. What problems, if any, do you see in maintaining a pure-line variety of wheat in its original state? What steps might be taken to maintain such a variety?

14-45. Many forage plants have multiple-allelic series for self-incompatibility (*S* alleles). Discuss whether this is advantageous or disadvantageous from the breeder's point of view.

14-46. Suppose that you, as a plant breeder, have on hand a variety of tomato that is satisfactory in every regard except that it is highly susceptible to a certain disease. You also have available a variety with several objectionable features, from the growers' standpoint, but that is disease resistant. You wish to transfer the disease resistance to the first variety through backcrossing. Will it make any difference in your plans, or in the ease of carrying them out, whether the disease resistance character is inherited as a simple dominant or as a simple recessive? Explain.

14-47. Which of these characters do you think would be simpler to transfer through the backcross method: (a) winter hardiness, or (b) plant color? Why?

14-48. Which is the more realistic criterion for establishing the practical value of a corn hybrid: (a) to compare its performance with the performances of its parental inbreds, or (b) to compare its performance with that of open-pollinated varieties grown in the same area?

14-49. Why, in general, are single-cross progenies more uniform than double-cross progenies?

14-50. Suggest how monoploids might be utilized in a corn-breeding program.

14-51. What are some of the various factors that limit yield in crop plants?

14-52. In connection with plant improvement programs, expeditions are sometimes organized for the purpose of collecting wild relatives of our cultivated plants. What practical purposes may such expeditions serve?

14-53. Are there any "improvements" you would like to see made in plants you are familiar with? What improvements? Would it be feasible to achieve them through breeding methods?

14-54. It has been suggested that experiments in plant improvement might well utilize complicated greenhouses that make possible the close control of many environmental factors such as light, temperature, and humidity. What advantages and disadvantages can you see in this suggestion as contrasted with improvement programs practiced in open field culture?

Assume a characteristic controlled by five pairs of alleles, acting in equal and additive fashion and uncomplicated by environmental effects. Assign a value of 50 units to the complete recessive, and assume that each substitution of an "active" allele (represented by a large letter) gives a phenotypic increment of 5 units. For example,

$$
\begin{array}{llllll}
aa & bb & cc & dd & ee & = \ \ 50 \\
Aa & bb & cc & dd & ee & = \ \ 55 \\
Aa & Bb & cc & DD & ee & = \ \ 70 \\
AA & BB & CC & DD & EE & = 100 \\
\end{array}
$$

√ *14-55.* Suppose that in a population with a phenotypic range from 50 to 70 units and an average value of 60, you selected to be parents of the next generation only individuals with the top rating of 70.

  a. List at random five different genotypes that might be represented among the individuals selected.
  b. Confirm by inspection that the average gamete produced by the selected individuals listed in (a) would contain two active alleles, and that the gametes might range from 0 to 4 active alleles.
  c. Confirm from (b) above that the average zygote in the next generation would have a phenotypic value of 70, and the range would be from 50 to 90.

14-56. Repeat the arbitrary analysis in Question 14-55 for another generation of selection, choosing the individuals with a phenotypic value of 90 to be parents of this next generation. What would be the average and the range of phenotypic values in the progeny?

14-57. Under the simple, additive inheritance postulated in Problems 14-55 and 14-56:

  a. How does the mean of the progeny compare with the mean of the selected parents?
  b. Make a general statement about the effectiveness of selection under these circumstances, using the term *selection differential*.
  c. What sets the limits to be achieved by selection under these conditions, and what determines the rapidity with which these limits are approached?

14-58. Now, postulate an arbitrary environmental effect on the expression of the characteristics discussed in Problems 14-55 to 14-57. Assume that the conditions described prevail for an "average" environment, but that a "poor" environment subtracts ten units from the phenotypic value of any particular genotype, and a "good" environment adds ten units. Describe, in a general way, the effects of this circumstance on the achievements of selection, if allowance for environmental effects cannot be made in the selection of individuals for breeding.

14-59. How would complete dominance of the "active" alleles affect the selection discussed above? Why?

14-60. Suppose that the situation is that described in the introduction to this series of problems (equal and additive gene effects, no environmental modification or dominance), except that the substitution of B for b has no effect unless A is also present in either the homozygous or heterozygous condition. For example,

$$
\begin{array}{llllll}
Aa & bb & Cc & Dd & ee & = 65, \text{ and} \\
Aa & Bb & Cc & Dd & ee & = 70, \text{ but} \\
aa & bb & Cc & Dd & ee \\
aa & Bb & Cc & Dd & ee \\
aa & BB & Cc & Dd & ee
\end{array} \right\} \text{all} = 60.
$$

Describe in a general way how this epistatic interaction may affect the progress of selection.

14-61. Frequently, in a comparison of two genotypes, one is found to be more effective in a given environment while the other is better in a different environment. How might this kind of interaction between genotype and environment affect a selection program?

14-62. How do the effects of environment, dominance, and epistasis, illustrated in Questions 14-58 to 14-61, explain the phenomenon of regression?

14-63. Why would you expect progeny tests to be particularly useful in the improvement of poultry and dairy cattle? In each of these classes of livestock, what advantages can you discern in estimating the breeding value of a male by indexes based on his dam and sisters, rather than on his progeny?

14-64. Following are dam-daughter comparisons in butterfat yields (corrected to maturity, 305 days, twice-daily milkings) for five Guernsey bulls in use in a single large herd. Considering butterfat production alone, and on the basis of the information given, which bull would you be likely to select for use in your Guernsey herd?

BULL	CORRECTED AVERAGE, FIVE DAUGHTERS	CORRECTED AVERAGE, CORRESPONDING DAMS
1	380	390
2	385	360
3	405	385
4	405	415
5	425	440

14-65. Compute coefficients of relationship for individuals A and B on the basis of the diagram below. If these two individuals are mated, what will be the inbreeding coefficient of their progeny?

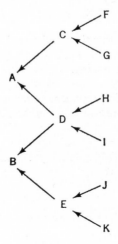

14-66. Studying the men and women of Pitcairn Island, Shapiro attempted to estimate the degree of inbreeding that had occurred by dividing the actual number of ancestors in a given generation of the pedigree of an individual by the total number he would have had in that generation if there had been no inbreeding. For example, a marriage between first cousins otherwise unrelated would have children with a coefficient of $\frac{6}{8} = 0.75$, if the computation is based on great-grandparental generation.

Compute the inbreeding coefficients for individuals P and P′ in the diagrams given below, assuming that the individuals in the most remote generation shown (great-grandparents) were unrelated.

a. Use the method described in this chapter.
b. Use the measure Shapiro utilized, counting ancestors in the great-grand-parental generation. (Note that P is the result of a mating between half-sibs, while P′ is the result of mating between first cousins.)
c. Which of the two measures describe the genetic situation better? Why?

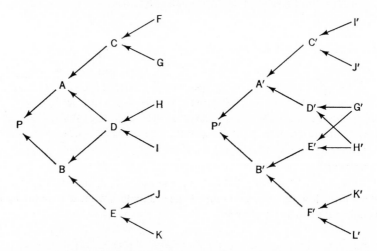

14-67. Characterize in general terms the usual genetic effects of the following systems of mating: (a) inbreeding, (b) outbreeding, (c) assortative mating, (d) disassortative mating, (e) linebreeding.

14-68. Work out a plan for the improvement of some animal population in which you are interested—*e.g.*, a breed of dogs used as hunters, a herd of Jerseys, or a population of earthworms to be used as fish bait. How might you select your foundation stock and your breeding stock from generation to generation? What mating patterns might you adopt? What information would you need for most efficient progress?

CHAPTER 15

Genetics and Man

Quite naturally, and legitimately, most of us have a special interest in ourselves and in members of our species. This is the basic reason for devoting a chapter of this book particularly to genetics in relation to man; for the fact is, the basic genetic mechanisms of man, mice, higher plants, flies, and molds appear to be the same. And, in principle, the genetics of man is not unlike the genetics of a virus. You will remember that in previous chapters, examples from human heredity have been used together with other material in the exposition of general genetic principles.

For a long time, man was considered a peculiarly awkward object of genetic investigation. His long life cycle and comparatively small individual progenies are, indeed, unfavorable to the use of certain standard research techniques of the geneticist. Moreover, we are neither able to subject ourselves to rigorous experimental conditions nor do we desire to do so. In the standard genetic approach, closely controlled matings and standardized environment are prerequisite. Where man is concerned these experimental ideals are not feasible.

Admitting that man is not the ideal organism for genetic research, certain circumstances ameliorate the difficulties. Thanks largely to decades of intensive medical research, techniques are available for examining individual human beings with a thoroughness seldom if ever applied to an individual Drosophila fly, maize plant, or virus particle. Thus a very large number of subtle but significant variations in phenotype are recognizable in man. And because vast numbers of people are subjected to such intensive examination each year, new and interesting variations constantly emerge.

Again because of interest in ourselves, much of it medically oriented, a very great deal is known about the biochemistry, morphology, anatomy, and physiology of man. Since genetics has important relationships with these subject matters, the student of human genetics can work with an unusually firm background of related knowledge.

Finally, technical ingenuity has enabled close analysis in areas of human genetics formerly thought largely closed to effective investigation. For example, human cytology, once extremely difficult and unrewarding to the investigator, is now a lively area of research, thanks to the development of methods for culturing human somatic cells *in vitro*. Similarly, the population genetics of man and even mutation in man are fruitful areas of study because of the development of elegant mathematical tools, some of them effective only because of the incredible speed and efficiency of electronic computers. These mathematical tools are beyond the level of sophistication of this book, but you should at least be aware that such tools exist and are contributing to a rapidly increasing knowledge of the genetics of man.

Pedigree Analysis in Man

Genes are generally recognized by the characters determined by them. This statement holds for man, where the recognition that particular traits are gene-controlled has depended primarily on the analysis of family pedigrees. In favorable cases, pedigree analysis permits the recognition of a familiar pattern of inheritance and the exclusion of rational alternatives.

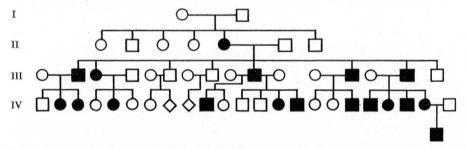

Figure 15-1. *Black symbols in this pedigree indicate individuals affected with severe foot blistering. The circles designate females; the squares, males. (From Haldane and Poole,* J. Heredity, **33:**17, 1942.)

Figure 15-1 is a pedigree showing the distribution of severe blistering of the feet in four generations of persons in families related by descent. According to the conventions for such representations, a circle represents a female individual and a square a male. Unshaded circles or squares designate people who are normal for the characteristic being studied. Shaded circles or squares represent "affected" individuals, in this instance persons showing severe foot blistering.

Each horizontal row, or tier, of circles and squares designates a generation, with the most recent generations shown lowest on the page. The generations are identified in sequence by Roman numerals.

In generation I of this pedigree, the horizontal line connecting the circle (woman) and square (man) is the indication of a mating between them. Counting from the left, individuals 1, 2, 3, 4, 5, and 7 in generation II are children of the man and wife above. Individual 6, not directly connected by lines with 1 and 2 above, is the husband of 5 in generation II. Individuals 9 and 11 in generation III are two different wives of 10. The diamonds in generation IV are used to indicate a number of individuals of unknown sex. Not all pedigrees in the literature of genetics conform entirely to fixed usage, but if you learn to read this one easily, variant forms will give you no real trouble.

What genetic meaning may we derive from the pedigree? It is readily seen that once the character for blistering appears in the ancestral line, no individual shows the character unless one of his parents is affected. Also, in families with an affected parent, both normal and affected progeny may occur, and usually do. If we consider all matings between a normal and an affected parent, we find that out of such matings 11 normal and 17 affected offspring occur, numbers consistent with a genetic ratio of 1:1. The foregoing observations are consistent with an interpretation that severe foot blistering is conditioned by an autosomal dominant gene. See whether you can exclude the obvious simple alternative hypotheses. For example, the hypothesis of a dominant gene linked to the X-chromosome is excluded by instances where an affected father has a normal daughter. Note that the hypothesis of an autosomal dominant gene suggests that the foot blistering trait occurs first in this pedigree because of mutation.

If the foregoing pedigree described some laboratory animal, the hypothesis of autosomal dominance in reference to foot blistering could be further tested by appropriate matings. In man the pedigree can only be left as it stands, interpreted in a satisfactory but not totally compelling way. Unfortunately, not all human pedigrees contain even as much information as the one presented here, and for many of them the interpretation is more tenuous, if an interpretation can be made at all. Nevertheless, persistent working out of pedigrees that refer to the same trait or to different traits has provided a working knowledge of Mendelian heredity for a large number of human characteristics. Some of the references at the end of this chapter make available to you a fair sample of the results of pedigree analysis in man.

Cytology and Cytogenetics in Man

As with other organisms, cytological analysis in man reveals a typical, or "normal," karyotype. Most somatic cells of most human beings have 46

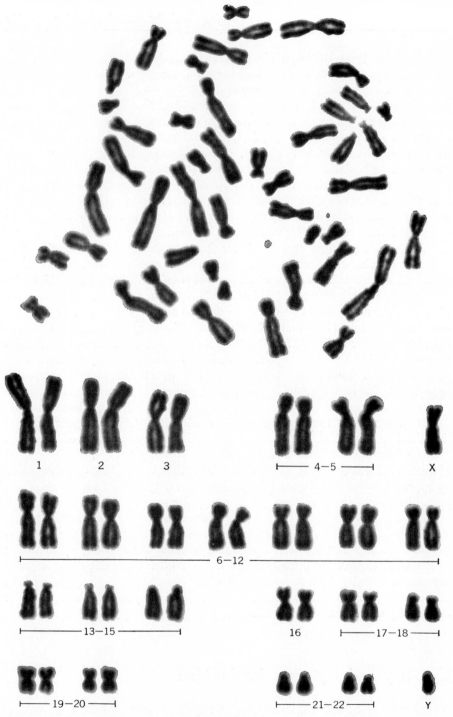

Figure 15-2. *Photograph of the chromosomes of a human male, with karyotype identifications shown below. (Courtesy of Margery Shaw.)*

chromosomes in their nuclei (Fig. 15-2). In the somatic nuclei, as expected for a diploid organism, the chromosomes exist in pairs. Members of a pair are morphologically alike, but different pairs are for the most part morphologically distinguishable on the basis of size, relative length of the arms of the chromosome, presence or absence of satellites, and so on. There is one, not unexpected exception to the rule that the chromosomes of the typical somatic nucleus can be arranged in pairs whose members are identical. The nuclei of males have two chromosomes that cannot be arranged into a like-membered pair. The larger of these chromosomes has its homologues in a like-membered pair found in females; the smaller is found only in males. The larger chromosome is the X-chromosome, the smaller the Y-chromosome. Thus typical females have 22 pairs of autosomes and an XX pair; typical males have 22 pairs of autosomes and a heteromorphic XY pair.

CHROMOSOMAL ABNORMALITIES IN MAN

Man is not exempt from the kind of cytological abnormality discussed in detail in Chapter 7. His chromosomal abnormalities reveal nothing new in principle but are of great interest, particularly to medicine. The newly emergent study of human cytology has in fact provided some real surprises, in that certain long-standing puzzles in human variation have suddenly become analyzable because of their ties with chromosomal anomalies.

Down's Syndrome. Among the relatively frequent abnormal variants in man are those called mongolian idiots (Down's syndrome). Affected individuals are mentally deficient and physically retarded in a variety of ways. Certain physical attributes such as broad face, stubby or flat nose, apparently slanted eyes are characteristic of most if not all mongolian idiots and in aggregate present a recognizable abnormal phenotype. Many affected individuals die in early childhood.

Mongolism is a syndrome that has been recognized for quite some time, and has been particularly interesting because its appearance is strikingly correlated with the age of the mother. Mongolism is about 100 times more frequent in the children of mothers over 45 than in the children of mothers between 16 and 28 years of age. Generally speaking, if one makes a classification by maternal age, the incidence of mongolism increases with increasing age of the mother. This correlation has stimulated various hypotheses to account for the appearance of an abnormality that occurs frequently enough to warrant concern.

In 1959, Lejeune and his co-workers in France found that the nuclei of somatic cells of mongols carry 47 chromosomes, one of the chromosomes of the genome being present in triplicate rather than duplicate. More recent studies by others confirm that mongolian idiots are trisomic for one of the chromosomes of the normal human complement.

How do the trisomics that we recognize as mongoloids arise? Most plausibly, the trisomic condition is ultimately attributable to nondisjunction, as described in Chapter 7. The fertilization that gives rise to a trisomic, then, is one involving a normal haploid gamete (*n*) and a gamete that is *n* + 1, this latter abnormal chromosomal complement occurring as the result of nondisjunction.

Two questions remain unanswered. First, why should trisomy for a particular chromosome produce the abnormal phenotype mongolian idiocy? We can only say that genic imbalance often results in malfunction and distorted phenotype. Imbalance for the particular loci in the chromosome trisomic in mongols somehow produces the particular syndrome we have just described. Second, why should mongolian idiots be more frequent among the children of older mothers? An attempt to answer this question would be even more speculative and less meaningful.

Abnormal Chromosomal Constitution for the X-Y Pair. Once you have learned that chromosomal anomalies in man do occur, you might wonder particularly what happens in instances of abnormality for the chromosomal pair that determines sex. As already mentioned briefly in Chapter 2, various deviations from the normal X-Y pattern do occur. We shall deal with only a few of the simplest here. You will find it interesting to compare them with analogous deviations in Drosophila, described in Chapter 12.

A well known but rare condition of abnormal sexual development in humans is Turner's syndrome. Persons with this condition have female external and internal genitalia but infantile uterus and breasts, and either no gonads or only traces of them. Karyotype analysis has shown that the somatic nuclei of affected individuals contain only 45 chromosomes, among which only one X-chromosome but no Y is present. The chromosomal constitution of these Turner-type persons can be designated XO.

Another syndrome, Klinefelter's, is typified by male genitalia with abnormally small testes containing few if any mature sperm. Body hair is underdeveloped, and the breasts are often large by the usual standards for a male. A number of Klinefelter individuals on whom karyotype analysis has been done are trisomics, having two X-chromosomes and a Y, the autosomal complement being normal. They can be designated XXY.

Not all abnormal humans that show the general characteristics of a Turner or a Klinefelter type necessarily possess the particular chromosomal anomalies we have designated. Perhaps some show no visible chromosomal aberrations of any kind. But for the simple, typical cases just described, the origin of the anomalies is clear enough. Again, nondisjunction of chromosomes seems to afford the explanation. Figure 15-3 summarizes the primary possibilities for nondisjunction of members of the sex-chromosome pair.

Consideration of the chromosomal constitution of Turner and Klinefelter types together with some study of Figure 15-3 will permit you to visualize a

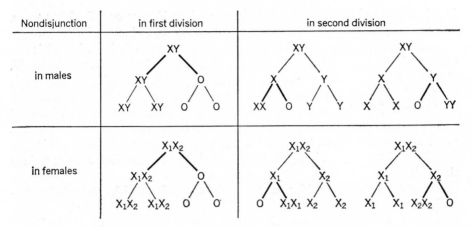

Nondisjunction	in first division	in second division
in males		
in females		

Figure 15-3. *Primary possibilities for nondisjunction of members of the sex-chromosome pair at meiosis. The nuclear division at which nondisjunction occurs is shown by heavy lines. (After Russell, Science, 133, 1961, p. 1799.)*

variety of ways in which the abnormal-type individuals might arise. For example, depending upon the meiotic division at which nondisjunction occurs, sperm that normally would be either X or Y may be either the abnormal types XY and O, XX and O, or YY and O. Fertilization of a normal egg (X) by an XY sperm gives a Klinefelter-type zygote (XXY); fertilization of a normal egg (X) by an O sperm gives a Turner-type zygote (XO). You can work out other possibilities. Somatic rather than meiotic nondisjunction may occur in early developmental stages and give rise to individuals mosaic for chromosomal anomalies.

Sex Chromatin. The nuclei of certain mammalian cells, including human cells, include a small stainable body called *sex chromatin*, or sometimes the *Barr body*—after the man who, with his associates, discovered it (Fig. 15-4). This body is regularly found in the cells of certain tissues in normal human females and is not found in the cells of normal males. Thus the Barr body can be used as a criterion to distinguish between cells that are XX and cells that are XY.

Extremely interesting relationships emerge when Barr bodies are considered

Figure 15-4. *A chromatin positive nucleus in a buccal smear from a human female. Notice the darkly stained spot (Barr body) at the periphery of the nucleus. (Courtesy of M. L. Barr.)*

in reference to chromosomal sex deviants such as Turner or Klinefelter types. It turns out, for example, that XXY (Klinefelter) individuals, who are male in terms of the presence of a Y-chromosome, have a Barr body in their nuclei; XO (Turner) individuals, who in many ways are phenotypic females, and lack a Y, show no Barr body. Rare XXX and XXXX females, trisomic and tetrasomic for the X-chromosome, show 2 and 3 Barr bodies respectively. Consideration of these relationships, and even more exotic ones not mentioned here, have led to the generalization that the maximum number of Barr bodies in any cell is one less than the number of X-chromosomes. Possible exceptions to this rule have been observed, but it is general enough that its implications deserve serious thought.

The numerical relationships summarized above, along with cytological observations, suggest that the heteropyknotic body called the Barr body is a heterochromatic X-chromosome, or is part of an X-chromosome. A significant extension of this interpretation is the hypothesis that only one X-chromosome, the one that is not heterochromatic, is genetically active in each nucleus. Underlying the hypothesis is an accumulation of evidence that heterochromatic material is genetically inert. (See also pages 363 and 364.)

If the hypothesis is correct, the formation of Barr bodies is a differentiation process whose effect is that only one of the X-chromosomes of the normal human female is active. One wonders, then, whether the hypothesis is not contradictory to the observed fact of regular patterns of inheritance for X-linked genes in the human female. Two kinds of observations permit escape from the apparent dilemma. First, cytological studies indicate that X-chromosome differentiation does not occur in the germ line. Second, even among somatic cells, differentiation of an X-chromosome is deferred until some time after the earliest stages of embryonic development.

What decides which member of a pair of X-chromosomes is to be inactivated? No answer to this question has been found, nor has present knowledge ruled out that it is essentially a random matter as to which of the X chromosomes in a nucleus is the active one. If essential randomness is the case, then, on the average, a human female has the same number of active X-chromosomes as a male, all of whose X-chromosomes (1 per nucleus) presumably are active. The plausibility of such a situation can be tested by referring it to a variety of kinds of observations. Let us consider it first in reference to appropriate observations on gene action.

Normal XX females homozygous for a sex-linked gene that directs production of glucose-6-phosphate dehydrogenase (G6PD) show no higher activity for this enzyme than do normal XY males, who have only one dose of the gene. The general case exemplified by this particular observation has been called dosage compensation. It is in contrast to the usual situation for autosomal loci that control the production of particular proteins that can be measured more or less quantitatively. What appears to be the typical case for autosomal loci can be illustrated by the comparison between homozygotes and heterozygotes for the gene that controls production of the enzyme galactose-1-P uridyl transferase. On the average, heterozygotes show only a little better than half the enzyme activity shown by homozygotes for the active allele.

Another test situation for the hypothesis of X-chromosome inactivation is provided by another kind of consideration of the phenotypes of heterozygotes for X-linked genes. If the hypothesis is correct, one might in suitable instances expect to find phenotypic mosaicism in heterozygotes, since in cells in which the dominant allele is in the inactive X the recessive would be expressed. This has been strongly suggested, for example, by studies of women heterozygous for G6PD deficiency. (Why, do you suppose, would women heterozygous for sex-linked recessive color blindness nevertheless have normal color vision?) Calico cats, you will remember, are females heterozygous for an X-linked allelic pair involved in the determination of coat color. The calicos are phenotypic mosaics that can be interpreted rationally as having certain patches of tissue with one of the X-chromosomes active and other patches of tissue showing activity for the genes in the homologous chromosome. A number of analogous cases are known to mammalian geneticists.

A final comment about sex chromatin seems worth making. We have referred to the Barr body as an instance of chromosomal differentiation through some process that is not yet understood. Without implying a mechanism, the process can be given some meaning if it is called heteropyknosis. So called, it seems perhaps to fall within a family of instances in which differentiation is achieved by the inactivation of chromosomal loci through heteropyknosis. One such instance, described in Chapter 7, is the variegated-type position effect.

Human Biochemical Genetics

In Chapter 10 you found that studies of the genetics of man provided early evidence that genes mediate biochemical reactions by controlling enzyme specificities. This evidence emerged from study of heritable metabolic-deficiency diseases like alcaptonuria and phenylketonuria. Since Garrod's time, and particularly in recent years, a molecular basis has been established for a large number of the heritable diseases of man. We can now be quite confident that the various chemical steps in man's metabolism are gene-controlled in the same fashion that the biochemistry of Neurospora or of *E. coli* is gene-controlled. Human individuals affected with a metabolic deficiency disease are the analogues of a biochemical-mutant strain of Neurospora. This fact has interest beyond its reaffirmation that human genetics is basically like that of other organisms.

Just as mutant blocks in microorganisms can sometimes be circumvented by nutritional supplements or other appropriate means, one can hope that understanding the molecular basis of heritable diseases will permit amelioration of the consequences of genetic defect. Most of you are already aware of an instance of circumvention of genetic defect, although you may not have thought of the matter in such terms. Human diabetes has a genetic basis, and when a diabetic is given insulin he is preserved by being supplied with an essential substance he cannot synthesize, just as an arginineless mutant of Neurospora is preserved by supplementing its diet with arginine. Medicine is still far from being able to control or ameliorate anything like the gamut of pathological phenotypes resulting from unfavorable heredity. But the following paragraphs illustrate that biochemical-genetic knowledge provides an increasingly sound basis for dealing with some of man's genetic defects.

Galactosemia. The sugar galactose is not required for the normal development of humans, but normal individuals can metabolize it and use it as an energy source. Galactose occurs commonly in the diet, particularly in the diet of children, because it is one of the components of the disaccharide lactose, the principal carbohydrate in milk. Individuals homozygous for a certain gene, *g*, are deficient in the enzyme gal-1-P uridyl transferase, referred to earlier in this chapter. This enzyme promotes a step in the reaction chain by which galactose is normally metabolized. Infants of genotype *gg* who drink milk or otherwise ingest a source of galactose accumulate galactose in the blood. A concomitant, and presumably a consequence, of this accumulation is a variety of unfortunate clinical symptoms, including enlarged liver, retarded mental development, and generally slow growth. Galactosemic individuals often die as children.

If newly born infants with the genotype for galactosemia are placed on a galactose-free diet, they are able to develop normally. Galactosemics whose clinical symptoms are detected early may be switched to an appropriate diet, minimizing the damage associated with accumulation of galactose. Clearly, it would be an advantage to be able to predict the probability of an unborn child being a galactosemic, so that in appropriate instances dietary precautions could be taken. Actually, the heterozygous carriers of the gene for galactosemia can often be identified as such, in that, on the average, their blood cells show a lesser activity for gal-1-P uridyl transferase than is found for the blood cells of homozygous normals.

Primaquine Sensitivity. A small fraction of the human population is adversely affected by treatment with the antimalaria drug primaquine. Treatment of these sensitive individuals is followed by severe hemolytic anemia. Sulfanilamide treatment seems to produce the same effect. Sensitivity has a familial pattern, with pedigree analysis indicating that a gene linked to the X-chromosome is the basis for the disorder.

Curiously, a condition very similar to primaquine sensitivity, which reaches a frequency of approximately 10 percent in a sample of American Negroes to whom the drug has been given, has been discovered under an apparently quite different guise. In certain Mediterranean areas, high family incidences of sensitivity to the fava bean have been known for some time. *Favism*, like the response to primaquine, is expressed as hemolytic anemia. The clinical symptom follows eating the raw bean or occasionally is detected after pollen from the bean plant has been inhaled. The distribution of sensitives within families is consistent with heredity controlled by an X-linked gene.

Not only are the sensitivities to the fava bean and primaquine inherited in the same manner, but in both instances affected individuals have a markedly low activity for the enzyme glucose-6-phosphate dehydrogenase. It now seems likely that favism and primaquine sensitivity are similar conditions and have a similar genetic basis.

Other Biochemical Differences in Man. It would be infeasible and fairly pointless to treat here all the various "molecular diseases" determined by genes in man. We shall mention a few more simply to give you a notion of their scope and variety. Among such diseases you should not forget sickle-cell anemia, traceable to a genetically determined alteration of the hemoglobin molecule. In fact, a large number of mutant hemoglobins are known and are distinguishable on the basis of particular amino acid substitutions of the kind described for S hemoglobin in Chapter 10. These are summarized in Table 15-1. The hemoglobins presently offer some of our finest examples of variant amino acid sequences that reflect variations in the genetic code.

Table 15-1. ABNORMAL HEMOGLOBINS IN MAN, DESCRIBED IN TERMS OF SINGLE AMINO ACID SUBSTITUTIONS.

Abnormal Hemoglobin Type	Amino Acid Substitution
S	valine for glutamic acid
G$_{San Jose}$	glycine for glutamic acid
I	asparagine for lysine
M$_{Saskatoon}$	tyrosine for histidine
M$_{Milwaukee}$	glutamic acid for valine
C, E, O	lysine for glutamic acid
Norfolk	asparagine for glycine
Zurich	arginine for histidine
G$_{Philadelphia}$	lysine for asparagine

Note: The different substitutions occur at various points in the alpha and beta chains. For example, the glutamic acid for which substitution is found in Hb G$_{San Jose}$ is not by position the same glutamic acid for which substitution is found in Hb S.

A substantial number of cases of hereditary deficiency in enzymatic activity is known in addition to those already discussed in some detail. Among the most interesting are *acatalasemia*, absence from the blood of catalase, an enzyme that promotes breakdown of peroxides, and deficiency for the enzyme cholinesterase, a deficiency recognized in relation to slow recovery from the effects of the muscle relaxant succinylcholine, which is administered in connection with electroshock therapy.

The number is even larger for heritable biochemical abnormalities not yet clearly identified with a deficiency in an enzyme system. These include rickets resistant to therapy with vitamin D; accumulation of the carbohydrate L-xyloketose in the urine; increased concentration of chlorides in the sweat, a concomitant of the disease cystic fibrosis of the pancreas; increased formation of free porphyrin compounds, giving rise to extreme sensitivity to light; accumulation of cystine in the urine, often accompanied by the formation of calculi, or stones, largely composed of this amino acid.

Finally, in the present context you should recall the large number of genetically controlled antigenic differences in man, and realize that they offer additional impressive evidence of heritable biochemical variability.

Human Heredity and Environment

Some of man's most significant attributes, such as intelligence, certain size characteristics, and personality traits, are complex in their manifestations. Sensitive to environmental influences and probably influenced by many genes, these characteristics are *quantitative*, and need to be referred to the kind of considerations presented in Chapter 14. For these characteristics, pedigree analysis in man is far less effective than experimental genetic procedures for studying controlled matings of plants and animals under defined environmental conditions.

THE USE OF TWINS IN STUDYING HUMAN GENETICS

Twin births provide a certain amount of compensation for some of the difficulties in studying quantitative characteristics in man. Twins have sometimes been called "experiments in nature," because they provide controls ordinarily lacking in human genetic studies.

One-Egg and Two-Egg Twins. About one in 88 births in the United States is a twin birth. Figure 15-5 shows the essential difference between two kinds of twins. *One-egg,* or *identical,* twins originate from the division and separation

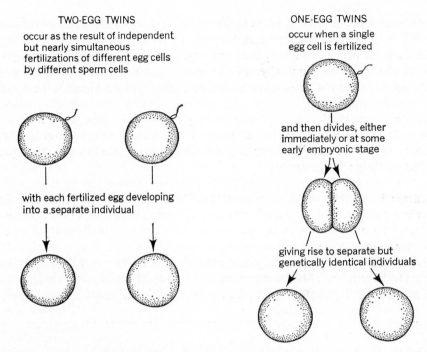

TWO-EGG TWINS

occur as the result of independent but nearly simultaneous fertilizations of different egg cells by different sperm cells

with each fertilized egg developing into a separate individual

ONE-EGG TWINS

occur when a single egg cell is fertilized

and then divides, either immediately or at some early embryonic stage

giving rise to separate but genetically identical individuals

Figure 15-5. *The origin of one-egg and of two-egg twins, showing the basic genetic difference between these two types.*

of a single fertilized egg cell, or very young embryo. Members of such a pair are genotypically identical, at least as to genes in chromosomes, since the nuclei of their cells have arisen by mitosis from a single common source. *Two-egg* twins, or *fraternal* twins, result from the nearly simultaneous fertilization of two different egg cells by different sperm. Members of such twin pairs are no closer genetic relatives than are sibs of different birth. *Sibs*, or *siblings*, are off-spring of the same parents. You can see that members of a one-egg twin pair must always be of the same sex, but members of a two-egg pair may be either of the same sex or of different sexes. Among human twin births in this country, about one in three on the average is a one-egg or monozygotic twin birth.

Methods for Utilizing Twins in Genetic Studies. A standard procedure in experimentation is to study the effects of one particular variable while other possible variables are nullified or controlled. Twins give control over the two important groups of variables that can be designated the *hereditary* and the *environmental.* One-egg twins provide genetic control, since each pair includes two individuals that are genotypically identical. Both kinds of twins may provide environmental control. Members of a twin pair are *in utero* at the same time, and while this does not insure uniformity of prenatal environment, it at least may reduce some of the variables associated with siblings born at different times. After birth, except where pair members are raised in separate homes, the environment for twins is as nearly the same as one might hope for without imposing experimental conditions. Certainly the environment of twins is usually much more alike than is the environment of two single-born individuals, even within the same family.

Triplets, quadruplets, and quintuplets may be one-egg, two-egg, or multiple-egg, and can be utilized for genetic studies in the same way twins are. But any widespread use of multiple-birth individuals in genetic studies has been precluded by their relative rarity.

Comparison of Physical Characteristics of Twin Pair Members. Identical twins are almost always strikingly similar in facial features, in hair and skin color and texture, in eye color, in various characteristics of the teeth, and in certain proportions of the body and limbs. Fraternal twins may also look very much alike. This is not surprising, since they are, after all, close relatives, and may be expected to have many of the same genes. But it is nearly impossible for fraternal twins to be genotypically identical for all their physical characteristics. Usually,

Figure 15-6. *Lois and Louise are strikingly similar one-egg twins. They were separated eight days after birth, and except for brief visits had no contact until they entered college at the age of eighteen. (From Gardner and Newman, J. Heredity, 31:120, 1940.)*

enough segregation has taken place that pair members, even of the same sex, are somewhat dissimilar in appearance.

Figure 15-6 shows a pair of identical twins that are remarkably alike. In fact, H. H. Newman, an outstanding authority on twins, states that he never learned to distinguish between members of this particular pair, even after many interviews and conversations. Their similarity is all the more impressive because they were separated eight days after they were born. They were reared apart, one in an urban community and one in a rural area, and, except for brief visits, spent no time together until they entered the same university as freshmen. Differences in environment clearly had no appreciable effects on the obvious physical features of these individuals. Louise and Lois, the girls in the picture, are perhaps more alike than most one-egg twins, but they are valid representatives of the proposition that heredity exerts a strong influence on many of our physical features.

When one turns to characters that can be described in precise quantitative terms, it becomes obvious that even the most similar of one-egg twins are not truly "identical." Actually, Lois and Louise, when examined by Dr. Newman, were both 5 feet 4¾ inches tall and had a head length of exactly 17 centimeters. But Lois had a head width 0.25 cm greater than the corresponding measurement for Louise, and Louise was a few pounds heavier. Weight comparisons do not always have permanent significance, however. At a date later than that of the first comparison, Lois and Louise weighed about the same. An analysis of the fingerprint patterns of the two girls gave ridge counts as follows:

	LEFT	RIGHT	TOTAL
Louise	72	67	139
Lois	75	69	144

Louise's left hand, in this regard, is more like Lois's than like her own right hand! Differences between these twins can be found, but the total impression is one of great similarity.

More meaningful to our general understanding of the influences of heredity and environment on physical characteristics are studies which compare one-egg twin groups with two-egg twin groups and with groups of paired siblings who are not twins. Table 15-2 summarizes such comparisons. The values are average pair differences for certain physical traits. Values for one-egg twins reared apart are also included. The two-egg twins and the paired sibs all represent instances in which the pair members are of the same sex.

The first point to be observed in the data is that differences between pair members occur within each biological group. From this it appears that environment has some effect on each of the traits studied. Secondly, the average pair

Table 15-2. AVERAGE PAIR DIFFERENCES IN CERTAIN PHYSICAL TRAITS FOR IDENTICAL AND FRATERNAL TWINS AND FOR SIBLINGS REARED TOGETHER, AND FOR IDENTICAL TWINS REARED APART.

Selected Traits	Average Pair Differences			
	Identical Twins (50 pairs)	Fraternal Twins (52 pairs)	Paired Siblings of Like Sex† (52 pairs)	Identical Twins Reared Apart (20 pairs)
Standing height	1.7 cm	4.4 cm	4.5 cm	1.8 cm
Weight	4.1 lb	10.0 lb	10.4 lb	9.90 lb
Head length	2.9 mm	6.2 mm	‡	2.20 mm
Head width	2.8 mm	4.2 mm	‡	2.85 mm
Cephalic index	0.016	0.028	‡	‡

Source: Data from Newman, Freeman, and Holzinger, *Twins: A Study of Heredity and Environment,* University of Chicago Press, 1937.
† Values represent brothers or sisters whose height and weight had been measured at the same age.
‡ The investigators report no values in these categories.

differences are not the same for the various biological groups. Take height, for example. Pair differences for fraternal twins and for sibs are about the same, and in each case the value is appreciably greater than the average difference between identical twins. The greater divergence between fraternal twins and between paired siblings, as compared to identical twins, can be attributed to heredity. Moreover, whatever may be the comparative differences of environment between fraternals as against sibs, these differences are not reflected in the mean pair-height differences for the two groups. There is another potent reason for thinking that ordinary environmental circumstances do not greatly affect the expression of height. Sets of identical twins reared apart show almost the same mean pair differences for height as do identical twins reared together.

Similarly, width and length of head are not sensitive to the kind of environment over which the twin method of study gives control. The situation is otherwise in regard to body weight. Identical twins reared apart are much more different than identical twins reared together. In fact, the former group shows an average pair difference much like that for fraternal twins or sibs. We can conclude that body weight responds to environmental change more readily than height or the dimensions of the head.

Finally, more recent studies have shown that after maturity the physical measurements of members of one-egg twin pairs show greater divergence. These observations indicate that growth rate is under more strict genetic control than is the amount of final growth.

Heredity and I.Q. Doubtless it would be wrong to refer to human intelligence as if it represented some single, well defined characteristic. Nevertheless, we agree that intelligence is important. We think of it as being the foundation for rational behavior and as enabling us to make competent judgments on all sorts of matters that affect our well-being. Differences in intelligence no doubt account for many of the misunderstandings among individuals and for many of the maladjustments of individuals to society. A knowledge of the bases of this important but elusive trait would do much to promote the understanding of man by man.

In a few cases of grossly subnormal intelligence, as in *amaurotic idiocy* or *phenylpyruvic idiocy*, we know that simple genic alternatives account quite directly for the differences between defectives and so-called normals. But among the vast bulk of people whom we consider "normal" there appears to be a wide range of continuous differences in intelligence. What is the basis of this variation? Pedigree studies have been of little aid in answering this question, because intelligence appears to involve highly complex relationships between heredity and environment. This is the kind of situation in which the twin method offers special advantages over other methods available to human geneticists.

Before reviewing what twin studies imply about differences in intelligence, we need to consider briefly the essential matter of measuring these differences. The best and most usable measures of intelligence so far available are the I.Q. (*intelligence quotient*) ratings devised by psychologists. These ratings depend upon scores made by individuals on certain standardized tests. The tests are so contrived that the test behavior, insofar as possible, reflects certain aspects of rational behavior. It should be emphasized, however, that the tests cannot be entirely removed from cultural and other kinds of environmental influence. While I.Q.'s are not considered wholly reliable or all-inclusive measures of "innate" intelligence, they represent a fairly reproducible estimate of capacity for some kinds of rational behavior.

Table 15-3 provides the same kind of comparison of average pair differences in intelligence as was given in Table 15-2 for certain physical traits. You will notice at once that identical twins reared apart show an appreciably greater average difference than identical twins reared together, indicating an effective

Table 15-3. AVERAGE PAIR DIFFERENCES FOR BINET I.Q. SCORES.

Identical Twins (50 pairs)	Fraternal Twins (52 pairs)	Paired Siblings of Like Sex (47 pairs)	Identical Twins Reared Apart (19 pairs)
5.9	9.9	9.8	8.2

Source: Data from Newman, Freeman, and Holzinger, *Twins: A Study of Heredity and Environment*, University of Chicago Press, 1937.

environmental influence on I.Q. The former group shows a smaller average pair difference than that found in fraternal twins or sibs, suggesting that heredity is also important. This difference between group averages is relatively small, however, and must be interpreted with caution, especially since I.Q. tests are sampling methods, so that scoring is subject to errors of chance. The fact that mean pair differences between fraternal twins and sibs are essentially the same implies that the kind of environmental influences that affect I.Q. are not the sort that vary significantly between individuals born at the same time as against those born at different times within a family.

Further insight into the effects of environment on I.Q. may be obtained from the case studies of identical twins reared apart. One thing the unsupplemented average pair difference value does not tell is that, for the majority of the 19 separated pairs that were studied, the individual differences were about the same as those found for unseparated pairs. But most of these separated pair members were not reared under drastically different environments. The few instances of large discrepancies in formal schooling, or in opportunities for education, could be correlated with differences in I.Q. test performance. The case histories of Gladys and Helen provide an illustration.

> Gladys and Helen were separated at the age of 18 months. They did not meet again until they were 28 years old. *Helen* was adopted out of an orphanage into a farm family living in Michigan. After going through high school she did four years of college work and took a bachelor's degree. She was then employed as a school teacher. *Gladys* was adopted into a Canadian family. Her foster father, a railway conductor, became ill, and in the interests of health moved to the Canadian Rockies for a two-year period. This was just at the time when *Gladys* was ready for third grade. No school was available at her new home, and when the family returned to Ontario she did not resume her formal education. From the age of 17 years on she worked at several occupations. At the time she and her sister were examined, she was employed in a printing house. On the Stanford-Binet test for I. Q. *Helen* scored 116 and *Gladys* 92.*

This difference in I.Q. scores is the largest in the group of identical twin pairs tested by Newman and his associates. In fact, about 69 percent of the scores of the general population lie between the scores of Gladys and Helen. It seems significant that this striking difference in I.Q. was found for the pair members having the greatest difference in educational experience of any of the twins studied.

On the basis of similar findings in other case histories, a number of authorities have drawn the conclusion that fairly large environmental differences may produce proportional differences in the I.Q. of genetically identical individuals, but that the effects of smaller environmental differences may not be detectable.

Any present judgment as to the genetic basis of I.Q. must be tentative, for

---

* Condensed from Newman, Freeman, and Holzinger, *Twins: A Study of Heredity and Environment*. Chicago: University of Chicago Press, 1937.

these reasons: (1) Twin studies have been carried out with only small numbers of individuals. (2) We can question whether fraternal twins reared together may not experience greater environmental differences than do identical twins. (3) The inability of the twin method to clarify which factors in heredity and which in environment are effective in modifying I.Q. leaves an unfortunate gap in our information. (4) Problems of the relative potency of postnatal and prenatal environments in producing certain kinds of differences have not been resolved.

When all difficulties, these and others, are taken into account, we are left with a very general picture of the relationships of genetics to intelligence. Heredity seems to provide the individual with upper and lower limits, between which some level of intelligence is realized. The realization of a particular I.Q. depends upon a number, probably large, of environmental factors. If this concept is at all correct, it has significant implications for us, even though it is in the stage of broad generality. One of the foremost of these implications is that a person with relatively favorable heredity for intelligence may fall short, in the achievement of intelligent behavior, of a person with relatively unfavorable heredity. The idea is illustrated in Figure 15-7. We may presume, then, that even if the genetic component of intelligence is large, there is a real point in working to achieve favorable environments for the expression of intelligence. In fact, the genetic component of intelligence in different individuals will continue to be difficult to define until environment is so manipulated that for all individuals the most favorable expression of intelligence is possible. It is highly probable that the

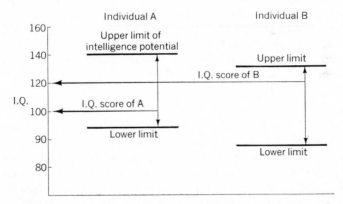

Figure 15-7. *One interpretation of the relationship of genetics to intelligence is that each individual has a fairly wide range of potentialities for intelligence, with the upper and lower limits of this range being determined by heredity. The realization of intelligence, as measured for example by an I.Q. score, is determined in complex fashion by a variety of factors of the internal and external environment. Thus the performance of individual A, with a greater potentiality for intelligent behavior, may fall short of that of individual B, whose I.Q. test score may lie much nearer the upper limit of his capacity.*

conditions most favorable for the utmost realization of one set of genetic potentialities will not be those most favorable for realization of the potentialities of some other genotype.

Genetics and Human Welfare

Man and the other organisms with which he interacts, from viruses to dairy cattle, are to a very great extent creatures of their genetic mechanisms. Almost automatically, everything that is found out and understood about genetics has implications—some more immediate than others—for human welfare. In the next few pages we will point out a few of the instances where genetic knowledge has evident relevance to human affairs.

MEDICAL GENETICS

The obvious point of contact between genetics and medicine is the fact that many impairments of normal human function falling into the category of "disease" are heritable characters. These may be relatively inconsequential departures from the normal, like color blindness, albinism, or polydactyly, or they may be much more serious, as hemophilia, juvenile amaurotic idiocy, sickle-cell anemia, or galactosemia. Some of the classical human-anomaly syndromes, like the Turner syndrome, represent aberrations in the genetic apparatus. A good deal that has already been said in preceding pages, then, refers to medical genetics. In this section, however, we will take a slightly different point of view—the point of view of applying this knowledge.

Counseling. The medical practitioner is often in a position to make forecasts, or prognoses, that can be utilized for the benefit of his patients. There is obvious value in being able to tell the parents of a galactosemic child the probability that another child born to them will also be a galactosemic. In other instances it is equally valuable to tell parents or parents-to-be that there is essentially no chance of their transmitting some particular unfavorable human attribute. A large number of people have legitimate and urgent genetic questions, ranging in topic from skin color to cancer. All such questions deserve the best possible answers, and the answers get better as knowledge of genetics increases.

Diagnosis. The fundamental problem in diagnosis of disease is to arrive at the correct causation of symptoms; otherwise, treatment is likely to be ineffective or possibly damaging. M. T. Macklin has recorded a striking example of how genetic knowledge can be utilized to determine the causation of pathological symptoms that are otherwise difficult to diagnose. The patient was a child whose symptoms were scanty hair, undeveloped teeth, dry skin, and a tendency to fever. An early diagnosis suggested deficiency in the functioning of the thyroid gland, but this was reconsidered when thyroid treatment accentuated the symptoms. Further inquiry revealed that one parent and other relatives of the child showed similar pathological symptoms. Converging trails of genetic and physiological approach led to the correct conclusion that the patient's difficulty was a dominant hereditary condition, *ectodermal dysplasia*, which, although rare, is well known. It is characterized by absence of the sweat glands as well as by deficiencies in the development of hair and teeth. With accurate diagnosis, the reason for trouble with temperature control became clear. At the same time, the undesirable effects of the original thyroid medication became explainable. Since the thyroid treatment stimulated metabolic activity, the patient's body temperature became even higher, and more difficult to control.

The Detection of Carriers of Hereditary Disease. Individuals who do not exhibit the symptoms of an hereditary disease but nevertheless are able to transmit it are called genetic *carriers*. Persons heterozygous for a recessive character, as for example the mothers of sons with hemophilia, are carriers. The term is also used when genes manifest themselves late in development, or are incompletely penetrant, or are otherwise irregular in expression.

The ability of medical genetics to deal with problems is greatly enhanced when carriers can be detected. Considering galactosemia as an example, we can appreciate how the detection of carriers might both refine prognosis and lead to the prevention of expression of the disease. Given simply the fact that a husband and wife are phenotypically normal, any estimate of the probability that they will have a galactosemic child is highly unrefined, depending wholly on considerations of population genetics. Given in addition a pedigree for husband and wife, one may or may not be able to refine the estimate, depending on the nature of the pedigree. But given a test that will detect carriers, one can tell whether the probability is one in four that they will have a galactosemic child or whether for practical purposes the probability is zero. Moreover, if both parents are detected as carriers, steps can be taken to nurture their child on galactose-free diet if necessary, to prevent the unfortunate consequences of clinical galactosemia.

With varying degrees of certainty, carriers for a variety of heritable diseases can now be detected In particular cases, detection may depend upon the observation that heterozygotes, as compared with the two kinds of homozygotes, show an intermediate amount of some chemical product of gene action. Thus, on the average, heterozygotes for galactosemia fall between normals and galactosemics when measurements for the activity of gal-1-P uridyl transferase are made. Unfortunately, those normals who show the lowest enzyme activity within their group overlap with the carriers who show the highest enzyme activity within their group. Carriers for acatalasemia can be detected with more certainty, since in studies thus far no overlap has been found between hetero-zygotes and homozygotes of either class, when measurements are made for blood catalase activity (Fig. 15-8). And, as you know, carriers for sickle-cell

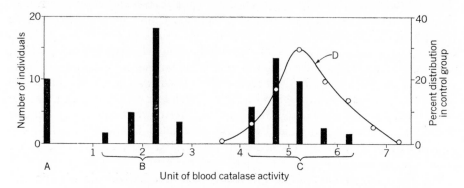

Figure 15-8. *Catalase activity in* (A) *10 individuals affected with acatalasemia,* (B) *30 carriers related to the acatalasemics,* (C) *36 normal relatives. The curve* (D) *represents the percentage distribution of activity in 206 controls. (After Nishimura, Hamilton, Kobara, Takahara, Ogura, and Doi; from Stern,* Principles of Human Genetics, *2nd ed., W. H. Freeman, 1960, p. 673.)*

anemia are readily distinguishable on the basis that they have both hemoglobin A and hemoglobin S. Likewise analysis of the uric acid content of blood helps to reveal carriers for gout. Not all detection of carriers, however, is made at the molecular level. It appears that carriers for Huntington's chorea, a severe nervous disorder, may be singled out on the basis of "brain wave" patterns revealed by electroencephalography.

Prevention. For the future, one hopes that detection of carriers will make increasingly feasible the anticipation and amelioration of the unfavorable expressions of hereditary defect. We have already indicated how the detection of carrier parents can enable the earliest possible, and therefore most favorable, use of dietary measures to avoid the severe expressions of galactosemia. For another kind of example, this time an instance of a gene that comes to maximum expression late in development, we may consider *xanthoma tuberosum*. The overt symptoms of this condition, which is transmitted on the basis of a dominant gene, are numerous nodules and tumors, largely present on the elbows, knees, fingers, heels, buttocks, and over certain tendons. Lesions of the heart sometimes occur, and implication of the cardiovascular system often culminates in death. The metabolic error, however, is expressed as *hypercholesterolemia*, an excess of cholesterol in the blood. Indeed increased blood cholesterol presages the appearance of the lesions described above. If this warning signal is detected in time, an affected person can be placed on a suitable diet, and years may be added to his life.

In cases like this, an awareness of genetic principles may lead to the detection of individuals with hypercholesterolemia before lesions appear. Although it is scarcely feasible to test periodically the level of cholesterol in the blood of each of us, it is practical and justifiable to apply such tests to relatives of persons with xanthoma tuberosum.

In the future, even infectious diseases are likely to be considered in the light of heredity. Man's resistance or susceptibility to certain pathogens has a genetic basis. Identical twins, for example, show a much greater concordance for tuberculosis than do fraternals. The ability to recognize persons with genetic susceptibility would be of enormous value in the prevention of infectious diseases for which immunizations or other preventive means are difficult to apply on a mass scale.

The Development of Resistance to Antibiotics by Bacteria. Some of man's most effective weapons in his struggle against infectious disease have been antibiotics such as penicillin, aureomycin, streptomycin, and the sulfas. There was early promise that antibiotics might neutralize completely the attacks by certain groups of bacteria on man. However, in an increasing number of cases where patients are treated, for example with penicillin, treatment is ineffective, even though the pathogenic organisms are of a kind once sensitive to the antibiotic. Now we recognize that strains of bacteria originally sensitive to an antibiotic can become resistant to it.

Laboratory investigations have shown that bacterial resistance to antibiotics originates through mutation. The detection of mutants of this kind is exceptionally easy, even when the mutation rate is low. Suppose, for example, that an investigator wishes to look for streptomycin-resistant mutants. Essentially all he needs do is to distribute bacteria of a sensitive type into Petri dishes containing a medium that includes streptomycin at a concentration that normally will not permit survival of the bacteria. If resistant mutants occur, they grow and form colonies. Under this system the mutants literally detect themselves. Since as many as several billion bacteria can be accommodated comfortably in a single Petri dish, the investigator can test almost unlimited numbers of individuals with relatively small effort.

M. Demerec, after a series of studies with important implications for medicine, has pointed out the interesting circumstance that bacteria show different patterns in acquiring resistance to different antibiotics. He has found that a high degree of resistance to streptomycin can be acquired in a single mutational step. On the other hand, a high degree of resistance to penicillin is attained only as the cumulative result of a series of mutations. Under this pattern of resistance, first-step mutants show a tolerance to penicillin that is only a modest advance over the lack-of-resistance characteristic of the original sensitive bacteria.

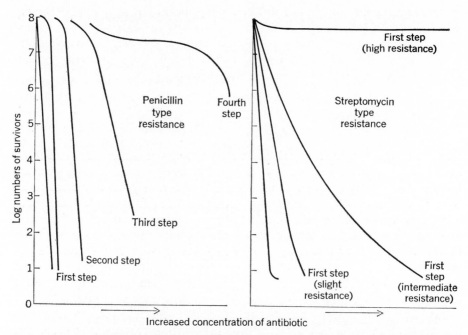

Figure 15-9. *Survival curves, showing the penicillin and streptomycin types of resistance pattern. Note that fewer cells survive at the higher concentrations of antibiotic. But within populations, differences in antibiotic resistance occur. If survivors from a population initially grown in the presence of an antibiotic are selected for reproduction, one may obtain a culture with an average resistance higher than that of the original population. By continuing to select mutants, through a series of steps of this kind, strains of bacteria quite tolerant to penicillin may be obtained. The curves show, however, that it is possible for bacteria to attain a high degree of resistance to streptomycin in what appears to be a single mutational step. (After Bryson and Demerec, Annals N.Y. Acad. Sci., 53:285, 1950.)*

However, further mutation in the new, mildly resistant strain can give somewhat greater tolerance. Subsequent accumulation of mutations may eventually give rise to strains that are highly resistant to penicillin. The penicillin and streptomycin patterns of resistance are compared in Figure 15-9.

The findings of Demerec suggest certain clinical applications. At least for the bacteria he studied, the development of penicillin-resistant strains seems avoidable if initial concentrations of the antibiotic are high enough to eliminate first-step mutants, and if these effective concentrations are maintained as long as any of the bacteria persist. On the other hand, since treatment even with massive initial doses of streptomycin does not preclude the occurrence of resistant bacteria, the clinician must be prepared for the occasional appearance of resistant mutants. In view of this possibility, he may wish to reserve the use of streptomycin for cases he cannot treat effectively in other ways. Knowledge of bacterial genetics, of course, provides the basis for a variety of medical applications. You will readily appreciate the medical significance of an understanding of the multiple resistance transfer factors discussed in Chapter 11.

PLANT AND ANIMAL IMPROVEMENT

Few of us appreciate the extent to which our physical well-being, certain aesthetic satisfactions, and, in general, our way of life depend upon the fact that the genetics of a variety of plants and animals has been manipulated to meet man's wishes and needs. Much of this manipulation has been unconscious, or, at best, empirical. The domestication of plants and animals and their improvement for man's needs had gone a long way before Mendel performed his classical experiments and founded the science of genetics. But as soon as the principles of genetics became widely appreciated, plant and animal breeders began more and more to put them into use, until today plant and animal breeding are highly specialized technologies, applied branches of the science of genetics. From a rather early point in man's history until now, with varying degrees of awareness and sophistication, he has taken advantage of inbreeding, hybridization, and particularly selection to alter the characteristics of organisms in ways better to suit his purposes. The story of plant and animal improvement is certainly one of the most dramatic and significant in the history of man. We cannot sketch that story here, or in any way do it justice, but examples of a few facets of the story may give you some sense of its importance.

If we were to choose but a single example to dramatize the effects of plant or animal breeding, we could scarcely make a better choice than hybrid corn. Part of its impact on our economy appears in this statement which came from Merle T. Jenkins, Principal Agronomist in charge of corn investigation, Bureau of Plant Industry, United States Department of Agriculture.

> During the three war years of 1917, 1918, and 1919, we produced 8 billion bushels of corn on a total of 311 million acres. During the three war years of 1942, 1943, and 1944, we produced $9\frac{1}{3}$ billion bushels on 281 million acres—1,366,201,000 more bushels than in the earlier period, on 30,522,000 fewer acres. This is equivalent to 5 billion pounds more meat a year—38 pounds more per person a year. When you look back to the fact that there were times during the second world war when meat rationing got down to as low as 115 pounds a person a year, you can appreciate the importance of this extra production.

What this increased efficiency of production of corn meant in terms of release of material, energies, and manpower to other activities essential to the national welfare is difficult to assess. Obviously, the contribution of hybrid corn was significant. When added to other contributions of plant breeding, it constituted one of our most powerful implements of defense.

The effects of hybrid corn on our standard of living in peacetime, on farming practice, and on the sociology of large segments of the population cannot be properly estimated even yet, but they are vast. We know that the superior standability of hybrid corn, which permits the effective use of mechanical pickers, has resulted in increased all-around mechanization for the farm. The need for hired men to help with corn picking has been significantly reduced, and a variety of more or less subtle socioeconomic consequences have followed. Some of these have to do with matters of labor supply, others with the composition of the rural community. Perhaps not the least significant effects have been on the farm housewife who does the cooking. Hybrid corn may prove to be a boon to soil conservation as well as to human conservation. The superior productiveness of hybrid corn is making it feasible for farmers to institute better cropping practices, to rotate their crops, and to restore fertility to the soil by adding fertilizers. The meteoric rise of hybrid corn as a factor in agricultural economy can be followed in percentages of total corn acreage planted with hybrid seed over the years following 1933, as shown in Table 15-4.

In the case of hybrid corn, improvements came as a sudden, powerful surge. Just as important in the long run are the improvements in such important matters as milk production, egg laying, and yield of a variety of plants (Table 15-5). Progress has not come easily, since productivity is usually a quantitative char-

Table 15-4. CORN HYBRIDS: PERCENTAGE OF TOTAL CORN ACREAGE IN THE UNITED STATES PLANTED WITH HYBRID SEED.

Year	Percent Acreage
1933	0.1
1938	14.9
1943	52.4
1948	76.0
1953	86.5
1958	94.1
1960	95.9

Source: Data from *Agricultural Statistics* 1962, published by the U.S. Department of Agriculture

Table 15-5. YIELDS OF WHEAT OBTAINED IN NEW YORK STATE SINCE 1866.

Period	Average Yield, Bushels Per Acre
1866–1875	14.1
1876–1885	15.5
1886–1895	15.4
1896–1905	17.5
1906–1915	20.2
1916–1925	19.9
1926–1935	19.3
1936–1945	23.8
1946–1955	28.1
1956–1964	33.2

Source: From Love, *Cornell Univ. Agr. Exp. Sta. Bull, 828, 22,* 1946, and from data provided by N. F. Jensen.

acter and not easily dealt with genetically. But correlated efforts to improve environments, for example by means of nutrition, and to manipulate genetic material in favorable ways promise a continuation of progress. It is significant that a characteristic need of underdeveloped countries is for appropriate plant and animal improvement.

As noted before, hybridization, inbreeding, and selection have had the major roles in plant and animal improvement. Since the rise of genetics as a science, somewhat more exotic means have been used as well. Properties of polyploidy have been utilized to advantage, particularly in ornamental horticulture. And even induced mutation has been utilized occasionally. In the early 1940's, when the tremendous clinical importance of penicillin was recognized, particularly in relation to the demands of wartime, great efforts were made to increase its production. One plan of attack was to irradiate the mold Penicillium, which produces penicillin, in the hope of inducing a mutation favorable to a larger output. You can see a basis for this hope if you recall instances where genetic blocks in metabolism have led to the accumulation of particular substances.

At least one lucky hit was made. From spores that were X-rayed, a culture of *Penicillium chrysogenum* was isolated that yielded approximately twice the amount of penicillin obtained from the original parent strain. Further irradiation experiments, in which conidia of the new mutant type were treated with ultraviolet light, produced a strain that yielded about 900 units of penicillin per milliliter in comparison with the 500-unit average obtained from the earlier X-ray mutant.

GENETICS AND MAN'S UNDERSTANDING OF MAN

We can argue that life would be dull indeed if differences among men did not exist, or even if their differences were greatly reduced. We can also speculate, if we wish, on the ways in which uniformity would affect our politics, economics, social organization, and cultural activities. But human variability is a fact. Variability is a characteristic of our species, and of organisms in general. As humans we are defined by our differences as well as our similarities. Genetics, by giving us insight into the nature of our similarities and differences, provides a base for rational appreciation of each other and for rational interaction.

The greater part of this book has been concerned with individual differences and their genetic bases. At this point we return to the subject of Chapter 13 and consider individual differences in relation to group differences, referring our remarks primarily to the so-called races of man.

Individual Diversity and Racial Diversity. The primary implication of *race* is that people have similarities and differences that permit them to be put into some sort of meaningful classification. We know, of course, that people have similarities and differences. Are these of a kind and distribution that permit classification? And, if so, is there any possible meaningful classification that fits at all with widely held notions of race?

First, let us examine two attributes common to the usual concepts of race. One of these is that by racial differences we mean heritable biological differences, not cultural differences as might be found in language, dress, or economic systems. The other is that race has something to do with geography, and with common origin. We might, for example, separate, quite accurately, all the people of the world as color blind or not color blind. But such a classification would have little to do with our concepts of race. There are color-blind individuals among the Japanese, among American whites, Indians, and Negroes, and among different groups of Europeans, Asiatics, and Africans. If you think about the matter, you will find that your ideas of race involve natural populations rather than artificial groupings of scattered individuals.

With the foregoing in mind, it may be asked: Are there heritable traits that distinctively characterize groups of people who have originated within a more or less limited geographic area? Despite commonly held beliefs, we will be hard pressed to point to more than a few traits representing anything like absolute differences among groups of people that fit the usual concepts of race. The combination of dark skin, thick lips, and kinky hair serves to set apart most Negroes into a distinct racial group. But within the group so set apart, there is a

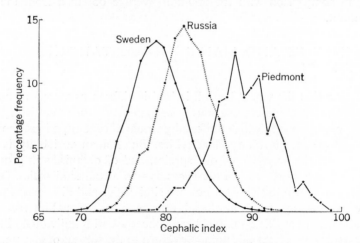

Figure 15-10. *Frequency distributions of cephalic index (breadth as percentage of length). Represented are a Swedish population (after Lamborg and Linders), a Russian population from the region of Smolensk (after Tchepowkovsky), and a population of Piedmontese (after Livi). (From Haldane,* Heredity and Politics, *1938. By permission of George Allen & Unwin Ltd. and of W. W. Norton.)*

tremendous range of variation for other heritable characteristics. And when we turn to different groups that by some definition or other have qualified as races, for example Caucasoids, we can scarcely find a criterion to which all members of the group conform and which excludes all members of different groups.

The key to the problem of defining race differences is that popular stereotypes of race are wrong in expecting absolute differences among racial groups. The most frequent real situation is that represented in Figure 15-10, which shows frequency distributions of cephalic index in three rather diverse European groups—Swedes, Russians, and Piedmontese. We see that there are significant differences among the groups, but that they are differences of average. With reference to individuals, the groups show a large overlap. In other words, the skull measurements of an individual are not diagnostic as to his "racial" origin. (Lacking other information, in what group would you place, for example, an individual whose cephalic index is 80?)

A Genetic Definition of Race. What is true of distributions of cephalic index is true for most other characteristics of the kind that we think of as comprising race differences. Assuming these characteristics to be gene controlled, we see that races come down to something very close to the populations described in Chapter 13. We might say, in fact, that races are population isolates with characteristic gene frequencies that distinguish them from other groups of the same general kind. Admittedly, as a definition this is extremely broad. It makes no attempt to fix the size or the character of the isolates. On the other hand, our concepts of race are not very fixed either, and a definition of race at this time needs considerable flexibility.

Fallacious Concepts of Race. Many fallacious concepts of race are based on fallacious concepts of heredity, some of them going back to pre-Mendelian times. The most common error is failure to take into account the particulate nature of hereditary material. The tacit assumption is that there are, or were, pure races, whose hereditary stuff is some kind of pure substance. Matings between different races are supposed to produce mixed races, in about the same fashion as mixing two different liquids, such as grape juice and ginger ale, gives rise to a mixed drink with definite but "hybrid" characteristics of its own. A further implication is that a series of matings between members of two different races will always give the same results. On such a basis, the character of the race assumes great importance in determining the character of its individual members. Persons belonging to races with "superior blood" are presumed to be predictably better than people who are members of races with less desirable hereditary stuff.

You realize at once, from your knowledge of genetics, that concepts of the sort we have just outlined are nonsensical. If in addition you realize that group differences are differences in frequencies of genes, not absolute differences, many widely accepted race concepts are laid bare as fallacies. If all of us can come to recognize these fallacies, and behave toward our fellow men in ways that recognition of these fallacies suggests, then human welfare and understanding will be truly promoted. To comment more particularly:

1. Pure races are fictional. A *pure* race would have to be one in which all individuals were homozygous and had the same allelic substitutions at each genetic locus. Except for one-egg twins, genetic identity is not approached even within families, and still less within the much larger groups called races. Complete homozygosity probably never occurs.

There is no reason to suppose that pure races existed even in the remote past. The available fossil evidence tends to refute rather than support the idea of a biological past in which the earth was peopled with pure races.

2. Judging an individual on the basis of some type or ideal that has been set up for his racial group is unfair and frequently leads to false conclusions. Races, for the most part, are distinguishable only on the basis of gene frequencies. An individual member of a group may have heritable traits considerably different from those characteristic of the most frequent type or of the average type for the group.

3. Evidence is lacking for associations between particular mental or psychological characteristics and the common heritable physical traits by which certain racial groups are identified. Black skins are sometimes supposed to go with shiftlessness and lack of ambition, or with uncanny musical talent. At the same time it may be forgotten that white skins may go with these traits as well, or that black skins are often not associated with these characteristics. Incorrect thinking along these lines is in some measure due to failure to understand the particulate nature of hereditary materials.

4. It is not established that there are racial differences in frequencies of genes determining mental and psychological traits. In fact, such genes have not yet been accurately identified in individuals, let alone studied in racial groups. Moreover, the considerable difficulty of obtaining precise measurements of such traits as I. Q. is magnified when comparisons between racial groups are made. For example, can mental tests based on the culture of western Europeans and white Americans give anything like a true picture of the mental capacities of individual American Indians? From several points of view, judgments as to racial differences in mental and psychological traits are premature.

On the other hand, to assert that all racial groups have an identical mental and psychological heritage is also at present unjustified. If genes have some influence in determining the various facets of intelligence, it would be remarkable if the frequencies of such genes were identical in different racial groups. *But to speak of differences is not to imply, at the same time, a scale of superiority and inferiority.*

5. On the basis of what has been said before, it follows that ideas of race superiority are ill founded. No race has a monopoly on "good" genes or is

entirely free from "bad" genes. And what are "good" and "bad" genes anyway? Genes that are unfavorable under one geographical and cultural environment may be relatively favorable under another. Finally, many of the so-called "superior" and "inferior" characteristics sometimes presumed to be associated with particular racial groups may be largely or entirely the effect of environment. Shiftlessness and lack of ambition have been said to characterize certain racial groups. These are the same groups that have lived under circumstances providing small incentive for the exercise of initiative or ambition. A primary aim of democratic society, to provide each individual with opportunities for the realization of his unique pattern of potentialities, makes sense on genetic as well as social grounds.

Keys to the Significance of This Chapter

Man has a past, present, and future, all of them dependent on his genetic endowment and on the genetic endowment of other living things. The understanding of this genetic endowment and of the mechanisms that operate it is far from complete but is substantial. As the science of genetics progresses, it offers the hope of increased self-understanding and well-being.

Progress in genetics is linked with advances in other branches of science, not only with presently obvious relatives such as cytology, biochemistry, physical chemistry, mathematics, physiology, and morphology, but also with areas such as information theory, psychology, communication, and behavioral science. Man's understanding of himself and of the living world is a matter of the unfolding and integration of all aspects of biology, among which genetics is fundamental.

REFERENCES

Burdette, W. J., editor, *Methodology in Human Genetics*. San Francisco: Holden-Day, 1962. (Although it is written primarily for the professional, students should be aware of this book's existence. It contains a series of articles on method, ranging in topic from tissue culture to mathematical analysis.)

Davidson, W. M., and Smith, D. R., editors, *Proceedings of the Conference on Human Chromosomal Abnormalities*. Springfield, Illinois: C. C. Thomas, 1961. (A series of short articles on various cytological anomalies in man.)

Dobzhansky, Th., *Mankind Evolving*. New Haven: Yale University Press, 1962. (Lively and broad in scope. A consideration of human evolution by a geneticist.)

Haldane, J. B. S., *Heredity and Politics*. New York: Norton, 1938. (Interesting both to the professional and to the casual reader. Perceptive treatment of race differences and eugenics proposals.)

Harris, H., *Human Biochemical Genetics*. Cambridge: Cambridge University Press, 1962. (Detailed. An excellent reference, particularly for the metabolic disorders of man.)

Ingram, V. M., *The Hemoglobins in Genetics and Evolution*. New York: Columbia University Press, 1963. (Brief but solid treatment of one of the most fascinating stories in human genetics.)

Knudson, A. G., Jr., *Genetics and Disease*. New York: McGraw-Hill, 1965. (A small book you will enjoy reading, written from a modern viewpoint and relating genetics to medicine.)

Lange, J., *Verbrechen als Schicksal*. Leipzig: Thieme, 1929. Translation: *Crime and Destiny*. New York: Charles Boni, 1930. (Early twin studies indicating a possible hereditary basis for criminality.)

Muller, H. J., "Our Load of Mutations." *Am. J. Human Genetics*, **1**:1–18, 1949. (Thought-provoking. Deals with possible deleterious consequences of increased mutation rate in man due to exposure to irradiants and other mutagens.)

Neel, J. V., "The Detection of the Genetic Carriers of Hereditary Disease." *Am. J. Human Genetics*, **1**:19–36, 1949. (Gives details of several important and interesting examples of this subject.)

Newman, H. H., *Multiple Human Births*. New York: Doubleday, Doran & Co., 1940. (This very readable book contains a wealth of information covering various aspects of multiple human births.)

Newman, H. H., Freeman, F. N., and Holzinger, K. J., *Twins: A Study of Heredity and Environment*. Chicago: University of Chicago Press, 1937. (The classical publication on twin studies in this country. Chap. 5 is most likely to be useful to the general student. Chap. 10 summarizes the fascinating case studies of identical twins reared apart.)

Osborn, F., *Preface to Eugenics*. New York: Harper & Bros., 1951. (Broad and temperate treatment of the case for applying genetic principles for the betterment of mankind.)

Race, R. R., and Sanger, R., *Blood Groups in Man*. Oxford: Blackwell, 4th ed., 1962. (Authoritative source book for a field in which information accumulates rapidly.)

Reed, S. C., *Counseling in Medical Genetics*. Philadelphia: W. B. Saunders Co., 1955. (Gives insight into the application of genetic principles in counseling. Includes actual examples of the problems presented to heredity clinics.)

Stern, C., *Principles of Human Genetics*. San Francisco: W. H. Freeman and Co., 2nd ed., 1960. (A textbook devoted to human genetics. Clearly and soundly written.)

QUESTIONS AND PROBLEMS

15-1. What do the following signify?

antibiotic	one-egg twins
average pair differences	prenatal environment
carrier	race
diagnosis	sex chromatin
identical twins	siblings
I.Q.	two-egg quadruplets
one-egg quintuplets	two-egg twins

15-2. What are some factors of the prenatal environment that might produce differences in one-egg twins?

15-3. In the extensive studies involving comparisons of one-egg and two-egg twins carried out by Newman, Freeman, and Holzinger, all of the two-egg pairs investigated were like-sexed. Why do you suppose that no boy-girl pairs were included in the comparisons?

15-4. An investigation involving a fairly large number of twin pairs revealed that, among one-egg twins, in 95 percent of all cases where one pair member had been infected with measles, the partner twin had a case history for measles too. Discuss whether this finding implies that susceptibility to measles is strongly hereditary.

15-5. Among two-egg twins where one pair member had a case history for measles in 87 percent of all the instances the partner member also had a case history for measles. Adding this information to that given in the preceding question, what can you say about a possible hereditary basis for susceptibility to measles?

15-6. J. B. Nichols studied the sex ratios of twins recorded in the 1900 census report for the U.S. and found that 717,907 pairs of twins were born between 1890 and 1900. Of these, 234,497 were boy-boy pairs, 219,312 were girl-girl pairs, and 264,098 were boy-girl pairs. Assuming for practical purposes a 1:1 sex ratio in man, estimate the number of one-egg twin pairs included in these data. Now, calculate the probability of male and female births from the data above, and estimate the number of one-egg twin pairs on this basis.

15-7. Give reasons why it would be a good idea for every medical practitioner to have training in genetics.

15-8. What are some of the practical advantages of being able to detect the carriers of hereditary diseases?

15-9. There is a tendency among many people to view heritable diseases as somehow more hopeless and horrible than infectious diseases or accidental mutilations. What arguments and examples could you muster to refute this idea?

15-10. Are there any kinds of genetic studies for which man, as compared with a standard experimental organism such as Drosophila, is a more favorable object?

15-11. Within certain social groups it appears that members of families that are somewhat larger than average for the group tend in turn to have relatively large families themselves. Is this convincing evidence for a hereditary basis for human fertility? What alternative explanations might the situation have?

15-12. Xeroderma pigmentosum is a severe recessive disease. Affected persons are photosensitive, and portions of their skin that have been exposed to the light show intensive pigmentation, freckling, and warty growths that often become malignant. Those afflicted with this heritable disease seldom live beyond the age of 15. Pedigree studies have led to the discovery that individuals heterozygous for the gene for xeroderma pigmentosum are characterized by very heavy freckling. This is quite independent of the freckling associated with red hair. Haldane has suggested that it may be undesirable for two heavily freckled persons to get married unless at least one is a red head. Explain the basis of his proposal.

15-13. Point out fundamental similarities in these two phenomena. (a) Plant varieties with established disease resistance against a pathogen sometimes suddenly fall prey to it. (b) Occasionally, antibiotics unexpectedly fail to work against human pathogens that previously were sensitive to them.

15-14. Why is it fallacious to evaluate an individual according to a racial stereotype?

15-15. W. C. Boyd and others have proposed tentative classifications of mankind on the basis of frequencies of the blood-group alleles. As compared with skin color or morphological characteristics, what genetic advantages do the blood groups have for purposes of race classification?

15-16. Does your knowledge of genetics supply any reason for believing that interracial marriages are biologically undesirable? Discuss.

15-17. On the basis of differing electrophoretic mobilities, two types (A and B) of the enzyme glucose-6-phosphate dehydrogenase can be distinguished in man. Individual clones of cultured cells from a female heterozygous for the alleles determining A and B respectively show one type of glucose-6-phosphate dehydrogenase or the other, but not both. Explain the implications of this observation.

15-18. Some human individuals with the typical male karyotype (XY 2A) appear externally to be female, with feminine breast development, female external genitalia, and luxuriant growth of hair on the head. At the same time, they lack a uterus and have abdominal or inguinal testes. Such individuals carry a gene *Tr*. Standard female (XX) carriers of the gene are only slightly, if at all, affected by it. Assume *Tr* to have an autosomal linkage. If a carrier female marries a normal male, what kinds of progeny are expected and in what proportions? Assuming now sex-linked transmission of *Tr*, would your expectation as to progeny be different? If so, how?

15-19. Assume that you are a genetic counselor. Two parents of normal phenotype have an albino child. They plan to have two or more children. Assuming the albinism is autosomal recessive, what can you tell the parents as to the probability that both children will be normal? Both albino? One normal and one albino, without regard to the order in which they are born? One normal and one albino, born in the order given?

15-20. Assume the pedigree presented below to be straightforward, with no complications such as mutation or illegitimacy. Trait *W*, found in individuals represented by the shaded symbols, is rare in the population at large. Tell which of the following patterns of transmission for *W* are consistent with and which are excluded by the pedigree:

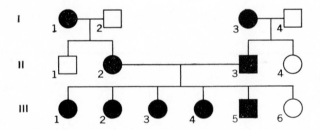

(a) autosomal recessive,   (b) autosomal dominant,   (c) sex-linked recessive,
(d) sex-linked dominant,   (e) Y-linkage.

Note: You may be able to analyze the pedigree more readily if you choose suitable gene symbols and then, under each hypothesis as to heritable pattern, assign appropriate genotypes to the individuals represented. For example, under the hypothesis of autosomal recessiveness, individuals II-2 and II-3 should both be double recessive genotype. Therefore, they could not have child III-6, who must carry a dominant gene. Under the terms of the question the hypothesis is effectively excluded.

15-21. If in your answer to question 15-20 you find more than one hypothesis not excluded, is one of the possible hypotheses more likely than the other(s)? Which?

15-22. For the increased welfare of man, what are some of the directions research in human genetics should take?

15-23. A husband and wife are both blood type AB. If they have a pair of dizygotic (two-egg) twins what is the probability that the twins will have identical blood type? That the twin pair members each will be blood type A? Answer the same questions but assuming the twins to be monozygotic.

15-24. A husband and wife are both heterozygous for an autosomal recessive gene for albinism. If they have a pair of dizygotic twins, what is the probability that the twins will be of the same phenotype with respect to pigmentation?

Index